Methods in Molecular Biology™

Series Editor

John M. Walker
School of Life Sciences
University of Hertfordshire
Hatfield, Hertfordshire, AL10 9AB, UK

For further volumes:
http://www.springer.com/series/7651

Network Biology

Methods and Applications

Edited by

Gerard Cagney

Conway Institute of Biomolecular and Biomedical Research,
University College Dublin, Dublin, Ireland

Andrew Emili

Banting and Best Department of Medical Research, Donnelly Centre for Cellular
and Biomolecular Research, University of Toronto, Toronto, ON, Canada and
Department of Molecular Genetics and Microbiology, University of Toronto, Toronto, ON, Canada

Humana Press

Editors
Gerard Cagney, Ph.D.
Conway Institute of Biomolecular
and Biomedical Research
University College Dublin
Dublin, Ireland
gcagney@gmail.com

Andrew Emili, Ph.D.
Banting and Best Department of Medical
Research, Donnelly Centre for Cellular
and Biomolecular Research
University of Toronto, Toronto
ON, Canada
and
Department of Molecular Genetics and
Microbiology, University of Toronto, Toronto
ON, Canada
andrew.emili@utoronto.ca

ISSN 1064-3745 e-ISSN 1940-6029
ISBN 978-1-4939-5855-9 ISBN 978-1-61779-276-2 (eBook)
DOI 10.1007/978-1-61779-276-2
Springer New York Dordrecht Heidelberg London

Printed on acid-free paper

Humana Press is part of Springer Science+Business Media (www.springer.com)

Preface

Researchers are all familiar with the spectacular success of reductionist approaches for elucidating biological mechanisms over the last century or so. Nevertheless, recent technological advances, particularly among the leading sub-specialities of molecular biology, present challenges to reductionism on at least two fronts.

First, the explosive growth in the scale of data generated by modern, highly parallel, computerized, and increasingly automated lab methods has become a major impediment. For instance, next-generation sequencing technology is capable of producing terabytes of sequence information per day. It is also now mainstream to report global expression patterns for thousands of genes in multiple strains or cell lines subjected to different conditions. Similarly, phenotypic screens analyzing singly or multiply gene-disrupted cells and organisms have recently become prominent. But how will such data be analyzed, interpreted, and presented when extended to patients where thousands of gene isoforms have unique functions in individual tissues? Secondly, it has become apparent that biological processes do not trace back to individual molecules, and that the discrete pathway diagrams of text books do not accurately reflect the more extensive interaction networks likely present in cells.

Yet, while extremely large datasets describing gene sequences, mRNA transcripts, protein interactions, and metabolite concentrations are increasingly commonplace, these represent only starting "parts lists" that are usually insufficient to unlock mechanistic insights on their own right. Notable examples include a marked increase in organism complexity that does not correspond to aggregate genome size (e.g. yeast versus fly versus human), the formation of individual proteins into macromolecular assemblies (i.e. protein complexes) in order to perform intricate biochemical tasks, and the combinatorial action of sequence-specific DNA-binding transcription factors, microRNAs, and chromatin regulators on transcriptional responses.

Fortunately, despite the clumsiness of our observations, nature herself has the clear objectives of survival and propagation in mind, so concepts emerging from the study of biological entities, such as networks (e.g. functional interactions linking genes, proteins, metabolites, etc.) suggest that order rather than chaos prevails. These principles include modular and hierarchical organization, reactive information-driven causal-response behaviours, systems robustness, co-evolution, and self-organization. Moreover, the experience of the last decade argues that continued success in the field of network biology will rely increasingly on accurate high-throughput methods for generating quantitative "omic" datasets in combination with innovative integrative analytical tools designed to discover organizing patterns and principles underlying the myriad data.

This book attempts to bring both approaches together in a single volume. It includes practical descriptions of the experimental and computational approaches currently prevalent in network biology.

Among the 10 experimental methods, three of the major protein interaction mapping approaches are described over three chapters. Seesandra Rajagopala and Peter Uetz describe how different versions of the yeast two hybrid assay serve as complementary protein interaction mapping tools (Chapter 1) while Zuyao Ni and co-workers provide detailed instructions on mapping mammalian protein interactions using a lentiviral affinity tagging and expression system (Chapter 2). In a description of the protein arrays methodology, Mikael Bauer and co-workers highlight the flexibility and power of this technique (Chapter 3). Next, Oliver Schilling and co-workers (Chapter 4) describe a quantitative mass spectrometry-based method for mapping proteome networks through C-termini; Mihaela Sardiu and Michael Washburn describe a set of procedures for quantifying probabilistic protein interaction networks using label-free proteomics (Chapter 5); Haiming Huang and Sachdev Sidhu describe approaches for mapping interactions between peptide recognition modules and short linear motifs to build protein interaction networks (Chapter 6). Mohan Babu and colleagues present methodology for carrying out large-scale genetic interaction (epistasis) screens in *E. coli* (Chapter 7); Yifat Cohen and Maya Schuldiner show how integrated use of high-content microscopy and gene disruption libraries provide unprecedented visual descriptions of dynamic cell processes (Chapter 8); Kim Blakely and co-workers describe the application of related gene knockdown approaches to mammalian cells based on shRNA libraries (Chapter 9). Finally, Susana Neves and Laura Forrest provide an in-depth overview of DNA sequence analysis technologies for molecular phylogenetics (Chapter 10).

A variety of cutting-edge computational approaches currently used in network biology are then surveyed over the next 11 chapters. Gregory Clark and co-workers describe the use of co-evolutionary methods for predicting protein interactions that rival alternative approaches, such as abundance based methods (Chapter 11); Daniele Merico and colleagues describe a method for finding functionally coherent sets of genes using a Plugin for the Cytoscape network visualization application applicable for many different data types (Chapter 12); Alexei Vazquez and co-workers outline statistical methods for determining error rates in protein interaction networks (Chapter 13) while a complementing method by Gabriel Musso and co-workers reports on methods for evaluating the quality of published datasets (Chapter 14); David Fung and co-workers tackle the widespread problem of classifying network data from clinical samples based on pathway and network topology analysis (Chapter 15); Jennifer Smith and co-workers describe methods for mapping transcriptional regulatory networks from chromatin immunoprecipitation data using models of gene regulation with combinatorial DNA binding activity (Chapter 16); Colm Ryan and colleagues describe a method for predicting epistatic interactions based on nearest neighbour imputation (Chapter 17); Iain Wallace, Corey Nislow and coworkers highlight an application of the widely used Cytoscape application for displaying chemogenomic protein networks (Chapter 18); One of the aims of studying protein networks is to gain a better understanding of the coordinated role of protein assemblies, or complexes, in the biology of the cell; Benjamin Webb, Andrej Sali, and co-workers (Chapter 19) show how combining such network data with constraints for biophysical experiments, sophisticated integrated models of protein complexes may be assembled. Sara Mostafavi and co-workers describe new algorithms for characterizing node attributes and assigning function based on annotated data as a core set of tools applicable to many networks (Chapter 20). In a key final chapter, Alexander Ratushny and colleagues discuss building models of biomolecular network processes based on reaction kinetics and other parameters (Chapter 21).

This book emphasizes the practical application of these approaches – protocols along with troubleshooting guides and benchmarking criteria – and allows for a more comprehensive description than is typically encountered in primary research papers. We are indebted to all the many contributing authors who have strived to make the procedures in this volume concise and accessible, and hope the ensemble will be of value to a broad assortment of readers, ranging from graduate students new to the art to seasoned professionals looking to polish their skill sets. We also encourage the emerging systems biology community to apply these methods regularly in the lab or at the computer. After all, the limitation of past reductionist approaches demand that we cast the net more widely.

Dublin, Ireland *Gerard Cagney, Ph.D.*
Toronto, ON, Canada *Andrew Emili, Ph.D.*

Contents

Contributors

JOHN D. AITCHISON • *Institute for Systems Biology, Seattle, WA, USA*
MOHAN BABU • *Banting and Best Department of Medical Research, Donnelly Centre for Cellular and Biomolecular Research, University of Toronto, Toronto, ON, Canada*
GARY D. BADER • *Banting and Best Department of Medical Research, Centre for Cellular and Biomolecular Research (CCBR), Toronto, ON, Canada; Department of Molecular Genetics, University of Toronto, Toronto, ON, Canada*
OLIVER BARRÉ • *Department of Oral Biological and Medical Sciences, Centre for Blood Research, University of British Columbia, Vancouver, BC, Canada*
MIKAEL BAUER • *Biophysical Chemistry and Biochemistry, Lund University, Lund, Sweden*
ALEXANDR BEZGINOV • *Department of Medical Biophysics, University of Toronto, Campbell Family Institute for Cancer Research, Ontario Cancer Institute, University Health Network, Toronto, ON, Canada*
KIM BLAKELY • *Donnelly Centre and Banting & Best Department of Medical Research, University of Toronto, Toronto, ON, Canada; Department of Molecular Genetics, University of Toronto, Toronto, ON, Canada*
GERARD CAGNEY • *Conway Institute of Biomolecular and Biomedical Research, University College Dublin, Dublin, Ireland*
DOLORES J. CAHILL • *School of Medicine & Medical Science, Conway Institute of Biomolecular and Biomedical Research, University College Dublin, Belfield, Dublin, Ireland*
ROBERT L. CHARLEBOIS • *Department of Medical Biophysics, University of Toronto, Campbell Family Institute for Cancer Research, Ontario Cancer Institute, University Health Network, Toronto, ON, Canada*
GREGORY W. CLARK • *Department of Medical Biophysics, University of Toronto, Campbell Family Institute for Cancer Research, Ontario Cancer Institute, University Health Network, Toronto, ON, Canada*
STEPHAN J. CLARKE • *Department of Medical Oncology, Sydney Cancer Centre, Sydney, NSW, Australia*
YIFAT COHEN • *Department of Molecular Genetics, Weizmann Institute of Science, Rehovot, Israel*
PÁDRAIG CUNNINGHAM • *School of Computer Science and Informatics, University College Dublin, Dublin, Ireland*
VAQAAR-UN-NISA DAR • *Department of Medical Biophysics, University of Toronto, Campbell Family Institute for Cancer Research, Ontario Cancer Institute, University Health Network, Toronto, ON, Canada*

ANDREW EMILI • *Banting and Best Department of Medical Research, Donnelly Centre for Cellular and Biomolecular Research, University of Toronto, Toronto, ON, Canada; Department of Molecular Genetics and Microbiology, University of Toronto, Toronto, ON, Canada*
LAURA L. FORREST • *Department of Ecology and Evolutionary Biology, University of Connecticut, Storrs, CT, USA*
DAVID C.Y. FUNG • *School of Biotechnology and Biomolecular Sciences, New South Wales Systems Biology Initiative, The University of New South Wales, Sydney, NSW, Australia*
ALLA GAGARINOVA • *Department of Molecular Genetics, University of Toronto, Toronto, ON, Canada*
GURI GIAEVER • *Department of Molecular Genetics, University of Toronto, Toronto, ON, Canada; Donnelly Centre for Cellular and Biomolecular Research, University of Toronto, Toronto, ON, Canada; Department of Pharmaceutical Sciences, University of Toronto, Toronto, ON, Canada*
ANNA GOLDENBERG • *Banting and Best Department of Medical Research, Centre for Cellular and Biomolecular Research, University of Toronto, Toronto, ON, Canada*
JACK F. GREENBLATT • *Banting and Best Department of Medical Research, Donnelly Centre for Cellular and Biomolecular Research, University of Toronto, Toronto, ON, Canada; Department of Molecular Genetics, University of Toronto, Toronto, ON, Canada*
DEREK GREENE • *School of Computer Science and Informatics, University College Dublin, Dublin, Ireland*
HAIMING HUANG • *Donnelly Centre for Cellular and Biomolecular Research, University of Toronto, Toronto, ON, Canada*
PITTER F. HUESGEN • *Department of Oral Biological and Medical Sciences, Centre for Blood Research, University of British Columbia, Vancouver, BC, Canada*
RUTH ISSERLIN • *Banting and Best Department of Medical Research, Centre for Cellular and Biomolecular Research (CCBR), Toronto, ON, Canada; Department of Molecular Genetics, University of Toronto, Toronto, ON, Canada*
LUCY JANKOVA • *Cancer Pharmacology Unit, ANZAC Research Institute, Concord Repatriation General Hospital, Sydney, NSW, Australia*
TROY KETELA • *Donnelly Centre and Banting & Best Department of Medical Research, University of Toronto, Toronto, ON, Canada*
SEUNG JOONG KIM • *Department of Bioengineering and Therapeutic Sciences, California Institute for Quantitative Biosciences, University of California, San Francisco, CA, USA; Department of Pharmaceutical Chemistry, California Institute for Quantitative Biosciences, University of California, San Francisco, CA, USA*
NEVAN KROGAN • *Department of Cellular and Molecular Pharmacology, California Institute for Quantitative Biomedical Research, University of California, San Francisco, CA, USA*

KEREN LASKER • *Department of Bioengineering and Therapeutic Sciences, California Institute for Quantitative Biosciences, University of California, San Francisco, CA, USA; Department of Pharmaceutical Chemistry, California Institute for Quantitative Biosciences, University of California, San Francisco, CA, USA*

SARA LINSE • *Biophysical Chemistry and Biochemistry, Lund University, Lund, Sweden*

AMY LO • *School of Biotechnology and Biomolecular Sciences, New South Wales Systems Biology Initiative, The University of New South Wales, Sydney, NSW, Australia*

MAGDALENA MAJ • *Department of Medicine III, Medical University of Vienna, Vienna, Austria*

DANIELE MERICO • *Banting and Best Department of Medical Research, Centre for Cellular and Biomolecular Research (CCBR), Toronto, ON, Canada; Department of Molecular Genetics, University of Toronto, Toronto, ON, Canada*

JASON MOFFAT • *Donnelly Centre and Banting & Best Department of Medical Research, University of Toronto, Toronto, ON, Canada; Department of Molecular Genetics, University of Toronto, Toronto, ON, Canada*

MARK MOLLOY • *Australian Proteome Analysis Facility, Department of Chemistry and Biomolecular Sciences, Macquarie University, Sydney, NSW, Australia*

QUAID MORRIS • *Department of Computer Science, Banting and Best Department of Medical Research, Centre for Cellular and Biomolecular Research, Toronto, ON, Canada; Department of Molecular Genetics, University of Toronto, Toronto, ON, Canada*

SARA MOSTAFAVI • *Department of Computer Science, Centre for Cellular and Biomolecular Research (CCBR), University of Toronto, Toronto, ON, Canada*

GABRIEL MUSSO • *Cardiovascular Division, Brigham & Women's Hospital, Boston, MA, USA; Harvard Medical School, Boston, MA, USA*

SUSANA S. NEVES • *Plant Cell Biotechnology Laboratory, ITQB Instituto de Tecnologia Química e Biológica, Universidade Nova de Lisboa, Oeiras, Portugal*

ZUYAO NI • *Banting and Best Department of Medical Research, Donnelly Centre for Cellular and Biomolecular Research, University of Toronto, Toronto, ON, Canada*

COREY NISLOW • *Department of Molecular Genetics, University of Toronto, Toronto, ON, Canada; Donnelly Centre for Cellular and Biomolecular Research, University of Toronto, Toronto, ON, Canada; Department of Pharmaceutical Sciences, University of Toronto, Toronto, ON, Canada*

DAVID J. O'CONNELL • *School of Medicine & Medical Science, Conway Institute of Biomolecular and Biomedical Research, University College Dublin, Belfield, Dublin, Ireland*

JONATHAN B. OLSEN • *Banting and Best Department of Medical Research, Donnelly Centre for Cellular and Biomolecular Research, University of Toronto, Toronto, ON, Canada; Department of Molecular Genetics, University of Toronto, Toronto, ON, Canada*

CHRISTOPHER M. OVERALL • *Department of Biochemistry and Molecular Biology, Centre for Blood Research, University of British Columbia, Vancouver, BC, Canada; Department of Oral Biological and Medical Sciences, Centre for Blood Research, University of British Columbia, Vancouver, BC, Canada*

JEREMY PHILLIPS • *Department of Bioengineering and Therapeutic Sciences, California Institute for Quantitative Biosciences, University of California, San Francisco, CA, USA; Department of Pharmaceutical Chemistry, California Institute for Quantitative Biosciences, University of California, San Francisco, CA, USA*
SEESANDRA V. RAJAGOPALA • *J. Craig Venter Institute, Rockville, MD, USA*
STEPHEN A. RAMSEY • *Institute for Systems Biology, Seattle, WA, USA*
ALEXANDER V. RATUSHNY • *Institute for Systems Biology, Seattle, WA, USA*
GRAHAM R. ROBERTSON • *Cancer Pharmacology Unit, ANZAC Research Institute, Concord Repatriation General Hospital, Sydney, NSW, Australia*
JEAN-FRANÇOIS RUAL • *Department of Pathology, University of Michigan, Ann Arbor, MI, USA*
DANIEL RUSSEL • *Department of Bioengineering and Therapeutic Sciences, California Institute for Quantitative Biosciences, University of California, San Francisco, CA, USA; Department of Pharmaceutical Chemistry, California Institute for Quantitative Biosciences, University of California, San Francisco, CA, USA*
COLM RYAN • *School of Computer Science and Informatics, University College Dublin, Dublin, Ireland*
RAMSEY A. SALEEM • *Institute for Systems Biology, Seattle, WA, USA*
ANDREJ SALI • *Department of Bioengineering and Therapeutic Sciences, California Institute for Quantitative Biosciences, University of California, San Francisco, CA, USA; Department of Pharmaceutical Chemistry, California Institute for Quantitative Biosciences, University of California, San Francisco, CA, USA*
MIHAELA E. SARDIU • *Stowers Institute for Medical Research, Kansas City, MO, USA*
OLIVER SCHILLING • *Institute for Molecular Medicine and Cell Research, University of Freiburg, Freiburg, Germany*
DINA SCHNEIDMAN-DUHOVNY • *Department of Bioengineering and Therapeutic Sciences, California Institute for Quantitative Biosciences, University of California, San Francisco, CA, USA; Department of Pharmaceutical Chemistry, California Institute for Quantitative Biosciences, University of California, San Francisco, CA, USA*
MAYA SCHULDINER • *Department of Molecular Genetics, Weizmann Institute of Science, Rehovot, Israel*
SACHDEV S. SIDHU • *Donnelly Centre for Cellular and Biomolecular Research, University of Toronto, Toronto, ON, Canada*
JENNIFER J. SMITH • *Institute for Systems Biology, Seattle, WA, USA*
ELISABETH R.M. TILLIER • *Department of Medical Biophysics, University of Toronto, Campbell Family Institute for Cancer Research, Ontario Cancer Institute, University Health Network, Toronto, ON, Canada*
ELINA TJIOE • *Department of Bioengineering and Therapeutic Sciences, California Institute for Quantitative Biosciences, University of California, San Francisco, CA, USA; Department of Pharmaceutical Chemistry, California Institute for Quantitative Biosciences, University of California, San Francisco, CA, USA*

PETER UETZ • *Proteros Biostructures, Martinsried, Germany and Gaithersburg, MD, USA*
ALEXEI VAZQUEZ • *Department of Radiation Oncology, The Cancer Institute of New Jersey and UMDNJ-Robert Wood Johnson Medical School, New Brunswick, NJ, USA*
JAVIER VELÁZQUEZ-MURIEL • *Department of Bioengineering and Therapeutic Sciences, California Institute for Quantitative Biosciences, University of California, San Francisco, CA, USA; Department of Pharmaceutical Chemistry, California Institute for Quantitative Biosciences, University of California, San Francisco, CA, USA*
KAVITHA VENKATESAN • *Novartis Institutes for Biomedical Research, Cambridge, MA, USA*
LUDWIG WAGNER • *Department of Medicine III, Medical University of Vienna, Vienna, Austria*
IAIN M. WALLACE • *Department of Molecular Genetics, University of Toronto, Toronto, ON, Canada; Donnelly Centre for Cellular and Biomolecular Research, University of Toronto, Toronto, ON, Canada; Department of Pharmaceutical Sciences, University of Toronto, Toronto, ON, Canada*
MICHAEL P. WASHBURN • *Stowers Institute for Medical Research, Kansas City, MO, USA; Department of Pathology and Laboratory Medicine, The University of Kansas Medical Center, Kansas City, KS, USA*
BENJAMIN WEBB • *Department of Bioengineering and Therapeutic Sciences, California Institute for Quantitative Biosciences, University of California, San Francisco, CA, USA; Department of Pharmaceutical Chemistry, California Institute for Quantitative Biosciences, University of California, San Francisco, CA, USA*
MARC R. WILKINS • *School of Biotechnology and Biomolecular Sciences, New South Wales Systems Biology Initiative, The University of New South Wales, Sydney, NSW, Australia*
JINGHAO M. YANG • *Department of Medical Biophysics, University of Toronto, Campbell Family Institute for Cancer Research, Ontario Cancer Institute, University Health Network, Toronto, ON, Canada*
ZHAOLEI ZHANG • *Banting and Best Department of Medical Research, Donnelly Centre for Cellular and Biomolecular Research, University of Toronto, Toronto, ON, Canada; Department of Molecular Genetics and Microbiology, University of Toronto, Toronto, ON, Canada*

Chapter 1

Analysis of Protein–Protein Interactions Using High-Throughput Yeast Two-Hybrid Screens

Seesandra V. Rajagopala and Peter Uetz

Abstract

The yeast two-hybrid (Y2H) system is a powerful tool to identify binary protein–protein interactions. Here, we describe array-based two-hybrid methods that use defined libraries of open reading frames (ORFs) and pooled prey library screenings that use random genomic or cDNA libraries. The array-based Y2H system is well-suited for interactome studies of existing ORFeomes or subsets thereof, preferentially in a recombination-based cloning system. Array-based Y2H screens efficiently reduce false positives by using built-in controls, retesting, and evaluation of background activation. Hands-on time and the amount of used resources grow exponentially with the number of tested proteins; this is a disadvantage for large genome sizes. For large genomes, random library screen may be more efficient in terms of time and resources, but not as comprehensive as array screens, and it requires significant sequencing capacity. Furthermore, multiple variants of the Y2H vector systems detect markedly different subsets of interactions in the same interactome. Hence, only multiple variations of the Y2H systems ensure comprehensive coverage of an interactome.

Key words: Yeast two-hybrid system, Protein–protein interactions, Two-hybrid array

1. Introduction

Specific interactions between proteins form the basis of most biological processes. Comprehensive analysis of protein–protein interactions on a genome scale is a challenging task of proteomics and has been best explored in budding yeast, human, and other model organisms. Protein interactome analysis on a genome scale was first achieved by using yeast two-hybrid (Y2H) screens (1) and next by large-scale mass spectrometric analysis of affinity-purified protein complexes (2, 3). The Y2H system is a genetic method that detects binary protein–protein interactions in vivo. Classical two-hybrid screens used random libraries (genomic or cDNA) to identify

Gerard Cagney and Andrew Emili (eds.), *Network Biology: Methods and Applications*, Methods in Molecular Biology, vol. 781, DOI 10.1007/978-1-61779-276-2_1,

novel interactions for a protein of interest. However, more recently, an array-based variation of this original principle has been increasingly used (Fig. 1). This approach can be applied not only to a few proteins, but also to whole genomes. Advantages of arrays are their built-in controls and their systematic nature. However, random library screens are more efficient for large genomes, but not as comprehensive as array screens and may yield more false negatives. Here, we describe the protocols for the array-based and random library two-hybrid screens.

1.1. The Principle of the Yeast Two-Hybrid System

The Y2H system is a genetic method extensively used to detect binary protein–protein interactions in vivo (in yeast cells). The system was developed by Stanley Fields (4) based on the observation that protein domains can be separated and recombined and can retain their properties. In particular, transcription factors can frequently be split into the DNA-binding domain (DBD) and activation domains (ADs). In the two-hybrid system, a DNA-binding domain (e.g., from the yeast Gal4 protein) is fused to a protein "B" (for bait) for which one wants to find interacting partners (Fig. 1). A transcriptional activation domain is then fused to some or all the predicted open reading frames (ORFs or "preys") of an organism. Bait and prey fusion proteins are then coexpressed in the same yeast cell. Usually, both protein fusions are expressed from plasmids that can be manipulated easily and then transformed into yeast cells. If the bait and prey proteins interact, a transcription factor is reconstituted which in turn activates one or more reporter genes. The expression of the reporter gene(s) allows the cell to grow only under certain conditions. For example, the HIS3 reporter encodes imidazoleglycerolphosphate (IGP) dehydratase, a critical enzyme in histidine biosynthesis. In the Y2H screening strain (lacking an endogenous copy of HIS3), expression of a HIS3 reporter gene is driven by a promoter that contains a Gal4p-binding site so that the bait protein fusion can bind to it. However, since the bait fusion should not contain a transcriptional activation domain, it remains inactive. If a protein ORF with an attached activation domain binds to the bait, this activation domain can recruit the basal transcription machinery, and expression of the reporter gene ensues. These cells can now grow in the absence of histidine because they can synthesize their own.

1.2. Applications

Originally, the two-hybrid system was invented to demonstrate the association of two proteins (4). Later, it was demonstrated that completely new protein interactions can be identified with this system, even when there are no candidates for an interaction with a given bait. Over time, it has become clear that the ability to conveniently perform unbiased library screens is the most powerful application of the system. With whole-genome arrays, such unbiased screens can be expanded to complete nonredundant sets of proteins. Arrays, like traditional two-hybrid screens, can also be

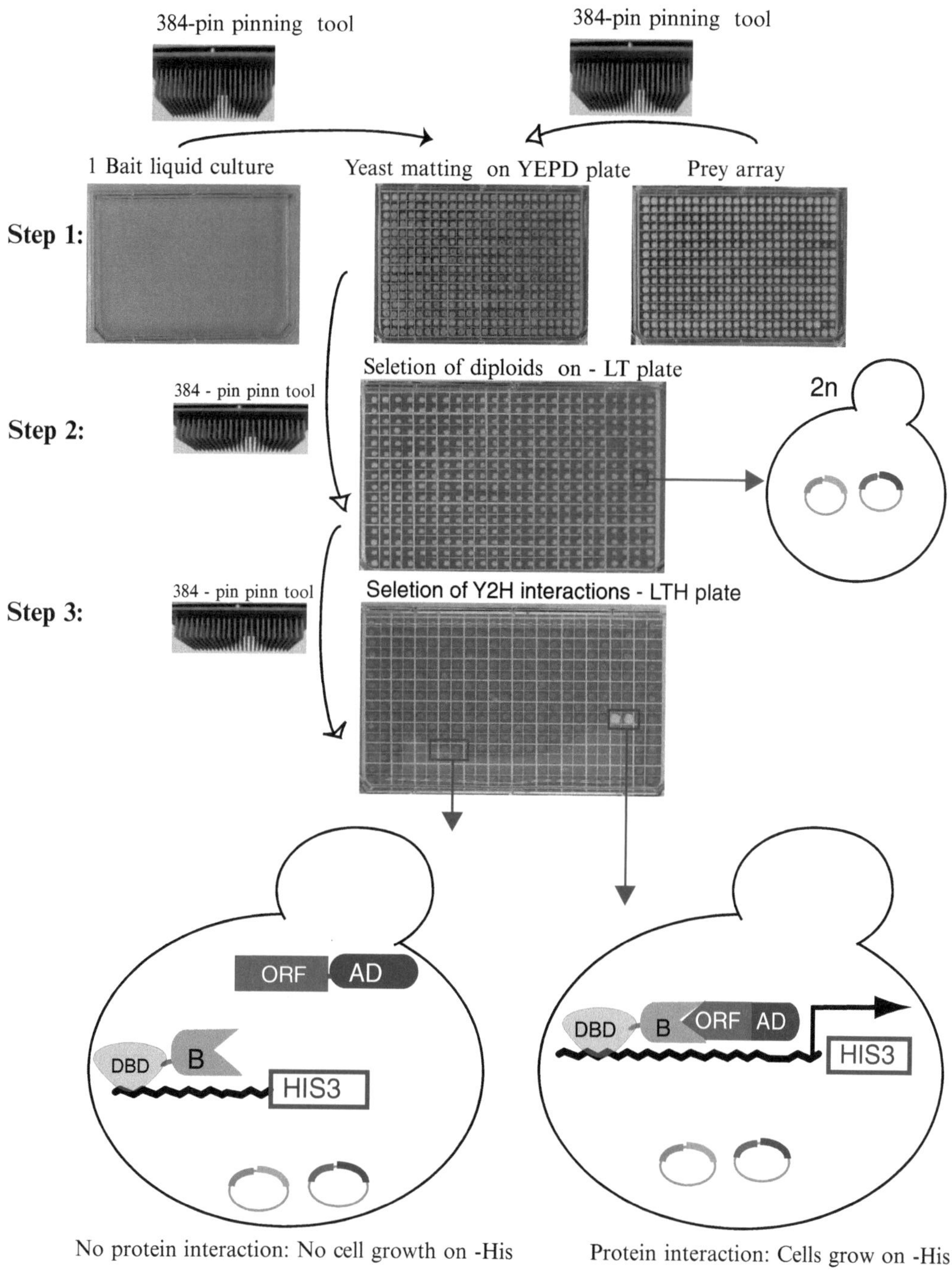

Fig. 1. Scheme of an array-based two-hybrid screen. *Step 1*: Yeast mating combines the bait and prey plasmids. First, the bait (DNA-binding domain (DBD) fusion) liquid culture is pinned onto YEPDA agar plates using a 384-pin pinning tool, and then the prey array (activation domain (AD) fusion) is pinned on top of the baits using the sterile pinning tool. Then, the mating plates are incubated at 30°C for 16 h. *Step 2*: The yeast-mating plates are pinned onto –Leu –Trp medium plates using a sterile 384-pin pinning tool. On –LT plates, only diploid cells grow. Selection on –LT media ensures that both the prey and bait plasmids are combined in the diploid yeast cells. *Step 3*: The diploid cells are pinned onto –Leu –Trp –His medium plates for protein interaction detection. Only if bait and prey proteins interact, an active transcription factor is reconstituted and transcription of a reporter gene is activated (*lower right panel*). The *rectangles* on the selective plate mark negative and positive interactions between the bait and the prey at these specific positions of the array (test is done in duplicates). After Uetz et al. (1).

adapted to a variety of related questions, such as the identification of mutants that prevent or allow interactions (5), screening for drugs that affect protein interactions (6, 7), identification of RNA-binding proteins (8), or semiquantitative determination of binding affinities (9). The system can also be exploited to map binding domains (10), to study protein folding (11), or to map interactions within a protein complex (proteasome (12), flagellum (13)). Finally, recent large-scale projects have been successful in systematically mapping interactions within whole proteomes or subsets thereof (yeast (1, 14); worm (15); fly (16); human (17, 18)). These studies have shown for the first time that most proteins in a cell are actually connected to each other (19).

In combination with structural genomics, gene expression data, and metabolic profiling, the enormous amount of data in these interaction networks should allow us eventually to model complex biological phenomena in molecular detail. An ultimate goal of this work is to understand the interplay of DNA, RNA, and proteins, together with small molecules, in a dynamic and realistic way.

1.3. Array-Based Screens

In an array, a number of defined prey proteins are tested for interactions with a bait protein (Fig. 1). Usually, the bait protein is expressed in one yeast strain and the prey is expressed in another yeast strain of different mating type. The two strains are then mated so that the two proteins are expressed in the resulting diploid cell (Fig. 1). The assays are done side-by-side under identical conditions, so they can be well-controlled, i.e., compared. As the identity of the preys is usually known, no sequencing is required after positives have been identified. However, the prey clones need to be obtained or made upfront. This can be done for a few genes or for a whole genome, e.g., an ORFeome (i.e., all ORFs of a genome).

In an array, each element has a known identity, and therefore it is immediately clear which two proteins are interacting when positives are selected. In addition, it is often immediately clear if an interaction is stronger than another one (but see below). Most importantly, since all these assays are done in an ordered array, background signals can be easily distinguished from true signals (Fig. 1, step 3). Until recently, it was much easier to construct a random library and screen it rather than to construct many individual clones and screen them individually. However, now whole genomes become increasingly available as ordered clone sets in a variety of vectors. Modern cloning systems also allow direct transfer of entry clones into many specialized vectors (20). For most model organisms, such genome-scale clone collections are already available (e.g., 21) or will be soon. One of the first applications of such clone collections is often a protein interaction screen.

In fact, in some cases, only an array screen may do the job. For example, if you have a bait protein that activates transcription on its own, a carefully controlled array may be the only way to distinguish between signal and background (see Fig. 1, step 3). Similarly, weak interactors may be detectable only when compared with a uniformly weak or no background.

1.4. Pooled Array Screening

A completely different screening strategy, the pooling strategy, has the potential to accelerate screening significantly, but might also have the disadvantage of increasing the number of false negatives. This may have been a reason why pooled screens in *Campylobacter jejuni* resulted in more false negatives than in one-by-one screens of *Treponema pallidum* (13, 22).

In the pooled array screening, preys of known identity (systematically cloned or sequenced cDNA library clones) are combined and tested as pools against bait strains. The identification of the interacting protein pair commonly requires either sequencing or retesting of all members of the respective pool. Zhong et al. established a method which allows for pooling up to 96 preys (23). It was estimated that this pooling scheme reduces the number of interaction tests required to 1/8–1/24 in the case of the yeast proteome. Two recent large-scale interaction mapping approaches for human proteins employed such a pooling strategy: Rual et al. tested baits against pools of 188 preys and identified individual interactions by sequencing (17); Stelzl et al. tested pools of 8 baits against a systematic library of individual preys and identified interactions by a second interaction mating (18). Recently, a "smart-pool-array" system was proposed, which allows the deconvolution of the interacting pairs through the definition of overlapping bait pools (24), and thus usually does not depends on sequencing or a second pair-wise mating procedure.

1.5. Random Library Screening (Genomic or cDNA)

Random library screens do not require systematic cloning of all prey constructs; however, the prey library must be created. Therefore, the complete DNA sequence of the genome of interest is no prerequisite. Random prey libraries can be made using genomic DNA- or cDNA-based libraries. For genomic libraries, the genomic DNA of interest is randomly cut, size-selected, and the resulting fragments ligated into one or more two-hybrid prey vector(s). Many two-hybrid screening projects used random genomic DNA libraries (10, 25). A cDNA library is made through reverse transcription of mRNA collected from specific cell types or whole organisms. To simplify the task even more, many cDNA libraries are commercially available. For example, Clontech has a collection of human and tissue-specific cDNA libraries. However, the bait clones that need to be screened with a random library need to be made independently as for an array screen.

In a random library screen, a library of prey proteins is tested for interactions with a bait protein (Fig. 2). Similar to array screens, the bait protein is expressed in one yeast strain and the preys are expressed in another yeast strain of different mating type. The two

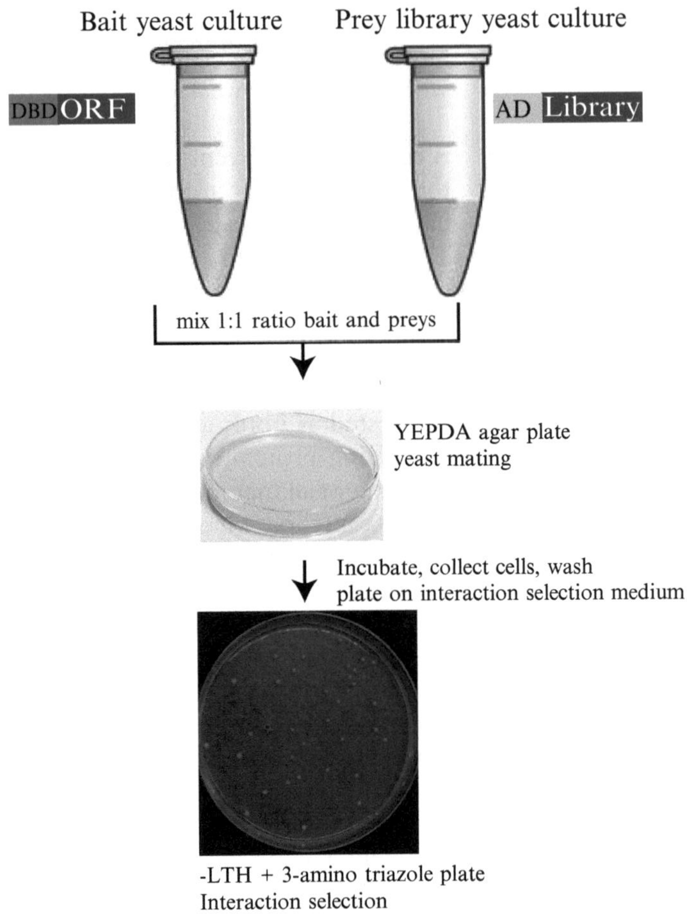

Fig. 2. Scheme of a random library two-hybrid screen: A haploid yeast strain expressing a single protein as a DBD fusion is mixed with the yeast haploid strains expressing cDNA or random genomic library (prey library). The bait and prey (1:1 ratio) culture is plated on YEPDA agar plate. The mating plates are incubated at 30°C for 6 h or overnight at room temperature. During this process, both the prey and bait plasmids are combined in the diploid yeast cells. The cells from the mating plates are collected and transferred onto –Leu –Trp –His medium plates (supplemented with different concentrations of 3-amino-triazole) for protein interaction detection, and plates are incubated at 30°C for 4–6 days. The identity of interacting prey is identified by yeast colony PCR of positive yeast colonies, followed by DNA sequencing of the PCR product.

strains are then mated so that the two proteins are expressed in the resulting diploid cell (Fig. 2). The diploids are plated on interaction selective medium, where only yeast cells having bait and its interacting prey grow. The prey is identified by isolating the prey plasmids, PCR amplification of the insert, and sequencing.

1.6. Multiple Variants of Y2H System

Y2H screens often produce vastly nonoverlapping interaction data when the screens are conducted using different vectors, strains,

Table 1
What you need for a yeast two-hybrid screen (examples)

		Gal4-Fusion		Selection			
Vector	**Promoter**	**DBD**	**AD**	**Yeast**	**Bacterial**	**Ori**	**Source**
Bait and prey vectors							
pDEST22	fl-ADH1	–	N-term	Trp1	Amp.	CEN	Invitrogen
pDEST32	fl-ADH1	N-term	–	Leu2	Gent.	CEN	Invitrogen
pGBKT7g	t-ADH1	N-term	–	Trp1	Kan.	2μ	(35)
pGBGT7g	t-ADH1	N-term	–	Trp1	Gent.	2μ	(29)
pGADT7g	fl-ADH1	–	N-term	Leu2	Amp.	2μ	(35)
pGBKCg	t-ADH1	C-term	–	Trp1	Kan.	2μ	(27)
pGADCg	fl-ADH1	–	C-term	Leu2	Amp.	2μ	(27)
Yeast strains							
Bait yeast strain	AH109						
Prey yeast strain	Y187						
Media and instruments							
Yeast media	YEPDA, selective liquid media, and agar plates						
Pin tool	Optional, but necessary, when large numbers are tested						

Fl full-length, *t* truncated, *N/C-term* N/C-terminal (fusion), *Amp* Ampicillin, *Kan* Kanamycin, *Gen* Gentamicin

and reporter genes or conducted in different laboratories. Rajagopala et al. investigated the underlying reasons for such inconsistencies and compared the effect of different vectors on their Y2H interactions. Low-copy Y2H vectors (yeast CEN ori) produce a higher fraction of interactions that are conserved and biologically relevant when compared to the high-copy vectors (yeast 2μ ori), but the latter appear to be more sensitive and thus detect more interactions (26).

Over the years, several Y2H vectors have been engineered and used for the interaction screening (Table 1). Traditionally, Y2H screens have been performed using N-terminal fusion proteins of DNA-binding and activation domains. Stellberger et al. constructed two new vectors that allow us to make both C-terminal fusion proteins of DNA-binding and activation domains and showed that permutations of C- and N-terminal Y2H vectors detect different subsets of interactions (27).

A recent study by Chen et al. benchmarked a number of two-hybrid vectors using a human positive reference set and a random reference set (92 protein pairs each) from Braun et al. (28). Chen et al. (29) cloned the positive reference set and the random reference set into five pairs of bait–prey vectors (pGBGT7g–pGADCg, pGBGT7g–pGADT7g, pDEST32–pDEST22, pGBKCg–pGADT7g, and pGBKCg–pGADCg). In addition to

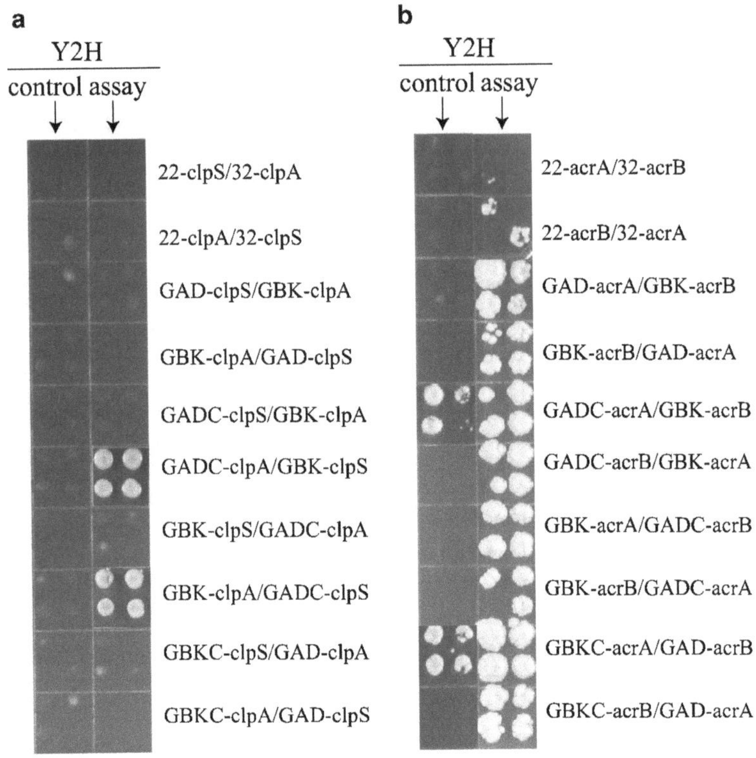

Fig. 3. An example to show the variations of the different two-hybrid vector systems. (**a**) The known protein interaction between *E. coli* AcrA and AcrB proteins tested in five two-hybrid vector pairs. Because each protein is tested as both bait and prey fusion, the interaction is tested in ten different configurations. The control column is tested with the bait and the empty prey vector which serves as a negative control. Only two out of ten tested combinations detected the interaction. (**b**) The known interaction between *E. coli* ClpS and ClpA proteins is tested in ten different configurations (as in (**a**)). Seven out of ten tested combinations show a positive interaction. Hence, protein interactions should be tested with multiple Y2H systems to reduce the number of false negatives. Note that vector names are abbreviated, but correspond to those listed in Table 1 (e.g., "22" = pDEST22 etc.).

each vector pair, they tested each protein both as activation (prey) and DBD fusion (bait), including C-terminal fusions in pGBKCg and pGADCg. This way, they tested each protein pair in ten different configurations (Figs. 3 and 4). The remarkable outcome of this study was that different Y2H variants detect markedly different subsets of interactions in the same interactome. All ten different configurations of bait–prey fusions were required to detect 73 of 92 interactions (79.3%), whereas individual vector pairs detected only 23.3 out of 92 interactions (25.3%) on average (29). Having multiple variants of Y2H that detect different subsets of interactions is of great value to generate more comprehensive protein interaction dataset; thus, future interactome projects must incorporate multiple Y2H vector systems with proper controls and adequate stringency.

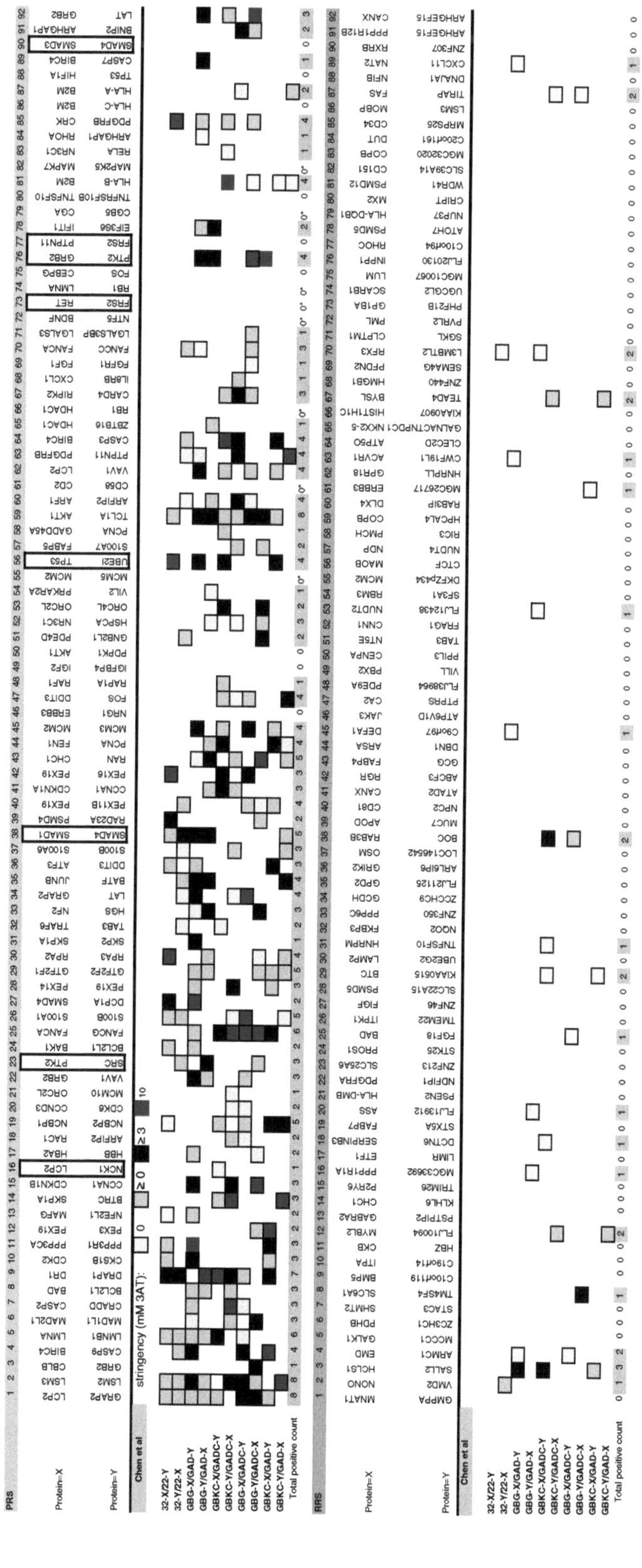

Fig. 4. Verification of known protein interactions by different two-hybrid vector systems: PRS is the Positive Reference Set of 92 known interactions among human proteins. The RRS panel shows the Random Reference Set of (most likely) noninteracting protein pairs from Braun et al. (28). X and Y indicate bait and prey proteins, respectively. Each of the 92 interaction pairs of PRS and RRS is tested in ten different configurations. The total positive count is the sum of vector pairs that detect the interaction (*gray boxes*). All ten different configurations of bait–prey fusions detect 73 out of 92 interactions (79.3%), whereas individual vector systems on average detected only 23.3 out of 92 interactions (25.3%). *Shades of gray* indicate levels of 3-AT (ranging from 0 [*light*] to 10 mM [*dark*]). Vector names are abbreviated but correspond to those listed in Table 1. After Chen et al. (29).

1.7. Evaluation of Raw Y2H Screening Data

Filtering of raw results significantly improves the data quality of the protein interaction set. In array screens, at least three parameters should be considered. *First*, protein interactions that are not reproduced in a retest experiment should be discarded. *Second*, for each prey, the number of different interacting baits (prey count) is counted; preys interacting with a large number of baits are nonspecific ("sticky" preys) and, thus, may have no biological relevance. The cutoff number depends also on the nature of baits and the number of baits screened: if a large family of related proteins is screened, it is not surprising that many of them find the same prey. As a general guideline, the number of baits interacting with a certain prey should not be larger than 5% of the bait number in genome-wide screenings. The *third* parameter is the background self-activation strength of the tested bait. The activation strength of interacting pairs must be significantly higher than with all other (background) pairs. Ideally, no activation (i.e., no colony growth) should be observed in noninteracting pairs. In the random library screening, the interacting protein is identified by sequencing. The interacting prey should be picked up at least two or more times with a bait.

In addition to the above parameters, more sophisticated statistical evaluations of the raw results have been suggested. For instance, filtering the raw interaction dataset by Logistic Regression (which uses positive and negative training sets of interactions) can help to identify the most reliable data (30, 31).

1.8. Quality of Large-Scale Two-Hybrid Interaction Data

Two-hybrid screens are not perfect. It is quite unlikely that you will detect all physiologically relevant interactors of your bait protein. *False negatives* may arise from steric hindrance of the two fusion proteins so that physical interaction or subsequent transcriptional activation is prevented. Other explanations for false negatives include instability of proteins or failure of nuclear localization, absence of a prey protein from a library, and inappropriate posttranslational modification of a bait or a prey, prohibiting an interaction. Earlier studies estimate that the false negative rate in array-based two-hybrid screens is on the order of 75%; i.e., up to 75% of all "true" interactions may be missed (13). This large number can be reduced by several strategies. For example, we investigated the interactome of bacterial flagella by using ORFs from both *T. pallidum* and *Campylobacter jejuni* which had estimated false negative rates of 76 and 77%, respectively. However, a combination of both datasets recovered 33% of all known flagellar interactions and, thus, had a false negative rate of 67%. When protein domains and fragments are used, this number can be further reduced. A recent study by Chen et al. has shown that different two-hybrid systems detect markedly different subsets of interactions in the same interactome (29). Using multiple variants of Y2H vectors (Figs. 4 and 5) for the same interactome can result in much

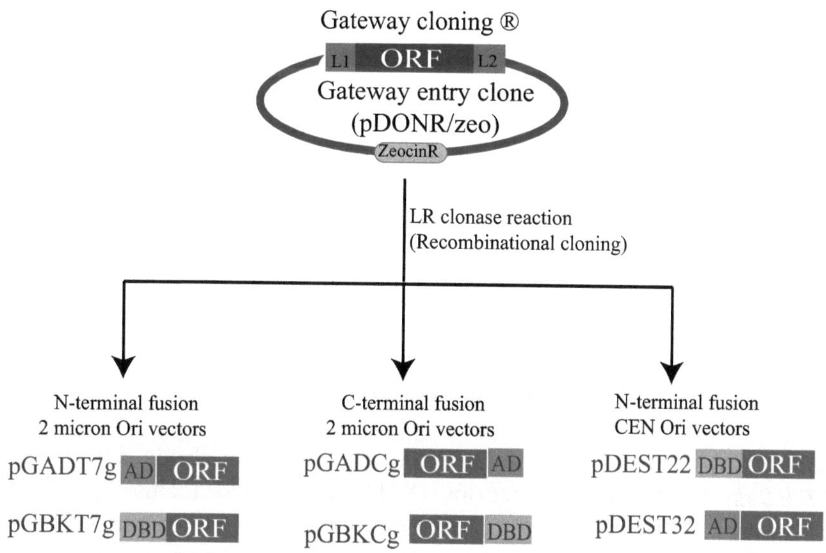

Fig. 5. Gateway cloning strategy for creating baits and preys. The Gateway-based Y2H expression clones are made by combining the ORFs of interest from a Gateway entry vector (such as pDONR/zeo or pDONR201) and the Y2H expression vectors (such as pDEST22 and pDEST32, Table 1) in the so-called Gateway LR reaction (20).

fewer false negatives, possibly on the order of 25–20%, compared to an average of 75% with single-vector systems (Fig. 4).

False positives: As with most assay systems, the two-hybrid system has the potential to produce false positives. "False positives" may be of technical or biological nature. A "technical" *false positive* is an apparent two-hybrid interaction that is not based on the assembly of two-hybrid proteins (i.e., the reporter gene(s) gets activated without a protein–protein interaction between bait and prey). Frequently, such false positives are associated with bait proteins that act as transcriptional activators. Some bait or prey proteins may affect general colony viability and, hence, enhance the ability of a cell to grow under selective conditions and activate the reporter gene. Mutations or other random events of unknown nature may be invoked as potential explanations as well. A number of procedures have been developed to identify or avoid false positives, including the utilization of multiple reporters, independent methods of specificity testing, or simply repeating assays to make sure that a result is reproducible (32–34) (described below).

A biological false positive involves a bona fide two-hybrid interaction with no physiological relevance. This includes the partners that can physically interact but that are never in close proximity to one another in the cell because of distinct subcellular localization or expression at different times during the life cycle. Examples may include paralogs that are expressed in different tissues or at different developmental stages. The problem is that the "false positive" nature can rarely be proven, as there may be unknown conditions under which these proteins do interact with a biological purpose. Overall, few technical false positives can be explained mechanistically.

While it often remains difficult to prove the biological significance of an interaction, many studies have attempted to validate them by independent methods. Finding an interaction by several methods certainly increases the probability that it is biologically significant. Recently, Uetz et al. (35) validated Y2H interactions of Kaposi Sarcoma Herpes virus (KSHV) by CoIP and found that about 50% of them can be confirmed. Similarly, when subsets of the large-scale human Y2H interactomes were evaluated, 78 and 65% of them could be verified by independent methods (17, 18).

1.9. Combination of AP/MS and Y2H Data and Protein Complex Topology

It is important to note that most methods detect either direct binary interactions (Y2H and related technologies) or indirect interactions (e.g., mass spectrometry) without knowing which proteins are interacting. Protein complexes are often interpreted as if the proteins that copurifiy are interacting in a particular manner consistent with either a spoke or a matrix model (36). The Y2H and other fragment complementation systems detect direct binary interactions. Combination of both methods gives a better picture of protein complex topology and an experimentally derived confidence score for each interaction. However, even the combination of both methods is usually not sufficient to establish accurate topology as some interactions may be too weak to be detected individually.

1.10. General Requirements for a Screen and Alternatives

Although the protocols in this chapter are based on the DNA-binding and activation domains of the yeast Gal4 protein, other DBDs and activation domains can be used.

In the LexA two-hybrid system, the DBD is provided by the entire prokaryotic LexA protein, which normally functions as a repressor in *Escherichia coli* when it binds to LexA operators. In the Y2H system, the LexA protein does not act as a repressor. An activation domain often used in the LexA two-hybrid system is the heterologous 88-residue acidic peptide B42 that strongly activates transcription in yeast. An interaction between the target protein (fused to the DBD) and a library-encoded protein (fused to an AD) creates a novel transcriptional activator with binding affinity for LexA operators.

In general, every component of the "classic" two-hybrid system can be replaced by different components: For example, the reporter gene does not need to be *HIS3*. Alternatively, *LEU2*, an enzyme involved in leucine biosynthesis, can be used. The reporter does not have to be a biosynthetic enzyme at all; green fluorescent protein (GFP) has been successfully used as a reporter gene (37), beta-galactosidase (lacZ) is common (38), and many others are under investigation. Finally, the two-hybrid system does not need to be based on transcription. Johnsson and Varshavsky (1994) developed a related system that is based on reconstituting artificially split ubiquitin, a protein that tags other proteins for degradation. As long as the function of a protein can be used as a selective marker, it is theoretically possible to divide it into fragments and drive the reassociation of the two fragments by exogenous "bait and prey" proteins, which are attached to each half. Several other variations have been developed and are described elsewhere (39, 40).

1.11. Genome-Wide Yeast Two-Hybrid Screening

The construction of an entire proteome array of an organism that can be screened in vivo under uniform conditions is a challenge. When proteins are screened on a genome scale, automated robotic procedures are necessary (see below). The procedure can be modified for manual use or for use with alternative screening strategies, such as synthetic lethal screens. With minor modifications, the array can be used to screen for protein interactions with DNA, RNA, or even small-molecule inhibitors of the Y2H interactions.

The protocols described here have been tested with yeast, bacterial, and viral proteins, but they can be applied to any other genome or subset thereof. Different high-throughput cloning methods used to generate two-hybrid clones, i.e., proteins with AD fusions (preys) and DBD fusions (baits), are therefore included below. The process involves the construction of the prey and bait array (Subheading 3.2) and screening of the array by either manual or robotic manipulation (Subheadings 3.5–3.8) screening the random cDNA or genomic library (Subheadings 3.7–3.9), including the selection of positives and scoring of results.

High-throughput screening projects deal with a large number of proteins; therefore, hands-on time and amount of resources become important issues. Options to reduce the screening effort are discussed. A prerequisite for array-based genome-wide screens is the existence of a cloned ORFeome; we briefly mention strategies to create such ORFeomes. Many ORFeome projects are currently being done (see ORFeomes below). We expect readily available complete ORFeomes for all major model organisms in the near future. Large-scale random library screening requires sequencing; having sequencing capacity becomes an important issue.

2. Materials

2.1. Yeast Media

1. *YEPD liquid medium*: 10 g yeast extract, 20 g peptone, 20 g glucose. Make up to 1 L with sterile water and autoclave.
2. *YEPDA liquid medium*: 10 g yeast extract, 20 g peptone, 20 g glucose. Make up to 1 L with sterile water and autoclave. After autoclaving, cool the medium to 60–70°C and then add 4 ml of 1% adenine solution (1% in 0.1 M NaOH).

3. *YEPDA solid medium*: 10 g yeast extract, 20 g peptone, 20 g glucose, 16 g agar. Make up to 1 L with sterile water and autoclave. After autoclaving, cool the medium to 60–70°C, then add 4 ml of 1% adenine solution (1% in 0.1 M NaOH), pour 40 ml into each sterile Omnitray plate (Nunc) under sterile hood, and let them solidify.
4. *Medium concentrate*: 8.5 g yeast nitrogen base, 25 g ammonium sulfate, 100 g glucose, 7 g dropout mix (see below). Make up to 1 L with sterile water and filter sterilize (Millipore).

2.2. Yeast Minimal Media (Selective) Plates

1. For 1 L of selective medium, autoclave 16 g agar in 800 ml water, cool the medium to 60–70°C, and then add 200 ml medium concentrate. Depending on the required selective plates, add the missing amino acids (see below) and/or 3-amino-1,2,4-triazole (3-AT) and pour plates.
2. –Trp plates (media-lacking tryptophan): Add 8.3 ml leucine and 8.3 ml histidine stock solution (see below).
3. –Leu plates (media-lacking leucine): Add 8.3 ml tryptophan and 8.3 ml histidine solution.
4. –Leu –Trp plates (media-lacking tryptophan and leucine): Add 8.3 ml histidine stock solution.
5. –Leu –Trp –His plates (media-lacking tryptophan, leucine, and histidine): Nothing needs to be added.
6. –Leu –Trp –His+ 3 mM 3-AT plates: Add 6 ml of 3-AT (0.5 M) to a final concentration of 3 mM.
7. *Dropout mix* (–His, –Leu, –Trp): Mix 1 g methionine, 1 g arginine, 2.5 g phenylalanine, 3 g lysine, 3 g tyrosine, 4 g isoleucine, 5 g glutamic acid, 5 g aspartic acid, 7.5 g valine, 10 g threonine, 20 g serine, 1 g adenine, and 1 g uracil and store under dry, sterile conditions.
8. *Amino acid stock solutions*: **Histidine** (His): Dissolve 4 g of histidine in 1 L sterile water and filter sterilize. **Leucine** (Leu): Dissolve 7.2 g of leucine in 1 L sterile water and sterile filter. **Tryptophan** (Trp): Dissolve 4.8 g of tryptophan in 1 L sterile water and filter sterilize.

2.3. Yeast Transformation

1. Salmon sperm DNA (carrier DNA): Dissolve 7.75 mg/ml salmon sperm DNA (Sigma) in sterile water, autoclave for 15 min at 121°C, and store at –20°C.
2. Dimethylsufoxide (DMSO, Sigma).
3. Competent host yeast strains, e.g., AH109 (for baits) and Y187 (for preys).
4. Lithium acetate (LiOAc) (0.1 M).

5. Selective plates (depending on the selective markers, described in Subheading 2.2).
6. 96 PEG solution: Mix 45.6 g PEG (Sigma), 6.1 ml of 2 M LiOAc, 1.14 ml of 1 M Tris, pH 7.5, and 232 μl 0.5 M EDTA. Make up to 100 ml with sterile water and autoclave.
7. Plasmid clones or linearized vector DNA and PCR product (for homologous recombination).

2.4. Bait Self-Activation Test

1. YEPDA liquid medium and selective media agar in single-well microtiter plates (Omnitray plates, Nunc).
2. –Trp –Leu ("–LT") plates (see Subheading 2.2).
3. Selective plates without Trp, Leu, and His ("–LTH"), but with different concentrations of 3-AT, e.g., 0 mM, 1 mM, 3 mM, 10 mM, 50 mM, and 100 mM (–LTH/3-AT plates).
4. Bait strains and the prey strain carrying the empty prey plasmid, e.g., Y187 strain with pDEST22 plasmid (Invitrogen).

2.5. Two-Hybrid Screening Protocol

1. 20% (v/v) bleach (1% sodium hypochlorite).
2. 95% (v/v) ethanol.
3. Single-well microtiter plate (e.g., OmniTray; Nalge Nunc) containing solid YEPD + adenine medium (see Subheading 2.1), –Leu –Trp, –His –Leu –Trp, and –His –Leu –Trp+ different concentrations of 3-AT.
4. 384-Pin replicator for manual screening or robot (e.g., Beckman Biomek FX).
5. Bait liquid culture (DBD fusion-expression yeast strain).
6. Yeast prey array on solid YEPDA plates.

2.6. Retest of Protein Interactions

1. 96-well microtiter plates (U- or V-shaped).
2. YEPDA medium and YEPDA agar in Omnitrays (Nunc).
3. Selective agar plates (–LT, –LTH with 3-AT).
4. Prey yeast strain carrying empty prey plasmid, e.g., pDEST22 in Y187 strain.
5. Bait and prey strains to be retested.

2.7. Beta-Galactosidase Filter Lift Assay

1. Selective plate (–LT) with diploid yeast colonies (from Subheading 3.6). The diploid cells carry the bait and prey combinations to be tested for activation of the beta-galactosidase reporter.
2. Omnitray plate.
3. Nitrocellulose membrane and Whatman paper.
4. Z-buffer: 60 mM Na_2HPO_4 (anhyd.), 60 mM NaH_2PO_4, 10 mM KCl, 1 mM $MgSO_4$.
5. X-GAL solution: 40 mg/ml in dimethylformamide (DMF).

3. Methods

3.1. Strategic Planning

Before starting an array-based screen, the size and character of the array must be designed and the ultimate aims of the experiment need to be considered. Factors that may be varied include the form of protein array (e.g., full-length protein or single domain, choice of epitope tags, etc.). Similarly, the arrayed proteins may be related (e.g., a family or pathway of related proteins, orthologs of a protein from different species, the entire protein complement of a model organism). In our experience, certain protein families work extremely well (e.g., splicing proteins) while others do not appear to work at all (e.g., many metabolic enzymes). We recommend to carry out a small-scale pilot study incorporating positive and negative controls before committing to a full-scale project.

Although high-throughput screening projects can be performed manually, automation is strongly recommended. Highly repetitive tasks are not only boring and straining, but also error-prone when done manually. If you do not have local access to robotics, you may have to collaborate with a laboratory that has.

3.2. Generation of a Protein Array Suitable for High-Throughput Screening

Once the set of proteins to be included in the array is defined, the coding genes need to be PCR-amplified and cloned into Y2H bait and prey vectors. In order to facilitate the cloning of a large number to proteins, site-specific recombination-based systems are commonly used (e.g., Gateway (20), see Fig. 5). Gateway cloning requires expensive enzymes and vectors, although both may be produced in the lab.

3.2.1. Gateway Cloning

Gateway (Invitrogen) cloning provides a fast and efficient way of cloning the ORFs (20). It is based on the site-specific recombination properties of bacteriophage lambda (41); recombination is mediated between the so-called attachment sites (att) of DNA molecules: between attB and attP sites or between attL and attR sites. The first step to Gateway cloning is inserting the gene of interest into a specific entry vector. One way of obtaining the initial entry clones is by recombining a PCR product of the ORF flanked by attB sites with the attP sites of a pDONR vector (Invitrogen). The resulting entry clone plasmid contains the gene of interest flanked by attL recombination sites. These attL sites can be recombined with attR sites on a destination vector, resulting in a plasmid for functional protein expression in a specific host. For example, a Gateway entry clone (in a pDONR vector) can be subsequently transferred to multiple Y2H expression vectors (such as pGADT7g, pGBKT7g, pGADCg, pGBKCg, pDEST22, and pDEST32, Fig. 5).

3.2.2. The ORFeome

The starting point of a systematic array-based Y2H screening is the construction of an ORFeome. An ORFeome represents all ORFs of a genome – in our case, the selected gene set individually cloned into entry vectors. More and more ORFeomes are available and can be directly used for generating the Y2H bait and prey constructs. These ORFeomes range from small viral genomes, e.g., KSHV and VZV (35), to several bacterial genomes, such as *E. coli* (21), *Bacillus anthracis*, or *Yersinia pestis*. These bacterial ORFeomes are available from BEI Resources (http://www.beiresources.org/). Clone sets of multicellular eukaryotes, e.g., *Caenorhabditis elegans* (42), human (43), or plant (44), have also been described. However, not all genes of interest are already available in entry vectors. Both entry vector construction and the subsequent destination vector cloning can be done for multiple ORFs in parallel. The whole procedure can be automated using 96-well plates so that whole ORFeomes can be processed in parallel.

3.3. The Prey Array

The Y2H array is set up from an ordered set of AD-containing strains (preys) rather than BD-containing strains (baits) because the former do not generally result in self-activation of transcription. The prey constructs are assembled by transfer of the ORFs from entry vectors into specific prey vectors by recombination. Several prey vectors for the Gateway system are available. In our lab, we primarily use the Gateway-compatible pGADT7g vector, a derivative of pGADT7 (Clontech), or pDEST22 (Invitrogen) (Fig. 5). These prey constructs are transformed into haploid yeast cells (Subheading 3.4), e.g., the Y187 strain (mating-type alpha) (Table 2). Finally, individual yeast colonies, each carrying one specific prey construct, are arrayed on agar plates in a 96- or 384-format in duplicates or quadruplicates.

3.3.1. Bait Construction

Baits are also constructed by recombination-based transfer of the ORFs into specific bait vectors. Bait vectors used in our lab are the Gateway-adapted pGBKT7g, pGBKCg (Clontech), and pDEST32

Table 2
Yeast strains and their genotypes

Yeast strains	Genotypes
Y187	MATα, ura3-52, his3-200, ade2-101, trp1-901, leu2-3, 112, gal4Δ, met–, gal80Δ,URA3::GAL1UAS-GAL1-TATA-lacZ (after Harper et al. (47))
AH109	MATa, trp1-901, leu2-3, 112, ura3-52, his3-200, gal4Δ, gal80Δ, LYS2::GAL1UAS-GAL1TATA-HIS3, GAL2UAS-GAL2TATA-ADE2, URA3::MEL1UAS-MEL1 TATA-lacZ (after James et al. (48))

(Invitrogen) (Fig. 5). The bait constructs are also transformed into haploid yeast cells (described in Subheading 3.4), e.g., the AH109 strain (mating-type a) (Table 2). After self-activation testing, the baits can be tested for interactions against the Y2H prey array or random prey library (see Note 1).

3.4. High-Throughput Yeast Transformation

This method is recommended for the high-throughput transformation of the bait or prey plasmid clones into respective yeast strains, and is based on the method of Cagney et al. (45). This protocol is suitable for 1,000 transformations, it can be scaled up and down as required, and most of the steps can be automated. Selection of the transformed yeast cells requires leucine- or tryptophan-free media (–Leu or –Trp depending on the selective marker on the plasmid). Moreover, at least one of the haploid strains must contain a two-hybrid reporter gene (here, HIS3 under GAL4 control).

1. Prepare competent yeast cells: Inoculate 250 ml YEPD liquid medium with freshly grown yeast strains on YEPD agar medium in a 2-L flask and grow in a shaker (shaking at 200 rpm) at 30°C. Remove the yeast culture from the shaker when the cell density reaches OD 1.0–1.3. This usually takes 12–16 h.
2. Spin out the cells at 2,000 × *g* for 5 min at room temperature; pour off the supernatant.
3. Dissolve the cell pellet in 30 ml of LiOAc (0.1 M); make sure that pellet is completely dissolved and there are no cell clumps.
4. Spin the cells in a 50-ml Falcon tube at 2,000 × *g* for 5 min at room temperature, pour off the supernatant, and dissolve the cell pellet in a total volume of 10 ml LiOAc (0.1 M).
5. Prepare the *yeast transformation mix* without yeast cells by mixing the following components in a 200-ml sterile bottle:

Component	For 1,000 reactions (ml)
96-PEG	100
Salmon sperm DNA	3.2
DMSO	3.4

6. Add the competent yeast cells prepared above (steps 1–4) to the yeast transformation mix; shake the bottle vigorously by hand or vortex for 1 min.
7. Pipette 100 μl of the yeast transformation mix into a 96-well transformation plate (we generally use Costar 3596 plates) by using a robotic liquid handler (e.g., Biomek FX) or a multistep pipette.
8. Now add 25–50 ng of plasmid; keep one negative control (i.e., only yeast transformation mix).

9. Seal the 96-well plates with plastic or aluminum tape and vortex for 2–3 min. Care should be taken to seal the plates properly; vigorous vortexing might cause cross-contamination.
10. Incubate the plates at 42°C for 30 min.
11. Spin the 96-well plate for 5 min at 2,000 × *g*; discard the supernatant and aspirate by tapping on a cotton napkin for a couple of times.
12. Add 150 μl of selective liquid media to each well, depending on the selective marker on the plasmid construct (for example, trytophan- or leucine-free liquid media). Seal the plates with AirPore tape (Qiagen) to protect from evaporation.
13. Incubate at 30°C for 36–48 h.
14. Pellet the cells by spinning at 2,000 × *g* for 5 min, discard the supernatant, and add 10 μl sterile H_2O to each well.
15. Transfer the cells to selective agar plate to select yeast with transformed plasmid (single-well Omnitrays, 128 × 86 mm, from Nunc are well-suited for robotic automation). Typically, we use a 96-pin tool (see reagent setup for the sterilization of the pin tool). As an alternative to the pin tool, one can use a multichannel pipette to transfer the cells. Allow the yeast spots to dry on the plates.
16. Incubate at 30°C for 2 days. Colonies start appearing after 24 h.

3.5. Self-Activation Test

Prior to the two-hybrid analyses, the bait yeast strains should be examined for self-activation. Self-activation is defined as detectable bait-dependent reporter gene activation in the absence of any prey interaction partner. Weak- to intermediate-strength self-activator baits can be used in two-hybrid array screens because the corresponding bait–prey interactions confer stronger signals than the self-activation background. In case of the *HIS3* reporter gene, the self-activation background can be suppressed by adding different concentrations of 3-AT, a competitive inhibitor of *HIS3*. Self-activation of all the baits is examined on plates containing different concentrations of 3-AT. The lowest concentration of 3-AT that suppresses growth in this test is used for the interaction screen (see below) because it avoids background growth while still detecting true interactions.

The aim of this test is to measure the background reporter activity (here, *HIS3*) of bait proteins in the absence of an interacting prey protein. This measurement is used for choosing the selection conditions used for Y2H screening described in Subheading 3.6.

1. Bait strains are arrayed onto a single-well Omnitray agar plate; either the standard 96-spot format or the 384-spot format is used (see Note 2).

2. The arrayed bait strains are mated with a prey strain carrying the empty prey plasmid, e.g., Y187 strain with pDEST22 (Invitrogen). Mating is conducted according to the standard screening protocol as described in Subheading 3.6. Note that here an array of baits is tested, whereas in a "real" screen (Subheading 3.6) an array of preys is tested.
3. After selecting for diploid yeast cells (on –LT agar), the cells are transferred to media selecting for the HIS3 reporter gene activity as described in Subheading 3.6. The –LTH transfer may be done to multiple plates with increasing concentrations of 3-AT. Suggested 3-AT concentrations are 0, 1, 3, 10, 25, 50, and 100 mM.
4. These –LTH+ 3-AT plates are incubated for 1 week at 30°C. The self-activation level of each bait is assessed: the lowest 3-AT concentration that completely prevents colony growth is noted. As this concentration of 3-AT suppresses reporter activation in the absence of an interacting prey, this 3-AT concentration is added to –LTH plates in the actual interaction screens as described in Subheading 3.6.

3.6. Screening for Protein Interactions Using a Yeast Protein Array

The Y2H prey array can be screened for protein interactions by a mating procedure that can be carried out manually or using robotics. A yeast strain expressing a single candidate protein as a DBD fusion is mated to all the colonies in the prey array (Fig. 1, step 1: shown for one prey plate). After mating, the colonies are transferred to a diploid-specific medium and then to the two-hybrid interaction selective medium. To manually screen with more than one bait, replicate copies of the array are used.

In many cases, a handheld 384-pin replicating tool can be used for routine transfer of colonies for screening. For large projects, however, a robotic workstation (e.g., Biomek 2000 or Biomek FX, Beckman Coulter) may be used to speed up the screening procedures and to maximize reproducibility. A 384-pin steel replicating tool (e.g., High-Density Replication Tool; V&P Scientific) can be used to transfer the colonies form one plate to another. Between the transfer steps, the pinning tool must be sterilized (see below).

Note that not all plasticware is compatible with robotic devices, although most modern robots can be reprogrammed to accept different consumables. In the procedure described here, the prey array is gridded on 86 × 128-mm single-well microtiter plates (e.g., OmniTray, Nalge Nunc International) in a 384-colony format (see Fig. 1).

1. *Sterilization*: Sterilize a 384-pin replicator by dipping the pins into 20% bleach for 20 s, sterile water for 1 s and 95% ethanol for 20 s, and sterile water again for 1 s. Repeat this sterilization after each transfer. Note: Immersion of the pins into these

solutions must be sufficient to ensure complete sterilization. When automatic pinning devices are used, the solutions need to be checked and refilled occasionally (especially ethanol, which evaporates faster than others).

Day 1

2. *Preparing prey array for screening*: Use the sterile replicator to transfer the yeast prey array from selective plates to single-well microtiter plates containing solid YEPD medium and grow the array overnight in a 30°C incubator (see Note 3).

3. *Preparing bait liquid culture (DBD fusion-expressing yeast strain):* Inoculate 20 ml of liquid YEPD medium in a 250-ml conical flask with a bait strain and grow overnight in a 30°C shaker (see Note 4).

Day 2

4. *Mating procedure*: Pour the overnight liquid bait culture into a sterile Omnitray plate. Dip the sterilized pins of the pin replicator (thick pins of ~1.5 mm diameter should be used to pin baits) into the bait liquid culture and place directly onto a fresh single-well microtiter plate containing YEPDA agar media. Repeat with the required number of plates and allow the yeast spots to dry onto the plates for 10–20 min.

5. Pick up the fresh prey array (i.e., AD) yeast colonies with sterilized pins (thin pins of ~1 mm diameter should be used to pin the preys) and transfer them directly onto the baits on the YEPDA plate so that each of the 384 bait spots per plate receives different prey yeast cells (i.e., a different AD fusion protein). Incubate overnight at 30°C to allow mating (Fig. 1, step 1, see Note 5).

6. *Seletion of diploids*: For the selection of diploids, transfer the colonies from YEPDA mating plates to plates containing –Leu –Trp medium using the sterilized pinning tool (thin pins should be used in this step). Grow for 2–3 days at 30°C until the colonies are >1 mm in diameter (Fig. 1, step 2, see Note 6).

7. *Interaction selection:* Transfer the colonies from –Leu –Trp plates to a single-well microtiter plate containing solid –His –Leu –Trp agar using the sterilized pinning tool. If the baits are self-activating, they have to be transferred to –His –Leu –Trp+ a specific concentration of 3-AT (Subheading 3.5). Incubate at 30°C for 6–8 days.

8. Score the interactions by looking for growing colonies that are significantly above background by size and are present as duplicate colonies.

9. The plates should be examined every day. Most two-hybrid positive colonies appear within 3–5 days, but occasionally positive interactions can be observed later. Very small colonies are

usually designated as background; however, there is no absolute measure to distinguish between the background and real positives. When there are many (e.g., >20) large colonies per array of 1,000 positions, we consider these baits as "random" activators. In this case, the screening should be repeated or the interactions should be retested.

10. Scoring can be done manually or using automated image analysis procedures. When using image analysis, care must be taken not to score contaminated colonies as positives.

3.7. Screening for Protein Interactions Using Random cDNA Libraries

A yeast strain expressing a single protein as a DBD fusion is mated to all the cells in the prey library (Fig. 2, step 1: shown for one bait sample). After mating, the resulting diploid cells are transferred to the Y2H selective medium, and the interacting prey is identified by yeast colony PCR of the resulting positive yeast colony, followed by DNA sequencing of the PCR product.

Day 1

1. Prepare prey library for screening: Inoculate 200 ml of selective medium (medium lacking leucine or tryptophan, depending on the selective marker on the prey plasmid) with an aliquot of yeast prey library (e.g., an *E. coli* prey library plasmid prep) to an O.D of 0.05. Grow at 30°C with shaking at 180 rpm for 16 h. Cultures should have an OD600 of 0.9–1.0 when they are harvested for the mating (the amount of prey library culture required depends on the number of baits you want to screen and the complexity of the prey library, see step 4).
2. Inoculate the empty prey vector in 200 ml selective medium (Y2H negative control).
3. Preparing bait liquid culture (DBD fusion-expressing yeast strain): Inoculate 10 ml of selective medium (medium lacking leucine or tryptophan, depending on the selective marker on the bait plasmid) with bait fusion-expressing yeast strain and grow the yeast overnight in a 30°C incubator (see Note 4).

Day 2

4. Mating procedure: Mix bait and prey at a 1:1 ratio, for example, 4 OD bait (4 ml of OD = 1) and 4 OD prey (4 ml of OD = 1) culture in 15-ml Falcon tubes (see Note 7).
5. For each bait, include one negative control, mix bait, and empty prey vector (1:1).
6. Centrifuge for 2 min at 3,000 × *g* at room temperature, and discard supernatant.
7. Resuspend pellet in 500 μl YPDA, plate on YEPDA agar plate (60 mm × 15 mm), and air dry the plates.

8. Incubate the plates at 30°C for 6 h or overnight at room temperature.
9. After incubation, collect the cells by washing the plate with 2 ml of sterile water.
10. Spin down the cells, remove the supernatant, and wash one time with 2 ml of sterile water.
11. Resuspend in 2 ml of selective medium (media lacking tryptophan, leucine, and histidine).
12. Plate 500 μl on the interaction selective agar plates –Leu –Trp –His+ 3 mM 3-AT (media lacking tryptophan, leucine, and histidine+ 3 mM 3-AT). To measure the number of diploids, make an aliquot of 1:100 dilution and plate the cells on –Leu –Trp plates (see Note 6); the screening depth in millions should be >0.1 million up to 1 million diploids in case of *E. coli* library screening, i.e., at least 20 times the number of library size. Store the remaining sample at 4°C for further use.
13. Interaction selection: Incubate the –Leu –Trp –His+ 3 mM 3-AT for 4–6 days at 30°C until the colonies are ~1 mm in diameter. If the baits are self-activating, they have to be transferred to –His –Leu –Trp+ a specific concentration of 3-AT (Subheading 3.5). Incubate at 30°C for 6–8 days.
14. Two-hybrid positives: The plates that show colony growth (but no colonies on control plates, i.e., bait mated to empty prey vector) identify the two-hybrid positive yeast clones.
15. Identity of interacting preys: The positive yeast colonies are picked either manually or using robotics and subjected to yeast colony PCR (Subheading 3.9), followed by DNA sequencing to identify the preys.

3.8. Protein Interaction Retesting

A major consideration when using the Y2H system is the number of false positives. The major sources for false positives are nonreproducible signals that arise through little-understood mechanisms. In Y2H screens, more than 90% of all interactions can be nonreproducible background (46). Thus, simple retesting by repeated mating can identify most false positives. We routinely use at least duplicate tests, although quadruplicates should be used if possible (see Fig. 1). Retesting is done by manually mating the interaction pair to be tested and by comparing the activation strength of this pair with the activation strength of a control, usually the bait mated with the strain that contains the empty prey vector.

Testing for reproducibility of interactions greatly increases the reliability of the Y2H interaction data. This method is used for specifically retesting interaction pairs detected in an array screen.

1. Re-array bait and prey strains of each interaction pair to be tested into 96-well microtiter plates. Use separate 96-well plates for baits and preys. For each retested interaction, fill one well of the bait plate and one corresponding well of the prey plate with 150 μl YEPD liquid medium.
2. For each retested interaction, inoculate the bait strain into a well of the 96-well bait.
3. Bait and prey strains should be at the corresponding position of the 96-well prey plate, for example, bait at position B2 of the bait plate and prey at position B2 of the prey plate. In addition, inoculate the prey strain with the empty prey vector (e.g., strain Y187 with plasmid pDEST22) into 20 ml YEPD liquid medium.
4. Incubate the plates overnight at 30°C.
5. Mate the baits grown in the bait plate with their corresponding preys in the prey plate. In addition, mate each bait with the prey strain carrying an empty prey vector as a background activation control. The mating is done according to Subheading 3.6 using the bait and prey 96-well plates directly as the source plates (see Note 8).
6. The transfers to selective plates and incubations are done as described in Subheading 3.6. As before, test different baits with different activation strengths on a single plate and pin the diploid cells onto –LTH plates with different concentrations of 3-AT. For choosing the 3-AT range, the activation strengths (Subheading 3.5) serve as a guideline.
7. After incubating for ~1 week at 30°C on –LTH/3-AT plates, the interactions are scored; positive interactions show a clear colony growth at a certain level of 3-AT, whereas no growth should be seen in the control (bait mated with empty vector strain).

3.9. Yeast Colony PCR and SAP

This protocol is designed to amplify the insert of the preys in the two-hybrid positive yeast clones using primers that bind to the upstream and downstream region of the insert. The PCR is optimized for 30 μl reaction; the total volume of the reaction may be increased if necessary. To clean up PCR products before sequencing, the PCR reaction is subjected to exonuclease I which removes leftover primers while the Shrimp Alkaline Phosphatase (SAP) removes the dNTPs.

1. Pick the yeast colony from interaction selective plate into 100 μl of sterile H_2O in 96-well plate (for longer storage, store at –80°C).
2. Take a 96-well PCR plate and pipette 5 U of zymolyase (1 μl) enzyme to each well.

3. Pipette 9 μl of the above yeast (1.) and incubate at 30°C for 60 min.
4. After incubation, add 20 μl PCR master mix with forward and reverse primers specific to prey vector used in the two-hybrid screening.
5. Run PCR cycles as recommended by the enzyme provider manual.
6. After PCR, load 5 μl of PCR reaction into agarose gel to check PCR products.
7. Purify the PCR reaction by SAP (see below).

SAP Master Mix

Components	100 samples (μl)
10× SAP buffer	50
Water	890
SAP (1 U/μl)	50
Exonuclease I (10 U/μl)	10

8. Spin the yeast colony PCR plate at 2,000 × *g* for 3 min (to sediment yeast debris).
9. Pipette 8 μl of PCR sample without touching the bottom yeast pellet into new PCR plate.
10. Add 10 μl of SAP master mix to 8 μl of PCR sample.
11. Incubate in the thermocycler as follows: 37°C for 60 min, 72°C for 15 min, and then put on hold at 4°C.
12. Use the sample for DNA sequencing using primers specific to prey vector.

3.10. Beta-Galactosidase Filter Lift Assay (Alternative Reporter Genes)

Y2H interactions can be reproduced using other reporter genes in addition to the one used in the actual screen depending on the different reporter genes present in the yeast strains used. Examples include beta-galactosidase or *ADE2* (for selection on adenine-deficient medium). Because of the use of different promoters, these reporter genes have different activation requirements, and Y2H interactions reproduced with different reporter genes are assumed to be more reliable. However, the use of multiple reporters may result in the loss of weaker Y2H positives. The beta-galactosidase reporter has the advantage of giving a semiquantitative output of the activation strength. Other reporters might be advantageous and can be transformed into yeast as additional plasmids or by using alternative strains, which contain the reporter as integrated construct. For example, the strain AH109 carries an alpha-galactosidase reporter gene which produces an enzyme that is secreted into the medium. Therefore, these cells do not require cell lysis for detection. The following method was adapted from the Breeden lab (http://labs.fhcrc.org/breeden/Methods/yeast%20 methods/B-Gal.doc).

1. Use the same diploid plate as in Subheading 3.6. As a control, the bait strains are mated with a prey strain containing an empty vector (following mating steps of Subheading 3.6).
2. Cut a nitrocellulose membrane to the dimensions of an Omnitray plate (Nunc). Place the nitrocellulose membrane on top of diploid yeast colonies and leave for 10 s.
3. Use tweezers to lift the filter and slowly submerge in liquid nitrogen for 1 min.
4. Place the membrane on an empty Omnitray plate (Nunc) to thaw.
5. Cut a Whatman paper to the same size as nitrocellulose membrane. Soak the Whatman paper with 2 ml Z-buffer to which 35 μl X-solution had been added.
6. Overlay the nitrocellulose filter with the Whatman paper and remove air bubbles.
7. Incubate at 30°C for 10–60 min.
8. Evaluate: A blue stain indicates the activation of the beta-galactosidase reporter and, therefore, a positive interaction.

4. Notes

1. Bait and prey must be transformed into yeast strains of opposite mating types to combine bait and prey plasmids by mating and to coexpress the fusion proteins in diploids. Bait and prey plasmids can go into either mating type. However, this decision also depends on existing bait or prey libraries to which the new library may be mated later.
2. Multiple baits are first inoculated at the different positions of a 96-well plate as liquid culture, and then cells are transferred (manually or with the use of a robot) to solid agar single-well plates (Omnitray plates). In this step, the 96-well format can also be converted into the 384-well format. This positions each bait in quadruplicates on the 384-well formatted plate. Full media agar (YEPDA agar) can be used; however, for long-term storage of the array, selective agar is suggested to prevent loss of plasmids.
3. In a systematic array-based Y2H screening, duplicate or quadruplicate prey arrays are usually used. In a random genomic

library screening, the entire experiment should be done in two copies to ensure reproducibility. Ideally, the master prey array should be kept on selective plates. The master copy of the array should only be used to make "working" copies on YEPDA agar plates for mating. These templates can be used for 1–2 weeks; after 2 weeks, it is recommended to copy the array onto fresh selective plates. Preys or bait clones tend to lose the plasmid if stored on YEPDA for longer periods, which may reduce the mating and screening efficiency.

4. If the bait strains are frozen, they are streaked or pinned on selective solid medium plates and grown for 1–2 days at 30°C. Baits from this plate are then used to inoculate the liquid YEPD medium. It is important to make a fresh bait culture for Y2H mating, as keeping the bait culture on reach medium (YEPD) for a long time may cause loss of plasmids. Usually, we grow baits overnight for mating.
5. Mating usually takes place in <15 h, but a longer period is recommended because some bait strains show poor mating efficiency. Adding adenine into the bait culture before mating increases the mating efficiency of some baits.
6. This is an essential control step to ensure successful mating because only diploid cells containing the Leu2 and Trp1 markers on the prey and bait vectors, respectively, grow in this medium. This step also helps the recovery of the colonies and increases the efficiency of the next interaction selection step.
7. The amount of OD units of bait and prey depends on the complexity of prey library; in case of *E. coli* which contains about 4,300 ORFs, we use 4 OD units of baits and preys. In case of human cDNA library screening, we recommend to use 12 OD units of baits and preys each.
8. First, the baits are transferred from their 96-well plate to two YEPDA plates (interaction test and control plate) using a 96-well replication tool. Let the plate dry for 10–20 min. Then, transfer the preys from their 96-well plate onto the first YEPDA plate and the empty prey vector control strain onto the second YEPDA plate.

Acknowledgments

Yu-Chi Chen and Thorsten Stellberger helped with the development and testing of new two-hybrid vectors mentioned in this chapter. Work on this chapter was funded by NIH grant RO1GM79710 and EU grant HEALTH-F3-2009-223101.

References

1. Uetz, P., Giot, L., Cagney, G., Mansfield, T.A., Judson, R.S., Knight, J.R., Lockshon, D., Narayan, V., Srinivasan, M., Pochart, P., et al. (2000). A comprehensive analysis of protein–protein interactions in Saccharomyces cerevisiae. Nature *403*, 623–627.
2. Gavin, A.C., Bosche, M., Krause, R., Grandi, P., Marzioch, M., Bauer, A., Schultz, J., Rick, J.M., Michon, A.M., Cruciat, C.M., et al. (2002). Functional organization of the yeast proteome by systematic analysis of protein complexes. Nature *415*, 141–147.
3. Ho, Y., Gruhler, A., Heilbut, A., Bader, G.D., Moore, L., Adams, S.L., Millar, A., Taylor, P., Bennett, K., Boutilier, K., et al. (2002). Systematic identification of protein complexes in Saccharomyces cerevisiae by mass spectrometry. Nature *415*, 180–183.
4. Fields, S., and Song, O. (1989). A novel genetic system to detect protein-protein interactions. Nature *340*, 245–246.
5. Schwartz, H., Alvares, C.P., White, M.B., and Fields, S. (1998). Mutation detection by a two-hybrid assay. Hum Mol Genet *7*, 1029–1032.
6. Vidal, M., and Endoh, H. (1999). Prospects for drug screening using the reverse two-hybrid system. Trends Biotechnol *17*, 374–381.
7. Vidal, M., and Legrain, P. (1999). Yeast forward and reverse 'n'-hybrid systems. Nucleic acids research *27*, 919–929.
8. SenGupta, D.J., Zhang, B., Kraemer, B., Pochart, P., Fields, S., and Wickens, M. (1996). A three-hybrid system to detect RNA-protein interactions in vivo. Proceedings of the National Academy of Sciences of the United States of America *93*, 8496–8501.
9. Estojak, J., Brent, R., and Golemis, E.A. (1995). Correlation of two-hybrid affinity data with in vitro measurements. Mol Cell Biol *15*, 5820–5829.
10. Rain, J.C., Selig, L., De Reuse, H., Battaglia, V., Reverdy, C., Simon, S., Lenzen, G., Petel, F., Wojcik, J., Schachter, V., et al. (2001). The protein-protein interaction map of Helicobacter pylori. Nature *409*, 211–215.
11. Raquet, X., Eckert, J.H., Muller, S., and Johnsson, N. (2001). Detection of altered protein conformations in living cells. J Mol Biol *305*, 927–938.
12. Cagney, G., Uetz, P., and Fields, S. (2001). Two-hybrid analysis of the Saccharomyces cerevisiae 26S proteasome. Physiol Genomics *7*, 27–34.
13. Rajagopala, S.V., Titz, B., Goll, J., Parrish, J.R., Wohlbold, K., McKevitt, M.T., Palzkill, T., Mori, H., Finley, R.L., Jr., and Uetz, P. (2007). The protein network of bacterial motility. Molecular systems biology *3*, 128.
14. Ito, T., Chiba, T., Ozawa, R., Yoshida, M., Hattori, M., and Sakaki, Y. (2001). A comprehensive two-hybrid analysis to explore the yeast protein interactome. Proceedings of the National Academy of Sciences of the United States of America *98*, 4569–4574.
15. Li, S., Armstrong, C.M., Bertin, N., Ge, H., Milstein, S., Boxem, M., Vidalain, P.O., Han, J.D., Chesneau, A., Hao, T., et al. (2004). A map of the interactome network of the metazoan C. elegans. Science *303*, 540–543.
16. Giot, L., Bader, J.S., Brouwer, C., Chaudhuri, A., Kuang, B., Li, Y., Hao, Y.L., Ooi, C.E., Godwin, B., Vitols, E., et al. (2003). A protein interaction map of Drosophila melanogaster. Science *302*, 1727–1736.
17. Rual, J.F., Venkatesan, K., Hao, T., Hirozane-Kishikawa, T., Dricot, A., Li, N., Berriz, G.F., Gibbons, F.D., Dreze, M., Ayivi-Guedehoussou, N., et al. (2005). Towards a proteome-scale map of the human protein-protein interaction network. Nature *437*, 1173–1178.
18. Stelzl, U., Worm, U., Lalowski, M., Haenig, C., Brembeck, F.H., Goehler, H., Stroedicke, M., Zenkner, M., Schoenherr, A., Koeppen, S., et al. (2005). A human protein-protein interaction network: a resource for annotating the proteome. Cell *122*, 957–968.
19. Schwikowski, B., Uetz, P., and Fields, S. (2000). A network of protein-protein interactions in yeast. Nature biotechnology *18*, 1257–1261.
20. Walhout, A.J., Temple, G.F., Brasch, M.A., Hartley, J.L., Lorson, M.A., van den Heuvel, S., and Vidal, M. (2000). GATEWAY recombinational cloning: application to the cloning of large numbers of open reading frames or ORFeomes. Methods Enzymol *328*, 575–592.
21. Rajagopala, S.V., Yamamoto, N., Zweifel, A.E., Nakamichi, T., Huang, H.K., Mendez-Rios, J.D., Franca-Koh, J., Boorgula, M.P., Fujita, K., Suzuki, K., et al. (2010). The Escherichia coli K-12 ORFeome: a resource for comparative molecular microbiology. BMC genomics *11*, 470.
22. Parrish, J.R., Yu, J., Liu, G., Hines, J.A., Chan, J.E., Mangiola, B.A., Zhang, H., Pacifico, S., Fotouhi, F., DiRita, V.J., et al. (2007). A proteome-wide protein interaction map for Campylobacter jejuni. Genome biology *8*, R130.
23. Zhong, J., Zhang, H., Stanyon, C.A., Tromp, G., and Finley, R.L., Jr. (2003). A strategy for constructing large protein interaction maps

using the yeast two-hybrid system: regulated expression arrays and two-phase mating. Genome Res *13*, 2691–2699.

24. Jin, F., Avramova, L., Huang, J., and Hazbun, T. (2007). A yeast two-hybrid smart-pool-array system for protein-interaction mapping. Nature methods *4*, 405–407.
25. Joung, J.K., Ramm, E.I., and Pabo, C.O. (2000). A bacterial two-hybrid selection system for studying protein-DNA and protein-protein interactions. Proceedings of the National Academy of Sciences of the United States of America *97*, 7382–7387.
26. Rajagopala, S.V., Hughes, K.T., and Uetz, P. (2009). Benchmarking yeast two-hybrid systems using the interactions of bacterial motility proteins. Proteomics *9*, 5296–5302.
27. Stellberger, T., Hauser, R., Baiker, A., Pothineni, V.R., Haas, J., and Uetz, P. (2010). Improving the yeast two-hybrid system with permutated fusions proteins: the Varicella Zoster Virus interactome. Proteome science *8*, 8.
28. Braun, P., Tasan, M., Dreze, M., Barrios-Rodiles, M., Lemmens, I., Yu, H., Sahalie, J.M., Murray, R.R., Roncari, L., de Smet, A.S., et al. (2009). An experimentally derived confidence score for binary protein-protein interactions. Nature methods *6*, 91–97.
29. Chen, Y.C., Rajagopala, S.V., Stellberger, T., and Uetz, P. (2010). Exhaustive benchmarking of the yeast two-hybrid system. Nature methods *7*, 667–668.
30. Bader, J.S., Chaudhuri, A., Rothberg, J.M., and Chant, J. (2004). Gaining confidence in high-throughput protein interaction networks. Nature biotechnology *22*, 78–85.
31. von Mering, C., Jensen, L.J., Kuhn, M., Chaffron, S., Doerks, T., Kruger, B., Snel, B., and Bork, P. (2007). STRING 7--recent developments in the integration and prediction of protein interactions. Nucleic acids research *35*, D358–362.
32. Koegl, M., and Uetz, P. (2007). Improving yeast two-hybrid screening systems. Briefings in functional genomics & proteomics *6*, 302–312.
33. Serebriiskii, I., Estojak, J., Berman, M., and Golemis, E.A. (2000). Approaches to detecting false positives in yeast two-hybrid systems. BioTechniques *28*, 328–330, 332–326.
34. Serebriiskii, I.G., and Golemis, E.A. (2001). Two-hybrid system and false positives. Approaches to detection and elimination. Methods in molecular biology (Clifton, N.J *177*, 123–134.
35. Uetz, P., Dong, Y.A., Zeretzke, C., Atzler, C., Baiker, A., Berger, B., Rajagopala, S.V., Roupelieva, M., Rose, D., Fossum, E., et al. (2006). Herpesviral protein networks and their interaction with the human proteome. Science *311*, 239–242.
36. Goll, J., and Uetz, P. (2006). The elusive yeast interactome. Genome biology *7*, 223.
37. Cormack, R.S., Hahlbrock, K., and Somssich, I.E. (1998). Isolation of putative plant transcriptional coactivators using a modified two-hybrid system incorporating a GFP reporter gene. Plant J *14*, 685–692.
38. Rossi, F., Charlton, C.A., and Blau, H.M. (1997). Monitoring protein-protein interactions in intact eukaryotic cells by beta-galactosidase complementation. Proceedings of the National Academy of Sciences of the United States of America *94*, 8405–8410.
39. Drees, B.L. (1999). Progress and variations in two-hybrid and three-hybrid technologies. Curr Opin Chem Biol *3*, 64–70.
40. Frederickson, R.M. (1998). Macromolecular matchmaking: advances in two-hybrid and related technologies. Curr Opin Biotechnol *9*, 90–96.
41. Landy, A. (1989). Dynamic, structural, and regulatory aspects of lambda site-specific recombination. Annu Rev Biochem *58*, 913–949.
42. Lamesch, P., Milstein, S., Hao, T., Rosenberg, J., Li, N., Sequerra, R., Bosak, S., Doucette-Stamm, L., Vandenhaute, J., Hill, D.E., et al. (2004). C. elegans ORFeome version 3.1: increasing the coverage of ORFeome resources with improved gene predictions. Genome Res *14*, 2064–2069.
43. Rual, J.F., Hirozane-Kishikawa, T., Hao, T., Bertin, N., Li, S., Dricot, A., Li, N., Rosenberg, J., Lamesch, P., Vidalain, P.O., et al. (2004). Human ORFeome version 1.1: a platform for reverse proteomics. Genome Res *14*, 2128–2135.
44. Gong, W., Shen, Y.P., Ma, L.G., Pan, Y., Du, Y.L., Wang, D.H., Yang, J.Y., Hu, L.D., Liu, X.F., Dong, C.X., et al. (2004). Genome-wide ORFeome cloning and analysis of Arabidopsis transcription factor genes. Plant Physiol *135*, 773–782.
45. Cagney, G., Uetz, P., and Fields, S. (2000). High-throughput screening for protein-protein interactions using two-hybrid assay. Methods Enzymol *328*, 3–14.
46. Uetz, P. (2002). Two-hybrid arrays. Curr Opin Chem Biol *6*, 57–62.
47. Harper, J.W., Adami, G.R., Wei, N., Keyomarsi, K., and Elledge, S.J. (1993). The p21 Cdk-interacting protein Cip1 is a potent inhibitor of G1 cyclin-dependent kinases. Cell *75*, 805–816.
48. James, P., Halladay, J., and Craig, E.A. (1996). Genomic libraries and a host strain designed for highly efficient two-hybrid selection in yeast. Genetics *144*, 1425–1436.

Chapter 2

Identification of Mammalian Protein Complexes by Lentiviral-Based Affinity Purification and Mass Spectrometry

Zuyao Ni, Jonathan B. Olsen, Andrew Emili, and Jack F. Greenblatt

Abstract

Protein complexes and protein–protein interactions (PPIs) are fundamental for most biological functions. Deciphering the extensive protein interaction networks that occur within cellular contexts has become a logical extension to the human genome project. Proteome-scale interactome analysis of mammalian systems requires efficient methods for accurately detecting PPIs with specific considerations for the intrinsic technical challenges of mammalian genome manipulation. In this chapter, we outline in detail an innovative lentiviral-based functional proteomic approach that can be used to rapidly characterize protein complexes from a broad range of mammalian cell lines. This method integrates the following key features: (1) lentiviral elements for efficient delivery of tagged constructs into mammalian cell lines; (2) site-specific Gateway™ recombination sites for easy cloning; (3) versatile epitope-tagging system for flexible affinity purification strategies; and (4) LC-MS-based protein identification using tandem mass spectrometry.

Key words: Protein–protein interactions, Protein complexes, Affinity purification, Lentivirus, Gateway™ cloning, Mass spectrometry

1. Introduction

Most cell type-specific functions and phenotypes are mediated by the activities of multiprotein complexes as well as other types of protein–protein interactions (PPIs) and posttranslational modifications. Accordingly, the formation and function of macromolecular protein complexes underpin the whole of cell biology, making the identification of PPI networks central – even fundamental – to understanding most molecular pathways and processes. Analysis of the variations of protein complex composition in different cell and tissue types is an important step toward understanding the relationship between gene products and cellular function and

Gerard Cagney and Andrew Emili (eds.), *Network Biology: Methods and Applications*, Methods in Molecular Biology, vol. 781, DOI 10.1007/978-1-61779-276-2_2, © Springer Science+Business Media, LLC 2011

phenotype in diverse physiological contexts. To date, most large-scale PPI studies have been performed in model organisms, primarily the budding yeast *Saccharomyces cervisiae*, including analysis by yeast two-hybrid (Y2H) and tandem affinity purification (TAP). While the former approach provides information relating to interactions between two proteins, typically of binary nature, the latter is a type of affinity purification and mass spectrometry (AP-MS) approach, which can detect multimeric protein complexes. In this chapter, we focus on AP-MS for the identification of protein complexes in mammalian contexts and outline a detailed lentiviral-based AP-MS protocol.

AP-MS is one of the more commonly used techniques in studies of PPIs and protein complex compositions (1–4). AP-MS relies on the fusion of a protein-coding sequence of interest with an epitope tag that allows protein purification to be conducted in a generic and standardized fashion (4–8). We recently developed a lentivirus-based tagging system for the affinity purification of mammalian proteins from tissue culture cells (9). This system contains an N- or C-terminal triple affinity tag, termed the versatile affinity (VA) tag, which consists of 3×FLAG, 6×His, and 2×StrepII epitopes with the dual tobacco etch virus (TEV) protease cleavage site located between 3×FLAG and 6×His. This lentivirus-based tagging system also incorporates an efficient system for cloning open reading frames (ORFs) from Gateway™ entry ORF clones into the vectors (10–12). The VA tag and Gateway™ recombination elements were reconstructed into lentiviral vectors that can infect most human cell types, including both dividing and nondividing cells (13–15). Multiple promoters are available in this system to control tagged gene expression, including a doxycycline-inducible promoter for adjusting protein expression levels. A vector map and brief outline of the lentiviral-based tagging method are found in Fig. 1.

Collectively, this innovative lentiviral-based AP-MS approach for mammalian protein complex identification addresses key bottlenecks related to large-scale mammalian AP-MS experiments. Important features include the following: (1) applicability to most mammalian cell types; (2) compatibility with publicly and commercially available cDNA libraries; (3) an efficient epitope-tagging system; and (4) compatibility with multiple purification strategies. As an overview, the workflow involved in such a pipeline entails (1) cloning target ORFs into destination vectors containing lentiviral elements, (2) production and harvesting of lentiviruses encapsulating the tagged constructs, (3) infection and scale-up of mammalian cell lines stably expressing the tagged constructs, (4) two-step purification for enrichment of target protein complexes, and (5) proteolytic digestion and sequencing of the peptide mixtures by shot-gun tandem mass spectrometry.

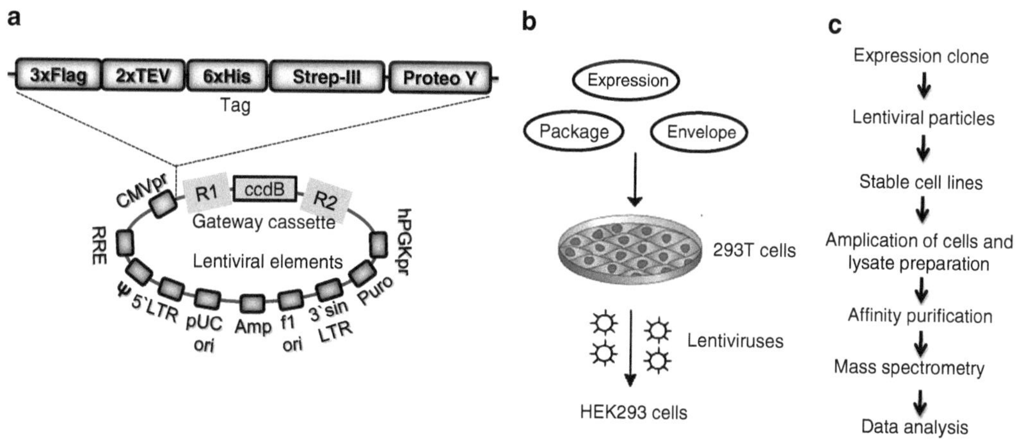

Fig. 1. MAPLE system. (**a**) Triple tag in gateway lentiviral system. The versatile affinity purification (VAP) tag consists of the following epitopes: 3×FLAG, 2× tobacco etch virus (TEV) protease cleavage site, 6×His, and 2× streptactin (Strep III). A yeast proteotyptic peptide (Proteo Y) is included for bait quantitation. The VAP tag is in-frame with the Gateway cassette in a lentiviral transfer vector. Only N-terminal tag is shown here, and the tags at the C terminus are in reverse order. The key lentiviral elements are as follows: *RRE* Rev-responsive element; Ψ packaging element (Psi); *5′-LTR* 5′-long terminal repeats; *3′-sin LTR* HIV 3′-LTR with self-inactivating U3 deletion; *Puro* Puromycin resistance; *Amp* Ampicillin resistance. (**b**) Lentiviral infection. Expression plasmids with tagged constructs are transfected into HEK293T cells together with the package and envelop plasmids to produce lentivirus. Harvested lentiviruses are then infected into HEK293 cells. (**c**) MAPLE system pipeline. Adapted from (9).

2. Materials

2.1. BP Reaction

1. BP clonase enzyme mix (Invitrogen).
2. Gateway™ DONR plasmids pDONR221 or pDONR223 (Invitrogen).
3. GeneJET plasmid miniprep kit (Fermentas).
4. GeneJET PCR purification kit (Fermantas).
5. Kanamycin (Sigma) dissolved into H_2O and stored at −20°C.
6. M13 sequencing primers. Forward: 5′-GTT TTC CCA GTC ACG AC-3′, reverse: 5′-CAG GAA ACA GCT ATG AC-3′.
7. NucleoBond Xtra Maxi Plus (Macherey-Nagel).
8. Protease K (Sigma).
9. SOC Medium (Bioshop).
10. Subcloning Efficiency DH5α cells (Invitrogen).

2.2. LR Reaction

1. Ampicillin and spectinomycin (Sigma).
2. Destination vector: VA tag destination vector in lentivirus cassette can be obtained from Dr. Jason Moffat (University of Toronto).

3. Gateway™-compatible entry clones for human and other species are obtained from (1) Open Biosystems, (2) UltimateORF collection (Invitrogen), or (3) BP reaction (outlined below).
4. Gateway™ LR clonase II enzyme mix (Invitrogen).

2.3. Lentivirus Infection

1. Dulbecco's modified Eagle's medium (DMEM; Wisent or LONZA).
2. Fetal Bovine Serum (FBS; Wisent or PAA).
3. FuGene (Roche).
4. Penicillin–Streptomycin (Wisent).
5. Glycerol for the storage of cell line (10% with DMEM media).
6. Opti-MEM (GIBCO).
7. Lentivirus packaging (Pax2) and envelope (MD2G) plasmids.
8. Polybrene (Sigma). Dissolve to a final concentration of 2 mg/ml in water and store at −20°C.
9. Puromycine dihydrochloride (Sigma). Dissolve to a final concentration of 2 mg/ml in water and store at −20°C.
10. Trypsin (0.05%)/EDTA (0.53 mM) (Wisent).

2.4. Protein Purification

1. 3×FLAG peptide (Sigma).
2. AFC (high salt) buffer: 10 mM Tris–HCl (pH 7.9), 420 mM NaCl, 0.1% NP-40.
3. AFC (low salt) buffer: 10 mM Tris–HCl (pH 7.9), 100 mM NaCl, 0.1% NP-40.
4. Anti-FLAG M2 agarose from mouse (Sigma). In a 1:1 slurry with glycerol. Store at −20°C.
5. Benzonase (Sigma).
6. β-Biotin (Sigma).
7. Bio-Spin Disposable Chromatography Column (small) (Bio-Rad).
8. Elution buffer (for Biotin): 50 mM ammonium bicarbonate, 2 mM biotin.
9. "Flat top" 1.5-ml microtube siliconized (BioCan Scientific).
10. Micro Bio-Spin Chromatography Columns (large, Bio-RAD).
11. Ni-NTA Agarose (Qiagen). Store at 2°C.
12. Phosphatase inhibitors: Make the stock solutions as follows: (1) 1 M sodium fluoride (NaF, Sigma). Dissolve 4.2 g into 100 ml H_2O and store at room temperature. (2) 100 mM sodium orthovanadate (NaOrth, Sigma). Dissolve 1.8392 g in 100 ml H_2O (pH is 12 when dissolved, and adjust to pH 10 with HCl). Then, place in boiling water and cool on ice. Repeat boiling and cooling cycles until no color remains. pH is

much lower than pH 10, so adjust with NaOH. Repeat boiling and cooling cycles until color has gone. Stabilize at pH 10. Aliquot and store at −20°C. (3) 100 mM Sodium Pyrophosphate decahydrate (Fisher Scientific).

13. Protease inhibitors: First, make mixes A and B. Mix A: 250 ml DMSO, 240 mg chymostatin, 150 mg pepstatin A, and 8.7 g PMSF. Mix B: 200 ml H_2O, 50 mg leupeptin, 1.7 g aprotinin, and 31 g benzamidine. Next, mix A and B as follows: 1 ml lysis buffer, 2.5 μl mix A, and 2 μl mix B. Alternatively, the use of commercially available protease inhibitor cocktails is acceptable.
14. Rinsing buffer: 50 mM ammonium bicarbonate, 125 mM NaCl.
15. Silver staining kit: ProteoSilver plus (Sigma).
16. Strep-Tactin Sepharose, 50% suspension, 20 ml (IBA).
17. TEV buffer (streptavidin-binding buffer): 10 mM Tris–HCl (pH 7.9), 125 mM NaCl, 0.1% NP-40.
18. TEV protease (2 mg/ml).

2.5. Materials for LC-MS

1. Orbitrap-Velos mass spectrometer (Thermo) and XCalibur software (Thermo), which is a software package and user interface for instrument operation and acquisition of mass spectra.
2. Trichloroacetic acid (TCA) (PIERCE) and acetone (PIERCE) for protein precipitation and washes, respectively. Keep ice cold prior to use.
3. Digestion buffer: 50 mM ammonium bicarbonate (NH_4HCO_3), pH 8.0, 1 mM calcium chloride ($CaCl_2$). Store at 4°C prior to use.
4. 100 mM Tris(2-carboxyethyl) phosphine (TCEP–HCl) (PIERCE) for cysteine reduction.
5. 0.5 M iodoacetamide (IAM) (Sigma) for alkylation of reduced cysteine residues.
6. Immobilized trypsin solution: Mix 18.7 μl digestion buffer, 1.8 μl PIERCE immobilized trypsin beads (PIERCE), 0.9 μl immobilized trypsin beads (Applied Biosciences), and 0.06 μl 1 M $CaCl_2$. Make sure that the pH of the immobilized trypsin is 8.0 for optimal digestion.
7. 75-μm fused silica (Polymicro Technologies).
8. Luna C18 reverse-phase packaging material (Phenomenex).
9. Solvent A: 5% ACN, 95% H_2O, 0.1% formic acid (all solutions HPLC grade or better).
10. Solvent B: 5% H_2O, 95% ACN, 0.1% formic acid (all solutions HPLC grade or better).
11. EASY-nLC nano LC pump (Proxeon).
12. P-2000 laser puller (Sutter Instrument Company).

3. Methods

3.1. BP Reaction to Make Gateway™-Compatible ORF Clones

The first step in lentivirus-based protein purification is to make the Gateway™ entry clone (see Note 1). Design PCR primers with attB1 forward primer (5′-GGGG-ACA-AGT-TTG-TAC-AAA-AAA-GCA-GGC-TNN-(gene-specific sequence)-3′, attB1 site underlined) and attB2 reverse primer (5′-GGGG-AC-CAC-TTT-GTA-CAA-GAA-AGC-TGG-GTN-(gene-specific sequence)-3′, attB2 site underlined). Gene-specific sequences should map to the 5′- and 3′-ends of the ORF of interest.

1. Amplify the PCR product from cDNA templates and use PCR purification kit to purify the PCR product.
2. Prepare Gateway™ DONR plasmids (pDONR221 or pDONR223).
3. Mix 2 μl 75 ng/μl PCR fragment, 1 μl 75 ng/μl pD221 plasmids, 2 μl BP clonase, and 5 μl TE buffer in a total volume of 10 μl and incubate at room temperature for 1–2 h.
4. Add 1 μl protease K and incubate at 37°C for 10 min, followed by cooling on ice for 2–5 min.
5. Transform the BP reaction products into DH5-α or TOP10 competent bacteria by adding 5–10 μl BP reaction mixture into 50 μl competent bacteria, keeping on ice for 15–30 min, followed by heat shock at 42°C for 1 min. Next, put on ice for 10 min and add 0.75 ml of SOC media to incubate at 37°C for 1 h. Finally, streak the bacteria on 100 μg ml kanamycin LB plate and incubate at 37°C overnight.
6. Pick four colonies and grow each overnight in 2 ml of LB with 100 μg/ml kanamycin at 37°C.
7. Isolate DNA by using miniprep kits.
8. Verify the cloned gene by either PCR, restriction enzyme digestion (e.g., by EcoRV), or sequencing (using M13 primers).

3.2. Gene (ORF) Tagging by LR Reaction

1. Streak bacteria that contain the entry ORF clone into LB agar plates with proper antibiotics (e.g., 150 μg/μl spectinomycin for the entry clones from OpenBiosystems in pDONR223 or 100 μg/ml kanamycin for the clones from Invitrogen in pENTR™ 221 or from OpenBiosystems in pDONR221). Incubate at 37°C overnight.
2. Pick two colonies and make DNA minipreps for each of the colonies.
3. Sequence the DNA (200–300 ng of each) using 10 μM of gene-specific or universal (such as M13) primers in the reaction mixture of 7.7 μl.

4. Check the sequencing results to ensure proper gene identity, length, and sequence.
5. Select correct sequence colonies for the LR reaction for tagging (see Note 3).
6. Make a mixture containing 75 ng entry clone DNA, 75 ng VA tag destination vector DNA, 6 μl TE buffer, and 2 μl Gateway™ LR clonase II enzyme mix in a total volume of 20 μl.
7. Incubate the mixture at 25°C for 1–2 h.
8. Add 1 μl protease K and incubate at 37°C for 10 min.
9. Transform the LR reaction products into 50 μl DH5α cells by incubating the DNA and 50 μl DH5α on ice for 30 min, then at 42°C for 90 s, and finally on ice for another 2 min. Then, add 250 μl SOC and incubate at 37°C for 1 h followed by centrifugation at 12,000 × *g* for 1 min.
10. Spread all or half the mixture on an ampicillin (100 μg/ml) LB agar plate and incubate overnight at 32 or 37°C (see Note 4).
11. Pick two colonies, grow each in 3 ml of LB with 100 μg/ml ampicillin, and prepare miniprep DNA. Run the DNA on a 0.8% agarose gel to check for correctly recombineered clones and do diagnostic enzymatic digestions (the latter is optional).
12. Make maxi-, midi-, or minipreps of the DNA.

3.3. Making Lentiviruses and Cell Infection

1. Split 1×10^6 293T cells into each 60-mm dish with 6 ml low-antibiotic media (DMEM with 10% heat-inhibited FBS and 0.1× penicillin–streptomycin) and incubate at 37°C overnight and 5% CO_2.
2. Make transfection mixes A and B: Mix A contains 6 μl FuGene and 90 μl Opti-MEM. Mix B contains 900 ng packaging plasmid (Pax2), 100 ng envelope plasmid (MD2G), 2,000 ng lentivirus expression plasmid, and Opti-MEM to a total volume of 20 μl.
3. Mix A and B dropwise, swirl the tube, and incubate the mix for 15–30 min at room temperature.
4. Carefully transfer (dropwise) the mix into the 293T cells (in the original 6 ml of 0.1× penicillin–streptomycin DMEM) and incubate at 37°C in 5% CO_2 for 18 h or overnight.
5. Replace the medium with 6 ml high-serum medium (DMEM + 30% iFS + 1 penicillin–streptomycin) and incubate for 24 h.
6. Harvest the viral supernatant (~6 ml), add new media, and incubate for 24 h.
7. Harvest the supernatant again, combine the supernatants, and centrifuge at 2,000 × *g* for 5 min at 4°C. The virus in the supernatants can be stored at −80°C.

8. Transduce cells by first seeding 3×10^6 HEK293 cells (or other cells) in 10-cm dishes with 10 ml medium (or 0.5×10^6 cells in each well of 6-well plate) on day 0 and then culturing them at 37°C in 5% CO_2 overnight. On day 1, remove the medium and add 5 ml (2 ml for 6-well plate) of the virus-containing medium plus 8 μg/ml of polybrene. The virus-containing medium is replaced by regular medium (DMEM, 10% FBS, and 1× penicillin–streptomycin) on day 2. Finally, add 2 μg/μl puromycin on day 3 and grow the cells under selection for 2–3 days (see Notes 15 and 16).
9. Stably transduced cells can be stored in liquid nitrogen or at −80°C in 10% glycerol DMEM medium. Cell lysates are collected for anti-FLAG Western blots to examine the expression of tagged proteins.
10. Expand the cells to five 15-cm dishes, grow for approximately 12 days, and harvest the cell pellets (approximately 1 g) for purification (see Note 11).

3.4. FLAG-Strep Two-Step Protein Purification

Two-step purifications usually yield cleaner products than do one-step purifications (4), although at the possible cost of losing some more weakly interacting proteins. In the lentivirus-based VA tag system, either FLAG-His or FLAG-Strep dual purification can be chosen for protein purification. Here, the latter protocol is described because it generally produces less background than is the case for FLAG-His purifications. All steps are carried out on ice or at 4°C unless indicated otherwise.

1. Thaw cell pellets with 5 ml of 1× high-salt AFC buffer for cell pellets from five 15-cm dishes (1 ml/dish), supplemented with protease inhibitors (2.5 μl/ml protease inhibitor A and 2 μl/ml protease inhibitor B) and phosphatase inhibitors (10 mM NaF, 1 mM sodium orthovanadate, 2 mM sodium pyrophosphate) (see Notes 5 and 6).
2. Perform three freeze–thaw cycles by quickly freezing cells on dry ice followed by thawing in a 37°C water bath, inverting frequently to prevent the temperature from rising above 4°C.
3. Add benzonase to a final concentration of 12.5–25 units/ml and incubate at 4°C for 30 min with rotation to digest RNA and DNA (see Note 7).
4. Transfer cell lysates to centrifuge tubes and spin at $100{,}000 \times g$ for 1 h at 4°C to remove cell debris and precipitates.
5. Make 20 μl slurry (10 μl beads and 10 μl liquid) of M2 agarose beads.
6. Wash beads twice with 0.5–1 ml of 1× high-salt AFC buffer and aspirate off the supernatant.
7. Resuspend beads in one volume of 1× high-salt AFC buffer.

8. Add 20 μl M2 agarose beads (1:1 slurry) to 5-ml extract and incubate for 4 h with rotation at 4°C.
9. Transfer the bead slurry into a column (Bio-Spin Disposable Chromatography Column).
10. Wash with 1 ml of 1× low-salt AFC buffer through the column three times, waiting for 10 min between each wash.
11. Wash with 1 ml of TEV buffer through the column two times, waiting for 10 min between each wash.
12. Add 100 μl TEV buffer and 5 μl TEV protease (2 mg/ml) (final concentration of TEV 100 μg/ml), and 2 μl 3×FLAG peptide, and rotate overnight at 4°C.
13. Wash 20 μl Strep-Tactin bead slurry with 1 ml TEV buffer three times in a column.
14. Collect, through the column, the 100 μl TEV buffer–protein mix in a siliconized microtube (the use of siliconized microtubes is optional). Otherwise, samples can be directly eluted into the next column that has the washed Strep-Tactin beads.
15. Add 100 μl TEV buffer to the beads twice, each time incubate for 2–4 (3) h at 4°C with rotation, and wait for 10 min between each collection (for a total volume of 300 μl).
16. Wash three times with 1 ml of TEV buffer, waiting for 10 min between washes.
17. Wash three times with 1 ml of rinsing buffer, waiting for 10 min between washes.
18. Resuspend beads in 150 μl elution buffer and incubate for 15 min–1 h at 4°C with rotation. Repeat the elution with 150 μl elution buffer for total volume of 300 μl eluate.

3.5. TCA Precipitation

1. Add 1 volume of ice-cold 100% TCA to 4 volumes of protein sample and incubate at 4°C overnight.
2. Centrifuge at 12,000 × *g* for 30 min at 4°C.
3. Discard supernatant.
4. Suspend pellet in 200 μl ice-cold acetone, incubate for 30 min at −20°C, and centrifuge at 14,000 rpm for 30 min at 4°C.
5. Repeat step 4.
6. Dry pellet by placing open tube at room temperature for 5–10 min. Cover with a kimwipe.
7. Dissolve the sample in 40 μl 50 mM NH_4HCO_3, pH 8.0.

3.6. Trypsin Digestion (Reduction/Alkylation)

1. Add TCEP–HCl to a final concentration of 2 mM and incubate at room temperature for 45 min.
2. Add IAM to a final concentration of 10 mM and incubate at room temperature for 40 min. Protect from light.
3. Add $CaCl_2$ to a final concentration of 1 mM.

4. Add trypsin (1:20–1:100 for trypsin:protein) and incubate at 37°C with shaking (1,000 × *g*) overnight.
5. Add FA to a final concentration of 1% to terminate the enzymatic digestion, and store digests at 4°C.

3.7. LC-MS/MS Analysis

1. For AP-MS experiments, we recommend using single-dimension reverse-phase chromatography. Prepare a spray tip using a 75-μm inner diameter/350-μm outer diameter fused-silica capillary column and a laser puller (program 19: heat=280, FIL=0, VEL=30, DEL=200). Pack 10–12 cm of C18 reverse-phase beads (Luna, 3 μ; solubilized in MeOH) by pressure bomb.
2. Connect the packed spray column to an EASY-nLC quaternary pump (Proxeon) and assemble in-line with the LTQ Orbitrap-Velos. Ensure that the nLC pumps maintain a stable pressure read-out and are free of leaks. The HPLC gradient programs for alternating data acquisition and wash cycles should be as follows:

Data acquisition (total runtime = 105 min)

Time	Duration	Flow (nl/min)	% B
0:00	N/A	300	2
1:00	1:00	300	2
3:00	2:00	300	6
75:00	72:00	300	24
91:00	16:00	300	90
96:00	5:00	300	90
97:00	1:00	300	2
105:00	8:00	300	2

MeOH wash (total runtime = 30 min)

Time	Duration	Flow (nl/min)	% B
0:00	N/A	300	2
1:00	1:00	300	2
2:00	1:00	300	6
17:00	15:00	300	24
18:00	1:00	300	90
21:00	3:00	300	90
22:00	1:00	300	2
30:00	8:00	300	2

3. The data-dependent data acquisition parameters include the following: 1 centroid MS (mass range 400–2,000) followed by MS/MS on the 10 most abundant product ions. General parameters include the following:

General parameters	*Setting*
Activation type	CID
Isolation width	3
Normalized collision energy	32
Activation Q	0.25
Activation time	30 ms
Data-dependent parameters	*Setting*
Minimum threshold	1,000
Repeat count	1
Repeat duration	30 s
Exclusion list size	500
Exclusion duration	30 s
Exclusion mass width (by mass)	Low = 1.2, high = 1.5

4. Ionize peptides from the nanocolumn by applying a 2.5-kV spray voltage. In data-dependent acquisition mode, the mass spectrometer acquires MS and MS/MS spectra through consecutive cycles of 11 scans throughout the duration of the gradient. Specifically, an initial full mass scan of precursor ions which detects the m/z values of intact peptide ions is followed by ten successive tandem mass spectra acquisitions of the ten most intense precursor ions.
5. The acquired tandem mass spectra are then matched to theoretical spectra based on an in silico digestion of a human proteome reference database, such as UniProtKB/Swiss-Prot, which can be downloaded in FASTA format from http://www.uniprot.org/downloads.
6. Peptide spectra acquired during the LC-MS run are identified using the SEQUEST search engine and validated using the STATQUEST probabilistic scoring algorithm. Alternatively, MASCOT or X!Tandem, among other search engines, may be used. Confidence scores are assigned to all putative matches using the probabilistic STATQUEST scoring program (or comparable statistical tools with other search algorithm suites). Proteins detected with two or more high-confidence peptide matches (i.e., 90–99%) and with a minimum likelihood threshold cutoff of 90% or greater probability are considered positive identifications.

4. Notes

1. *Gateway™ ORFs.* Large numbers of human, mouse, and rat ORFs have been cloned into Gateway™ cassettes and are commercially available from various sources, such as Invitrogen, OpenBiosystems, and GeneCopoeia. For these genes, the BP reaction can be skipped and the LR reaction can be directly carried out.
2. *ORF quality.* It is important to sequence the ORFs that are obtained to ensure that the sequence is correct and to know which isoform was cloned. Ensure that the ORF does not possess undesirable mutations, such as premature stop codons. For larger ORFs, multiple overlapping sequencing primers should be used for complete coverage of the coding sequence.
3. *LR reactions.* Run a 0.8% agarose gel to verify insertion of the gene and avoid incorrect clones from self-recombination. Since the lentiviral DNA contains repeats that are sometimes unstable and may recombine between repeats, we recommend using the following growth conditions to reduce the chances of recombination between direct repeats: (1) culture transformants at 25–30°C; (2) do not use "rich" bacterial media as they tend to give rise to a greater number of unwanted recombinants; and (3) if the plasmid confers chloramphenicol resistance, select and culture transformants using LB medium containing 15–30 μg/ml chloramphenicol in addition to the antibiotic appropriate for selection of plasmids.
4. *Selection.* It is important to determine the optimal concentrations of selection drugs for different cell lines. For example, the transduced cells were selected with puromycin at a concentration of 1 μg/ml for mouse R1 embryonic stem cells and 2 μg/ml for human HEK293 cells for a minimum of 48 h.
5. *High- and low-salt cell lysis buffers.* The advantage of carrying out cell lysis in a buffer with a high salt concentration (420 mM NaCl) is that it efficiently disrupts the nuclear membrane, thus permitting better identification of nuclear complexes. However, its disadvantage is the possible loss of some salt-sensitive interactions. Therefore, a buffer with a low salt concentration (approximately 100 mM NaCl) is the best choice for preparing cell lysates when studying cytoplasmic proteins or weak interactions.
6. *Protein sample dialysis.* To rescue and recover protein interactions disrupted by high salt concentrations, sample dialysis could be carried out to reduce the salt concentration after the cell lysis step. Dialysis of samples with a low-salt buffer reduces the salt concentration, resulting in the potential recovery of some interactions. However, dialysis may also increase the purification

background. Dialysis is carried out with 10 mM HEPES pH 7.9, 100 mM NaCl, 0.1 mM ETDA, and 20% glycerol in a Spectra/Por Membrane (MWCO 3500) for 4 h at 4°C, followed by centrifugation at 100,000 × *g* for 30 min at 4°C.

7. *Treatment of cell extracts with nuclease.* For protein–RNA/DNA interaction studies, avoid the use of benzonase. Instead, when studying protein–RNA interactions, for example, add RNase inhibitors and carry out the purification in DEPC-treated water to keep the RNA intact.
8. *Tags interfering with protein interactions.* In some cases, the tag may interfere with protein interactions (9). One of the possible solutions is to add the tag at the other terminus of the ORFs (i.e., test clones with both N- and C-terminal tags). Alternatively, multiple subunits of a protein complex can be tagged.
9. *Tag choice for purifications.* The VA tag is compatible with single, dual, or triple affinity purification schemes. We have found that dual purification from HEK293 cell lysates by FLAG followed by Strep III yields less background and higher spectral counts for purified proteins than do the FLAG-His purifications.
10. *Tag affecting cell viability.* In rare cases, the tag itself could affect cell viability. For example, a tag that interferes with the essential function of a protein, resulting in loss of function, could have a dominant negative effect that reduces cell viability. Such cases may require the use of alternative tagging strategies, such as fusing the tag to the opposite end of the protein or using an inducible promoter system.
11. *Number of cells used for purification.* The choice of purification method affects the requirement of starting materials for purification. As well, the expression level of the tagged protein also affects the number of cells that are required, (i.e., the higher the expression, the fewer cells that are needed). The lentivirus-based system typically uses approximately 10^8 cells for starting material, similar to other methodologies described elsewhere (5).
12. *Inducible promoter.* One possible effect of bait overexpression is the formation of nonbiologically relevant interactions, that is, interactions that do not occur under endogenous contexts. Moreover, overexpression of certain genes can affect cell viability (e.g., negative regulators of the cell cycle). In such instances, an inducible promoter is a viable option, where the bait expression level can be modulated by the addition of an inducer, such as tetracycline for the tetracycline-inducible promoter (9). This type of system permits experimental control of bait expression, both in terms of amount and timing.
13. *Endogenous tagging.* As an alternative to an inducible promoter, an ideal solution is to insert the tag into the endogenous gene by homologous recombination (16). Although endogenous

tagging links the tagged construct to the endogenous promoter, resulting in more or less endogenous expression levels, the success rate for endogenous tagging in mammalian systems is extremely low.

14. *Cell type-specific promoter activities.* Certain promoters exhibit cell/tissue type-specific activation or restriction (17). For example, the cytomegalovirus (CMV) promoter is active in many mammalian cells, but not in human and mouse ES cells. In contrast, phosphoglycerate kinase (PGK) and human elongation factor-1a (EF1a) promoters are transcriptionally active in ES cells (9, 17). Therefore, care should be taken to choose the optimal promoter.
15. Integration site, copy number, and lentivirus biosafety (see lentiviral shRNA screening).
16. *Multiplicity of infection (MOI).* (See lentiviral shRNA screening.) Try to transduce a minimum number of lentiviruses into target cells. Low MOI keeps the expression levels lower and reduces the cell-to-cell variation of expression.

Acknowledgments

We thank Anthony B. Mak and Denitza Roudeva for critical reading of the manuscript. This project is supported by grants from the Canadian Institutes of Health Research (CIHR) and Ontario Research Fund to JG and AE. ZN is supported by a CIHR fellowship.

References

1. Gavin, A.C., et al., *Proteome survey reveals modularity of the yeast cell machinery.* Nature, 2006. **440**(7084): p. 631–6.
2. Krogan, N.J., et al., *Global landscape of protein complexes in the yeast Saccharomyces cerevisiae.* Nature, 2006. **440**(7084): p. 637–43.
3. Butland, G., et al., *Interaction network containing conserved and essential protein complexes in Escherichia coli.* Nature, 2005. **433**(7025): p. 531–7.
4. Rigaut, G., et al., *A generic protein purification method for protein complex characterization and proteome exploration.* Nat Biotechnol, 1999. **17**(10): p. 1030–2.
5. Burckstummer, T., et al., *An efficient tandem affinity purification procedure for interaction proteomics in mammalian cells.* Nat Methods, 2006. **3**(12): p. 1013–9.
6. Forler, D., et al., *An efficient protein complex purification method for functional proteomics in higher eukaryotes.* Nat Biotechnol, 2003. **21**(1): p. 89–92.
7. Blagoev, B., et al., *A proteomics strategy to elucidate functional protein–protein interactions applied to EGF signaling.* Nat Biotechnol, 2003. **21**(3): p. 315–8.
8. Schmidt, T.G. and A. Skerra, *The Strep-tag system for one-step purification and high-affinity detection or capturing of proteins.* Nat Protoc, 2007. **2**(6): p. 1528–35.
9. Mak, A.B., et al., *A lentiviral functional proteomics approach identifies chromatin remodeling complexes important for the induction of pluripotency.* Mol Cell Proteomics, 2010. **9**(5): p. 811–23.
10. Walhout, A.J., et al., *GATEWAY recombinational cloning: application to the cloning of*

large numbers of open reading frames or ORFeomes. Methods Enzymol, 2000. **328**: p. 575–92.

11. Hartley, J.L., G.F. Temple, and M.A. Brasch, *DNA cloning using in vitro site-specific recombination.* Genome Res, 2000. **10**(11): p. 1788–95.
12. Freuler, F., et al., *Development of a novel Gateway-based vector system for efficient, multiparallel protein expression in Escherichia coli.* Protein Expr Purif, 2008. **59**(2): p. 232–41.
13. Dull, T., et al., *A third-generation lentivirus vector with a conditional packaging system.* J Virol, 1998. **72**(11): p. 8463–71.
14. Root, D.E., et al., *Genome-scale loss-of-function screening with a lentiviral RNAi library.* Nat Methods, 2006. **3**(9): p. 715–9.
15. Moffat, J., et al., *A lentiviral RNAi library for human and mouse genes applied to an arrayed viral high-content screen.* Cell, 2006. **124**(6): p. 1283–98.
16. Kim, J.S., et al., *Epitope tagging of endogenous genes in diverse human cell lines.* Nucleic Acids Res, 2008. **36**(19): p. e127.
17. Xia, X., et al., *Transgenes delivered by lentiviral vector are suppressed in human embryonic stem cells in a promoter-dependent manner.* Stem Cells Dev, 2007. **16**(1): p. 167–76.

Chapter 3

Protein Networks Involved in Vesicle Fusion, Transport, and Storage Revealed by Array-Based Proteomics

Mikael Bauer, Magdalena Maj, Ludwig Wagner, Dolores J. Cahill, Sara Linse, and David J. O'Connell

Abstract

Secretagogin is a calcium-binding protein whose expression is characterised in neuroendocrine, pancreatic, and retinal cells. We have used an array-based proteomic approach with the prokaryotically expressed human protein array (hEx1) and the eukaryotically expressed human protein array (Protoarray) to identify novel calcium-regulated interaction networks of secretagogin. Screening of these arrays with fluorophore-labelled secretagogin in the presence of Ca^{2+} ions led to the identification of 12 (hEx1) and 6 (Protoarray) putative targets. A number of targets were identified in both array screens. The putative targets from the hEx1 array were expressed, purified, and subjected to binding analysis using surface plasmon resonance. This identified binding affinities for nine novel secretagogin targets with equilibrium dissociation constants in the 100 pM to 10 nM range. Six of the novel target proteins have important roles in vesicle trafficking; SNAP-23, ARFGAP2, and DOC2alpha are involved in regulating fusion of vesicles to membranes, kinesin 5B and tubulin are essential for transport of vesicles in the cell, and rootletin builds up the rootlet, which is believed to function as scaffold for vesicles. Among the targets are two enzymes, DDAH-2 and ATP-synthase, and one oncoprotein, myeloid leukaemia factor 2. This screening method identifies a role for secretagogin in secretion and vesicle trafficking interacting with several proteins integral to these processes.

Key words: Secretagogin, Calcium-regulated, Vesicle, Exocytosis, Endocytosis

1. Introduction

Secretagogin is a recently discovered calcium-sensing protein found in the Langerhans islet cells in pancreas (1), in neural and neuroendocrine cells (2), as well as in cone cells in the retina (3). Secretagogin contains six EF-hands, and the 3D structure is similar to the homologue calbindin D28k with some distinct differences in the packing of the EF-hands and secretagogin has a deeper cleft in the structure. In the human protein, four of the EF-hands bind Ca^{2+} resulting in conformational changes and exposure of hydrophobic surfaces (4).

Gerard Cagney and Andrew Emili (eds.), *Network Biology: Methods and Applications*, Methods in Molecular Biology, vol. 781, DOI 10.1007/978-1-61779-276-2_3, © Springer Science+Business Media, LLC 2011

Secretagogin has been found to interact with SNAP-25, a protein involved in vesicle trafficking in exocytosis. This interaction is stronger with the calcium-loaded compared to the calcium-free form indicating that secretagogin could be a protein that controls Ca^{2+}-dependent secretion (5). A Ca^{2+}-dependent interaction has also been found between secretagogin and the TAU protein which was identified using GST pull-down assays (6). Few studies of the biological role have been reported; however, secretagogin-expressing neurons in the hippocampus were found to be resistant to neurodegenerative damage associated with Alzheimer's disease (7). We have previously reported the successful implementation of an array-based screen for protein:protein interactions with another Ca^{2+}-binding protein, calmodulin (8, 9). The screen contains 37,200 redundant proteins expressed from a human fetal brain cDNA library spotted onto a PVDF membrane, which is screened for interactions using fluorescently labelled proteins. Identified putative targets are then expressed and the interaction is verified and quantified using independent methods, for example surface plasmon resonance (SPR) assays, which verified over 70 novel human targets of calmodulin (8, 9). In the present study, we have applied the same array-based technique for screening protein:protein interactions with fluorophore-labelled secretagogin, yielding 12 potential targets, 9 of which were validated using SPR technology (Table 1). For comparison and to extend the repertoire of putative targets, a parallel screen was performed using fluorophore-labelled secretagogin and the Protoarray (Invitrogen) system.

Table1
Secretagogin-binding proteins and kinetic parameters and dissociation constants from the fits of the targets validated with SPR

Target protein	Log k_{on}	Log k_{off}	K_d
Rootletin	6	-4	100 pM
N(G),N(G)-dimethylarginine dimethylaminohydrolase 2 (DDAH-2)	4	-4	10 nM
Synaptosomal-associated protein-23 (SNAP-23)	5	-5	100 pM
Myeloid leukaemia factor 2	5	-5	100 pM
ADP-ribosylation factor GTPase-activating protein 2 (ARFGAP2)	4	-3	100 nM
Kinesin heavy chain (KIF5B)	5	-4	1 nM
Double C2-like domain-containing protein alpha (DOC2alpha)	5	-5	100 pM
ATP synthase O subunit	5	-4	1 nM
Tubulin, beta	6	-4	100 pM

2. Materials

2.1. Screening of High-Content Arrays

1. High-density protein arrays of the human brain library (hEx1) (German Resource Centre for Genome Research – RZPD). Protoarray (Invitrogen).
2. Tris-buffered saline (TBS) (10 mM Tris–HCl pH 7.5, 150 mM NaCl).
3. TBS-Tween (TBST) (20 mM Tris–HCl pH 7.5, 500 mM NaCl, 0.05% Tween (Reidel de Hein)).
4. TBST-Triton X-100 (20 mM Tris–HCl pH 7.5, 500 mM NaCl, 0.05% Tween, 0.5% Triton X-100 (Sigma)).
5. Blocking solution (2% non-fat milk (Fluka), TBS). Protoarray blocking solution (50 mM Hepes, pH 7.5, 200 mM NaCl, 0.08% Triton X-100, 25% glycerol, 1 mM DTT, 1% milk powder).

2.2. Fluorophore Labelling of Secretagogin

1. One molar equivalent of AlexaFluor488 C5-aminooxyacetamide was added to 5 mg/ml secretagogin in 20 mM phosphate buffer with 150 mM NaCl and 1 mM EDTA, pH 7.3, and left to react at room temperature in the dark for 1 h.
2. Excessive unbound dye was removed by passing the reaction twice through an NAP-10 size-exclusion column and collecting the protein fraction.

2.3. Expression and Purification of Recombinant Proteins from hEx1 Library

1. 2XTY medium (Sigma), 40% Glucose (Sigma), Ampicillin (100 mg/ml) (Sigma), Kanamycin (15 mg/ml) (Sigma).
2. Overnight Express (Novagen), 10% Glycerol (Sigma).
3. Lysis buffer (100 mM NaH_2PO_4, 10 mM Tris–HCl, 6 M guanidine HCl, adjust pH to 8.0 using NaOH).
4. Wash buffer (100 mM NaH_2PO_4, 10 mM Tris–HCl, 8 M urea, adjust pH to 6.3 using HCl).
5. Elution buffer (100 mM NaH_2PO_4, 10 mM Tris–HCl, 8 M urea, adjust pH to 4.5 using HCl).
6. Ni-NTA agarose (Qiagen), 5-ml polypropylene columns (Qiagen).

2.4. SDS-Polyacrylamide Gel Electrophoresis

1. Separating buffer (4×): 1.5 M Tris–HCl, pH 8.7, 0.4% SDS. Store at room temperature.
2. Stacking buffer (4×): 0.5 M Tris–HCl, pH 6.8, 0.4% SDS. Store at room temperature.
3. Thirty percent acrylamide/bis solution (37.5:1 with 2.6% C) (this is a neurotoxin when unpolymerised, so care should be taken not to receive exposure) and *N,N,N,N′*-Tetramethyl-ethylenediamine (TEMED, Sigma).

4. Ammonium persulfate: Prepare 10% solution in water and immediately freeze in single-use (200 μL) aliquots at 20°C.
5. Water-saturated isobutanol: Shake equal volumes of water and isobutanol in a glass bottle and allow to separate. Use the top layer. Store at room temperature.
6. Running buffer (5×): 125 mM Tris, 960 mM glycine, 0.5% (w/v) SDS. Store at room temperature.
7. Prestained molecular weight markers: Kaleidoscope markers (Bio-Rad, Hercules, CA).

2.5. Surface Plasmon Resonance Measurement Validation of Binding Affinity

1. Lyophilised secretagogin was dissolved in water at a concentration of 4 μg/ml and covalently coupled onto the carboxymethyl matrix of CM5 sensor chips using standard amine coupling 0.05 M EDC and 0.2 M NHS.
2. Flow buffer contained 10 mM Tris, pH 7.5, 150 mM KCl, 2 mM $CaCl_2$, and 0.005% Tween 20.

3. Methods

In these experiments, we use high-content human protein arrays on 22.2 × 22.2 cm polyvinylidene fluoride (PVDF) membranes and high-content micro-arrays on nitrocellulose-coated microscope slides (Protoarray, Invitrogen). Each PVDF array is robotically spotted in a standard 5 × 5 spotting pattern, with each array containing 27,648 protein spots in duplicate, making a total of 55,296 protein spots per array (Fig. 1). The protein collection of 37,200 clones is spotted as follows: one with 27,648 individual clones spotted in duplicate and one with 9,552 individual clones spotted in duplicate multiple times. Together, the two membranes hold 37,200 redundant proteins, of which over 10,000 are estimated to be unique (non-redundant) human proteins based on sequencing of a subset of the clones. Twelve proteins are arrayed in duplicate in the 5 × 5 spotting pattern around a central ink-guiding point (Fig. 1). Before use, the membranes are washed clean of residual bacterial colonies leaving the expressed proteins on the membrane. These arrays were screened for secretagogin-binding proteins in the presence of 1 mM Ca^{2+} using AlexaFluor488-labelled secretagogin. The Protoarray has >9,500 full-length human proteins expressed in sf9 cells spotted in duplicate with a number of control features (Fig. 2). These arrays were screened for secretagogin-binding proteins in the presence of 1 mM Ca^{2+} using AlexaFluor546-labelled secretagogin.

3.1. Preparation of High-Content Arrays

1. The PVDF arrays were soaked in 95% ethanol, rinsed in deionised water, and washed clean of residual bacterial colonies with 20 mM Tris–HCl, 500 mM NaCl, and 0.05% (v/v)

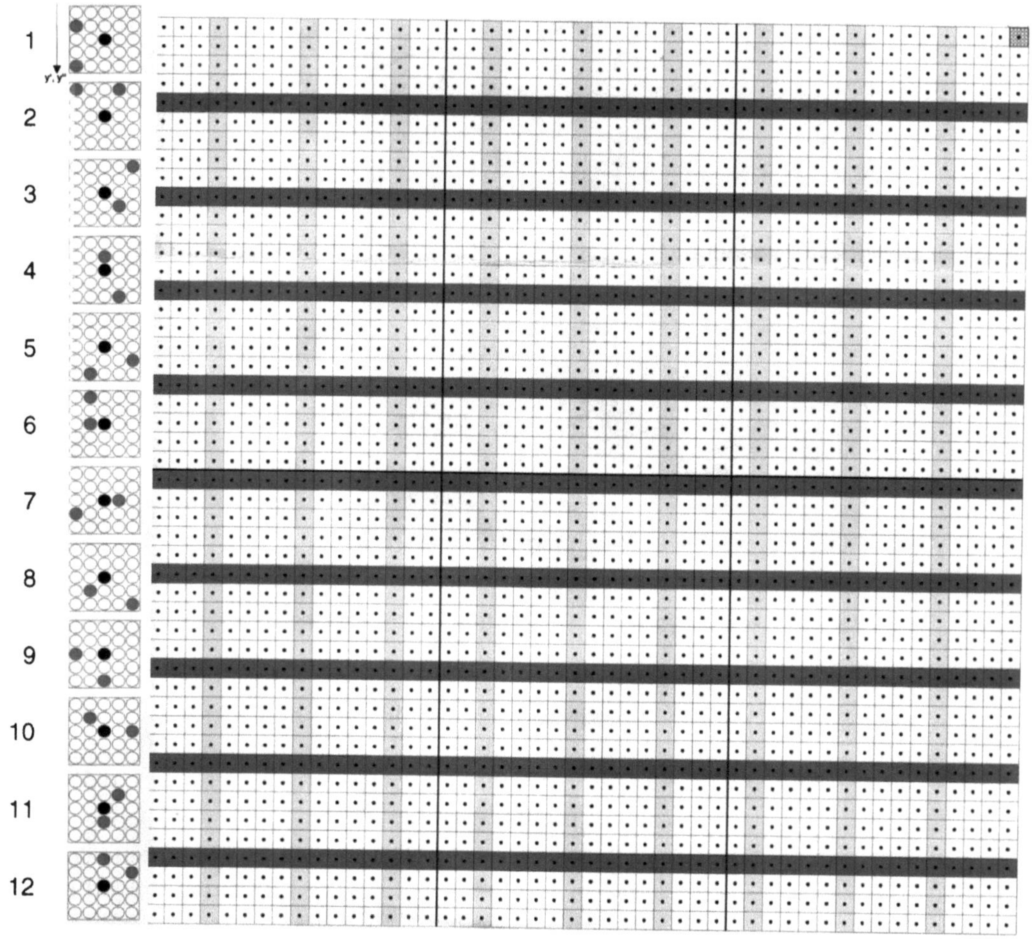

Fig. 1. hEX1 protein array format. Each duplicate protein pattern spotted in a 12 × 12 grid is indicated on the *left*. The 12 × 12 pattern is spotted around a central ink guide dot. The arrangement of these grids on the array is shown on the *right*.

Tween-20, pH 7.4 (TBST), with 0.5% (v/v) Triton X-100 (see Notes 1 and 2).

2. For secretagogin-binding protein profiling, the protein arrays were blocked in 2% (w/v) non-fat, dry milk powder in 20 mM Tris–HCl and 150 mM NaCl, pH 7.4 (TBS), for 2 h, washed twice in TBST, and subsequently incubated with Sec-Alexa488 at a concentration of 1 μM in TBS with 1 mM $CaCl_2$ for 24 h (see Notes 3–5).
3. The protein arrays were washed in TBST with 1 mM $CaCl_2$ six times for 10 min each.
4. The arrays were illuminated with blue epifluorescence, and the images were taken using a high-resolution CCD detection system (Fuji LAS3000).
5. Image analysis was performed with VisualGrid (GPC Biotech, Germany). Hits were counted as positive if both spots of a clone were significantly brighter than background (Fig. 2).

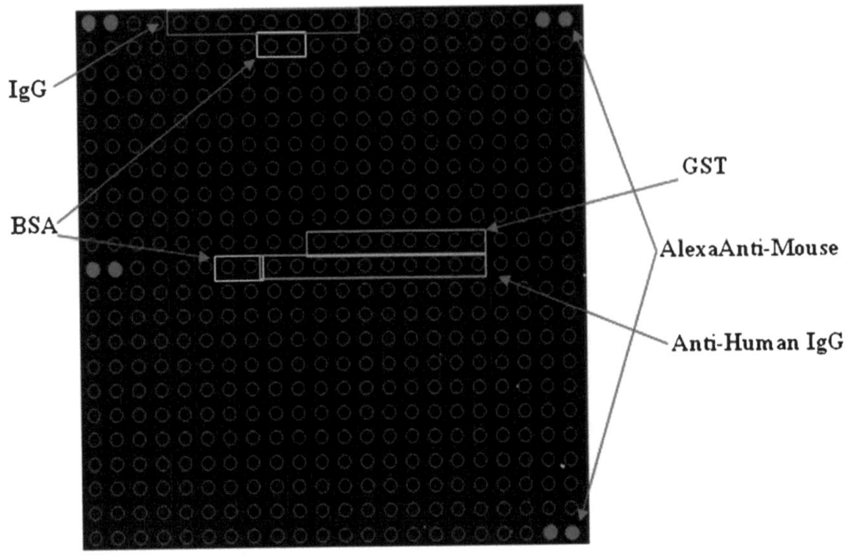

Fig. 2. Protoarray protein micro-array format. Pairs of proteins are spotted side by side in sub-arrays of 22 columns × 22 rows. Forty-eight sub-arrays make up the entire micro-array content. Control landmarks are indicated with *arrows*.

6. The Protoarrays were equilibrated to 4°C for 15 min before blocking in 50 mM HEPES buffer, pH 7.5, 200 mM NaCl, 0.08% Triton X-100, 25% glycerol, 1 mM DTT, and 1% milk powder for 60 min rotating at 4°C.
7. The Sec-Alexa546 probe was diluted to 1 μM in 150 μl of PBS 0.1% Tween and incubated on the Protoarrays for 60 min at 4°C beneath a 25 × 60-mm coverslip.
8. The Protoarrays were washed for 4 × 5 min in PBS 0.1% Tween.
9. The Protoarrays were rinsed three times in deionised water to remove salts and dried by centrifugation at 250 × *g* for 3 min.
10. The Protoarrays were scanned in a micro-array scanner (Genepix 4000b, Axon Instruments) at PMT gain settings of 635/500 and 532/250.

3.2. Fluorophore Labelling of Secretagogin

1. Secretagogin was expressed and purified according to the method published by Rogstam et al. (5) and was labelled with the fluorescent dye AlexaFluor488.
2. One molar equivalent of AlexaFluor488 C5-aminooxyacetamide was added to 5 mg/ml secretagogin in 20 mM phosphate buffer with 150 mM NaCl and 1 mM EDTA, pH 7.3, and left to react at room temperature in the dark for 1 h.
3. Excessive unbound dye was removed by passing the reaction twice through an NAP-10 size-exclusion column and collecting the protein fraction. The protein was immediately frozen in aliquots.
4. Labelling with AlexaFluor546 C5-aminooxyacetamide was performed in the same way.

3.3. Expression and Purification of Recombinant Proteins from hEx1 Library

1. Prepare single colonies of clones by streaking on selective media (2XTY, 2% glucose, 100 μg/ml ampicillin, 15 μg/ml kanamycin, 10% agar) and pick a single colony for inoculation into 1 ml of the same liquid media minus agar.
2. Inoculate 50 ml of starter culture into 10 ml of Overnight Express culture medium, 100 μg/ml ampicillin, and 15 μg/ml kanamycin for 20 h culture. This auto-induction system allows for efficient expression of recombinant proteins without the need for IPTG induction (see Note 6).
3. Cells are pelleted at 4,000 × *g*, supernatants poured off, and the pellets frozen at −80°C for 1 h to disrupt the cells (see Note 7).
4. The pellets are re-suspended in lysis buffer (100 mM NaH_2PO_4, 10 mM Tris–HCl, 6 M guanidine HCl, adjust pH to 8.0 using NaOH) and incubated for 30 min to solubilise cellular proteins. The cell debris is cleared from the lysate by centrifugation at 14,000 × *g* in a micro-centrifuge.
5. 1 ml of Ni-NTA purification resin is added to 5 ml polypropylene columns and equilibrated with addition of 2 × 1-ml lysis buffer.
6. The cleared cellular lysate is added to the column and allowed to purify by gravity flow. The 6 histidine tag on the expressed recombinant proteins binds to the nickel ions immobilised on the NTA matrix.
7. The specifically bound proteins are eluted from the column in 1 ml of elution buffer (100 mM NaH_2PO_4, 10 mM Tris–HCl, 8 M urea, adjust pH to 4.5 using HCl) and are visualised using polyacrylamide gel electrophoresis (PAGE) (Fig. 3).

3.4. SDS-Polyacrylamide Gel Electrophoresis

1. It is critical that the glass plates for the gels are scrubbed clean with 70% ethanol and rinsed extensively with distilled water prior to use (see Note 8).
2. Prepare a 1.0-mm thick, 10% gel by mixing 7.5 mL of 4× separating buffer with 10 mL acrylamide/bis solution, 12.5 mL water, 100 μL ammonium persulfate solution, and 20 μL TEMED. Pour the gel, leaving space for a stacking gel, and overlay with water-saturated isobutanol. The gel should polymerise in about 30 min.
3. Pour off the isobutanol and rinse the top of the gel twice with water.
4. Prepare the stacking gel by mixing 2.5 mL of 4× stacking buffer with 1.3 mL acrylamide/bis solution, 6.1 mL water, 50 μL ammonium persulfate solution, and 10 μL TEMED. Use about 0.5 mL of this to quickly rinse the top of the gel, then pour the stack, and insert the comb. The stacking gel should polymerise within 30 min.

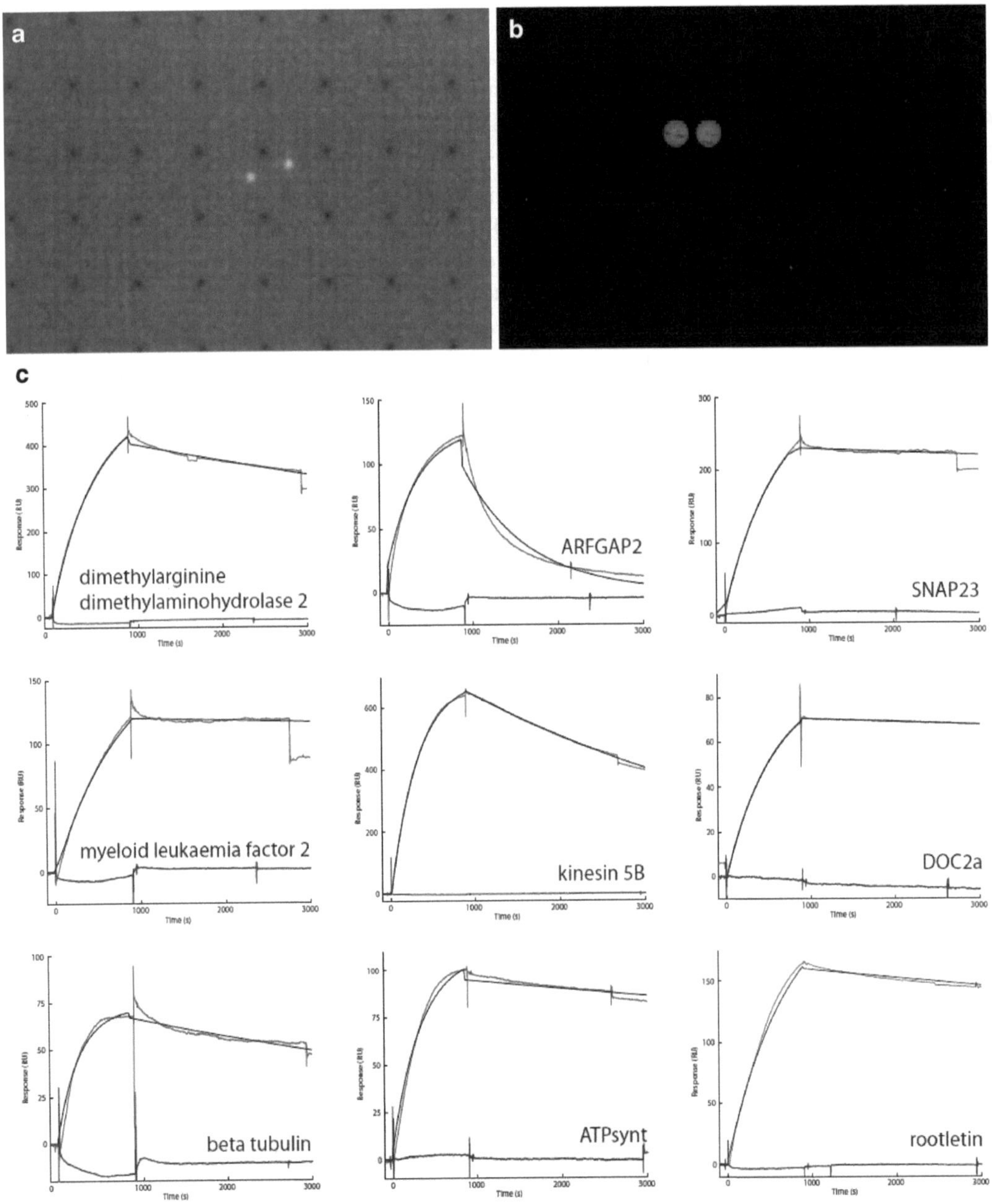

Fig. 3. Identification and characterisation of secretagogin-binding proteins. (**a**) SNAP-23 on the hEx1 array, (**b**) SNAP-23 on the Protoarray, and (**c**) sensorgrams identifying the specific binding kinetics of nine proteins identified in the array screening approach measured with surface plasmon resonance.

5. Prepare the running buffer by diluting 100 mL of the 4× running buffer with 400 mL of water in a measuring cylinder.
6. Once the stacking gel has set, carefully remove the comb and use a 3-mL syringe fitted with a 22-gauge needle to wash the wells with running buffer.

7. Add the running buffer to the upper and lower chambers of the gel unit and load the 50 μL of each sample in a well. Include one well for pre-stained molecular weight markers.
8. Complete the assembly of the gel unit and connect to a power supply. The gel can be run at a constant current of 20 mA per gel.

3.5. Surface Plasmon Resonance Measurement Validation of Binding Affinity

1. The SPR experiments were performed using a Biacore 3000 system (GE Healthcare) with four flow cells.
2. Lyophilised secretagogin was dissolved in water to a concentration of 4 μg/ml and covalently coupled onto the carboxymethyl matrix of CM5 sensor chips using standard amine coupling.
3. The chip was activated with an injection of 25 μl of a mixture of 0.05 M EDC and 0.2 M NHS, followed by the injection of approximately 25 μl 4 μg/ml secretagogin in one flow cell to yield about a 2000 RU response (the injection volume varied somewhat from time to time as the injection was stopped when the signal had increased by 2000 RU).
4. Remaining reactive groups in the channel with secretagogin and the channel preceding it were blocked with an injection of 100 μl ethanolamine; thus, the channel before the secretagogin channel serves as a blank.
5. Another pair of cells were used in the same manner to couple calbindin D28k (negative control) in one flow cell (2000 RU) again with the second cell as a blank.
6. Buffer was injected over the chip for 1 h to achieve a stable baseline.
7. Binding experiments were performed by injecting 150 μL of each target protein at 0.5 to 100 nM in the flow buffer over the chip surface to monitor association kinetics, followed by buffer flow for at least 2 h to measure dissociation kinetics. The flow buffer contained 10 mM Tris, pH 7.5, 150 mM KCl, 2 mM $CaCl_2$, and 0.005% Tween 20 (see Note 9).
8. All binding experiments were performed in duplicate. The concentration of each target was estimated based on SDS-PAGE of purified stock solutions.
9. Between experiments, the surface was regenerated by injecting 1 mM EDTA in 10 mM Tris–HCl, pH 7.5, 150 mM KCl, and 0.005% Tween 20 until the response level returned to the baseline level (see Note 10).
10. Binding experiments with the previously known secretagogin target SNAP-25 were used as positive controls.

3.6. Fitting of SPR Sensorgrams for Affinity Measurement

1. Curve fitting to all SPR data was performed using the BIAeval software (GE Healthcare). Dissociation-phase data were analysed by fitting a single exponential decay function to the data

$$R(t) = A\exp(-k_{\text{off}}t), \quad (1)$$

where R is the response, A the amplitude, and k_{off} the dissociation rate constant. The variable parameters were k_{off} and A.

2. Association-phase data were analysed by fitting the following function to the data with a fixed value of k_{off} as obtained above.

3. $R(t) = R_{\text{max}}(ck_{\text{on}} / (k_{\text{off}} + ck_{\text{on}}))(1 - \exp(-(k_{\text{off}} + ck_{\text{on}})t)) + R_0, \quad (2)$

   where R is the response, c the concentration of target protein in the flow, k_{on} the association rate constant, R_{max} the signal that would have been obtained at full saturation of the immobilised secretagogin, and R_0 the instantaneous response resulting from injection of the target protein due to a difference in refractive index between solutions. The variable parameters were k_{on}, R_{max}, and R_0.

4. An estimate of K_d was obtained from the rate constants as

$$K_d = k_{\text{off}} / k_{\text{on}}. \quad (3)$$

5. Since the association rate constant and, thus, also the equilibrium dissociation constant depend on the concentration of the target as estimated from SDS-PAGE, we have chosen to present values with an accuracy of one order of magnitude ($\pm 0.5 \times 10$ log units).

4. Notes

1. Handling the protein arrays requires plenty of space as they are quite large (similar to A4 in size) and while PVDF is quite robust, it is easy to tear the arrays if not handled carefully. Use a pair of forceps with flat gripping heads to grab arrays outside of the print area. There are wide margins on at least two sides of the array to facilitate the careful gripping of the membrane as it is moved around. Using two forceps reduces the risk of tearing the array. The Protoarrays are easily damaged by scratching their surface, and it is also important to visually inspect for damage prior to use as false negatives can be accounted for by scratching of the surface.
2. It is very important to allow the membranes to equilibrate well in TBST-Triton X-100 after the membranes have been activated in ethanol and rinsed. Allow to soak at least 5 min in this buffer before wiping bacterial colonies from the surface. If the membrane is not completely equilibrated, the dessicated bacterial colonies are difficult to remove and background is significantly increased. Also ensure that the tissue paper used to wipe

the arrays clean is thoroughly wet. We have best experience using Kimwipes for this task.

3. For the blocking solution, dissolve the milk powder with stirring at high speed for enough time to ensure that no milk powder aggregates are present. It is useful practice to heat this to 50°C when stirring to help dissolve the milk fully.
4. For blocking and incubation steps, it is sufficient to rock the arrays at a slow speed to ensure complete coverage of the membranes with blocking or probe solution. In washing steps, it is necessary to increase the speed of rocking to agitate the arrays and to completely remove all traces of non-specifically bound material.
5. All incubations with fluorophore-labelled proteins were covered in aluminium foil to prevent photo-bleaching, and the foil was carefully wrapped to ensure that the arrays were level on the rocking surface to provide complete coverage of the array surface. For the Protoarrays, it is especially important to take great care in trapping the incubation fluid beneath the coverslip. It is useful to pipette the 150 ml onto one end of the array and carefully lower the coverslip towards the other end while rolling the fluid gently forwards beneath to avoid trapping of air bubbles. Should air become trapped gently, press with a pipette tip to extrude bubbles.
6. While a 20-h culture for auto-induction works well for the majority of protein-expressing clones and facilitates the expression of many clones simultaneously, it is also advisable to titrate for protein expression for clones that have different expression characteristics. In general, proteins can be well-expressed from 9 h and may suffer damage if left for incubation times past 12 h. In this case, a titration series should be performed for individual protein-expressing clones.
7. For 10-ml cultures, the cell pellet is easily disrupted with pipetting and vortexing of the pellet in lysis buffer. For larger volumes, it is advisable to rock the pellet for 1 h to ensure that the cells are completely lysed before clearing the cell debris by centrifugation.
8. For PAGE, it is critical that the glass plates used to cast the gels are scrubbed clean with 70% ethanol and rinsed extensively with distilled water prior to use. Failure to clean the plates properly can lead to problems in separating the plates after electrophoresis.
9. Appropriate concentrations of the target proteins to use vary depending on the affinity and kinetics of the binding reaction. Aim for a concentration, where you get curvature of the signal during the injection phase. To get reliable fittings of your data, use concentrations where the binding reactions reach equilibrium.

10. At the end of the binding experiment, check whether the regeneration of the surface has worked well and ensure that all of the target proteins have been washed off before starting a new injection. If not, make another injection of the EDTA solution.

References

1. Wagner, L., Oliyarnyk, O., Gartner, W., Nowotny, P., Groeger, M., Kaserer, K., Waldhäusl, W., and Pasternack, M. (2000) Cloning and expression of secretagogin, a novel neuroendocrine and pancreatic islet of Langerhans-specific Ca2+-binding protein. J. Biol. Chem. **275**, 24740–24751
2. Mulder, J., Zilberter, M., Spence, L., Tortoriello, G., Uhlén, M., Yanagawa, Y., Aujard, F., Hökfelt, T., and Harkany, T. (2009) Secretagogin is a Ca2+-binding protein specifying subpopulations of Telencephalic neurons. Proc. Natl. Acad. Sci. USA. **106**, 22492–22497
3. Puthussery, T., Gayet-Primo, J., and Rowland Taylor, W. (2009) Localization of the Calciumbinding Protein Secretagogin in Cone Bipolar Cells in the Mammalian Retina. J. Comp. Neurol. **518**, 513–525
4. Bitto, E., Bingman, C.A., Bittova, L., Frederick, R.O., Fox, B.G., and Phillips G.N. Jr (2009) X-ray structure of Danio rerio secretagogin: A hexa-EF-hand calcium sensor. Proteins **76**, 477–83
5. Rogstam, A., Linse, S., Lindqvist, A., James, P., Wagner, L., and Berggård, T. (2007) Binding of calcium ions and SNAP-25 to the hexa EF hand protein secretagogin. Biochem. J. **401**, 353–363
6. Maj, M., Gartner, W., Ilhan, A., Neziri, D., Attems, J., and Wagner, L. (2010) Expression of TAU in insulin-secreting cells and its interation with the calcium-binding protein secretagogin. J. Endocrinol. **205**, 25–36
7. Attems, J., Preusser, M., GrosingerQuass, M., Wagner, L., Lintner, L., and Jellinger, J. (2008) Calcium-binding protein secretagogin-expressing neurons in the human hippocampus are largely resistant to neurodegeneration in Alzheimer's disease. Neuropathol. Appl. Neurobiol. **34**, 23–32
8. O'Connell, D.J., Bauer, M.C., O'Brien, J., Johnson, W.M., Divizio, C.A., O'Kane, S.L., Berggård, T., Merino, A., Akerfeldt, K.S., Linse, S., and Cahill, D.J. (2010) Integrated protein array screening and high throughput validation of 70 novel neural calmodulin binding proteins. Mol. Cell. Proteomics **9**, 1118–32
9. Bauer, M.C., O'Connell, D.J., Cahill, D.J., and Linse, S. (2008) Calmodulin binding to the polybasic C-termini of STIM proteins involved in store-operated calcium entry. Biochemistry **47**, 6089–91

Chapter 4

Identification and Relative Quantification of Native and Proteolytically Generated Protein C-Termini from Complex Proteomes: C-Terminome Analysis

Oliver Schilling, Pitter F. Huesgen, Olivier Barré, and Christopher M. Overall

Abstract

Proteome-wide analysis of protein C-termini has long been inaccessible, but is now enabled by a newly developed negative selection strategy we term C-terminomics. In this procedure, amine- and carboxyl groups of full-length proteins are chemically protected. After trypsin digestion, N-terminal and internal tryptic peptides – but not C-terminal peptides – posses newly formed, unprotected C-termini that are removed by coupling to the high-molecular-weight polymer poly-allylamine. Ultrafiltration separates the uncoupled, blocked C-terminal peptides that are subsequently analyzed by liquid chromatography-tandem mass spectrometry. On a proteome-wide scale, this strategy profiles native protein C-termini together with neo C-termini generated by endoproteolytic cleavage or processive C-terminal truncations ("ragging"). In bacterial proteomes, hundreds of protein C-termini were identified. Stable isotope labeling enables quantitative comparison of protein C-termini and C-terminal processing in different samples. Using formaldehyde-based chemical labeling, this quantitative approach termed "carboxy-terminal amine-based isotope labeling of substrates (C-TAILS)" identified >100 cleavage sites of exogenously applied GluC protease in an *Escherichia coli* proteome. C-TAILS complements recently developed N-terminomic techniques for endoprotease substrate discovery and is essential for the characterization of carboxyprotease processing.

Key words: Degradomics, C-terminus, Protease, TAILS, Carboxypeptidase

1. Introduction

The human genome encodes >560 proteases, which collectively sculpt the human proteome by irreversibly processing or truncating protein substrates. Although proteolysis has long been predominantly associated with protein degradation, the importance of fine-tuned proteolytic cleavage for protein fate and function has become increasingly apparent. Several proteomic techniques have been

Gerard Cagney and Andrew Emili (eds.), *Network Biology: Methods and Applications*, Methods in Molecular Biology, vol. 781, DOI 10.1007/978-1-61779-276-2_4, © Springer Science+Business Media, LLC 2011

invented to characterize proteolytic processing in the cellular or organism-wide context. System-wide approaches aimed at the elucidation of protease substrates are commonly referred to as degradomics (1) and mostly focus on protein N-termini (2, 3). These terminomic approaches typically deplete internal peptides in order to focus analysis, time, and sample consumption on terminal peptides and potential cleavage sites. One of the most powerful N-terminomic strategies – referred to as terminal amine isotopic labeling of substrates (TAILS) – (4–6) employs a negative, polymer-based selection strategy of N-terminal peptides and, hence, analyzes both modified (e.g., acetylated or cyclized) and unmodified N-termini. In addition, TAILS employs stable isotope labeling to distinguish induced cleavage events (e.g., through exogenous application of a recombinant protease) from background proteolysis in different biological samples. In contrast, proteome-wide analysis of protein C-termini has long been inaccessible. This has been a serious deficit since C-terminal analysis complements N-terminal cleavage site characterization and is crucial to characterize C-terminal processing by carboxypeptidases, a group of proteases with important roles in physiological and pathological processes (7).

We have developed a polymer-based, negative selection strategy for the proteome-wide analysis of protein C-termini which detects endoproteolytic as well as carboxyproteolytic processing in cellular proteomes (8) (Fig. 1). In this C-terminomic strategy, thiol groups as well as N-terminal α-amines and lysine ε-amines of full-length protein are chemically protected to prevent side reactions during the workflow. After cleanup, protein carboxyl groups, including protein carboxy-termini together with aspartate and glutamate side chains, are chemically protected by carbodiimide-mediated (1-ethyl-3-[3-dimethylaminopropyl]carbodiimide hydrochloride (EDC)) condensation with the primary amine compound ethanolamine. At a 1.0 M concentration, ethanolamine is used in large excess similar to other carboxyl-labeling approaches (9). Proteins are then trypsin-digested. However, cleavage occurs exclusively at arginine residues since trypsin does not recognize chemically modified lysine residues. This is advantageous as it generates longer peptides that generally improve peptide identification. After digestion, newly formed N-terminal α-amines are chemically protected to prevent side reactions during the remaining workflow. Internal and N-terminal tryptic peptides differ from C-terminal peptides in possessing a free C-terminus after trypsination. Through their free C-termini, internal and N-terminal tryptic peptides are covalently coupled to the high-molecular-weight polymer poly-allylamine (molecular weight approximately 56 kDa). C-terminal peptides lack free carboxyl groups and remain uncoupled. They are separated from the polymer–peptide mixture by ultrafiltration using a size cutoff of 10 kDa and analyzed by liquid chromatography-tandem mass spectrometry (LC-MS/MS).

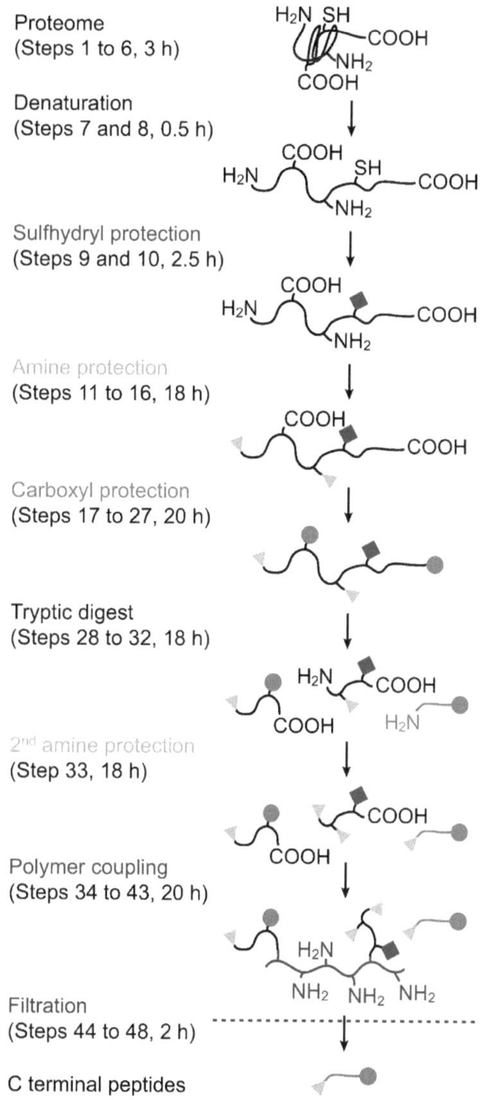

Fig. 1. Isolation of protein carboxy-terminal peptides. Full-length proteins are denatured, followed by protection of cysteine thiols (modification indicated by *squares*) and free amines at protein N-termini and lysine side chains (*triangles*). Next, carboxy-groups at protein C-termini and in acidic amino acid side chains are protected (*circles*). After digestion with trypsin, protein N-terminal and internal peptides possess free carboxy-groups that are used for specific removal by condensation with an amine-containing high-molecular-weight polymer. Protein C-terminal peptides, which do not contain free carboxyls, do not react with the polymer and are isolated by ultrafiltration. Corresponding steps in Subheading 3.1 and approximate reaction times are indicated and include cleanup steps that are not explicitly shown in this scheme.

In conjunction with metabolic or chemically introduced stable isotope labeling, our C-terminomic approach is well-suited to compare proteolytic processing in different biological samples by detecting neo C-termini (Fig. 2). The present workflow employs an orthogonal labeling strategy and combines compulsory amine

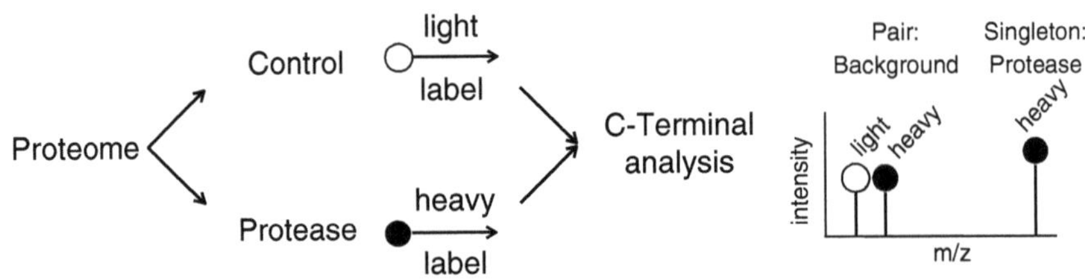

Fig. 2. Endoprotease or carboxyprotease substrate discovery by relative quantification of proteolytically generated C-termini. For this degradomic application of the C-terminome strategy, a proteome sample is incubated with a test protease and chemically labeled with a stable heavy isotope. An untreated control sample is labeled with the corresponding light isotope. The workflow presented in this chapter uses formaldehyde-based protection of protein and peptide primary amines for stable isotope labeling (steps 11 and 33). After labeling, samples are mixed at equal abundance and protein C-termini are isolated. In LC-MS/MS analysis, C-termini that are present in both samples occur in both heavy and light labeled states ("isotope pair"), whereas C-termini that are exclusively present in the protease-treated samples occur only in the heavy labeled form without the light counterpart. Hence, relative quantification of protein C-termini enables distinction of induced cleavage events from background proteolysis, a crucial feature for protease substrate discovery.

protection (formaldehyde-based reductive methylation, see below) with stable isotope introduction. It is referred to as "carboxy-terminal amine-based isotope labeling of substrates (C-TAILS)" and has been successfully validated by detection of limited GluC proteolysis in an *Escherichia coli* proteome (8).

In contrast to chromatographic fractionation schemes for C-terminal peptides (10), this protocol clearly distinguishes C-terminal peptides via the presence of the C-terminal ethanolamine group. A typical C-terminal analysis contains approximately 20% of noncaptured tryptic internal peptides. Due to the ethanolamine tag, C-terminal peptides with a C-terminal arginine residue are not confounded with noncaptured, internal tryptic peptides that also end with an arginine but without a label.

An important caveat of C-terminomic experiments is reduced proteome coverage in comparison to N-terminomic approaches, which feature less-complex workflows (4–6, 8). Hence, N-terminomic experiments are typically considered to be more adequate for the characterization of general endoprotease specificity while C-terminomic techniques are of particular interest for more specialized experiments that are not sufficiently covered by N-terminomic techniques, such as C-terminal truncations. Combining C-terminome and N-terminome analysis, such as C-TAILS and N-TAILS, has the potential to often identify both the C- and N-terminal semi-tryptic peptides of the same cleavage event. Alternately, C-TAILS can identify cleavage sites that are missed in N-TAILS analysis, e.g., as a result of insufficient peptide lengths for LC-MS/MS analysis.

Generally, our C-terminomics strategy focuses on cost-effective, commercially available reagents without the need for custom-synthesized reagents. In this protocol, we describe how to enrich protein C-terminal peptides from complex proteomes for LC-MS/MS analysis and we introduce stable isotope labeling for the relative quantitation of C-terminal processing in different biological samples. This workflow typically yields hundreds of C-terminal peptide sequences. Although beyond the scope of this protocol, the workflow can be adapted to characterize the C-termini of individual, purified proteins.

2. Materials

2.1. Enrichment of C-Terminal Peptides from Cellular Proteomes

1. EDC.
2. 10 kDa cut-off Microcon filter devices (Millipore).
3. 2-(N-morpholino)ethanesulfonic acid (MES).
4. 4-(2-hydroxyethyl)piperazine-1-ethanesulfonic acid (HEPES).
5. Acetonitrile.
6. Bicinchoninic acid (BCA) assay.
7. Bradford protein assay.
8. C18 solid-phase extraction columns.
9. Dimethylsulfoxide (DMSO).
10. Dithiothreitol (DTT).
11. DNAse I.
12. Ethylenediaminetetraacetic acid (EDTA).
13. Ethanolamine hydrochloride.
14. Formaldehyde (light isotope d(0)C12) for stable isotope labeling in comparative experiments, also heavy formaldehyde (d(2)C^{13}, Cambridge Isotopes).
15. Guanidine hydrochloride.
16. Iodoacetamide.
17. Methanol.
18. N-hydroxysuccinimide (NHS).
19. *Ortho*-phenanthroline.
20. Phenylmethanesulfonyl fluoride (PMSF).
21. Phosphate-buffered saline (PBS).
22. Poly-allylamine (average molecular weight ~56 kDa) (Sigma).
23. RNAse A.
24. Sample: Cellular proteome from an organism with a fully sequenced genome.
25. Sodium cyanoborohydride ($NaCNBH_3$, ALD coupling solution, Sterogene).
26. Standard laboratory equipment for cell culture, molecular biology, and protein chemistry.

27. Tabletop centrifuge.
28. Trans-epoxysuccinyl-L-leucylamido(4-guanidino)butane (E-64).
29. Trichloracetic acid (TCA).
30. Trypsin, sequencing grade.
31. Ultrasonication bath.

2.2. Identification of C-Terminal Peptides by Liquid Chromatography-Tandem Mass Spectrometry

1. Acetonitrile.
2. Formic acid.
3. C18 resin capillary column.
4. Conversion utility for mzXML data format (11).
5. Software for spectrum-to-sequence assignment, such as MASCOT (12) or X!TANDEM (13).
6. Tandem mass spectrometer in conjunction with a capillary liquid chromatography system.
7. Trans Proteomic Pipeline software (14).

3. Methods

3.1. Enrichment of C-Terminal Peptides from Cellular Proteomes

1. Grow and harvest cellular proteome
   (a) Adherent cell line:
   - Grow cells to desired confluence
   - Detach cells preferably with protease-free detachment methods
   - Centrifuge cells at 400 × *g*, 4°C, for 5 min
   - Decant supernatant and proceed with pellet.

   (b) Suspension cell culture, both eukaryotic and bacterial:
   - Grow cells to desired density
   - Centrifuge cells at 400 × *g*, 4°C, for 5 min
   - Decant supernatant and proceed with pellet
   - Wash cell pellet with PBS and repeat centrifugation.
2. Lyse cells in 100 mM HEPES, pH 7.5, in the presence of 1 μg/ml of DNAse I and RNAse A and the following protease inhibitors: 1 mM E-64, 10 mM PMSF, and 1 mM *ortho*-phenanthroline. Use mechanical lysis methods, such as ultrasonication, freeze–thaw cycles, or needle shearing. Avoid usage of detergents.

   Keep samples separate in experiments, where comparisons are made (see Note 1).

3. Centrifuge at 12,000 × *g* for 10 min.
4. Proceed with supernatant.
5. Determine protein concentration using Bradford assay.
6. Use 2 mg proteome as starting material.
7. Add guanidine hydrochloride to 2.0 M (final concentration).
8. Incubate with 10 mM DTT (final concentration) for 30 min at 70°C to reduce disulfide bonds.
9. Incubate with 25 mM iodoacetamide (final concentration) for 2 h at 25°C in the dark to protect free sulfhydryls.
10. Incubate with additional 5 mM DTT (total final DTT concentration is now 15 mM) for 30 min at 25°C to quench unreacted iodoacetamide.
11. Incubate with 20 mM formaldehyde (final concentration) and 20 mM $NaCNBH_3$ (ALD coupling solution, final concentration) for 16 h at 25°C to protect primary amines by reductive dimethylation. In case of a comparative experiment, label one sample with light formaldehyde (d(0)C^{12}) and one sample with heavy formaldehyde (d(2)C^{13}, Cambridge Isotopes).
12. Add 15% (v/v, final concentration) TCA to precipitate proteins.
13. Incubate on ice for 1 h.
14. Centrifuge at 12,000 × *g* for 10 min.
15. Proceed with pellet and discard supernatant.
16. Wash pellet four times with −20°C cold methanol (see Note 2). If pellet loosens, repeat centrifugation.
17. Resolubilize proteins in 200 mM MES, pH 5.0, 2.0 M guanidine hydrochloride, and 1.0 M ethanolamine. Incubate in an ultrasonication bath to enhance resolubilization.
18. Check that pH equals 5.0 and adjust if required.
19. Add 10 mM NHS and 20 mM EDC.
20. After a 1-min incubation, check that pH equals 5.0 and adjust if required.
21. Incubate for 1 h at 25°C.
22. Repeat EDC addition step, including pH adjustment, twice with incubations of 1–16 h.
23. Add 15% (v/v, final concentration) TCA to precipitate proteins.
24. Incubate on ice for 1 h.
25. Centrifuge at 12,000 × *g* for 10 min.
26. Proceed with pellet and discard supernatant.
27. Wash pellet three times with −20°C cold methanol. If pellet loosens, repeat centrifugation.

28. Resuspend proteins in a minimal volume of 20 mM HEPES, pH 7.5, and 2.0 M guanidine hydrochloride. Incubate in an ultrasonication bath to improve resolubilization.
29. Dilute guanidine hydrochloride concentration to 0.5 M by addition of 20 mM HEPES, pH 7.5.
30. Determine protein concentration using the Bradford assay.
31. Add sequencing-grade trypsin at a ratio of 1:50 (w/w, trypsin/total protein).
32. Incubate for 16 h at 37°C.
33. Incubate with 20 mM formaldehyde (final concentration) and 20 mM $NaCNBH_3$ (ALD coupling solution, final concentration) for 16 h at 25°C to protect newly generated N-termini of internal peptides by reductive dimethylation. In case of a comparative experiment, label one sample with light formaldehyde ($d(0)C^{12}$) and one sample with heavy formaldehyde ($d(2)C^{13}$) (Cambridge Isotopes). Use the same labeling scheme as in step 11.
34. In case of a comparative experiment employing formaldehyde-based labeling, mix samples at equal abundance.
35. Prepare a 2 mM stock solution of poly-allylamine (average molecular weight ~56 kDa) in 400 mM MES, pH 5.0, 2 M guanidine hydrochloride, and 20% (v/v) acetonitrile. Incubation in an ultrasonication bath can facilitate the dissolving of poly-allylamine.
36. Add one volume of poly-allylamine stock to the tryptic digest to reach a final poly-allylamine concentration of 1 mM.
37. Add 10 mM NHS (final concentration, see Note 3).
38. Control pH of proteome/poly-allylamine mixture and adjust to 5.0 if required.
39. Start coupling reaction by addition of 50 mM EDC (final concentration, see Note 4).
40. Incubate for 1 min at 25°C.
41. Check that pH equals 5.0 and adjust if required.
42. Incubate for 3 h at 25°C.
43. Repeat EDC step – including pH control – twice with the last incubation lasting 16 h.
44. Pass polymer/peptide mixture through a 10-kDa cutoff Microcon filter device (Millipore). Due to the hygroscopicity of poly-allylamine, the ultrafiltration step can be very lengthy. Use 25°C as centrifugation temperature.
45. Keep the filtrate, which contains uncoupled, C-terminal peptides.
46. Repeat ultrafiltration step twice with 500 μl water.

47. Pool the three filtrates and desalt using C18 solid-phase extraction columns.
48. Determine peptide concentration using BCA assay.

3.2. Identification of Cleavage Products by Liquid Chromatography-Tandem Mass Spectrometry

1. Analyze C-terminal by LC-MS/MS. Standard LC-MS/MS settings for peptide/proteome analysis are employed. These settings are machine specific and not within the scope of this protocol.
2. Identify peptide sequences by spectrum-to-sequence assignment using Mascot (12) and/or X!Tandem (13) (see Note 5). Since trypsin does not cleave at dimethylated lysines, set cleavage specificity as C-terminal to arginine (ArgC-like specificity). Apply semi-style searches with up to three missed cleavages. Set static modifications as carboxyamidomethylation of cysteine residues (+57.02 Da), dimethylation of lysines and amino termini (+28.03 Da for normal isotope abundance; 34.06 Da for d(2)C^{13} formaldehyde), and ethanolamine modification (+43.04 Da) for aspartate and glutamate residues as well as peptide C-termini (see Note 6). For quantitative comparison with stable isotope labeling, use software tools like XPRESS (15) to determine heavy:light (H/L) ratios (see Note 7). The mass difference between light and heavy dimethylated amines is 6.03 Da.

4. Notes

1. In a comparative C-terminomic experiment employing chemical labeling, samples are kept separate until after the second reductive methylation step (step 34). If metabolic labeling strategies are used, samples can be mixed prior to or immediately after lysis. When opting for a metabolic labeling strategy, note that C-terminal tryptic peptides lack a specific "anchor" – such as a C-terminal lysine or arginine residue – and metabolic labeling of every C-terminal peptide cannot be achieved with standard SILAC-style (16) approaches.
2. TCA coprecipitates guanidine hydrochloride, which is subsequently washed off with methanol.
3. Prepare a 1 M NHS stock solution in DMSO.
4. EDC rapidly degrades in water. Store at –20°C and bring to room temperature before opening. Prepare a fresh 1 M stock in DMSO immediately before usage.
5. In terminal proteomics, protein identifications are typically based on single peptide sequences, hence requiring a particularly low false discovery rate in spectrum-to-sequence assignment.

As described by Kleifeld et al. for the N-TAILS technique, this can be achieved by replicate experiments or by independently validating a peptide sequence with a second spectrum-to-sequence assignment software (5). Here, the combination of Mascot (12) and X!Tandem (13) has proven to be particularly useful.

6. Uncoupled, internal tryptic peptides can be recognized via a free, unprotected peptide C-terminus.
7. In an initial C-TAILS study, an H/L ratio cutoff value of 4 was found to accurately distinguish neo C-termini (8).

Acknowledgments

O.S. acknowledges support from the Deutsche Forschungsgemeinschaft (DFG, SCHI 871/2-1). P.F.H. is supported by Postdoctoral Fellowships of the German Academic Exchange Service (DAAD) and the Michael Smith Foundation for Health Research (MSFHR). O.B. acknowledges support from the Swiss National Science Foundation and Canadian Institutes of Health Research (CIHR). C.M.O is supported by a Canada Research Chair in Metalloproteinase Proteomics and Systems Biology. This work was supported by a grant from the CIHR and from a program project grant in Breast Cancer Metastases from the Canadian Breast Cancer Research Alliance (CBCRA) with funds from the Canadian Breast Cancer Foundation and the Cancer Research Society as well as with an Infrastructure Grant from MSHFR.

References

1. López-Otín, C., and Overall, C.M. (2002) Protease degradomics: a new challenge for proteomics, *Nat. Rev. Mol. Cell. Biol.* *3*, 509–519.
2. Auf dem Keller, U., and Schilling, O. (2010) Proteomic Techniques and Activity-based Probes for the System-wide Study of Proteolysis, *Biochimie.*
3. Doucet, A., Butler, G.S., Rodríguez, D., Prudova, A., and Overall, C.M. (2008) Metadegradomics: toward in vivo quantitative degradomics of proteolytic post-translational modifications of the cancer proteome, *Mol. Cell. Proteomics* *7*, 1925–1951.
4. Auf dem Keller, U., Prudova, A., Gioia, M., Butler, G. S., and Overall, C. M. (2010) A statistics-based platform for quantitative N-terminome analysis and identification of protease cleavage products, *Mol. Cell. Proteomics* *9*, 912–927.
5. Kleifeld, O., Doucet, A., Auf dem Keller, U., Prudova, A., Schilling, O., Kainthan, R. K., Starr, A. E., Foster, L. J., Kizhakkedathu, J. N., and Overall, C. M. (2010) Isotopic labeling of terminal amines in complex samples identifies protein N-termini and protease cleavage products, *Nat. Biotechnol.* *28*, 281–288.
6. Prudova, A., Auf dem Keller, U., Butler, G. S., and Overall, C.M. (2010) Multiplex N-terminome analysis of MMP-2 and MMP-9 substrate degradomes by iTRAQ-TAILS quantitative proteomics, *Mol. Cell. Proteomics* *9*, 894–911.
7. Reznik, S. E., and Fricker, L. D. (2001) Carboxypeptidases from A to Z: implications in

embryonic development and Wnt binding, *Cell. Mol. Life Sci.* *58*, 1790–1804.

8. Schilling, O., Barré, O., Huesgen, P. F., and Overall, C. M. (2010) Proteome-wide analysis of protein carboxy termini: C terminomics, *Nat. Methods* *7*, 508–511.
9. Panchaud, A., Hansson, J., Affolter, M., Bel Rhlid, R., Piu, S., Moreillon, P., and Kussmann, M. (2008) ANIBAL, stable isotope-based quantitative proteomics by aniline and benzoic acid labeling of amino and carboxylic groups, *Mol. Cell. Proteomics* *7*, 800–812.
10. Van Damme, P., Staes, A., Bronsoms, S., Helsens, K., Colaert, N., Timmerman, E., Aviles, F. X., Vandekerckhove, J., and Gevaert, K. (2010) Complementary positional proteomics for screening substrates of endo- and exoproteases, *Nat. Methods* *7*, 512–515.
11. Pedrioli, P. G. A., Eng, J. K., Hubley, R., Vogelzang, M., Deutsch, E. W., Raught, B., Pratt, B., Nilsson, E., Angeletti, R. H., Apweiler, R., Cheung, K., Costello, C. E., Hermjakob, H., Huang, S., Julian, R. K., Kapp, E., McComb, M. E., Oliver, S. G., Omenn, G., Paton, N. W., Simpson, R., Smith, R., Taylor, C. F., Zhu, W., and Aebersold, R. (2004) A common open representation of mass spectrometry data and its application to proteomics research, *Nat. Biotechnol.* *22*, 1459–1466.
12. Perkins, D. N., Pappin, D. J., Creasy, D. M., and Cottrell, J. S. (1999) Probability-based protein identification by searching sequence databases using mass spectrometry data, *Electrophoresis* *20*, 3551–3567.
13. Craig, R., and Beavis, R. C. (2004) TANDEM: matching proteins with tandem mass spectra, *Bioinformatics* *20*, 1466–1467.
14. Pedrioli, P. G. A. (2010) Trans-proteomic pipeline: a pipeline for proteomic analysis, *Methods Mol. Biol.* *604*, 213–238.
15. Han, D. K., Eng, J., Zhou, H., and Aebersold, R. (2001) Quantitative profiling of differentiation-induced microsomal proteins using isotope-coded affinity tags and mass spectrometry, *Nat. Biotechnol.* *19*, 946–951.
16. Ong, S.-E., Blagoev, B., Kratchmarova, I., Kristensen, D. B., Steen, H., Pandey, A., and Mann, M. (2002) Stable isotope labeling by amino acids in cell culture, SILAC, as a simple and accurate approach to expression proteomics, *Mol. Cell. Proteomics* *1*, 376–386.

Chapter 5

Construction of Protein Interaction Networks Based on the Label-Free Quantitative Proteomics

Mihaela E. Sardiu and Michael P. Washburn

Abstract

Multiprotein complexes are essential building blocks for many cellular processes in an organism. Taking the process of transcription as an example, the interplay of several chromatin-remodeling complexes is responsible for the tight regulation of gene expression. Knowing how those proteins associate into protein complexes not only helps to improve our understanding of these cellular processes, but can also lead to the discovery of the function of novel interacting proteins. Given the large number of proteins with little to no functional annotation throughout many organisms, including human, the identification and characterization of protein complexes has grown into a major focus of network biology. Toward this goal, we have developed several computational approaches based upon label-free quantitative proteomics approaches for the analysis of protein complexes and protein interaction networks. Here, we describe the computational approaches used to build probabilistic protein interaction networks, which are detailed in this chapter using the example of complexes involved in chromatin remodeling and transcription.

Key words: Proteomics, Quantitative, Proteins, Complexes, Networks, Spectral counting, Normalized spectral abundance factor

1. Introduction

Quantitative proteomics currently contributes to the identification of biological networks in two main areas with the identification and characterization of protein complexes and the prediction of protein interactions. The main experimental pipeline for this is the use of affinity purification of protein complexes followed by mass spectrometry analysis (APMS) (1–5). In this approach, specific proteins are tagged (subsequently called "baits") and expressed in cells, cell lines, or even whole organisms. Through affinity purification of the bait, the specifically binding proteins (which are called "preys") are isolated and subsequently analyzed and quantified using mass spectrometry approaches. APMS studies are typically carried out

Gerard Cagney and Andrew Emili (eds.), *Network Biology: Methods and Applications*, Methods in Molecular Biology, vol. 781, DOI 10.1007/978-1-61779-276-2_5, © Springer Science+Business Media, LLC 2011

without the use of stable isotopes, and spectral counting approaches are used for the quantitative analysis (2–5). In spectral counting, the sum of the total number of scans detecting and identifying a particular protein is used as the basis for quantitative analysis (6–8). We have successfully used spectral counting-based approaches to determine quantitative differences in protein complexes (9) and for the probabilistic analysis of protein interaction networks (4, 10).

In this chapter, we describe the different steps and bioinformatics analyses necessary to translate quantitative proteomics data generated from APMS experiments using multidimensional protein identification (MudPIT) (Fig. 1) into biological networks.

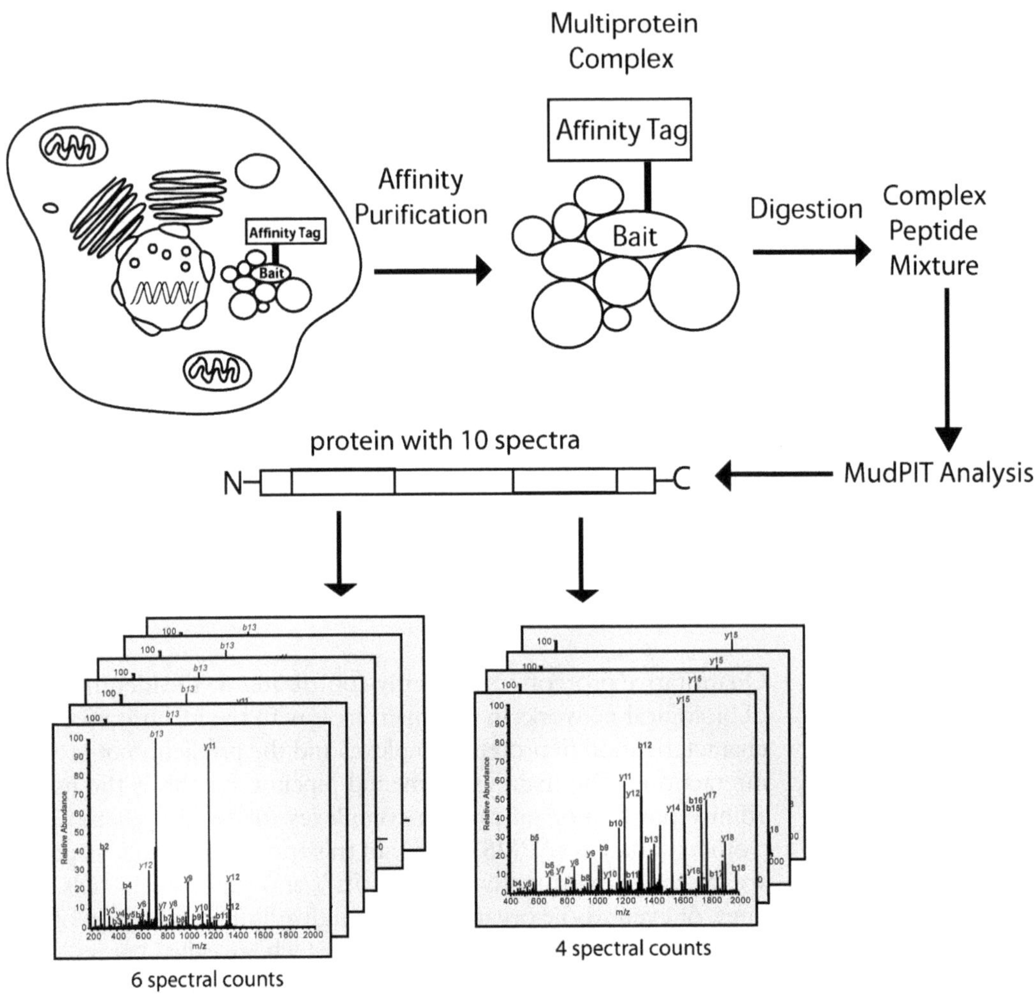

Fig. 1. Spectral count-based quantitative proteomic analysis of protein complexes. An affinity-tagged protein that is part of a complex is isolated from cells using an affinity purification procedure. The purified protein complex is then digested into peptides, which subsequently are analyzed by MudPIT. As an example, one of the proteins in the mixture, here, a short and low-abundance protein, is detected and identified by two unique peptides. One of these peptides has six spectral counts and the other has four spectral counts. Therefore, the protein is detected and identified by ten total spectral counts. Typical analyses of multiprotein complexes actually yield hundreds of spectral counts.

In short, there are four main methods applied after the identification and quantification of the proteins in a sample. The first is data preprocessing, which includes background elimination, normalization, and assessment of reproducibility. Second is discriminant analysis to identify proteins enriched in the sample using a vector algebra approach named singular value decomposition. The third step is hierarchical clustering and data visualization, which is followed by the fourth step, the construction of probabilistic interaction networks based on Bayes' theorem. We have previously used these approaches to analyze a human chromatin-remodeling network centered on Tip49a and Tip49b (4) and a yeast deletion network analysis of the Rpd3/Sin3 histone deacetylase complex (10).

2. Materials

2.1. Initial Data Sets

The identification of proteins in the samples is determined using MudPIT (11). The SEQUEST algorithm (12) is used in order to match MS/MS spectra to peptides. Peptide/spectrum matches are selected, compared, and filtered using DTASelect/CONTRAST (13) with the use of concatenated forward and reversed databases to determine false discovery rates and other standard approaches (11). The total number of the tandem mass spectra detected and identified for a given protein defines the spectral counting, which is used as a quantitative measure of protein abundance (Fig. 1) (9, 14). Since these steps have been described in detail elsewhere (11), they are not described in detail here.

3. Methods

3.1. Normalization

Historically, the measurement of the relative abundance of proteins in a protein sample was performed by labeling the samples with two different (light and heavy) isotopes followed by a simultaneous analysis in the same MS run (15). More recent work established the feasibility of using single, label-free MS runs for the relative quantification of proteins in a complex sample (15). This removed the cost and experimental effort necessary for isotopic labeling. In addition, the major advantage of this strategy is that it allows the simultaneous analysis of multiple protein samples from different MS runs using proper statistical and bioinformatics tools. The relative abundance of proteins in the complex protein samples analyzed in different MS runs is deduced from the spectra detected in the analyzer of the mass spectrometer. Several quantitative measures were proposed which either use the shape of the peaks (area under the curve or the summed intensities) or the number of the spectra ("spectral count") corresponding to a peptide (15, 16).

We routinely use spectral counts, which faithfully reproduce protein associations in protein complexes (4, 10).

Especially in the case of label-free approaches, the normalization of the spectral counts obtained in an experiment is crucial. The two main features that have to be taken into account for the normalization are the total number of peptides/proteins detected in the MS run and the length of a protein, since longer proteins comprise more peptides which therefore are more likely to be detected than shorter ones. We, therefore, introduced the normalized spectral abundance factor (NSAF) (14) for normalization in quantitative proteomics experiments.

In a typical experiment, the raw data is assembled into a matrix $A_{N\times M}$ in which the columns (M) represent the individual purifications (baits) and rows (N) represent the associated proteins (preys). Each element of the matrix $A_{N\times M}$ is denoted by the spectral count. The first step in preprocessing the data is a normalization procedure known as the distributed normalized spectral abundance factor (dNSAF) (17), which is defined as:

$$\mathrm{dNSAF} = \frac{(\mathrm{uSpC} + d \times \mathrm{sSpC}) / L}{\sum_{i=1}^{N} \left((\mathrm{uSpC} + d \times \mathrm{sSpC}) / L\right)_i}, \quad (1)$$

where d is defined as $d = \frac{\mathrm{uSpC}}{\sum_{i=1}^{N} \mathrm{uSpc}_i}$ and uSpC is the unique spectral counts for a given protein.

Basically, the number of spectra for every protein in a dataset is calculated based upon the sum of the spectra that are unique to the protein (uSpC) and the shared spectral counts (sSpC), which are weighted according to their distribution (d) among all proteins (17). Some peptides are not unique to a single protein or protein isoform; instead of either ignoring those peptides or counting the full spectra count for each protein, their spectra count is distributed over all proteins to which they can be assigned (17). This value is then first normalized by the length (L) to form the abundance factor, and then normalized by the sum of all abundance factors in the sample.

3.2. Extraction of Contaminants

Every method employing the principle of affinity purifications has the intrinsic feature that some proteins bind nonspecifically to the resin/column or the affinity tag used, resulting in the carryover of unspecific proteins in the eluted samples (4). Therefore, before performing any kind of analysis, it is crucial to remove those proteins. For that reason, control experiments are performed using cell extracts of wild-type cells passed through the affinity purification procedure (4), affinity-tagged mock proteins, or samples overexpressing only the affinity tag that is not fused to a protein (see Note 1). The simplest approach would be to disregard all proteins present in the control experiment(s); however, this would exclude proteins present in the controls that nevertheless are enriched in the true samples, and therefore can be considered as specific binders to the affinity-purified protein.

To achieve this, a matrix is prepared in a bait–prey format that consists of dNSAF values which is subjected to the extraction of contaminants. The nonspecific binding proteins are identified and removed from the dataset by comparing the dNSAF value of every protein in each of the individual purifications with the dNSAF value from the affinity purification control experiments, which are performed in the same yeast strain, cell line, and/or organism that the affinity-tagged proteins of interest were analyzed from (4). If the dNSAF value in the purification is higher than the dNSAF in the mock control, the protein is considered specific to that particular purification. Otherwise, the dNSAF value is replaced by zero and the protein is considered nonspecific to that particular purification (10). For an ideal experimental series which consists of as many negative controls as positive samples, the contaminant extraction is performed using a ratio vector comparing the vector of a protein in the purifications (**x**) versus the controls (**y**) (4, 18). The ratio vector (α) between two vectors x and y is defined as:

$$\alpha = \sqrt{\frac{< y, y >}{< x, x >}}. \tag{2}$$

α greater than 1 indicates that the value expected as **y** is a "greater" vector. A protein with a value of α greater than 1 is considered contaminant and is excluded from the data. Although both methods remove the majority of the contaminant proteins, one can also control for proteins that show up consistently in TAP purifications, for example, (see Note 2).

3.3. Assessment of Reproducibility and Quality Control

Depending on the scale of the experiment, it is not always feasible to perform multiple replicates for each bait. It is valid to omit the majority of replicates and rather focus on more baits, since APMS experiments are fairly reproducible. However, this reproducibility has to be verified using replicates for selected baits. To measure the reproducibility of the replicates, Pearson correlation (r) and/or ANOVA analysis is applied (10). Pearson correlation is employed for baits that contain two replicates while ANOVA is the method of choice for baits with three or more replicates (). The Pearson correlation (r), which is computed by a pairwise comparison of each protein in the two samples, is calculated using *R* statistical language (http://www.r-project.org) function cor(). For more details, see Note 3.

One-way ANOVA can be computed on the replicates using the R function lm(). ANOVA tests the hypothesis that each sample (each replicate) comes from a population with the same mean (H_0: $\mu_1 = \mu_2 = \mu_3$) versus the alternative hypothesis that the mean differs for at least two of the replicates (H_1 at least one inequality: $\mu_1 \neq \mu_2$ or $\mu_1 \neq \mu_3$ or $\mu_2 \neq \mu_3$) (see Note 4). The ANOVA results are then assessed using Tukey's test in order to pairwise test the mean of each replicate with other replicates. The *R* function TukeyHSD() can be used to compute Tukey's Honest Significant Difference test (see Note 5).

3.4. Singular Value Decomposition

To determine the proteins that are enriched in the purifications, singular vector decomposition (SVD) is applied on a matrix whose elements represent the normalized spectral count (i.e., dNSAF) for each prey in each bait (19, 20). SVD is applied to the matrix using the R function svd() (see Note 6). The output of the SVD is a list of vectors that corresponds to the left and the right singular vectors of a matrix. The information obtained from the first left singular vector (lsv) defines the proteins that are enriched in the purifications by using a rank-estimated method (4, 10). Basically, the distribution of the lsv is investigated in order to determine a proper cutoff for selecting the proteins that were the most associated with the baits (i.e., are components of the same complex as the bait). By plotting the lsv in a log–log (or a linear–log) scale, the proteins that are more directly associated with the bait can be discriminated from the indirect bait associations at the breaking point of the distribution. For instance, in the case of the Rpd3 complexes, the coefficients of the known subunits of the Rpd3(S) and Rpd3(L) complexes were situated at the top of the distribution while the rest of the proteins, which represent the indirect complex associations, were located on the lower part of the distribution (10).

3.5. Clustering Methods

The aim of cluster analysis is to separate the proteins identified in an APMS experiment based on their abundance level. In principle, the proteins that exhibit similar abundance levels across the baits are considered as being part of the same complex or subcomplex and, therefore, are expected to cluster together (21). Yet, different baits belonging to the same protein complex do not always purify identical preys (due to phenomena, like transient interactions, different isoforms of protein complexes, etc.) (4). There are proteins which are almost always present in each purification (core proteins), proteins which are identified only in a few purifications (attachment proteins), and proteins which are always present together (modules) (22, 23). In addition, in some cases, different complexes share the same core proteins. Therefore, cluster analysis of quantitative proteomics data can be very challenging.

There are two different kinds of hierarchical clustering techniques, agglomerative and divisive, and a variety of methods and distance measures (see Note 7) (21). Not every combination of them leads to satisfactory results when using quantitative values (dNSAF) as an input for clustering analysis. Generally, best results are obtained using agglomerative clustering with Pearson correlation as a distance matrix and UPGMA/Ward method as an algorithm (21). First, a matrix is generated after SVD analysis with dNSAF values for all enriched proteins as rows and the baits as columns. Next, the software PermutMatrix (http://www.lirmm.fr/~caraux/PermutMatrix/) can be employed for the assessment of the hierarchical agglomerative procedures (24). Using this method, a dendrogram and an area image of the clustering result are produced. Next, the dendrogram has to be inspected manually in order to delimit the proteins into different groups. After the groups are determined, the proteins can be represented in a Venn diagram. An example of the results of a UPGMA and Pearson cluster analysis of the Tip49a- and b-centered chromatin-remodeling and nutrient-signaling proteins is shown in Fig. 2 (4).

3.6. Probabilistic Network Analysis

The probability between each prey protein and the affinity-tagged bait is calculated based upon Bayes' theorem (4, 10). For each protein, the spectral counts of the observed dataset are transformed into probabilities in which the numerator is the product of a likelihood and a prior probability and the denominator is just the sum of all of the numerators among the different baits. The probabilistic measure for the connectivity between a bait and a prey protein is calculated for each prey based on the observed spectral counts in the different baits according to the following considerations.

To quantify the connectivity relationship between a prey protein i ($i=1,\ldots, N$) and a bait j ($j=1,\ldots, M$), the conditional probability is first estimated, that is, the probability of a prey i being in the sample given the fact that the bait j is in the sample by:

$$P(i \mid j) = \frac{P(i, j)}{P(j)}, \tag{3}$$

where $P(i, j)$ is the joint probability between protein i and protein j and is defined as:

$$P(i, j) = \frac{C_{i,j}}{\sum_{i',j'} C_{i'j'}}, \tag{4}$$

where $C_{i,j}$ is the number of distributed spectral abundance factor (dSAF) value of prey i in bait j while $\sum_{i',j'} C_{i',j'}$ sums the total number of all dSAFs in all samples. $P(j)$ is the likelihood of protein j and is estimated by:

$$P(j) = \frac{\sum_{i} C_{i,j}}{\sum_{i',j'} C_{i'j'}}, \tag{5}$$

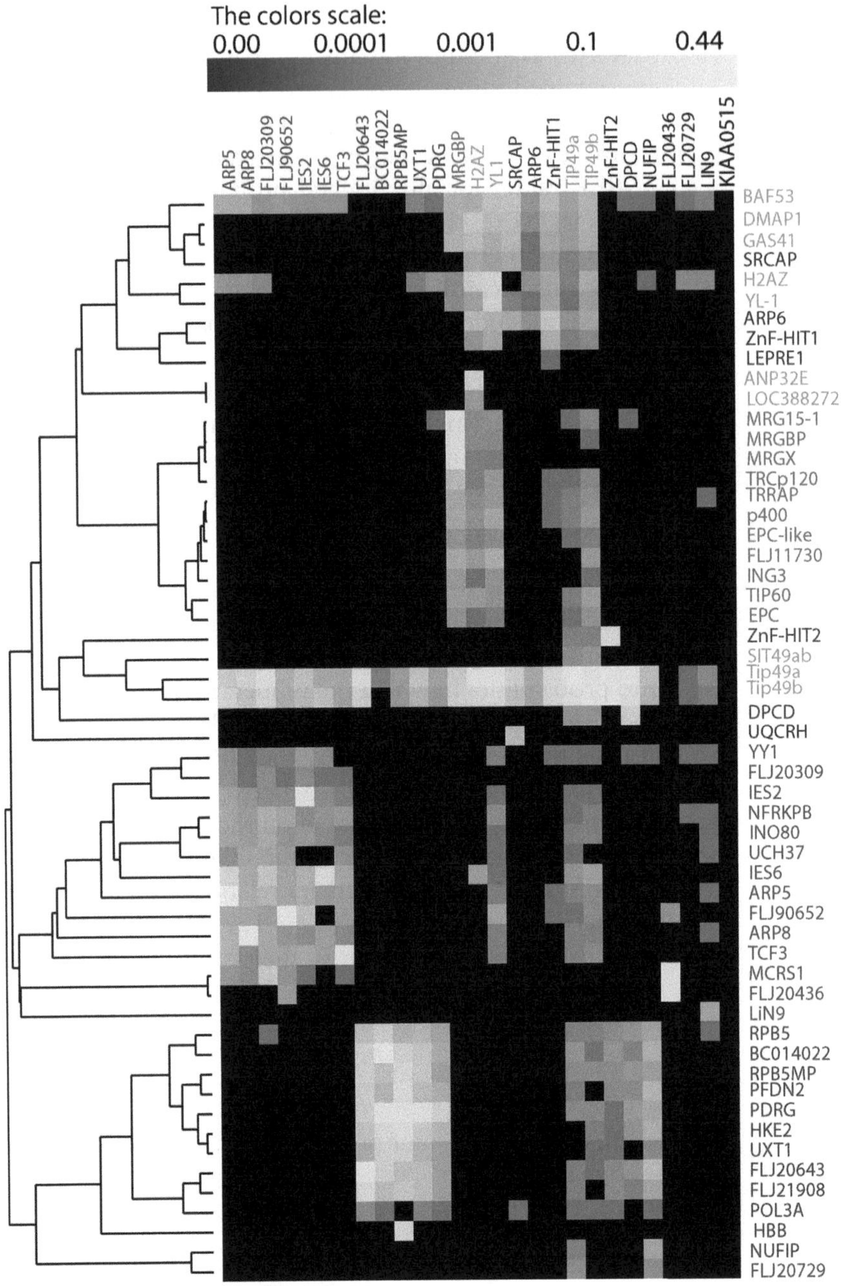

Fig. 2. Hierarchical clustering. A hierarchical cluster using UPGMA algorithm and Pearson correlation as distance metric was performed on the relative protein abundances expressed as NSAFs. Each *column* represents an isolated purification (bait), and each *row* represents an individual protein (prey). The color intensity represents protein abundance with the brightest indicating highest abundance and decreasing intensity indicating decreasing abundance. *Black* indicates that the protein was not detected in a particular sample. The protein complexes are hINO80, URI/Prefoldin, TRRAP/TIP60, and SRCAP. Reprinted with modification from original article Sardiu et al. (2008) Probabilistic assembly of human protein interaction networks from label-free quantitative proteomics. *Proc Natl Acad Sci USA* 105(5): 1454–9. doi: 10.1073/pnas.0706983105) (4) Copyright (2008) National Academy of Sciences, USA.

where $\sum_i Cij$ sums the dSAF values of all preys (summation over all possible values of i) in the bait j. When the conditional probability is known, the probability of prey i can be calculated using:

$$P(i) = \sum_j P(i \mid j)P(j), \tag{6}$$

where the summation is over all possible values of j.

For a bait j and prey i, the posterior probability $P(j|i)$ is defined by Bayes' rule:

$$P(j \mid i) = \frac{P(i \mid j)P(j)}{P(i)}. \tag{7}$$

It quantifies the preference of a bait to have a connection with a prey. Similar to previous studies, an arbitrary prior probability was chosen, that is, $1/N$.

Since probabilities correlate to the tendency of two proteins to associate with each other, we partition the components of a complex into three biologically meaningful groups, i.e., those exhibiting low (<0.02), medium (0.02–0.04), and high (>0.04) probabilities. These three different categories correspond to proteins with a low, intermediate, and high association to the bait. It is intuitive to expect that those preys with a higher probability are more likely to associate stronger with the bait. However, to answer this question, probabilistic deletion networks have to be generated (10). Finally, using the program Cytoscape (http://www.cytoscape.org) (25), images for probabilistic networks can be generated for the protein complex analyzed. In these networks, the nodes correspond to proteins and the connections are weighted according to the different probabilities (4). An example of a probabilistic network of the Tip49a and b complexes involved in chromatin-remodeling and nutrient-signaling complexes (4) can be seen in Fig. 3 in which the different probability levels are visualized using different colors.

3.7. Probabilistic Deletion Networks

A high probability between a prey protein and a bait could either indicate a strong pairwise interaction and/or that the two proteins are part of a stable complex. In order to obtain a score that is more likely to reflect the true direct interaction of a prey–bait pair, it is important to minimize the effect of other proteins affecting this interaction. One approach to accomplish this goal is to disrupt the integrity of the protein complex (10). To achieve this, APMS experiments are performed in yeast strains in which certain components of the complex studied are genetically deleted (10). The expectation is that after the deletion of a protein that is essential for the stability of a tight complex and/or for the interaction between prey and bait, the probability of the prey–bait pair will significantly decrease unless the two proteins exhibit a strong association. On the other hand, the change of probabilities also indicates whether

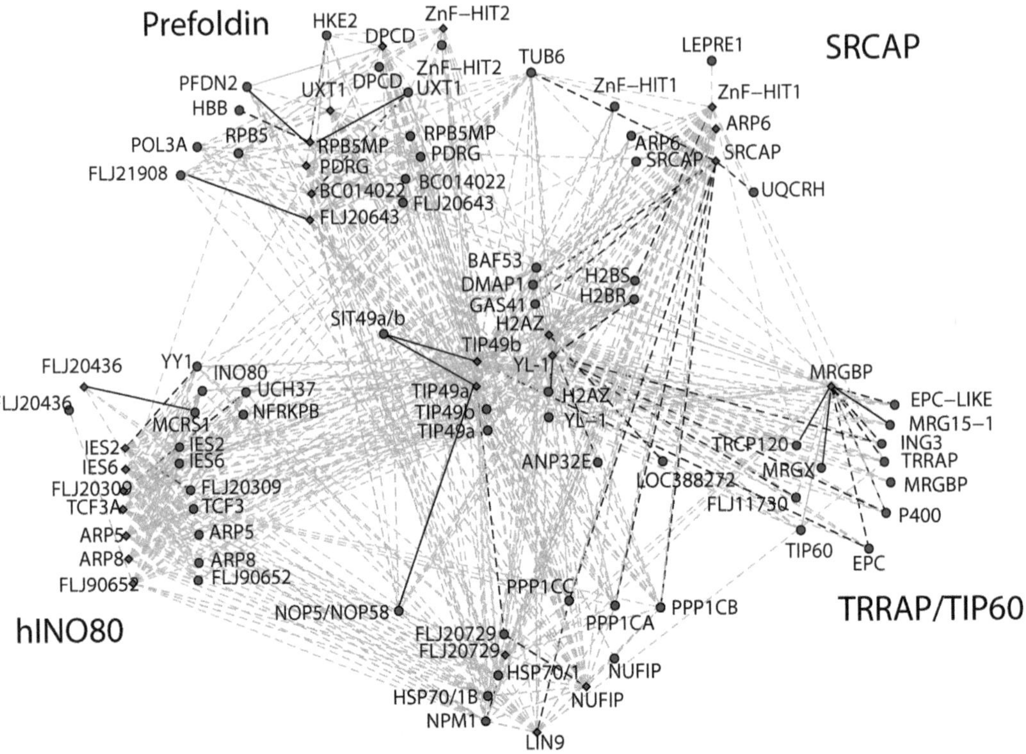

Fig. 3. Probabilistic network of TIP49a/TIP49b interactions. The entire network is shown, in which nodes represent protein baits (depicted by *diamonds*) or protein preys (*circles*) and the weighted edges represent the calculated posterior probabilities. *Black dashed lines* represent the interactions with the highest probability, the *grey dashed lines* represent interactions with a moderate and low probability. *Solid lines* represent protein–protein interactions validated by the literature or the present study. Reprinted with modification from original article Sardiu et al. (2008) Probabilistic assembly of human protein interaction networks from label-free quantitative proteomics. *Proc Natl Acad Sci USA* 105(5): 1454–9. doi: 10.1073/pnas.0706983105) (4) Copyright (2008) National Academy of Sciences, USA.

the deleted protein is essential for the integrity of the complex and/or for the interaction between the prey and the bait (i.e., if the interaction between prey and bait depends on the deleted subunit). Therefore, if a prey–bait pair retains its high probability after deletion of a protein (especially for those that affect the integrity of the complex; for an assessment of the effect on the complex, see below), this could indicate a strong association between the two proteins. If the probability decreases, the prey–bait pair requires the deleted protein to interact. An example of the hierarchical cluster used to build probabilistic deletion networks of the Rpd3 complex studied in (10) can be seen in Fig. 4.

In addition to determining the direct relationships between prey and bait, probabilistic deletion networks also allow the identification of subcomplex using cluster analysis. While the wild-type RPD3 complex is very tight and therefore leads to solid clusters corresponding to the large (Rpd3L) and small (Rpd3S)

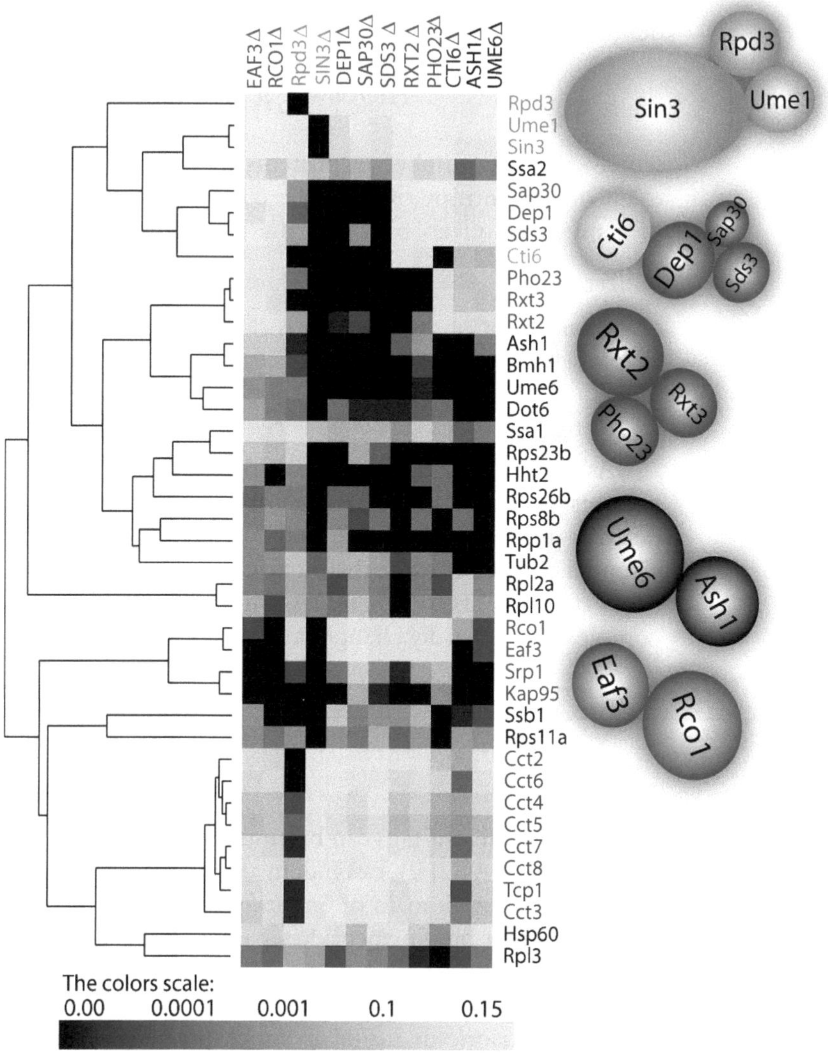

Fig. 4. Analysis of Rpd3-TAP deletion strains. Hierarchical clustering of the purifications of Rpd3-TAP in different deletion strains. Each *column* represents an isolated Rpd3-TAP in a different deletion strain, and each *row* represents an individual protein (prey). The color intensity represents the protein abundance (dNSAF) with the brightest indicating highest abundance and decreasing intensity depicting decreasing abundance. *Black* indicates that the protein was not detected in a particular purification. The formation of subcomplexes is illustrated on the *right side* of the cluster. Reprinted with modification from the original article Sardiu et al. (2009) Determining protein complex connectivity using a probabilistic network derived from quantitative proteomics. *PLoS ONE* 4(10): e7310. doi: 0.1371/journal.pone.0007310) (10).

subunits, after deletion analysis, the clusters segregate into smaller subcomplexes (10). The effect of a deleted protein on the whole complex can be estimated by calculating the information theory-based entropy of a complex. Like the entropy in physics which measures the “disorder” of a system, in information theory, the Shannon entropy measures the amount of information of a system. Hence, Shannon entropy can also be used to calculate the amount of information contained by a protein complex which is defined by:

$$H(p) = -\sum_{i=1}^{N} p_i \log p_i, \tag{8}$$

where N is the number of proteins in the complex (in order to calculate the entropy only for the complex) and p_i is the dNSAF of a prey in a given bait, and the summation is taken over the nonzero p_i ($\sum_{i=1}^{N} p_i = 1$). The *information* lost after genetic deletion of a component of the complex is calculated based on the entropy as follows:

$$I(x) = H_{\text{before}} - H_{\text{after}}. \tag{9}$$

As was seen for the example of the Rpd3 complex, some components strongly affect the integrity of the complex after being deleted (high H_{after}), low I, and can therefore be considered as essential components for the integrity of the complex, whereas for other components, which are less important for the stability of the complex, H_{after} hardly changes after deletion (10).

4. Notes

1. Several different tags can be employed for APMS experiments, like the FLAG, HA, or tandem affinity tag, which consist of the IgG-binding domain of *Staphylococcus aureus* protein A and a calmodulin-binding peptide separated by a TEV protease cleavage site (reviewed in ref. 26). However, it is crucial to perform the control experiments with samples overexpressing the same tag (tag alone or fusion to mock protein) as for the specific experiment, since every tag has its intrinsic unspecific binding proteins.
2. When working with data generated in the yeast strain BY4741, we use a list of proteins that appeared in greater than 3% of all TAP purifications in a proteome-wide study performed by Krogan et al. (27) which used the same library of TAP-tagged proteins and the same strain as our ongoing studies.
3. Pearson correlation is our method of choice of measuring the reproducibility between two replicates, i.e., vectors a and b. The numerical values of these vectors are the normalized abundance factors (dNSAF) of all proteins identified in each bait. The Pearson correlation is defined as:

$$r(a,b) = \frac{\sum_i (a_i - \bar{a})(b_i - \bar{b})}{(n-1)S_a S_b}, \tag{10}$$

in which *Sa* and *Sb* are the corresponding standard deviations of vectors *a* and *b*. Values of *r* close to 1 indicate a *strong correlation*, whereas values of *r* close to 0 indicate *little* or *no correlation.* To compute the Pearson correlation in R for two numerical vectors *a* and *b*, the function scan() is first used to read in the data which is computed as follows: out<–scan("data", list(" ",0,0)), followed by reading the columns of data frame as protein_label<–out[[1]], a<–out[[2]], and b<–out[[3]]. Finally, we run the Pearson correlation for the two vectors *a* and *b* using the function *cor()* as cor(a,b, method=c("Pearson")).

4. To test the reproducibility between multiple replicate vectors (i.e., number of vectors greater than two), we use the ANOVA method. The following contains an example for the analysis of variance of a data consisting of three vectors. The data and its corresponding columns are read as shown in Note 3. For a one-way analysis of variance, the data must be in a data frame containing two columns, one an index for the independent variable and the other for the dependent variable's value. Creating a data frame requires two steps: first, creating a data frame with columns for each independent variables using the function data.frame() which executes the command as result=data.frame(a,b,c). Second, the columns are stack into a single column using the function stack() which is run as output=stack(result). Next, the analysis of variance is performed using the syntax anova(lm(values~ind, data=output)). The result of ANOVA is a *P*-value that measures if there is a difference between the three replicates or not. A *P*-value >0.05 indicates that there is no significant difference between any of the populations' means, and a *P*-value ≤0.05 indicates that at least two of the populations have significantly different means.
5. If the *P*-value obtained from the analysis of variance indicates a significant difference between the three vectors, we next seek to identify where the difference lies. For this, we use Tukey's Honest Significant Difference Test and use the optional command: TukeyHSD(aov(values~ind, data=output)). The result from this analysis is a list of *P*-values between each pair of replicates (in our case, three vectors, three *P*-values are obtained).
6. Singular value decomposition method is applied to the data using the R function svd(). The data (called "matrix") is first read using the function read.table() with the syntax output<– read.table("matrix"). The svd is next applied to the table as svd(output). The lsv obtained from this method is then used to select the proteins enriched in the samples.
7. In a previous study, we compared three different partitioning clustering approaches for clustering, namely, the methods PAM, FANNY, and K-means, and used two distances,

i.e., Euclidean and Manhattan (21). For a detailed discussion of the practicability of the different methods and distances on quantitative proteomics data, see Sardiu et al. (21).

Acknowledgments

This work was supported by the Stowers Institute for Medical Research.

References

1. Ewing, R. M., Chu, P., Elisma, F., Li, H., Taylor, P., Climie, S., McBroom-Cerajewski, L., Robinson, M. D., O'Connor, L., Li, M., Taylor, R., Dharsee, M., Ho, Y., Heilbut, A., Moore, L., Zhang, S., Ornatsky, O., Bukhman, Y. V., Ethier, M., Sheng, Y., Vasilescu, J., Abu-Farha, M., Lambert, J. P., Duewel, H. S., Stewart, II, Kuehl, B., Hogue, K., Colwill, K., Gladwish, K., Muskat, B., Kinach, R., Adams, S. L., Moran, M. F., Morin, G. B., Topaloglou, T., and Figeys, D. (2007) Large-scale mapping of human protein-protein interactions by mass spectrometry, *Mol Syst Biol 3*, 89.
2. Behrends, C., Sowa, M. E., Gygi, S. P., and Harper, J. W. (2010) Network organization of the human autophagy system, *Nature 466*, 68–76.
3. Breitkreutz, A., Choi, H., Sharom, J. R., Boucher, L., Neduva, V., Larsen, B., Lin, Z. Y., Breitkreutz, B. J., Stark, C., Liu, G., Ahn, J., Dewar-Darch, D., Reguly, T., Tang, X., Almeida, R., Qin, Z. S., Pawson, T., Gingras, A. C., Nesvizhskii, A. I., and Tyers, M. (2010) A global protein kinase and phosphatase interaction network in yeast, *Science 328*, 1043–1046.
4. Sardiu, M. E., Cai, Y., Jin, J., Swanson, S. K., Conaway, R. C., Conaway, J. W., Florens, L., and Washburn, M. P. (2008) Probabilistic assembly of human protein interaction networks from label-free quantitative proteomics, *Proc Natl Acad Sci USA 105*, 1454–1459.
5. Sowa, M. E., Bennett, E. J., Gygi, S. P., and Harper, J. W. (2009) Defining the human deubiquitinating enzyme interaction landscape, *Cell 138*, 389–403.
6. Liu, H., Sadygov, R. G., and Yates, J. R., 3rd. (2004) A model for random sampling and estimation of relative protein abundance in shotgun proteomics, *Anal Chem 76*, 4193–4201.
7. Old, W. M., Meyer-Arendt, K., Aveline-Wolf, L., Pierce, K. G., Mendoza, A., Sevinsky, J. R., Resing, K. A., and Ahn, N. G. (2005) Comparison of label-free methods for quantifying human proteins by shotgun proteomics, *Mol Cell Proteomics 4*, 1487–1502.
8. Zybailov, B., Coleman, M. K., Florens, L., and Washburn, M. P. (2005) Correlation of relative abundance ratios derived from peptide ion chromatograms and spectrum counting for quantitative proteomic analysis using stable isotope labeling, *Anal Chem 77*, 6218–6224.
9. Paoletti, A. C., Parmely, T. J., Tomomori-Sato, C., Sato, S., Zhu, D., Conaway, R. C., Conaway, J. W., Florens, L., and Washburn, M. P. (2006) Quantitative proteomic analysis of distinct mammalian Mediator complexes using normalized spectral abundance factors, *Proc Natl Acad Sci USA 103*, 18928–18933.
10. Sardiu, M. E., Gilmore, J. M., Carrozza, M. J., Li, B., Workman, J. L., Florens, L., and Washburn, M. P. (2009) Determining protein complex connectivity using a probabilistic deletion network derived from quantitative proteomics, *PLoS One 4*, e7310.
11. Swanson, S. K., Florens, L., and Washburn, M. P. (2009) Generation and analysis of multidimensional protein identification technology datasets, *Methods Mol Biol 492*, 1–20.
12. Eng, J., McCormack, A. L., and Yates, J. R., 3rd. (1994) An approach to correlate tandem mass spectral data of peptides with amino acid sequences in a protein database., *J Am Soc Mass Spectrom 5*, 976–989.
13. Tabb, D. L., McDonald, W. H., and Yates, J. R., 3rd. (2002) DTASelect and Contrast: tools for assembling and comparing protein identifications from shotgun proteomics, *J Proteome Res 1*, 21–26.
14. Zybailov, B., Mosley, A. L., Sardiu, M. E., Coleman, M. K., Florens, L., and Washburn, M. P. (2006) Statistical analysis of membrane proteome expression changes in Saccharomyces cerevisiae, *J Proteome Res 5*, 2339–2347.

15. Kline, K. G., Finney, G. L., and Wu, C. C. (2009) Quantitative strategies to fuel the merger of discovery and hypothesis-driven shotgun proteomics, *Brief Funct Genomic Proteomic 8*, 114–125.
16. Lundgren, D. H., Hwang, S. I., Wu, L., and Han, D. K. (2010) Role of spectral counting in quantitative proteomics, *Expert Rev Proteomics 7*, 39–53.
17. Zhang, Y., Wen, Z., Washburn, M. P., and Florens, L. (2010) Refinements to label free proteome quantitation: how to deal with peptides shared by multiple proteins, *Anal Chem 82*, 2272–2281.
18. Kuruvilla, F. G., Park, P. J., and Schreiber, S. L. (2002) Vector algebra in the analysis of genome-wide expression data, *Genome Biol 3*, RESEARCH0011.
19. Alter, O., Brown, P. O., and Botstein, D. (2000) Singular value decomposition for genome-wide expression data processing and modeling, *Proc Natl Acad Sci USA 97*, 10101–10106.
20. Wall, M. E., Dyck, P. A., and Brettin, T. S. (2001) SVDMAN--singular value decomposition analysis of microarray data, *Bioinformatics 17*, 566–568.
21. Sardiu, M. E., Florens, L., and Washburn, M. P. (2009) Evaluation of clustering algorithms for protein complex and protein interaction network assembly, *J Proteome Res 8*, 2944–2952.
22. Gavin, A. C., Aloy, P., Grandi, P., Krause, R., Boesche, M., Marzioch, M., Rau, C., Jensen, L. J., Bastuck, S., Dumpelfeld, B., Edelmann, A., Heurtier, M. A., Hoffman, V., Hoefert, C., Klein, K., Hudak, M., Michon, A. M., Schelder, M., Schirle, M., Remor, M., Rudi, T., Hooper, S., Bauer, A., Bouwmeester, T., Casari, G., Drewes, G., Neubauer, G., Rick, J. M., Kuster, B., Bork, P., Russell, R. B., and Superti-Furga, G. (2006) Proteome survey reveals modularity of the yeast cell machinery, *Nature 440*, 631–636.
23. Gavin, A. C., Bosche, M., Krause, R., Grandi, P., Marzioch, M., Bauer, A., Schultz, J., Rick, J. M., Michon, A. M., Cruciat, C. M., Remor, M., Hofert, C., Schelder, M., Brajenovic, M., Ruffner, H., Merino, A., Klein, K., Hudak, M., Dickson, D., Rudi, T., Gnau, V., Bauch, A., Bastuck, S., Huhse, B., Leutwein, C., Heurtier, M. A., Copley, R. R., Edelmann, A., Querfurth, E., Rybin, V., Drewes, G., Raida, M., Bouwmeester, T., Bork, P., Seraphin, B., Kuster, B., Neubauer, G., and Superti-Furga, G. (2002) Functional organization of the yeast proteome by systematic analysis of protein complexes, *Nature 415*, 141–147.
24. Meunier, B., Dumas, E., Piec, I., Bechet, D., Hebraud, M., and Hocquette, J. F. (2007) Assessment of hierarchical clustering methodologies for proteomic data mining, *J Proteome Res 6*, 358–366.
25. Shannon, P., Markiel, A., Ozier, O., Baliga, N. S., Wang, J. T., Ramage, D., Amin, N., Schwikowski, B., and Ideker, T. (2003) Cytoscape: a software environment for integrated models of biomolecular interaction networks, *Genome Res 13*, 2498–2504.
26. Chang, I. F. (2006) Mass spectrometry-based proteomic analysis of the epitope-tag affinity purified protein complexes in eukaryotes, *Proteomics 6*, 6158–6166.
27. Krogan, N. J., Cagney, G., Yu, H., Zhong, G., Guo, X., Ignatchenko, A., Li, J., Pu, S., Datta, N., Tikuisis, A. P., Punna, T., Peregrin-Alvarez, J. M., Shales, M., Zhang, X., Davey, M., Robinson, M. D., Paccanaro, A., Bray, J. E., Sheung, A., Beattie, B., Richards, D. P., Canadien, V., Lalev, A., Mena, F., Wong, P., Starostine, A., Canete, M. M., Vlasblom, J., Wu, S., Orsi, C., Collins, S. R., Chandran, S., Haw, R., Rilstone, J. J., Gandi, K., Thompson, N. J., Musso, G., St Onge, P., Ghanny, S., Lam, M. H., Butland, G., Altaf-Ul, A. M., Kanaya, S., Shilatifard, A., O'Shea, E., Weissman, J. S., Ingles, C. J., Hughes, T. R., Parkinson, J., Gerstein, M., Wodak, S. J., Emili, A., and Greenblatt, J. F. (2006) Global landscape of protein complexes in the yeast Saccharomyces cerevisiae, *Nature 440*, 637–643.

Chapter 6

Studying Binding Specificities of Peptide Recognition Modules by High-Throughput Phage Display Selections

Haiming Huang and Sachdev S. Sidhu

Abstract

Peptide recognition modules (PRMs) play critical roles in cellular processes, including differentiation, proliferation and cytoskeleton organization. PRMs normally bind to short linear motifs in protein ligands, and by so doing recruit proteins into signaling complexes. Based on the binding specificity profile of a PRM, one can predict putative natural interaction partners by searching genome databases. Candidate interaction partners can in turn provide clues to assemble potential in vivo protein complexes that the PRM may be involved with. Combinatorial peptide libraries have proven to be effective tools for profiling the binding specificities of PRMs. Herein, we describe high-throughput methods for the expression and purification of PRM proteins and the use of peptide-phage libraries for PRM specificity profiling. These high-throughput methods greatly expedite the study of PRM families on a genome-wide scale.

Key words: Phage display, High-throughput selection, Peptide recognition modules, Protein expression and purification, ELISA, DNA sequencing

1. Introduction

Peptide recognition modules (PRMs), which normally reside in multi-domain proteins, serve as adaptors to bind to short peptide motifs in protein ligands (1). PRMs bind to their ligands in response to external stimuli or internal cues, and consequently trigger signaling pathways or cause the formation of protein complexes. For example, growth factor receptor-bound protein 2 (Grb2) binds to tyrosine-phosphorylated epidermal growth factor receptor (EGFR) through its Src Homology 2 (SH2) domain upon stimulation and then links to Ras signaling by its Src Homology 3 (SH3) domains (2).

Peptide libraries are powerful tools for exploring the binding specificity profiles of PRMs (3–5). The peptide binding profile of a PRM is a good clue to unveil its ligands in vivo through searching

Gerard Cagney and Andrew Emili (eds.), *Network Biology: Methods and Applications*, Methods in Molecular Biology, vol. 781, DOI 10.1007/978-1-61779-276-2_6, © Springer Science+Business Media, LLC 2011

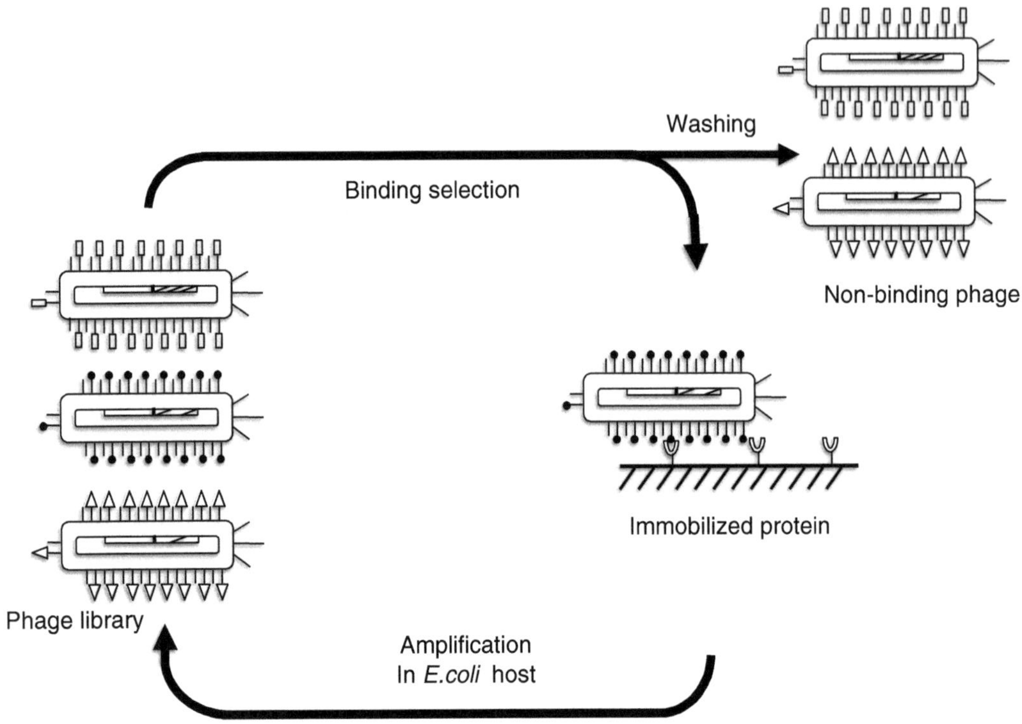

Fig. 1. Phage display selection process. Peptides (shown as *rectangles*, *dots*, or *triangles*) are displayed on phage particles that also contain the encoding DNA. Specific binding clones can be selectively enriched from library pools by binding to immobilized target protein and washing away nonbinding clones. Subsequently, the bound phage can be eluted and amplified for additional rounds of selection by passage through an *E. coli* host coinfected with helper phage. In a successful experiment, binding clones dominate the phage pool following four or five selection rounds. The sequences of phage-displayed peptides are decoded by DNA sequencing.

of genome databases (3, 6). Thus, this method serves as an alternative to other approaches that have been applied to the dissection of protein interaction networks mediated by PRMs (e.g., co-immunoprecipitation, mass spectrometry). Peptide libraries expressed in yeast (yeast two-hybrid) (7) or chemically synthesized on membranes (peptide arrays) (8) are methods that have been applied to the study of PRMs. Combining the advantages of yeast two-hybrid methods (genetic encoding) and peptide arrays (in vitro selections), phage-displayed peptide libraries have proven to be highly effective for the study of the binding specificities of PRMs (4, 9) (Fig. 1).

Recent studies have expanded the study of PRMs to a genome-wide scale (10–12). Rather than working with single domains, researchers are now able to purify and analyze hundreds of PRM family members in parallel. In this chapter, we describe high-throughput phage display methods to study the binding specificities of PRMs (e.g., SH3 and WW domains). First, we describe a method for the high-throughput expression and purification of PRM protein targets (Fig. 2a). Second, we describe a

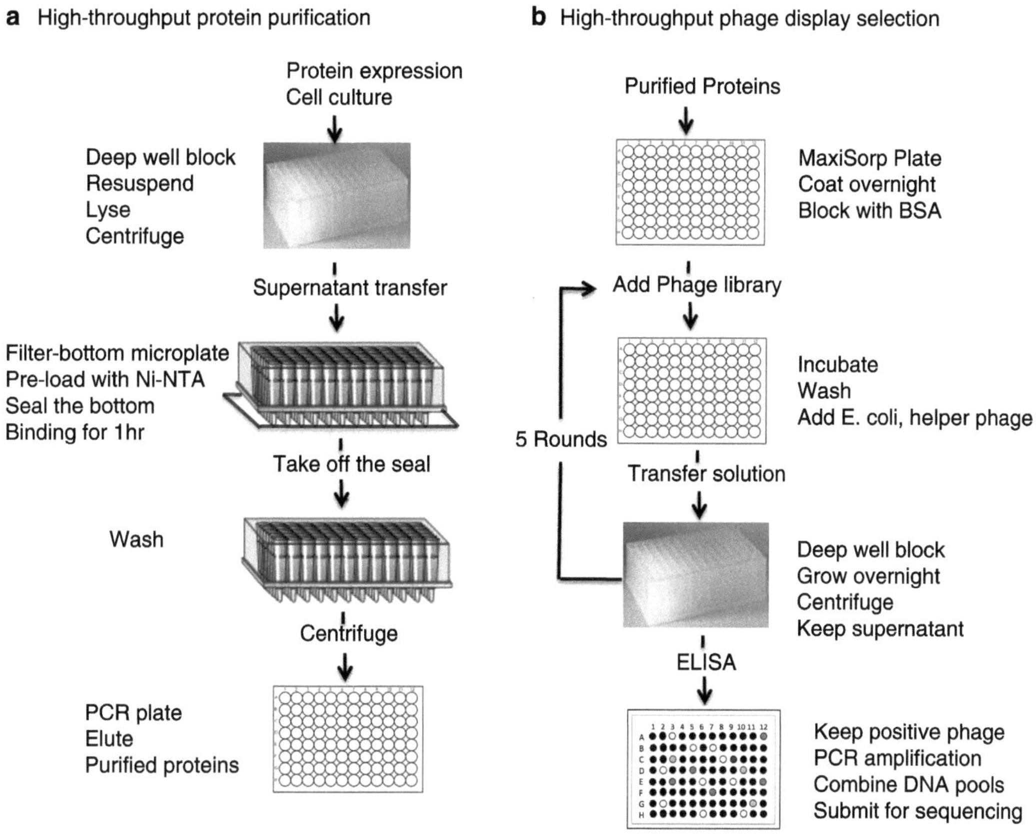

Fig. 2. High-throughput phage display. (**a**) High-throughput protein purification. Cell pellets are lysed in a 96-deep-well block. After centrifugation, the supernatants are transferred into a 96-well filter-bottom microplate preloaded with Ni-NTA agrose. After washing, the purified proteins are eluted into a 96-well PCR plate. (**b**) High-throughput phage display selection. Purified proteins are immobilized in a 96-well Maxisorp microplate overnight and phage-displayed peptide libraries are added. After washing away nonbinding phage, bound phage are eluted and amplified for further rounds of selection by passage through an *E. coli* host coinfected with helper phage. The progress of the selection process can be monitored by phage ELISA with the amplified phage supernatants.

method for high-throughput selections with peptide-phage libraries against the purified PRM proteins (Fig. 2b), and subsequently the use of deep sequencing technology to determine specificity profiles.

2. Materials

2.1. Protein Expression

1. 2YT medium: 10 g yeast extract, 16 g tryptone, 5 g NaCl. Add water to 1 l. Sterilize by autoclaving.
2. Carbenicillin (Carb): Make 1,000× stock in water (50 mg/ml), filter sterilize.
3. Kanamycin (Kan): Make 1,000× stock in water (25 mg/ml), filter sterilize.

4. Isopropyl-β-D-thiogalactopyranoside (IPTG): Make 1,000× stock (1.0 M) in water, filter sterilize.
5. 96-deep-well block (Whatman).
6. Axygen 96-well Mini tube system 0.6-ml (Blue box) (Axygen).

2.2. Protein Purification

1. Binding buffer: 50 mM HEPES, pH 7.5, 500 mM NaCl, 5 mM imidazole, 5% glycerol. Store at room temperature.
2. Pre-wash buffer: binding buffer, 1% TritonX-100. Store at room temperature.
3. Wash buffer: 50 mM HEPES, pH 7.5, 500 mM NaCl, 30 mM imidazole, 5% glycerol. Store at room temperature.
4. Elution buffer: 50 mM HEPES, pH 7.5, 500 mM NaCl, 250 mM imidazole, 5% glycerol. Store at room temperature.
5. Protease Inhibitor Cocktail (Sigma).
6. Benzonase nuclease (Novagen).
7. Lysis buffer: 98 ml binding buffer, 1 ml TritonX-100, 1 ml 100× protease inhibitor cocktail, 16.5 μl benzonase (4,130 units), 100 mg lysozyme, 77 mg DTT.
8. Ni-NTA agarose (Qiagen).
9. Filter bottom microplate (Seahorse Bioscience).
10. Bottom plate seal (Seahorse Bioscience).
11. Coomassie blue protein staining dye: 0.5 g Coomassie blue, 100 ml glacial acetic acid, add water to 1.0 l. Store at room temperature.

2.3. Phage Display Selection and ELISA

1. Nunc 96-well flat-bottom immuno maxisorp plate (Nalgene Nunc International).
2. Gas permeable film (VWR).
3. PBS buffer: 137 mM NaCl, 3 mM KCl, 8 mM Na_2HPO_4, and 1.5 mM KH_2PO_4. Adjust pH to 7.4, autoclave. Store at room temperature.
4. PBT buffer: PBS, 0.05% Tween 20, 0.5% BSA. Filter sterilize. Store at room temperature.
5. PT buffer: PBS, 0.05% Tween 20. Filter sterilize. Store at room temperature.
6. *E.coli* XL1-Blue (Stratagene).
7. M13KO7 helper phage (New England Biolabs).
8. Anti-M13 antibody-HRP conjugate (GE Healthcare).
9. TMB substrate (KPL).
10. 1.0 M H_3PO_4.

2.4. Preparation of Samples for Illumina Sequencing

1. PEG/NaCl, 20% PEG-8000 (w/v), 2.5 M NaCl. Mix and autoclave. Store at room temperature.
2. PCR primers with barcodes.
3. Qiaquick gel Extraction kit (Qiagen).
4. High fidelity PCR enzyme mix (Fermentas).

3. Methods

Obtaining high-quality protein targets is essential for phage display selections. One plate of proteins (96 samples) can be expressed and purified in a deep-well block by the methods provided here. Proteins can be purified by use of a fused affinity tag (e.g., hexa-His, GST, etc.). The yield of the proteins in each well can be up to 100 μg, which is sufficient for the following phage display selections and screening assays.

The peptide-phage display library used in this protocol was a random dodecamer library fused to the N-terminus of the M13 bacteriophage major coat protein. The phage display selections should be performed at low temperature (4°C). Normally, five rounds of selections are sufficient to enrich for binding phage-displayed peptides. From the fifth-round phage pool, which represents a pool of peptides binding to the PRM of interest, individual clones were assayed for binding to the PRM using a phage ELISA. It was found that greater than 95% of the individual phage clones bound specifically to the PRM. The fifth-round phage pool was used as the template for a PCR that amplified the region to be sequenced and added a unique bar code. The amplified DNA-representing binding clones from selections against different PRMs were pooled together and submitted for deep sequencing by Illumina.

3.1. High-Throughput Protein Expression and Purification

Day 1

1. Add 400 μl of 2YT/Carb (50 μg/ml) media to each well of a 96-well blue box.
2. Inoculate 10 μl of a glycerol stock of bacteria harboring the expression plasmid of interest into each well and grow overnight at 37°C with shaking at 200 rpm.

Day 2

1. Add 1.4 ml 2YT/Carb (50 μg/ml) media to each well of two duplicate 96-deep-well blocks.
2. Inoculate 50 μl overnight culture to each well (for each well of the overnight culture, inoculate the two corresponding wells of the duplicate blocks).
3. Grow for 5 h at 37°C with shaking at 200 rpm (OD_{600} ~0.6).

4. Add IPTG to a final concentration of 1.0 mM.
5. Incubate the cultures at 16°C for 24 h with shaking at 200 rpm.

Day 3

1. Centrifuge one block at 2,000 × *g* for 10 min at 4°C. Decant supernatant.
2. Transfer the culture from the corresponding wells of the duplicate block to the first block and centrifuge at 2,000 × *g* for 10 min at 4°C. Decant supernatant.
3. Store the pellets at −20°C overnight.

Day 4

1. Thaw the pellets at room temperature.
2. Make 30 ml of fresh suspension buffer (30 ml binding buffer and 0.3 ml 100× protease inhibitor cocktail).
3. Add 250 μl of suspension buffer to each well and shake at 200 rpm for 10 min at room temperature.
4. Make 100 ml of fresh lysis buffer.
5. Add 750 μl of lysis buffer to each well of a 96-well plate (lysis plate).
6. Resuspend the pellets in 250 μl of suspension buffer by pipetting and transfer the suspension to the lysis plate.
7. Shake the lysis plate at 200 rpm for 30 min at room temperature.
8. Prewash 7.0 ml of Ni-NTA agrose resin three times. For each prewash, add 15 ml of prewash buffer, centrifuge at 200 × *g* for 5 min, and aspirate the supernatant. After the last prewash, add 17.5 ml of prewash buffer.
9. Seal the filter-bottom microplate with four layers of parafilm and one bottom plate seal.
10. Add 150 μl of Ni-NTA agrose resin to each well.
11. Centrifuge the lysis plate at 2,000 × *g* for 10 min at 4°C.
12. Transfer 850 μl of supernatant from the lysis plate to the filter-bottom microplate.
13. Shake the filter-bottom microplate at 170 rpm for 1 h at room temperature.
14. Remove the parafilm and bottom plate seal from the filter-bottom microplate.
15. Place an empty 96-well block at the bottom of the binding block to collect the waste.
16. Centrifuge the filter-bottom microplate at 200 × *g* for 5 min.
17. Wash the filter-bottom microplate three times by adding 1.0 ml of wash buffer to each well and centrifuging at 200 × *g* for 2 min.

18. Add 100 μl of elution buffer and incubate at room temperature for 5 min.
19. Place a 96-well PCR plate under the filter block and elute the protein by centrifuging at 200 × *g* for 5 min at 4°C.

3.2. High-Throughput Phage Display Selection

Day 1

1. Coat a 96-well Maxisorp plate with target proteins from the high-throughput purification protocol (one well per protein). To each well, add 90 μl PBS followed by 10 μl eluted target protein (at least 0.5 μg of protein). In parallel, coat a negative control plate with GST (5 μg/ml).
2. Cover the plates with a clear adhesive film and incubate overnight at 4°C.

Day 2

1. Block each well with 200 μl PBS and 0.5% BSA for 1 h at 4°C.
2. Wash the GST-coated plate four times with PT buffer and add 100 μl of phage solution to each well. Incubate for 1 h at 4°C.
   (a) Round 1: Add 100 μl of the naïve peptide library.
   (b) Rounds 2–5: Add 100 μl of pH-adjusted phage supernatant from the overnight culture from the previous round (see step 13).
3. Wash the target-coated plate four times with PT buffer and transfer the phage solution from the GST-coated plate to the target-coated plate. Incubate for 2 h at 4°C with shaking at 200 rpm.
4. Remove the phage solution and wash the plate eight times with PT buffer.
5. Elute bound phage by adding 100 μl of actively growing *E.coli* XL1-blue culture to each well. Cover the plate with a gas-permeable film and incubate for 30 min at 37°C with shaking at 200 rpm.
6. Add 10 μl of M13KO7 helper phage (1×10^{11} p.f.u./ml) to each well. Incubate for 45 min at 37°C with shaking at 200 rpm.
7. Make frozen stocks by transferring 10 μl of bacterial culture from each well to a well containing 90 μl of 2YT, 15% glycerol. Freeze on dry ice or liquid nitrogen and store at −80°C.
8. Transfer the remaining culture in each well to 1.4 ml of 2YT (supplemented with 50 μg/ml Carb, 25 μg/ml Kan, 50 μM IPTG) in a 96-well deep-well block.
9. Seal the block with gas-permeable film and incubate overnight at 37°C with shaking at 200 rpm).

Day 3

1. Spin the 96-well block at 2,000 × *g* for 10 min and transfer the phage supernatant to a new 96-well block.
2. Adjust the pH of the phage supernatant by adding 1/10 volume of 10× PBS. Sterilize by heating at 65°C for 30 min.
3. Repeat selection steps using the pH-adjusted phage supernatant for the next round of binding selection. This procedure is repeated for a total of five rounds.

3.3. Phage ELISA

To monitor the selections for the enrichment of specific binding clones, a phage ELISA is performed at the second round and all subsequent rounds. The phage ELISAs are performed in parallel with the rounds of binding selections.

Day 1

1. Coat a 96-well Maxisorp plate with target proteins from the high-throughput purification protocol (one well per protein). To each well, add 45 μl PBS followed by 5 μl eluted target protein (at least 0.25 μg of protein). In parallel, coat a negative control plate with GST (5 μg/ml).
2. Cover the plates with a clear adhesive film and incubate overnight at 4°C.

Day 2

3. Block each well with 100 μl PBS and 0.5% BSA for 1 h at 4°C.
4. Wash the blocked plate four times with PT buffer.
5. Add 50 μl pH-adjusted phage supernatant from each well of the plate from the binding selections to the corresponding well of the target-coated plate and the GST-coated plate. Incubate the plates for 2 h at 4°C.
6. Wash six times with PT buffer.
7. To each well, add 50 μl anti-M13 antibody-HRP conjugate (1:5,000) in PBT buffer and incubate for 1 h at 4°C.
8. Wash eight times with PT buffer and two times with PBS.
9. To each well, add 50 μl TMB substrate. Allow color to develop for 5 min and stop the reaction by adding 50 μl 1.0 M H_3PO_4.
10. Read spectrophotometrically at 450 nm in a microtiter plate reader.
11. To assess the progress of the binding selections, compare the signal strength for each well coated with target protein to the corresponding well coated with GST. For a successful selection with specific binding clones, the signal strength for the target-coated well should be greater than that for the GST-coated well.

3.4. Preparation of Samples for Illumina Sequencing

The phage supernatant from a successful round of binding selection, as evidenced by phage ELISA, can be used as the template for a PCR to amplify a pool of DNA fragments for Illumina sequencing.

1. In a 96-well deep block, add 120 μl PEG/NaCl and 600 μl phage supernatant to each well. Mix well and incubate for 5 min at room temperature to precipitate phage.
2. Centrifuge for 30 min at 2,000 × *g* and 4°C. Decant the supernatant and remove the remaining supernatant by tapping the block gently on a stack of paper towels.
3. Resuspend the phage pellet in 40 μl PBS. The concentrated phage is used as the template in the following PCR to amplify DNA fragments for sequencing.
4. Prepare 100× PCR mix as follows:

Component	Amount (1 reaction)	Amount (100 reactions)
10× high-fidelity PCR buffer	2.5 μl	250 μl
dNTP mix (10 mM)	0.5 μl	50 μl
Forward primer (10 μM)	1 μl	100 μl
Reverse primer (10 μM)	1 μl	100 μl
High-fidelity PCR enzyme mix (5 u/μl)	0.2 μl	20 μl
H_2O	18.8 μl	1,880 μl
Total	24 μl	2,400 μl

5. Aliquot 24 μl PCR mix into each well of a 96-well PCR plate. Add 1 μl concentrated phage solution. Mix the solution by shaking the plate on a shaker for 5 min.
6. Amplify DNA fragments using the following program:

	Temperature (°C)	Time	Cycles
Initial denaturation	94	5 min	1×
Denaturation cycle	94	30 sec	18×
Annealing cycle	55	30 sec	18×
Elongation cycle	72	40 sec	18×
Final elongation	72	5 min	1×

7. Assess reaction yield by agarose gel electrophoresis.
8. Estimate the concentration of each PCR product and combine equal quantities of DNA from successful reactions.

9. Resolve the combined DNA sample by agarose gel electrophoresis to remove PCR primers. Cut the band from the gel and purify the band using a Qiagen gel extraction kit.
10. Measure the concentration and purity of the DNA.
11. Submit the purified DNA for Illumina sequencing.

4. Notes

1. The vector we use expresses the PRM fused to both a hexa-His tag and GST. We used Ni-NTA for purification and coated a GST-negative plate during the selection. Alternatively, the proteins could also be purified using glutathione resin.
2. To increase the yield of protein, we inoculated two identical expression blocks and combined the pellets prior to purification.
3. When preparing the lysis buffer for protein purification, add lysozyme powder directly to the buffer and ensure that the lysozyme dissolves completely. Adding 1% TritonX-100 improves bacterial lysis dramatically and does not seem to affect the purified proteins we have tested (e.g., SH2, SH3, WW domains). However, it is advisable to test the stability of other proteins in the presence of this detergent. Sonication is not necessary when using this lysis buffer.
4. Selections should be performed at 4°C to minimize denaturation of sensitive proteins and to improve peptide-phage binding.
5. If using GST fusion proteins, prebinding the library on a GST-coated plate is necessary to remove GST-binding peptide phage.
6. It is necessary to perform a phage ELISA to check the phage pool enrichment after each round of selection. Prior to DNA sequencing from a selection round, it is important to analyze a small number of randomly chosen clones by phage ELISA to ensure that a significant proportion of individual clones bind specifically to the target protein.
7. High-throughput sequencing with an Illumina instrument was used to sequence a pool representing phage from many individual selections. One should design an appropriate sequencing strategy and PCR primers with barcodes compatible with Illumina sequencing. Fewer cycles of PCR can reduce biases due to amplification. Concentration of the phage particles by precipitation with PEG/NaCl is required to obtain good PCR product yield with a reduced number of reaction cycles.

8. Because of the high yield of Illumina sequencing (up to 20 million reads in one run), one could combine more than one plate of selection results for sequencing. We have successfully combined up to 4 plates, representing 384 individual selection experiments.

Acknowledgments

This work was supported by grant from the CIHR (MOP-93684) to S.S.S. We thank Andreas Ernst for figure preparation.

References

1. Pawson, T. & Scott, J. D. (1997). Signaling through scaffold, anchoring, and adaptor proteins. Science 278, 2075–80.
2. Lowenstein, E. J., Daly, R. J., Batzer, A. G., Li, W., Margolis, B., Lammers, R., Ullrich, A., Skolnik, E. Y., Bar-Sagi, D. & Schlessinger, J. (1992). The SH2 and SH3 domain-containing protein GRB2 links receptor tyrosine kinases to ras signaling. Cell 70, 431–42.
3. Tong, A. H., Drees, B., Nardelli, G., Bader, G. D., Brannetti, B., Castagnoli, L., Evangelista, M., Ferracuti, S., Nelson, B., Paoluzi, S., Quondam, M., Zucconi, A., Hogue, C. W., Fields, S., Boone, C. & Cesareni, G. (2002). A combined experimental and computational strategy to define protein interaction networks for peptide recognition modules. Science 295, 321–4.
4. Tonikian, R., Zhang, Y., Sazinsky, S. L., Currell, B., Yeh, J. H., Reva, B., Held, H. A., Appleton, B. A., Evangelista, M., Wu, Y., Xin, X., Chan, A. C., Seshagiri, S., Lasky, L. A., Sander, C., Boone, C., Bader, G. D. & Sidhu, S. S. (2008). A specificity map for the PDZ domain family. PLoS Biol 6, e239.
5. Huang, H., Li, L., Wu, C., Schibli, D., Colwill, K., Ma, S., Li, C., Roy, P., Ho, K., Songyang, Z., Pawson, T., Gao, Y. & Li, S. S. (2008). Defining the specificity space of the human SRC homology 2 domain. Mol Cell Proteomics 7, 768–84.
6. Li, L., Wu, C., Huang, H., Zhang, K., Gan, J. & Li, S. S. (2008). Prediction of phosphotyrosine signaling networks using a scoring matrix-assisted ligand identification approach. Nucleic Acids Res 36, 3263–73.
7. Song, E., Gao, S., Tian, R., Ma, S., Huang, H., Guo, J., Li, Y., Zhang, L. & Gao, Y. (2006). A high efficiency strategy for binding property characterization of peptide-binding domains. Mol Cell Proteomics 5, 1368–81.
8. Rodriguez, M., Li, S. S., Harper, J. W. & Songyang, Z. (2004). An oriented peptide array library (OPAL) strategy to study protein-protein interactions. J Biol Chem 279, 8802–7.
9. Tonikian, R., Xin, X., Toret, C. P., Gfeller, D., Landgraf, C., Panni, S., Paoluzi, S., Castagnoli, L., Currell, B., Seshagiri, S., Yu, H., Winsor, B., Vidal, M., Gerstein, M. B., Bader, G. D., Volkmer, R., Cesareni, G., Drubin, D. G., Kim, P. M., Sidhu, S. S. & Boone, C. (2009). Bayesian modeling of the yeast SH3 domain interactome predicts spatiotemporal dynamics of endocytosis proteins. PLoS Biol 7, e1000218.
10. Machida, K., Thompson, C. M., Dierck, K., Jablonowski, K., Karkkainen, S., Liu, B., Zhang, H., Nash, P. D., Newman, D. K., Nollau, P., Pawson, T., Renkema, G. H., Saksela, K., Schiller, M. R., Shin, D. G. & Mayer, B. J. (2007). High-throughput phosphotyrosine profiling using SH2 domains. Mol Cell 26, 899–915.
11. Liu, B. A., Jablonowski, K., Raina, M., Arce, M., Pawson, T. & Nash, P. D. (2006). The human and mouse complement of SH2 domain proteins-establishing the boundaries of phosphotyrosine signaling. Mol Cell 22, 851–68.
12. Stiffler, M. A., Chen, J. R., Grantcharova, V. P., Lei, Y., Fuchs, D., Allen, J. E., Zaslavskaia, L. A. & MacBeath, G. (2007). PDZ domain binding selectivity is optimized across the mouse proteome. Science 317, 364–9.
13. Tonikian, R., Zhang, Y., Boone, C. & Sidhu, S. S. (2007). Identifying specificity profiles for peptide recognition modules from phage-displayed peptide libraries. Nat Protoc 2, 1368–86.

Chapter 7

Array-Based Synthetic Genetic Screens to Map Bacterial Pathways and Functional Networks in *Escherichia coli*

Mohan Babu, Alla Gagarinova, and Andrew Emili

Abstract

Cellular processes are carried out through a series of molecular interactions. Various experimental approaches can be used to investigate these functional relationships on a large-scale. Recently, the power of investigating biological systems from the perspective of genetic (gene–gene, or epistatic) interactions has been evidenced by the ability to elucidate novel functional relationships. Examples of functionally related genes include genes that buffer each other's function or impinge on the same biological process. Genetic interactions have traditionally been investigated in bacteria by combining pairs of mutations (for example, gene deletions) and assessing deviation of the phenotype of each double mutant from an expected neutral (or no interaction) phenotype. Fitness is a particularly convenient phenotype to measure: when the double mutant grows faster or slower than expected, the two mutated genes are said to show alleviating or aggravating interactions, respectively. The most commonly used neutral model assumes that the fitness of the double mutant is equal to the product of individual single mutant fitness. A striking genetic interaction is exemplified by the loss of two nonessential genes that buffer each other in performing an essential biological function: deleting only one of these genes produces no detectable fitness defect; however, loss of both genes simultaneously results in systems failure, leading to synthetic sickness or lethality. Systematic large-scale genetic interaction screens have been used to generate functional maps for model eukaryotic organisms, such as yeast, to describe the functional organization of gene products into pathways and protein complexes within a cell. They also reveal the modular arrangement and cross-talk of pathways and complexes within broader functional neighborhoods (Dixon et al. Annu Rev Genet 43:601–625, 2009).

Here, we present a high-throughput quantitative *Escherichia coli* synthetic genetic array (eSGA) screening procedure, which we developed to systematically infer genetic interactions by scoring growth defects among large numbers of double mutants in a classic gram-negative bacterium. The eSGA method exploits the rapid colony growth, ease of genetic manipulation, and natural efficient genetic exchange via conjugation of laboratory *E. coli* strains. Replica pinning is used to grow and mate arrayed sets of single-gene mutant strains as well as to select double mutants en mass. Strain fitness, which is used as the eSGA readout, is quantified by the digital imaging of the plates and subsequent measuring and comparing single and double mutant colony sizes.

While eSGA can be used to screen select mutants to probe the functions of individual genes; using eSGA more broadly to collect genetic interaction data for many combinations of genes can help reconstruct a functional interaction network to reveal novel links and components of biological pathways as well as unexpected connections between pathways. A variety of bacterial systems can be investigated, wherein the genes impinge on a essential biological process (e.g., cell wall assembly, ribosome biogenesis, chromosome

Gerard Cagney and Andrew Emili (eds.), *Network Biology: Methods and Applications*, Methods in Molecular Biology, vol. 781,
DOI 10.1007/978-1-61779-276-2_7,

replication) that are of interest from the perspective of drug development (Babu et al. Mol Biosyst 12:1439–1455, 2009). We also show how genetic interactions generated by high-throughput eSGA screens can be validated by manual small-scale genetic crosses and by genetic complementation and gene rescue experiments.

Key words: *Escherichia coli*, Conjugation, Double mutant, Hypomorphs, Epistasis, Genetic interaction, Network, Synthetic lethality or sickness, Aggravating, Alleviating, Suppression

1. Introduction

While physical interactions and computational inferences can indicate bacterial proteins that are associated into complexes or are evolutionarily linked (1, 2), these inferences do not necessarily reveal the nature of the functional relationship at a pathway level (3–5). With the exception of metabolic pathways (6), genetic dependencies and pathway architecture in bacteria have been traditionally explored in a step-wise manner by isolating second-site genetic modifier mutations in classic suppressor/enhancer screens. For example, the loss of functional alleles in two genes may cause "synthetic lethality" or "synthetic sickness" even if neither mutation alone significantly reduces cell viability. Such aggravating (or negative) genetic interactions usually result when the products of two genes either jointly control an essential process or encode subunits of an essential protein complex, or function within redundant or convergent pathways such that one can functionally compensate for, or buffer, defects in the other. Conversely, alleviating (or positive) genetic interactions (e.g., suppression) typically arise between genes operating at different steps within the same pathway (7–10).

Despite wide conservation, only ~300 protein coding genes are essential for *Escherichia coli* viability under standard laboratory growth conditions (11). The fact that the other ~3,900 genes are dispensable for viability presumably reflects, in part, the adaptability and robustness of microbial processes to environmental perturbations (12, 13). Obvious backup controls that "buffer" system failures include DNA repair systems, protein chaperones, and duplicated genes (i.e., paralogs) (14, 15). However, alternate types of functional dependencies and genetic redundancy likely exert more pervasive stabilizing effects on the molecular networks and phenotypes of cells (16, 17). Indeed, the organization of gene products into redundant parallel pathways and buffered functional modules can explain the dispensability of most genes (18).

1.1. *Escherichia coli* Synthetic Genetic Array and GIANT

The development of a high-throughput synthetic genetic array (SGA) method for detecting genetic dependencies by systematically assaying the fitness of digenic combinations by mating and meiotic assortment of unlinked nonessential gene deletions allowed for the genome-scale mapping of genetic interaction networks in *Saccharomyces cerevisiae* (19–22). A related screening strategy, termed epistatic miniarray profiling (23), or E-MAP, has likewise been used to generate comprehensive maps of alleviating and aggravating interactions by making quantitative measurements of fitness of strains deleted for pairs of genes representing a particular functional neighborhood of interest. For example, comprehensive genetic interaction profiling has revealed the biological roles of individual genes, the extent of pleiotropy and specialization among the subunits of protein complexes, and the extensive nature of pathway cross-talk and redundancy among components of the chromatin remodeling machinery (9), protein secretion apparatus (23), and RNA processing systems (24).

Although *E. coli* does not have a sexual lifecycle in the same manner as *S. cerevisiae* does, like many other bacterial species, it is naturally capable of conjugation and genetic exchange. This ability for genetic transfer and the efficiency of the ensuing homologous recombination are exploited for making double mutants in two analogous but independently developed methods for investigating *E. coli* genetic interactions in GIANT-coli (for *G*enetic *I*nteraction *AN*alysis *T*echnology for *E. coli*), and eSGA (for *E. coli* *s*ynthetic *g*enetic *a*rrays) (25, 26). Although the readout in both approaches, digenic mutant fitness, is determined by systematically generating and measuring the colony sizes of double mutants, we focus on the eSGA procedure in this chapter. In addition to scoring deletion mutants, analogous to what has been done in *S. cerevisiae*, hypomorphic alleles (e.g., temperature-sensitive conditional alleles with reduced gene function (27) or mRNA perturbation (DAmP) alleles (28)) can likewise be used to assess the genetic interaction patterns of essential bacterial genes. Genes in the same pathway typically display alleviating interactions with each other and highly correlated patterns of aggravating interactions with genes in other overlapping pathways (4, 19). Hierarchical clustering of the patterns of genetic interactions measured for various genes associated with a biological system of interest can therefore reveal the arrangement of genes into pathways (23).

1.2. Conjugation

In bacterial conjugation, DNA is unidirectionally transferred from a male (F^+ or Hfr) strain to a female (F^-) strain. Male *E. coli* donor strains harbor a self-transmissible low-copy plasmid (Fertility, or F factor) (29), which encodes functions necessary to promote its own transfer into a naïve (female) F^- bacterial cell via conjugation (29). Transfer of the F factor is catalyzed upon nicking of the double-stranded DNA at the origin of transfer, *oriT*, followed by rolling-circle replication and transfer of single-stranded DNA through a sex pilus into the recipient F^- cell (30). Occasionally, the F factor integrates into the host chromosome (31), giving rise to High frequency of recombination (Hfr) male strains. An Hfr donor transfers host DNA, concomitantly with the F factor, to a conjugated F⁻ recipient strain. Transfer starts with loci adjacent to *oriT* and continues in a unidirectional manner until mating is physically interrupted or the transfer is complete. Once in the recipient, the donor DNA is integrated into the corresponding genomic locus by homologous recombination (14). To achieve high-throughput, arrayed sets of single-gene deletion mutant recipient (F⁻) and donor (Hfr) strains are grown on solid media, conjugated by replica pinning, and the double mutants are selected.

Our high-throughput, quantitative eSGA assay is used to make large-numbers of double mutants in parallel, as ordered colonies on plates, to investigate genetic interactions on genomic scale (25). In eSGA, first, a male donor Hfr strain is constructed with a single marked "query" gene mutation. The donor is constructed by targeted introduction of a specific gene mutation, either representing a full deletion of the entire open reading frame (in the case of non-essential genes), or a point mutation (for essential genes), using a cassette marked with chloramphenicol-resistance (Cm^R). Temperature inducible expression of an integrated phage lambda Red-mediated homologous recombination system is used to dramatically improve the efficiency of homologous recombination (25, 32) (Fig. 1). Conjugation is then performed by pinning the donor onto an arrayed set of viable kanamycin (Kan)-resistance marked (Kan^R) single-gene deletion *E. coli* K-12 F⁻ recipient strains (11). A genome-wide single-gene deletion recipient collection representing 3,968 nonessential genes of *E. coli* is publicly available (11). Since essential genes cannot be deleted, we typically perturb the 3′UTR to destabilize transcript abundance to generate hypomorphic alleles (25). After genomic DNA transfer and homologous recombination into the recipient host, the resulting double mutants are selected by replica pinning onto solid medium containing both antibiotics. The screening process can be speed up using a robotic pinning device, but manual strain manipulation is readily performed with a handheld device. Growth fitness, the primary eSGA readout, is recorded by imaging the plates and measuring the respective colony sizes (25). A workflow diagram for the entire process is shown in Fig. 2.

As with yeast, the ability to perform eSGA screens easily and systematically allows one to examine the genetic interaction profiles of both individual genes and the components of entire systems in *E. coli* in an unbiased, comprehensive manner (3, 25), permitting the investigation of bacterial gene function at a higher order, pathway level. Functional relationships are usually identified by the non-multiplicative growth fitness of the double mutants. For example, combinations of mutations in enzymes participating in two alternate pathways (e.g., Isc, Suf) that generate iron–sulfur cluster prosthetic groups widely used in multiple bacterial processes (33) are synthetic lethal by eSGA. In such cases, genetic interactions

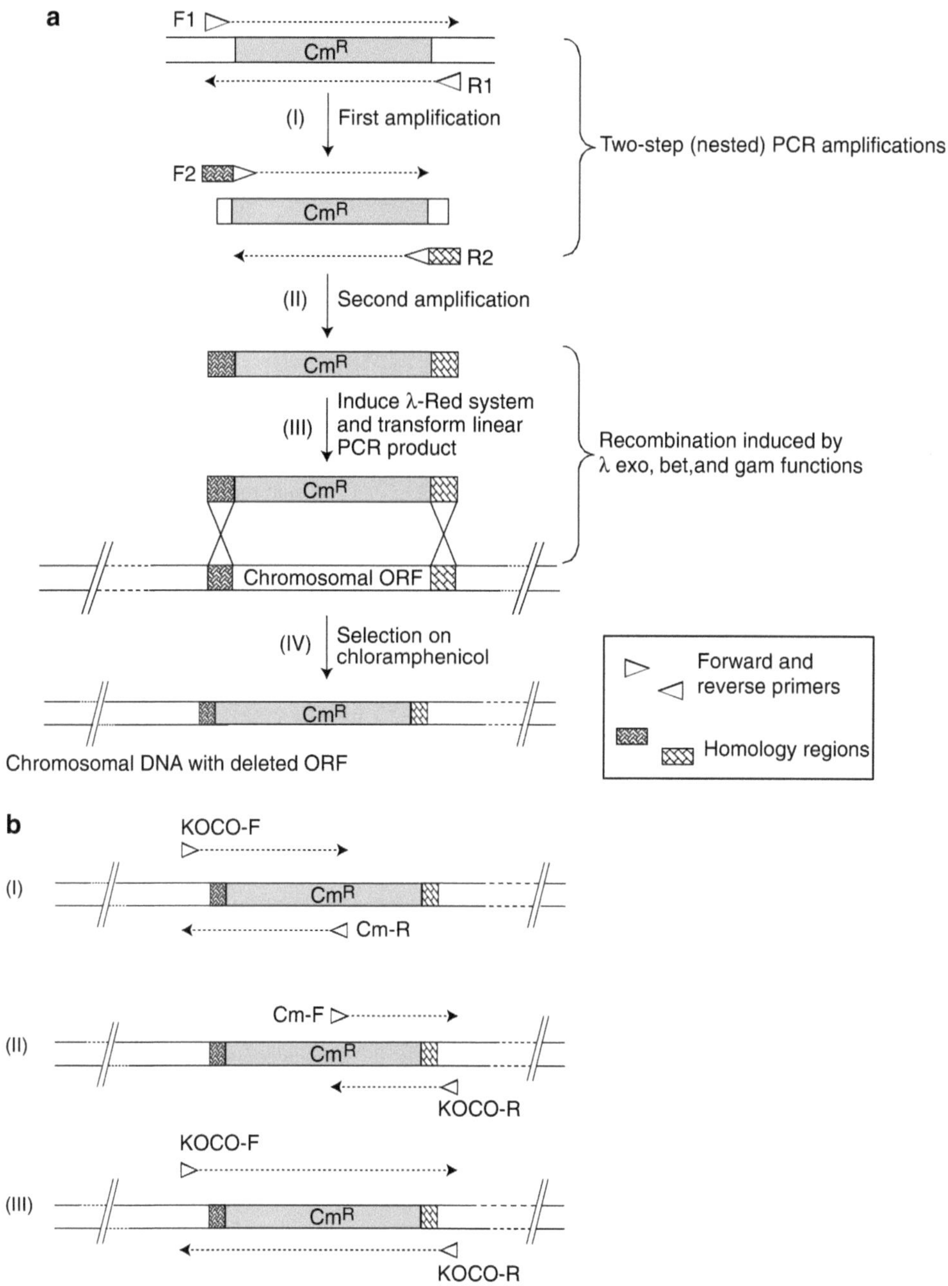

Fig. 1. Donor construction and confirmation. *Panel A*: Construction of eSGA donor mutant strains by deletion of *E. coli* chromosomal ORF in Hfr Cavalli. In the first step, the chloramphenicol resistance (Cm^R) cassette, with short adjacent regions, is amplified from the pKD3 plasmid using primers F1 and R1. In the second step, the Cm^R region of the plasmid is amplified and 45-nt homology regions, for site-specific recombination, are added using primers F2 and R2. In the third step, the product of the second amplification is transformed into the Hfr Cavalli strain after the λ-red system derepression to specifically replace the target ORF with the Cm^R. In the fourth step, the mutants having the Cm^R are selected on chloramphenicol. *Panel B*: The gene deletions in Hfr Cavalli are confirmed by separate PCRs with three primer sets. The first primer set consists of a 20-nt flanking primer, located 200 base pairs upstream of the targeted region (KOCO-F), and reverse (Cm-R) primer complementary to the Cm^R cassette sequence (*shown in top panel*). The second set includes a forward (Cm-F) primer, annealing to the Cm^R cassette sequence, and a reverse flanking confirmation primer (KOCO-R), which should be designed to anneal 200 bp downstream of the 3′ end of the deleted gene (*shown in middle panel*). The third PCR includes KOCO-F and KOCO-R primers (*shown in lower panel*). See Subheading 3 for details.

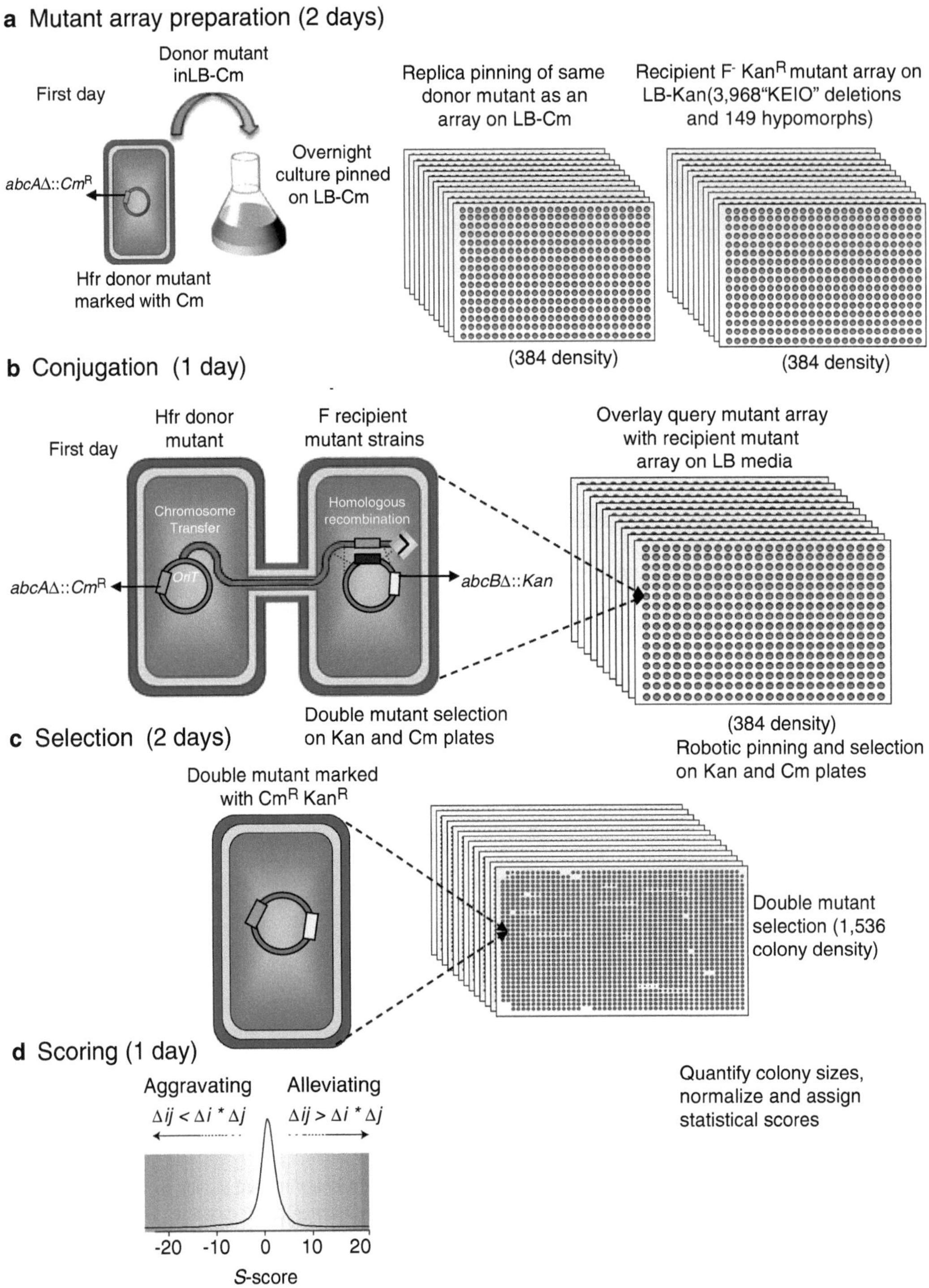

Fig. 2. Systematic double mutant construction in *E. coli* using eSGA. Schematic summary of key eSGA steps: (**a**) Overnight Hfr query mutant strain (marked with Cm^R), grown overnight in LB-Cm, is pinned onto LB-Cm plates in 384 format. Simultaneously, the recipient F^- mutant array strains (marked with Kan^R) are pinned onto LB-Kan plates. (**b**) After overnight growth the Hfr query strain colonies and recipient array colonies are pinned over each other on LB plates. Conjugation ensues, with DNA transfer initiating at a specific origin of transfer, *oriT*, and proceeding via a rolling circle mechanism of replication. The donor chromosome undergoes homologous recombination with the recipient chromosome (marked as "X in *broken lines*"). (**c**) The resulting colonies are pinned onto plates containing both Kan and Cm for selection of double mutants. (**d**) The double mutant plates are then imaged and colony sizes are scored to identify aggravating (synthetic lethal and synthetic sick) and alleviating (buffering) interactions.

between two alleles can be measured based on how the phenotype of an organism lacking both alleles (double mutant) differs from that expected when the phenotypes of single mutations are combined (13). Based on this proposition, the genetic interactions ε_{IJ} can be defined between mutations *I* and *J* in terms of any quantitative phenotype *P* as the difference between an observed ($P_{IJ,\ \text{observed}}$) and an expected ($P_{IJ,\ \text{expected}}$) phenotype of the double mutants if no interaction exists between the two mutations:

$$\varepsilon_{IJ} = P_{IJ}^{\text{observed}} - P_{IJ}^{\text{expected}}$$

This mathematical equation depends on our ability to compute P_{IJ}^{expected} as a function based on the combined effect of two individual mutations with phenotypes $P_{I,\ \text{observed}}$ and $P_{J,\ \text{observed}}$. For example, if loss of allele I ($P_{I,\ \text{observed}}$) results in a growth rate 0.8 times the wild-type growth rate, whereas loss of allele J ($P_{J,\ \text{observed}}$) results in a growth rate of 0.9, then the expected (P_{IJ}^{expected}) growth rate of the double mutant (lacking alleles *I* and *J*) would be 0.72 times that of the wild-type. This neutral model assumes that two genes do not normally impact each other, and in fact, experimental observations support the intuitive idea that genetic interactions are rare (21, 23). In cases, when deletion of two genes causes a more deadly effect than the fitness reduction expected from the combined loss of individual genes are said to have negative or aggravating interaction (e.g., synthetic sickness or lethality). Such interactions often identify proteins that function in distinct but parallel pathways in a given process (4). Alternatively, when double mutant has a better than expected fitness, the two genes are said to have positive, or alleviating interaction (e.g., suppression).

To date, we have performed over 160 genome-wide eSGA screens using donors bearing mutations in diverse query genes, including hypomorphic alleles of several essential *E. coli* genes. Since linkage may bias the interaction score for the pairs of genes within 30 kbp from each other, we eliminate the 30 kbp regions on either side of the query gene from analysis (25). This means that unrelated genes are not said to interact with the query gene simply because linkage prevented formation of the double mutant, producing an apparent synthetic lethal phenotype. On the other hand, this also means that relationships between genes with related function – such as often functionally related genes within the same operon – cannot be investigated (25). Nevertheless, our screens have revealed hundreds of novel genetic interactions suggestive of pathway relationships (34). Typically, the number of genetic interactions per screen, at an average of 30 synthetic lethal interactions per bacterial query gene, is roughly similar to the number of interactions found in yeast (22). Although genetic interactions reveal many interesting connections suggestive of novel mechanistic links, analyzing the network for concurrence with protein–protein interactions (PPI), membrane protein expression levels, co-expression and functional connections predicted by genomic-context methods, namely: the conservation of gene order (operons); gene fusions; operon recombination frequencies derived from intergenic distances of predicted operons across genomes and phylogenetic profiling – can be especially informative. For example, like the buffered pathways that are linked by genetic interactions, genes encoding subunits of protein complexes tend to buffer each other and will also be connected by genetic interactions. Previous E-MAP studies in yeast have shown that genes exhibiting alleviating genetic interactions are more likely to encode proteins that are physically associated (9, 35), and examination of our recent envelope data in *E. coli* revealed a similar overall tendency toward alleviating interactions between genes encoding protein subunits of the same bacterial complex (34). In addition, comparative genomic procedures can be performed in conjunction with genetic interaction data to investigate the evolutionary significance of the putative functional relationships detected in *E. coli* and in other proteobacterial species and prokaryotic taxa. Alterations to the components of the pathways and functional modules in virulent *E. coli* strains and other pathogens can illuminate divergent functional adaptations. Analogous comparisons of budding and fission yeast (~400 million years divergence) have shown that although there is a highly significant conservation of synthetic lethal genetic interactions, substantial rewiring of functional modules occurs between distantly related eukaryotes (35). The interaction data generated for *E. coli* may be similarly used to gain insight into the pathway architecture of other microbes for which functional annotations are largely lacking. Collectively, our results have confirmed the place of genetic interaction screens using eSGA, or conceptually similar alternate methods like GIANT-coli (26) and next-gen TnSeq (36), to illuminate novel functional interactions in *E. coli* often missed by other experimental or computational approaches. Since many of the *E. coli* genes are widely conserved across microbes (1), and because antibiotic sensitivity is often enhanced in combination with certain gene mutations (i.e., underlining the field of chemical genomics) (37), any functional relationships illuminated by eSGA may be exploited for designing innovative combination drug therapies.

In this chapter, we describe the key steps required to perform a single whole-genome eSGA screen and subsequent data analysis procedures that allow identifying the quantitative defects in double mutant fitness to reveal the functional dependencies and pathway redundancy. These steps include generating query gene donor strains, performing conjugation using high-density colony plates, and processing digital images for statistical scoring. We also briefly describe how genetic interaction scores can be used in formulating a testable biological hypothesis. Overall, these protocols are readily implemented in a lab with experience in basic microbiological techniques, can be scaled-up through the use of commercial robotic platforms, and make use of *E. coli* lab strains that are either publicly accessible (11) or are readily generated in house (1, 25).

2. Materials

2.1. Media, Stock Solutions, and Reagents

1. Luria-Bertani (LB) Medium: Solid medium is prepared by dissolving 25 g of LB powder and 20 g of agar in 1,000 mL of distilled water. Liquid LB medium is prepared without agar.
2. SOC medium (Invitrogen).
3. Kan antibiotic stock: Dissolve 50 mg/mL of Kan in double distilled water and filter-sterilize using a 0.22 μm millipore filter. The filter-sterilized Kan stock solution is stored in single use aliquots at −20°C.
4. Chloramphenicol (Cm) antibiotic stock: Dissolve 34 mg/mL of Cm in 95% ethanol and filter-sterilize using a 0.22 μm millipore filter. The filter-sterilized Cm stock solution is stored in single use aliquots at −20°C.
5. Ampicillin (Amp) antibiotic stock: Dissolve 100 mg/mL of Amp in double distilled water and filter-sterilize using a 0.22 μm millipore filter. The filter-sterilized Amp stock solution is stored in single use aliquots at −20°C.
6. 10% ice-cold sterile glycerol and ice-cold sterile distilled water for preparing competent cells.
7. 70% sterile glycerol stock for long-term strain storage.
8. Desalted knock out and confirmation primers (Subheading 3). These can be purchased from a commercial supplier.
9. Taq DNA polymerase (Fermentas) for PCR amplifying the Cm template to make the gene tagging or replacement cassettes.
10. Genomic DNA isolation kit (Promega).
11. Plasmid DNA and PCR purification kits (Qiagen).

2.2. Equipment

1. Thermal cycler for standard PCR (BioRad iCycler). The amplification conditions may need to be optimized if a different thermal cycler is used.
2. Equipment and supplies for standard agarose gel electrophoresis: gel box, power supply, buffer TBE, DNA ladder, as well as ethidium bromide and UV transilluminator for visualizing the DNA.
3. Electroporator (BioRad MicroPulser).
4. A centrifuge for pelleting cultures. We typically use a Beckman Coulter TJ-25 centrifuge with TS-5.1-500 rotor, 42°C water bath shaker, 0°C ice-slurry shaker, 32°C shaker, and 32°C plate incubator.
5. The 96-pin and 384-floating pin handheld pinning device (V & P Scientific, Inc.) for the manual pinning of single and double mutants. (Optional: To automate and speed up the pinning

process, we use the RoToR-HDA benchtop robot (Singer Instruments), fitted with 96 or 384 pin density pads for the replica-pinning procedures).

6. Digital camera with close-up (macro) imaging capabilities – for plate imaging (10 megapixels minimum resolution).
7. To allow for automatic colony size quantization, reduce variation in lighting between the plates, and to make imaging faster, use a camera stand (Kaiser) and a Canon digital camera with remote shoot and close-up (macro) imaging capabilities (10 megapixels minimum resolution). Affix the camera to the Kaiser, 50 cm above the surface of the stand. Make an adapter for the plates that, once attached to the Kaiser stand, will hold the plates in place and provide the background to allow automated plate and colony detection. For this, first, prepare the cardboard placeholder: trace a plate that you will be using on a ≥0.3 cm thick cardboard, leaving ca. 3–4 cm of cardboard on each side of the plate (trim the cardboard, if necessary). Cut the cardboard along the trace lines and discard the center piece. Second, make a narrow white edge that will aid in automatic location of the plate borders. For this, obtain a piece of thin white plastic; cut it to match the outer measurements of the cardboard. Attach the plastic to the cardboard with tape along the outer edges. Cut out the center of the plastic such that a ~0.3 cm edge is visible from the center opening of the cardboard. Third, to provide a background to the plates, obtain a piece of black velvet larger than the central opening in the plastic. Tape the velvet, around its outside edge, to the plastic (on the opposite side of the cardboard) so that the velvet does not have any wrinkles. Now, the narrow white plastic edge and the velvet pile (the reverse side of the velvet will not provide sufficiently even background) should be visible through the center opening in the cardboard. Place the entire plate adaptor in the middle of the Kaiser stand, with velvet side down and the cardboard side up so that the plate's borders are exactly vertical and horizontal in the photograph. Affix (using black electrical tape) the adaptor to the stand – completely and evenly cover the cardboard with tape, leaving the white plastic edges and the velvet free of tape. Once the adaptor assembly is complete and attached to the stand, place light boxes (two parallel Testrite 16 in. × 24 in. units; Freestyle Photographic supplies) on the sides of the Kaiser stand, 15–25 cm from the adaptor – to provide light during imaging. Cover the light boxes with white or blue bench liners to reduce the light intensity, if necessary. Cover the lower half of the light boxes with 1–2 (additional) layers of bench liners to distribute the light primarily upward. This makes the light distribution around the whole imaged plate more even, which aids in later plate quantization. Lastly, shield the set up from external light – to reduce day-to-day and

time-of-day variation in images – by covering the set up with black light-impermeable or nonreflective material. For taking a plate image, place the plate in the center opening of the cardboard and keep the set up shielded from outside light.

2.3. Pinning System, Plates and Accessories for Working with Cultures

1. Sterile 50 mL polypropylene tubes and 1.5 mL microcentrifuge tubes for preparing competent cells and transformation. These need to be chilled on ice for at least 10 min prior to use.
2. 250 mL flasks for growing cell cultures.
3. Sterile 0.2-cm electroporation cuvettes. These need to be chilled on ice for at least 10 min prior to use.
4. 15 mL sterile culture tubes for recovery of double mutants following a transformation.
5. Rectangular plates (Nunc) for manual pinning (or optional: rectangular plates (Singer Instruments) – for all robotic replica-pinning procedures).
6. 384 well microtiter plates (Nunc) for constructing a stock copy culture of the F^- recipient single-gene deletion mutants.
7. Distilled water, 70% ethanol (v/v), 95% ethanol, and cheesecloth for cleaning the reusable handheld pinning device. Three suitable rectangular liquid containers – to hold the liquids. These should be large enough to fit all pins of the handheld pinning device (for example, large enough solid tip box lids). Fold the cheesecloth and place it in the first container, cover it with ca. 0.5 cm distilled water. This first water station will be used to remove the cells from the pins. Pour 70% ethanol in the second container. During this step, the floating pins are sterilized after washing them in water (the height of the ethanol in this container should be about 0.7 cm – slightly higher than the liquid level in the first container). In the third container, place 95% ethanol (ca. 0.9 cm height). After a 70% ethanol wash, briefly dip the pins in the 95% ethanol and flame them to sterilize. The pins of the handheld pinning device are washed successively in water, 70% ethanol, 95% ethanol, and are flamed prior to each pinning step.
8. Optional: cleaning accessories for robot plastic pinning pads: short- and long-pin plastic pads can be reused up to three times. Sterilize the pads, following each use; by soaking them in 10% bleach overnight, rinsing with distilled water, washing in 70% ethanol, and air-drying the pads in a flow hood under UV. Store the sterile pads in sealed sterile plastic bags until use.

2.4. Bacterial Strains and Plasmids

1. Hfr Cavalli strain with an integrated temperature inducible λ_{red} system (JL238) (25) is used as the parental strain for the construction of Hfr query mutant donor strains.
2. pKD3 (38) as PCR template for chloramphenicol resistance marker.

3. The Keio *E. coli* F^- recipient nonessential single-gene deletion mutant collection (11) can be obtained from the National BioResource Project (NBRP) of Japan: (http://www.shigen.nig.ac.jp/ecoli/strain/top/top.jsp).
4. Potentially hypomorphic *E. coli* F^- recipient SPA-tagged essential gene strains (39) can be obtained from open biosystems: (https://www.openbiosystems.com/GeneExpression/NonMammalian/Bacteria/EcoliTaggedORFs/).

3. Methods

3.1. Construction of Query Deletion Mutants in an Hfr Cavalli Donor Strain

Donor strains are made by using homologous recombination to replace the query gene with a "Cm" resistance marker in an engineered *E. coli* Hfr Cavalli strain which bears an integrated temperature-inducible λ-Red recombination system to improve integration efficiency (25). In making the donors, the following steps are performed: first, the gene mutagenesis (e.g., deletion) cassette is amplified from pKD3 plasmid (Subheading 3.1.1). Second, competent cells are prepared from the Hfr Cavalli strain with the λ-Red system (Subheading 3.1.2). Third, the cassette is transformed into the competent cells and mutants are selected (Subheading 3.1.3). Fourth, the gene mutation is confirmed by PCR of individual clones (Subheading 3.1.4) and validated mutants are stored for future use (Subheading 3.1.5).

3.1.1. Preparation and Generation of the Linear DNA Mutagenesis Cassette

The linear DNA cassette to replace (i.e., delete) a target open reading frame with a "Cm" resistance marker is synthesized by a two-step (nested) PCR amplification from the pKD3 plasmid (38). Nested amplification (Fig. 1) is employed to minimize the chance of transforming and recovering an intact plasmid when making the donors.

1. In the first two rounds of amplifications, ~45 ng of pKD3 is used as a template in PCR with forward F1 (5′-AGATTGCAGCATTACACGTCTT-3′) and reverse R1 (5′-GGCTGACATGGGAATTAGC-3′) primers to produce a 1,070 bp product. PCR is carried out in 50 μL reactions containing 1× reaction buffer with $(NH_4)_2SO_4$, 4 mM $MgCl_2$, 200 μM dNTPs, 2 μM of each primer, and 1.25 U *Taq* DNA polymerase per 50 μL (Fermentas). The following cycling conditions are used: 3 min denaturation at 95°C is followed by 30 cycles of 1 min denaturation at 95°C, 1 min annealing at 56°C, and 2 min extension at 72°C. Subsequently, a 10 min final extension at 72°C and a final hold at 4°C are performed.
2. Perform agarose gel electrophoresis to confirm that the correct fragment was obtained.

3. PCR purify the obtained fragment (Qiagen) following standard manufacturer protocol and dilute the purified product (5 ng/μL) in sterile distilled H_2O.
4. The purified linear product is stored in −20°C for use as a template for all subsequent second-step amplifications.
5. Set up a second (nested) PCR with gene-specific (e.g., knockout) primers, F2 and R2, that have 45 nt of homology to the gene of interest at the 5′-ends (to allow for homologous recombination), immediately upstream and downstream of the target; and 20 nt at the 3′-ends, to prime the synthesis of the "Cm" marker. The size of the produced fragment is 1,123 bp. The reactions are set up with ~2.5 ng of template PCR (from step 4 above) per 50 μL reaction. The remaining conditions are as described in step 1, Subheading 3.1.1.
6. Confirm using agarose gel electrophoresis that the expected product is obtained.
7. PCR purify the obtained fragment (Qiagen) following standard manufacturer protocol, eluting into 30 μL sdH_2O. Dilute the product to ~50 ng/μL. The purified product can be stored in −20°C for long-term storage prior to use in transformation (Subheading 3.1.3).

3.1.2. Preparation of Competent Cells for Donor Mutant Construction

1. Inoculate a single Hfr Cavalli colony in 5 mL of LB medium with 2.5 μL of 100 μg/mL ampicillin. Incubate the strain overnight at 32°C with shaking at 220 rpm.
2. Inoculate 1 mL of the saturated overnight culture into 70 mL of fresh LB medium with 35 μL of 100 mg/mL ampicillin in a 250 mL flask.
3. Incubate the culture at 32°C with shaking until an optical density (OD) of ~0.5–0.6 is obtained (~2 h) (see Note 1).
4. When an OD reaches ~0.5–0.6, transfer the culture to a water bath for heat induction of the λ_{red} recombination system at 42°C for 15 min with shaking at 160 rpm (see Note 2).
5. Transfer the culture to a chilled ice-slurry water bath for 10–20 min at 160 rpm in order to stop the induction. Make sure to keep the cells cold from this point until after transformation (Subheading 3.1.3, step 2).
6. While keeping the culture on ice, divide it equally into two prechilled 50 mL polypropylene tubes (~35 mL culture per tube) and centrifuge at 4,400 ×*g* for 6 min at 4°C.
7. Decant the supernatant; resuspend the cell pellets in ~10 mL of ice-cold sterile distilled water. To ensure cell survival, do not vortex the cells past this stage! Gentle pipetting may be used if necessary to dislodge the cells. Add ice-cold sterile distilled water up to 40 mL. The cells should resuspend easily by gentle

inversion once they are dislodged. Perform centrifugation under the same conditions as mentioned in step 6.

8. Repeat step 7 using 10% glycerol instead of water.
9. Decant the supernatant, resuspend each cell pellet in 20 mL ice-cold sterile 10% glycerol and centrifuge again as described in step 6.
10. Decant the supernatant and resuspend the cell pellet in 500 μL of ice-cold 10% glycerol.
11. Aliquot 50 μL of the cell suspension into individual prechilled 1.5 mL microcentrifuge tubes and proceed with transformation (see Note 3).

3.1.3. Electroporation and Donor Mutant Selection

1. Add 100 ng of purified gene deletion cassette in 2 μL of ice-cold water (from Subheading 3.1.1, step 7) to the prepared competent cells (from Subheading 3.1.2, step 11). Flick the tube – do not pipette mix. Allow suspension to sit on ice for 5 min.
2. Transfer the suspension of ice-cold competent cells and DNA to a prechilled electroporation cuvette. Electroporate the cell mixture using 2.5 kV, 25 μF, 200 Ω setting (applies for 0.2 cm cuvettes) and immediately add 1 mL of room temperature SOC medium.
3. Transfer the electroporated cells in SOC medium with a sterile pasteur pipette into a 15 mL culture tube and incubate at 32°C for 1 h with orbital shaking at 220 rpm.
4. After incubation, centrifuge the cells in a Beckman at 4,400 ×*g* for 5 min (at room temperature).
5. Remove approximately 850 μL of the supernatant and resuspend the cell pellet in the remaining liquid.
6. Spread the cells on LB plates containing 34 μg/mL Cm.
7. Incubate the plates at 32°C overnight (see Note 4).
8. Pick and streak out two to three individual transformants on LB-Cm plates for mutant confirmation. Also, streak the same transformants on LB-Kan to make sure that the strains are not Kan resistant.

3.1.4. PCR Confirmation of Successful Gene Deletion

1. Grow overnight the individual knock out strains in liquid LB, complemented with 34 μg/mL Cm, and isolate the genomic DNA following the manufacturer's instructions (Promega) (see Note 5).
2. The DNA is amplified in three separate 25 μL reactions, with three different sets of knock out confirmation primers. All reactions are performed under conditions described in step 1 of Subheading 3.1.1, but with ~150 ng of genomic DNA template.

After the PCR products are synthesized, the products are run out on an agarose gel to confirm that the correct fragments were obtained.

3. The first primer set consists of a 20-nt flanking primer, located 200 base pairs upstream of the targeted region (KOCO-F), and Cm-R primer (5′-TTATACGCAAGGCGACAAGG-3′), which is complementary to the "Cm" cassette sequence. This amplification is expected to produce a 445-nt amplicon (Fig. 1).
4. The second set includes a forward primer (Cm-F, 5′-GATCTTCCGTCACAGGTAGG-3′), annealing to the "Cm" cassette sequence, and a reverse flanking confirmation primer (KOCO-R), which should be designed to anneal 200 bp downstream of the 3′ end of the deleted gene. This amplification reaction is expected to produce a 309-nt amplicon (Fig. 1).
5. The third PCR contains KOCO-F and KOCO-R primers. This reaction is required to verify that the selected strain is not a merodiploid, with one gene locus having been replaced by the cassette and another duplicated, but otherwise wild-type gene copy is still present. This amplification is expected to produce a 1.433 kb product (Fig. 1).

3.1.5. Storage of Confirmed Query Deletion Donor Mutant Strains Prior to Screening

1. Strains can be stored at 4°C for up to a month.
2. Overnight cultures of successfully confirmed recombinant donor strain clones for each query mutant are placed in individually labeled cryovials, supplemented with 15% glycerol (by volume), vortexed for 20 s, frozen on dry ice, and transferred to −80°C for long-term storage.

3.2. Arraying an *E. coli* F^- Recipient Strain Collection for Genome-Wide eSGA Screens

1. The entire Keio *E. coli* single-gene deletion mutant collection (3,968 strains; F^- BW25113) (11) is replicated by pinning to twenty-four 384-well microplates containing 80 μL of liquid LB medium supplemented with 50 μg/mL Kan per well. Each strain is pinned into one well, without replicates.
2. To make room for border control strain, which will aid in within and between plate normalization as well as in colony quantization, remove inoculated media from the outermost wells of each plate (from Subheading 3.2, step 1) and transfer to new plates, again leaving the outermost wells empty.
3. Similarly, make negative control spots by removing two strains from the inner wells of each plate to a new plate. The strains should be removed from a different location in each plate. Thus, these negative control spots are expected to be empty in the recipient and double mutant plates, ensuring that there were no processing errors when numbering the plates or in plate orientation during imaging or pinning.

4. Fill the empty wells, where cultures have been removed in steps 2 and 3 above, with LB containing 34 μg/mL of chloramphenicol.
5. Inoculate Keio deletion strain JW5028 (11) into a 500-mL flask with 150-mL LB, containing 50 μg/mL of Kan. In our whole genome eSGA experiments, this particular deletion mutant showed the smallest number of interactions and is thus used as a border control. Since the same recipient strain is used in all the border wells and all recipient plates, it is expected to grow equally well across all plates of the same screen. Thus, the measurement of the border control strain growth should help with between plate normalization. Further, since the left and the right as well as the top and the bottom border sports should grow and be detected by the imaging software equally well, comparing the border colony sizes provides a good control for efficient automatic quantization and is useful for within plate normalization.
6. Grow the arrayed strains as well as the JW5028 culture overnight at 32°C with 190 rpm orbital shaking to OD of ~0.4–0.6 at 600 nm.
7. After the overnight growth, remove completely the LB-Cm from all border wells and replace it with 80 μL of JW5028 overnight culture.
8. Supplement each well in the recipient plates with 15% glycerol by volume. Gently pipette-mix the liquids, without spilling, to uniformly distribute the glycerol, and store the plates at −80°C for long-term storage.

3.3. Construction of E. coli Double Mutants Using an Arrayed Strain Mating Procedure

Once a query donor mutant strain is available and the recipients have been arrayed, genome-wide eSGA screens can be performed (see Note 6). The construction and screening process for generating and evaluating large sets of double mutants is divided into four steps: (1) mutant array preparation, (2) conjugation, (3) selection of double mutants on "Cm" and "Kan" – containing plates, and (4) plate imaging and quantification of individual colony sizes. Construction of double mutants in high-density gridded arrays (Fig. 2) involves replica pinning and incubations over 6 days. We suggest performing at least three independent biological replicate experiments. One experiment, without replicates, is described below:

1. First day: The Hfr donor strain, bearing the query gene replacement (marked with Cm, from Subheading 3.1), is grown overnight at 32°C in rich LB liquid media (with 34 μg/mL of Cm).
2. Second day: The thawed collection of ordered recipient mutants is pinned in 384 format onto solid LB plates supplemented with 50 μg/mL Kan. Simultaneously, the donor query strain is pinned in 384 densities onto the same number of LB plates supplemented with 34 μg/mL of Cm.

3. Third day: Conjugation plates are made by (a) pinning the donor from the 384-spot overnight donor plates onto solid LB plate and (b) pinning the arrayed recipients, also grown overnight in 384-spot density, over the freshly pinned donor (see Note 7).
4. Fourth day: Conjugants are subjected to first round of selection: each 384-density conjugation plate is pinned onto one solid LB plate containing Cm and Kan (see Note 8).
5. Fifth day: Double mutant colonies from the first double drug selection plate are repinned onto a second double drug selection plate in 1,536-spot format, such that each first selection colony is represented on the second selection plate by four colonies. In each step of the above process, the plates are usually incubated for 16–36 h at 32°C (see Note 9). The double mutants generated by the plate-based assay can be randomly checked by PCR, as described in Fig. 3, to confirm the correctness of the mutations.
6. Sixth day: The second selection double mutant plates are photographed to record the growth of all colonies for quantitatively

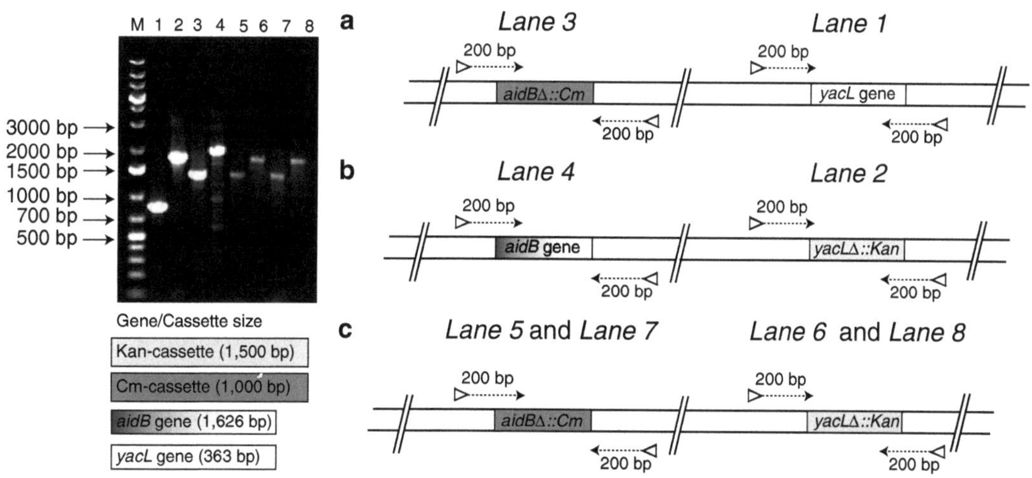

Fig. 3. PCR confirmation of deletions at two unrelated loci, combined through conjugation. Following 24 h conjugation of Δ*aidB*-Cm donor deletion strain (**a**) and Δ*yacL*-Kan recipient deletion strain (**b**), the selected double mutants (**c**) are confirmed by PCR. The sizes of the Cm and Kan cassettes, as well as *aidB* and *yacL* loci, in base pairs, are shown in parentheses. Amplifications with flanking primers, KOCO-F and KOCO-R (as shown in Fig. 1), from strains with the target gene replaced by Cm^R or Kan^R cassette are expected to produce 1,400 bp and 1,900 bp products, respectively. On the other hand, amplifications from the isolates containing a wild-type copy of the target gene result in a product equal to the size of the gene plus 400 bp. *Lanes from left to right*: M. DNA marker; Lane 1. Δ*aidB*-Cm donor deletion strain amplified with *yacL* knock out confirmation primers; Lane 2. Δ*yacL*-Kan recipient deletion strain amplified with *yacL* knock out confirmation primers; Lane 3. Δ*aidB*-Cm query deletion strain amplified with *aidB* knock out confirmation primers; Lane 4. Δ*yacL*-Kan recipient deletion strain amplified with *aidB* knock out confirmation primers; Lanes 5 and 7; and 6 and 8 show amplifications from two independently constructed double mutants (Δ*aidB*-Cm* Δ*yacL*-Kan), amplified with *aidB* and *yacL* knock out confirmation primers, respectively. The molecular weights, in bp, are shown beside arrows to the left of the gel. The lanes 1–4 serve as controls and show that donor and recipient are mutants for *aidB* and *yacL* only, respectively, each bearing a wild-type copy of the second locus. Lanes 5–8 show the same PCR amplifications using conjugant genomic DNA and confirm that PCR products corresponding to both mutant loci are present in both isolates.

assessing mutant fitness (Subheading 3.4) and analyzing the interactions between gene pairs (Subheading 3.5).

3.4. Data Processing and Score Generation

3.4.1. Quantitative Image Analysis and Plate Colony Size Normalization

1. The size of each colony is measured by processing plates in batch mode or individually using specialized colony imaging software (19, 25). The output is a raw colony size measurement for each of the colonies present on the plate.
2. The raw colony measurements can be normalized using multiple normalization and filtering steps to correct for systematic biases in colony growth and measurement within and between plates, such as plate edge effects, interplate variation effects, uneven image lighting, artifacts due to physical curvature of the agar surface, competition effects for neighboring colonies and possible pinning defects, as well as differences in growth time (25). These systematic artifacts are independent of the growth properties of *E. coli* deletion mutants used in the study, and should be corrected as they give rise to spurious growth fitness estimates.
3. A correction can be applied for the larger trend in colony sizes typically observed in the outermost colony rows and columns (due to lower competition for nutrients than found in the center of the plate). Thus, for each edge row or column, the colony sizes can be scaled such that the median size in that row or column is equal to the median size of the colonies in the center of the plate (see Note 10).
4. Further, normalization can be applied to account for the differences in the growth phenotype of the donor mutations. Since most mutations in the array have little or no growth defect, and most of the mutation combinations will not interact (4, 40), we typically normalized the colony sizes according to the peak of the histogram of colony sizes on a given plate (see Note 11).
5. Finally, the results are analyzed statistically to take into account the reproducibility and the deviation (standard variance) from the median sizes of the replicate colony measurements.

3.4.2. Generation of Genetic Interaction Scores

1. The normalized median colony growth sizes are used to generate a genetic interaction score (S) for each gene pair as follows:

$$S\text{ - score} = \frac{\mu_{\text{Exp}} - \mu_{\text{Cont}}}{\sqrt{\frac{S_{\text{var}}}{n_{\text{Exp}}} + \frac{S_{\text{var}}}{n_{\text{Cont}}}}}$$

where, $S_{\text{var}} = (\text{var}_{\text{Exp}} \times (n_{\text{Exp}} - 1) + \text{var}_{\text{Cont}} \times (n_{\text{Cont}} - 1))/(n_{\text{Exp}} + n_{\text{Cont}} - 2)$; var_{Exp} = the maximum variance of the normalized colony sizes for the double mutant; var_{Cont} = median of the variances in the normalized double mutant colony sizes from the reference set; nE_{xp} = number of measurements of double mutant colony sizes; nC_{ont} = the median number of experimental replicates over all the experiments; nE_{xp} = median normalized colony sizes of the double mutants; and nC_{ont} = median of normalized colony sizes for all double mutants arising from the single donor mutant strain. The *S*-score reflects both the statistical confidence of a putative digenic interaction as well as the biological strength of the interaction. Strong positive *S*-scores indicate alleviating (e.g., suppression) effects (i.e., growth rate of the double mutant is better than the product of the growth rates of the two respective single mutants), suggesting that the interacting genes participate in the same pathway (25) while significant negative *S*-scores reflect synthetic sickness or lethality (i.e., growth rate of the double mutant is worse than the product of the growth rates of the two single mutants), which is often suggestive of membership in parallel redundant pathways (25).

2. Closely linked genes should exhibit negative *S*-scores because recombination between these genes happens with greatly lower frequencies, which results in a failure to form double mutants. Based on our analysis in the previous study (25), the linkage suppression is evident and inversely proportional to the genetic distance, but is generally undetectable 30 kbp from the donor query gene loci. Although functionally linked genes are often located in the same operon on the *E. coli* chromosome, it is recommended to remove all recipient genes located within 30 kbp of a query locus from further assessment. Nonetheless, linkage suppression can be used to confirm the correctness of a particular query mutation (i.e., linked loci should exhibit a detectable linkage effect).
3. After taking into account of all the above steps, a final data set is created for functional analysis with a single *S*-score generated for each pair of tested genes (see Note 12).

3.4.3. Assessing the Genetic Interaction Scores

1. The interaction *S*-scores from the genome-wide screens can be plotted to examine the normality of the *S*-score distribution. If the *S*-scores approximate a normal distribution, then the next step would be to see what significant fraction of *S*-scores is observed in the two tails of the distribution of the experimental data set compared to chance alone (see Note 13).
2. The genetic interaction data can be randomized to evaluate whether the *S*-scores calculated for pairs of genes in the genome-wide screens reflect true genetic interactions or are due to residual position effects of the strains on the plate. For example, one can scramble the row and column coordinates of the dataset using the original raw colony sizes of each gene

extracted from the image extraction software. The *S*-scores are then recalculated.

3. *S*-scores deviating significantly from the mean represent candidates for functional associations. One can then evaluate the extent to which available knowledge – about functionally related gene pairs – is reflected at various *S*-score thresholds (see Note 14).
4. Candidate interacting gene pairs can be considered to be functionally bridged if they share a physical interaction or another known (annotated) functional relationship (25). With this assumption, one can generate a positive prediction rate [TP/(TP + FP)] for the genetic interaction data, where true positives (TP) can be calculated for functional associations predicted by eSGA that have links from other experimental studies while false positives (FP) can be defined as eSGA hits without any supporting evidence. Using this strategy, one can determine not only the optimal threshold score, but also the rate of positive prediction with increasing |*S*-score| value.
5. Since genetic interaction (*S*) scores, deviating significantly from the mean, represent candidates for functional associations (37), one can also make use of pathway level annotations to examine the genetic interaction network at a more global mechanistic level (23, 41). A wealth of curated information is contained in public functional annotation databases like KEGG (42), EcoCyc (43) and MultiFun (44), and other sources of functional associations (1), to determine if putative genetic interactors have related functions, which should reflect the overall accuracy of the filtered dataset. Specifically, one can evaluate the extent to which genes belonging to a particular functional category (i.e., annotation term) show statistical enrichment for genetic interactions at various *S*-score thresholds after correcting for multiple-hypothesis testing. To accomplish this, one can calculate enrichment between each pair of functional processes by performing a hypergeometric distribution analysis (45) with correction for multiple hypothesis testing (46). After accounting for the false discovery rate, the scored data should show a marked corresponding increase in enrichment for both aggravating and alleviating gene pairs (p-value ≤ 0.05) between select biological processes with increasing *S*-score values. This alternative strategy can also serve to define a confidence threshold for selecting the putative genetic interactions for further analysis.
6. The reliability of high-confidence genetic interactions can be confirmed by reanalyzing individual donor-recipient digenic mutant combinations using custom mini-array conjugation assays (25). For example, the specificity of the interaction of *pdxB* with *iscS* in the Fe-S cluster biosynthesis pathway from Butland et al. (25) can be explained by looking at the curated

functional annotations. The gene *pdxB* is involved in the biosynthesis of pyridoxal 5-phosphate (PLP), which is a cofactor for IscS cysteine desulfurases, suggesting a plausible link. The reliability of the putative interaction was confirmed by manually conjugating the *iscS*Δ*::Cm*^{R} query deletion mutant strain to the *pdxB*Δ*::Kan*^{R} recipient deletion mutant strain, and then by reciprocally conjugating *pdxB*Δ*::Cm*^{R} as query deletion strain to *iscS*Δ*::Kan*^{R} as the recipient deletion mutant strain. As shown in Fig. 4a, the *iscS* mutant displayed synthetic lethality with *pdxB*, and a similar aggravating interaction was observed when *pdxB*Δ*::Cm*^{R} was used as donor, implying *iscS* and *pdxB* are functionally related. Such small-scale validation experiments are useful for verifying putative genetic interactions generated from the high-throughput eSGA screens.

7. One can also perform rescue experiments with one or both mutant alleles by gene complementation using an inducible expression plasmid from the public *E. coli* ORFeome collection (47).

3.5. Discerning Pathway Level Relationships

1. Genes in the same pathway are expected to display closely correlated patterns of genetic interactions with genes in parallel, functionally redundant pathways.
2. Groupings reflecting such biological relationships can be visualized by two-dimensional hierarchical clustering of the *S*-scores to suggest the possible membership of novel genes in known pathways and biological processes. A representative example is shown in Fig. 4b, where the components of the functionally redundant Isc and Suf pathways form distinct clusters that are linked together by extensive aggravating interactions.
3. A hypergeometric distribution function (45) can be applied to evaluate pathway cross-talk (i.e., determine the significance of interactions between combinations of pathways). Likewise, nonthreshold-based enrichment algorithms, such as Gene Set Enrichment Analysis (48) may allow for a more sensitive detection of trends and modularity in the genetic interaction networks. Integration of genetic interaction networks with physical associations (i.e., protein–protein interactions) and other functional association data, such as genomic contexts, can potentially detect groupings reflective of higher order functional modules defining biological systems.
4. The similarity of the genetic interaction profiles should represent the congruency of the phenotypes of the two mutations. For instance, one would logically expect both measures to be indicative of whether the two genes act in the same pathway. In fact, gene pairs exhibiting alleviating interactions with highly correlated interaction profiles tend to encode proteins that are physically associated (9, 10, 23, 24). Additionally, in cases

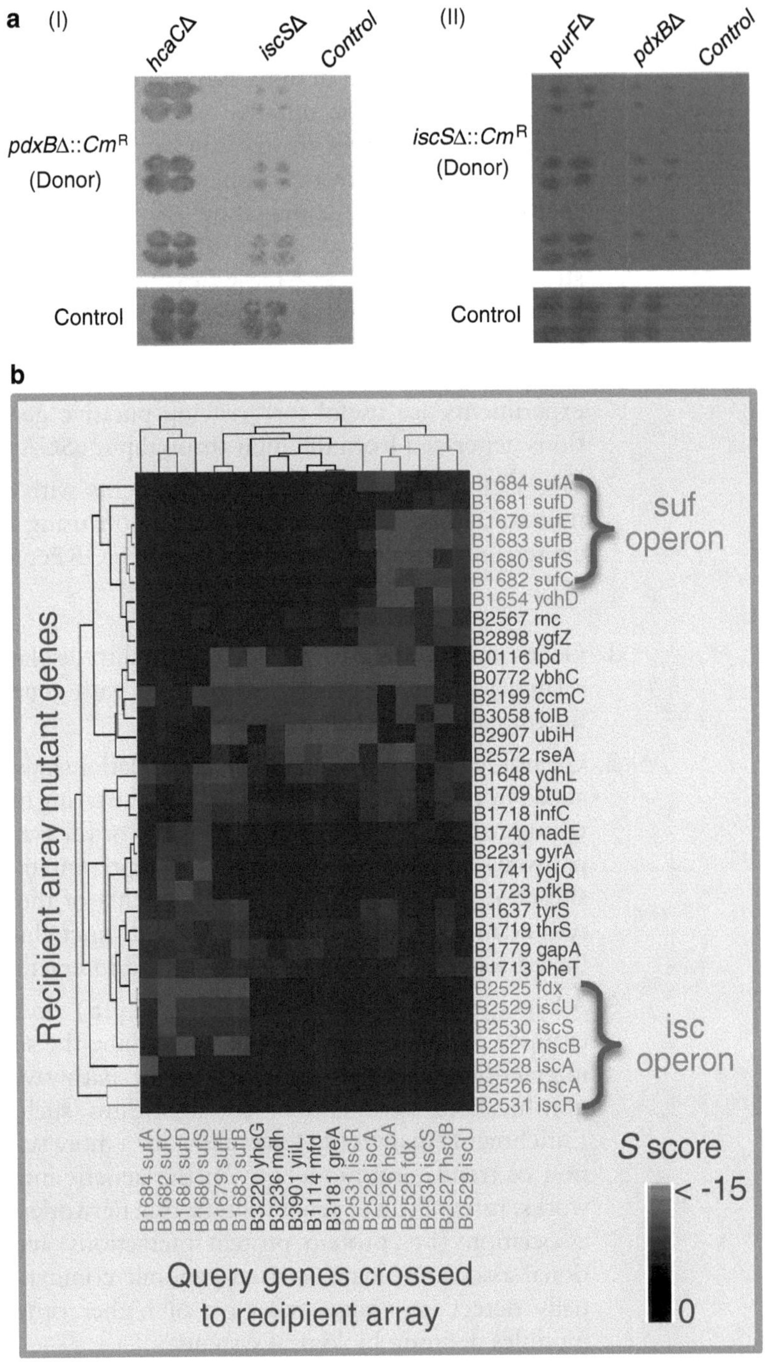

Fig. 4. Validation of genetic interactions using custom miniarrays and clustering of genetic interaction profiles. *Panel A*: Confirmation of genetic interactions of *iscS* and *pdxB* strains using custom miniarrays. The eSGA miniarray confirmations, shown as plate images, were performed by crossing a *pdxB* query mutant strain (marked with Cm^R) with an *iscS* recipient deletion strain (marked with Kan^R) and vice-versa. Both donor-recipient combinations, *pdxB* and *iscS* (I) as well as *iscS* and *pdxB* (II) displayed an SSL relationship. The *hcaC*Δ::*KanR* and *purF*Δ::Kan^R recipient mutants were used as negative controls (no SSL interactions expected), "C" represents recipient control and *ybaS*Δ represents a donor control. *Panel B*: A subset of the high-confidence genetic interaction network highlighting the known Isc (*gray print*) and Suf (*gray print*)

where the proteins do not physically associate, products of such gene pairs tend to act coherently in a biochemical pathway. Such cases are particularly informative as these very close functional relationships are difficult to detect by other methods. Cluster analysis can therefore be applied to genetic interaction networks to group genes according to profile similarity and to predict functions of unannotated genes (21–23, 25, 35, 49). Since clustering algorithms vary, putative functional relationships determined through clustering require independent experimental verification (see Note 15).

4. Notes

1. At an OD between 0.5 and 0.6, *E. coli* culture is at an exponential growth phase with cells dividing at a maximal rate, allowing for the preparation of highly competent cells.
2. At this temperature, the temperature-sensitive λ cI-repression is removed, which leads to the expression of λ *exo*, *bet*, and *gam* genes that in turn make the cells more recombinogenic. A maximum level of recombination is reached at induction times between 7.5 and 17.5 min, with reduction in efficiencies occurring at times longer than 17.5 min (32).
3. The competent cells, resuspended in 10% glycerol, can be stored for up to 6 months by first freezing the cells on dry ice and then transferring them into –80°C for storage. Prior to subsequent transformation (Subheading 3.1.3) the cells are thawed on ice for 10–15 min.
4. Include positive and negative transformation controls. For a negative control, we suggest a mock electroporation of competent cells without any PCR product to test for the presence of contaminating strains. As a positive control, use a PCR product (or a plasmid) that is known to generate transformants.
5. Colony PCR can also be performed in order to more rapidly confirm the replacement of the target gene with the Cm cassette. This method does not require the purification of genomic DNA; however, care has to be taken to pick a very small volume of cells (ca. 0.5 mm^2) using a plastic pipette tip. Furthermore, a 5-min initial denaturation step is used in this case.

Fig. 4. (continued) pathway genes with similar patterns of genetic interactions. Both pathways function in the same process and their patterns of genetic interactions cluster together. *Columns*: recipient array genes. *Gray* represents aggravating (negative *S*-score) interactions and *black* represents the absence of genetic interactions. The array strains with the supposed *iscR* and *hscA* mutations were defective. The *ydhD* gene displays interactions with the Isc pathway. The network clustering analysis shown is performed on data from Butland et al. (25).

6. The construction and screening process for generating and evaluating large sets of double mutants can be applied for E-MAP studies as well (23, 34). In this case, the donor query strain is crossed against an array of recipient genes selected to answer a specific biological question.
7. In the standard eSGA screening process, the donor and recipient strains are incubated for conjugation at an approximate cell ratio of ~1:1. We observed that the selection of conjugants using only one antibiotic (Kan or Cm), followed by selection using both antibiotics, results in highly variable double mutant colonies. This additional selective outgrowth step is deemed to be advantageous for two reasons. First, it has been previously reported that having >1 colony measurement greatly increases the accuracy of the fitness estimates allowing for the natural measure of variation as the standard deviation. Second, it has been noted during the development process that direct double antibiotic selection on the initial conjugation plate leads to highly variable colony size distributions which severely impairs the accuracy of strain fitness estimates.
8. For standard genome-wide eSGA screens, each individual Keio gene deletion mutant (11) is arrayed in a row of four replicates: two copies of each of the two biological replicates ("Isolate 1 and 2", which are independently confirmed deletion strains for the same gene). Further, the entire collection is screened twice on separate plates to establish score reproducibility and statistical significance. Therefore, in total, each combination of mutations is assayed using a total of eight replicate colony measurements.
9. We suggest allowing conjugation to take place over a 24 h period because only a few viable conjugants were obtained for short mating periods (<12 h) (25). Monitor growth of colonies on plates to determine appropriate incubation time for the other steps. The time may vary from screen to screen, since different mutants may have different growth rates.
10. There may be cases in which one is unable to detect or quantify genetic interactions because one of the mutations causes an *E. coli* strain to have an unusual mutant colony morphological phenotype (e.g., spreading or mucoid) that precludes colony transfer or conjugation, which in turn masks or exaggerates growth defects. Measurements for such colonies are removed from further analysis.
11. Systematic variability can also arise due to batch-to-batch effect, where the size of a colony estimated for one set of screens completed at approximately the same time using the same media preparation differs from the values estimated for another set of screens completed at a different time. If such

batch-to-batch variability is evident, it is preferable to compute the expected colony sizes independently for each batch.

12. In cases where two *S*-scores are produced from two independent biological experiments, we suggest calculating and substituting in either a median or an average score.
13. A randomized dataset can be created for each query mutant screen, where the row and column coordinates and the raw colony sizes of each recipient mutant are shuffled. A randomized dataset to control for score distributions obtained for each query mutant screen is permuted 1,000 times. *S*-scores are then calculated for each query mutant screen from this permuted randomized dataset. The difference in tails between the experimental and the randomized datasets can be then used to assess the statistical significance of aggravating or alleviating genetic interactions.
14. Predicted functional associations between *E. coli* genes can be obtained from public databases like STRING (50) or eNET (1).
15. Since *E. coli* pathways have evolved adaptive mechanisms to withstand environmental perturbations and since certain functional relationships are likely manifested only under particular physiological conditions (3), one can subject the array of viable double mutants to the additional stress of growth on minimal glucose medium or at high temperature (i.e., at 42°C). This strategy is likely to reveal condition-dependent genetic interactions. For example, Gross and colleagues have shown condition-dependent synthetic genetic interactions involving the outer membrane-associated lipoprotein *pal* in a media-dependent manner (e.g., *pal-yfgL*; *pal-ompA*; *pal-cpxR*) (26). The use of rich LB medium for the conjugation step typically results in more rapid and consistent growth of recipient mutants from the KEIO collection. Cells grown on minimal media require much longer incubation times (a total of approximately 2 weeks to complete a full genome screen in minimal medium versus 6 days on rich medium). Further, screens, where conjugation and selection are performed on minimal media, display far more variability in colony sizes and conjugation success compared to rich media screens. Thus, we incubate all plates at 32°C during the conjugation and selection steps. For experiments that require other conditions to study genetic interactions, conjugation and selection can be performed in standard conditions, and then the double mutants can be replicate pinned and grown in the condition of interest (for example, on different media, or at a different temperature). Thus, a new set of condition-specific genetic interactions can be generated.

Acknowledgments

We thank members of the Emili and Jack Greenblatt laboratories for assistance and advice. AG is a recipient of Vanier Canada Graduate Scholarship. This work was supported by a Canadian Institute of Health Research (CIHR) grant (MOB-82852) to A.E.

References

1. Hu, P., Janga, S. C., Babu, M., Diaz-Mejia, J. J., and Butland, G. (2009) Global functional atlas of *Escherichia coli* encompassing previously uncharacterized proteins. *PLoS Biol 7*, e1000096.
2. Pereira-Leal, J. B., Levy, E. D., and Teichmann, S. A. (2006) The origins and evolution of functional modules: lessons from protein complexes. *Philos Trans R Soc Lond B Biol Sci 361*, 507–17.
3. Babu, M., Musso, G., Diaz-Mejia, J. J., Butland, G., Greenblatt, J. F., and Emili, A. (2009) Systems-level approaches for identifying and analyzing genetic interaction networks in *Escherichia coli* and extensions to other prokaryotes. *Mol Biosyst 12*, 1439–55.
4. Beltrao, P., Cagney, G., and Krogan, N. J. (2010) Quantitative genetic interactions reveal biological modularity. *Cell 141*, 739–45.
5. Beyer, A., Bandyopadhyay, S., and Ideker, T. (2007) Integrating physical and genetic maps: from genomes to interaction networks. *Nat Rev Genet 8*, 699–710.
6. Motter, A. E., Gulbahce, N., Almaas, E., and Barabasi, A. L. (2008) Predicting synthetic rescues in metabolic networks. *Mol Syst Biol 4*, 168.
7. Bandyopadhyay, S., Kelley, R., Krogan, N. J., and Ideker, T. (2008) Functional maps of protein complexes from quantitative genetic interaction data. *PLoS Comput Biol 4*, e1000065.
8. Boone, C., Bussey, H., and Andrews, B. J. (2007) Exploring genetic interactions and networks with yeast. *Nat Rev Genet 8*, 437–49.
9. Collins, S. R., Miller, K. M., Maas, N. L., Roguev, A., Fillingham, J., Chu, C. S., Schuldiner, M., Gebbia, M., Recht, J., Shales, M., Ding, H., Xu, H., Han, J., Ingvarsdottir, K., Cheng, B., Andrews, B., Boone, C., Berger, S. L., Hieter, P., Zhang, Z., Brown, G. W., Ingles, C. J., Emili, A., Allis, C. D., Toczyski, D. P., Weissman, J. S., Greenblatt, J. F., and Krogan, N. J. (2007) Functional dissection of protein complexes involved in yeast chromosome biology using a genetic interaction map. *Nature 446*, 806–10.
10. Fiedler, D., Braberg, H., Mehta, M., Chechik, G., Cagney, G., Mukherjee, P., Silva, A. C., Shales, M., Collins, S. R., van Wageningen, S., Kemmeren, P., Holstege, F. C., Weissman, J. S., Keogh, M. C., Koller, D., Shokat, K. M., and Krogan, N. J. (2009) Functional organization of the *S. cerevisiae* phosphorylation network. *Cell 136*, 952–63.
11. Baba, T., Ara, T., Hasegawa, M., Takai, Y., Okumura, Y., Baba, M., Datsenko, K. A., Tomita, M., Wanner, B. L., and Mori, H. (2006) Construction of *Escherichia coli* K-12 in-frame, single-gene knockout mutants: the Keio collection. *Mol Syst Biol 2*, 2006.0008.
12. Hartwell, L. H., Hopfield, J. J., Leibler, S., and Murray, A. W. (1999) From molecular to modular cell biology. *Nature 402*, C47-52.
13. Mani, R., St Onge, R. P., Hartman, J. L. 4th., Giaever, G., and Roth, F. P. (2008) Defining genetic interaction. *Proc Natl Acad Sci USA 105*, 3461–6.
14. Kuzminov, A. (1999) Recombinational repair of DNA damage in *Escherichia coli* and bacteriophage lambda. *Microbiol Mol Biol Rev 63*, 751–813.
15. Seoighe, C., and Wolfe, K. H. (1998) Extent of genomic rearrangement after genome duplication in yeast. *Proc Natl Acad Sci USA 95*, 4447–52.
16. Dixon, S. J., Costanzo, M., Baryshnikova, A., Andrews, B., and Boone, C. (2009) Systematic mapping of genetic interaction networks. *Annu Rev Genet 43*, 601–25.
17. Dixon, S. J., Fedyshyn, Y., Koh, J. L., Prasad, T. S., Chahwan, C., Chua, G., Toufighi, K., Baryshnikova, A., Hayles, J., Hoe, K. L., Kim, D. U., Park, H. O., Myers, C. L., Pandey, A., Durocher, D., Andrews, B. J., and Boone, C. (2008) Significant conservation of synthetic lethal genetic interaction networks between distantly related eukaryotes. *Proc Natl Acad Sci USA 105*, 16653–8.

18. Cork, J. M., and Purugganan, M. D. (2004) The evolution of molecular genetic pathways and networks. *Bioessays 26*, 479–84.
19. Costanzo, M., Baryshnikova, A., Bellay, J., Kim, Y., Spear, E. D., Sevier, C. S., Ding, H., Koh, J. L., Toufighi, K., Mostafavi, S., Prinz, J., St Onge, R. P., VanderSluis, B., Makhnevych, T., Vizeacoumar, F. J., Alizadeh, S., Bahr, S., Brost, R. L., Chen, Y., Cokol, M., Deshpande, R., Li, Z., Lin, Z. Y., Liang, W., Marback, M., Paw, J., San Luis, B. J., Shuteriqi, E., Tong, A. H., van Dyk, N., Wallace, I. M., Whitney, J. A., Weirauch, M. T., Zhong, G., Zhu, H., Houry, W. A., Brudno, M., Ragibizadeh, S., Papp, B., Pal, C., Roth, F. P., Giaever, G., Nislow, C., Troyanskaya, O. G., Bussey, H., Bader, G. D., Gingras, A. C., Morris, Q. D., Kim, P. M., Kaiser, C. A., Myers, C. L., Andrews, B. J., and Boone, C. (2010) The genetic landscape of a cell. *Science 327*, 425–31.
20. Pan, X., Yuan, D. S., Xiang, D., Wang, X., Sookhai-Mahadeo, S., Bader, J. S., Hieter, P., Spencer, F., and Boeke, J. D. (2004) A robust toolkit for functional profiling of the yeast genome. *Mol Cell 16*, 487–96.
21. Tong, A. H., Evangelista, M., Parsons, A. B., Xu, H., Bader, G. D., Page, N., Robinson, M., Raghibizadeh, S., Hogue, C. W., Bussey, H., Andrews, B., Tyers, M., and Boone, C. (2001) Systematic genetic analysis with ordered arrays of yeast deletion mutants. *Science 294*, 2364–8.
22. Tong, A. H., Lesage, G., Bader, G. D., Ding, H., Xu, H., Xin, X., Young, J., Berriz, G. F., Brost, R. L., Chang, M., Chen, Y., Cheng, X., Chua, G., Friesen, H., Goldberg, D. S., Haynes, J., Humphries, C., He, G., Hussein, S., Ke, L., Krogan, N., Li, Z., Levinson, J. N., Lu, H., Menard, P., Munyana, C., Parsons, A. B., Ryan, O., Tonikian, R., Roberts, T., Sdicu, A. M., Shapiro, J., Sheikh, B., Suter, B., Wong, S. L., Zhang, L. V., Zhu, H., Burd, C. G., Munro, S., Sander, C., Rine, J., Greenblatt, J., Peter, M., Bretscher, A., Bell, G., Roth, F. P., Brown, G. W., Andrews, B., Bussey, H., and Boone, C. (2004) Global mapping of the yeast genetic interaction network. *Science 303*, 808–13.
23. Schuldiner, M., Collins, S. R., Thompson, N. J., Denic, V., Bhamidipati, A., Punna, T., Ihmels, J., Andrews, B., Boone, C., Greenblatt, J. F., Weissman, J. S., and Krogan, N. J. (2005) Exploration of the function and organization of the yeast early secretory pathway through an epistatic miniarray profile. *Cell 123*, 507–19.
24. Wilmes, G. M., Bergkessel, M., Bandyopadhyay, S., Shales, M., Braberg, H., Cagney, G., Collins, S. R., Whitworth, G. B., Kress, T. L., Weissman, J. S., Ideker, T., Guthrie, C., and Krogan, N. J. (2008) A genetic interaction map of RNA-processing factors reveals links between Sem1/Dss1-containing complexes and mRNA export and splicing. *Mol Cell 32*, 735–46.
25. Butland, G., Babu, M., Díaz-Mejía, J. J., Bohdana, F., Phanse, S., Gold, B., Yang, W., Li, J., Gagarinova, A. G., Pogoutse, O., Mori, H., Wanner, B. L., Lo, H., Wasniewski, J., Christopolous, C., Ali, M., Venn, P., Safavi-Naini, A., Sourour, N., Caron, S., Choi, J. Y., Laigle, L., Nazarians-Armavil, A., Deshpande, A., Joe, S., Datsenko, K. A., Yamamoto, N., Andrews, B. J., Boone, C., Ding, H., Sheikh, B., Moreno-Hagelseib, G., Greenblatt, J. F., and Emili, A. (2008) eSGA: *E. coli* synthetic genetic array analysis. *Nat Methods 5*, 789–95.
26. Typas, A., Nichols, R. J., Siegele, D. A., Shales, M., Collins, S. R., Lim, B., Braberg, H., Yamamoto, N., Takeuchi, R., Wanner, B. L., Mori, H., Weissman, J. S., Krogan, N. J., and Gross, C. A. (2008) High-throughput, quantitative analyses of genetic interactions in *E. coli*. *Nat Methods 5*, 781–87.
27. Davierwala, A. P., Haynes, J., Li, Z., Brost, R. L., Robinson, M. D., Yu, L., Mnaimneh, S., Ding, H., Zhu, H., Chen, Y., Cheng, X., Brown, G. W., Boone, C., Andrews, B. J., and Hughes, T. R. (2005) The synthetic genetic interaction spectrum of essential genes. *Nat Genet 37*, 1147–52.
28. Breslow, D. K., Cameron, D. M., Collins, S. R., Schuldiner, M., Stewart-Ornstein, J., Newman, H. W., Braun, S., Madhani, H. D., Krogan, N. J., and Weissman, J. S. (2008) A comprehensive strategy enabling high-resolution functional analysis of the yeast genome. *Nat Methods 5*, 711–18.
29. Firth, N., Ippen-Ihler, K. and Skurray, R.A. (1996) Structure and function of the F-factor and mechanism of conjugation. *Escherichia coli and Salmonella: Cellular and Molecular Biology, Vol. 1 (Neidhardt, F.C., Ed.), ASM Press, Washington, DC*, 2377–401.
30. Ippen-Ihler, K. A., and Minkley, E. G., Jr. (1986) The conjugation system of F, the fertility factor of *Escherichia coli*. *Annu Rev Genet 20*, 593–624.
31. Chumley, F. G., Menzel, R., and Roth, J. R. (1979) Hfr Formation Directed by Tn10. *Genetics 91*, 639–55.
32. Yu, D., Ellis, H. M., Lee, E. C., Jenkins, N. A., Copeland, N. G., and Court, D. L. (2000) An efficient recombination system for chromosome engineering in *Escherichia coli*. *Proc Natl Acad Sci USA 97*, 5978–83.

33. Johnson, D. C., Dean, D. R., Smith, A. D., and Johnson, M. K. (2005) Structure, function, and formation of biological iron-sulfur clusters. *Annu Rev Biochem 74*, 247–81.
34. Babu, M., Díaz-Mejía, J. J., Phanse, S., Vlasblom, J., Gagarinova, A., Graham, C., Ding, H., Hu, P., Yousif, F., Nazarians-Armavil, A., Pogoutse, O., Ali, M., Peer, A., Margalit, H., Wodak, S. J., Moreno-Hagelsieb, G., Greenblatt, J. F., and Emili, A. (2010) Genetic interaction profiling reveals the global functional organization of cell-envelope pathways in *Escherichia coli*. *Cell.*
35. Roguev, A., Bandyopadhyay, S., Zofall, M., Zhang, K., Fischer, T., Collins, S. R., Qu, H., Shales, M., Park, H. O., Hayles, J., Hoe, K. L., Kim, D. U., Ideker, T., Grewal, S. I., Weissman, J. S., and Krogan, N. J. (2008) Conservation and rewiring of functional modules revealed by an epistasis map in fission yeast. *Science 322*, 405–10.
36. van Opijnen, T., Bodi, K. L., and Camilli, A. (2009) Tn-seq: high-throughput parallel sequencing for fitness and genetic interaction studies in microorganisms. *Nat Methods 6*, 767–72.
37. St Onge, R. P., Mani, R., Oh, J., Proctor, M., Fung, E., Davis, R. W., Nislow, C., Roth, F. P., and Giaever, G. (2007) in *Nat Genet* pp 199–206.
38. Zeghouf, M., Li, J., Butland, G., Borkowska, A., Canadien, V., Richards, D., Beattie, B., Emili, A., and Greenblatt, J. F. (2004) in *J Proteome Res* pp 463–8.
39. Butland, G., Peregrín-Alvarez, J. M., Li, J., Yang, W., Yang, X., Canadien, V., Starostine, A., Richards, D., Beattie, B., Krogan, N., Davey, M., Parkinson, J., Greenblatt, J., and Emili, A. (2005) in *Nature* pp 531–7.
40. Collins, S. R., Schuldiner, M., Krogan, N. J., and Weissman, J. S. (2006) A strategy for extracting and analyzing large-scale quantitative epistatic interaction data. *Genome Biol 7*, R63.
41. Ma, X., Tarone, A. M., and Li, W. (2008) Mapping genetically compensatory pathways from synthetic lethal interactions in yeast. *PLoS One 3*, e1922.
42. Kanehisa, M. (2002) The KEGG database. *Novartis Found Symp 247*, 91–101; discussion 101–3, 119–28, 244–52.
43. Keseler, I. M., Bonavides-Martinez, C., Collado-Vides, J., Gama-Castro, S., Gunsalus, R. P., Johnson, D. A., Krummenacker, M., Nolan, L. M., Paley, S., Paulsen, I. T., Peralta-Gil, M., Santos-Zavaleta, A., Shearer, A. G., and Karp, P. D. (2009) EcoCyc: a comprehensive view of *Escherichia coli* biology. *Nucleic Acids Res 37*, D464-70.
44. Serres, M. H., and Riley, M. (2000) MultiFun, a multifunctional classification scheme for Escherichia coli K-12 gene products. *Microb Comp Genomics 5*, 205–22.
45. Le Meur, N., and Gentleman, R. (2008) Modeling synthetic lethality. *Genome Biol 9*, R135.
46. Benjamini, Y., and Yekutieli, D. (2001) The control of the false discovery rate in multiple testing under dependency. *Annals of Statistics 29*, 1165–88.
47. Saka, K., Tadenuma, M., Nakade, S., Tanaka, N., Sugawara, H., Nishikawa, K., Ichiyoshi, N., Kitagawa, M., Mori, H., Ogasawara, N., and Nishimura, A. (2005) A complete set of *Escherichia coli* open reading frames in mobile plasmids facilitating genetic studies. *DNA Res 12*, 63–8.
48. Subramanian, A., Tamayo, P., Mootha, V. K., Mukherjee, S., Ebert, B. L., Gillette, M. A., Paulovich, A., Pomeroy, S. L., Golub, T. R., Lander, E. S., and Mesirov, J. P. (2005) Gene set enrichment analysis: a knowledge-based approach for interpreting genome-wide expression profiles. *Proc Natl Acad Sci USA 102*, 15545–50.
49. Wong, S. L., Zhang, L. V., Tong, A. H., Li, Z., Goldberg, D. S., King, O. D., Lesage, G., Vidal, M., Andrews, B., Bussey, H., Boone, C., and Roth, F. P. (2004) Combining biological networks to predict genetic interactions. *Proc Natl Acad Sci USA 101*, 15682–7.
50. von Mering, C., Jensen, L. J., Kuhn, M., Chaffron, S., Doerks, T., Kruger, B., Snel, B., and Bork, P. (2007) STRING 7 – recent developments in the integration and prediction of protein interactions. *Nucleic Acids Res 35*, D358–62.

Chapter 8

Advanced Methods for High-Throughput Microscopy Screening of Genetically Modified Yeast Libraries

Yifat Cohen and Maya Schuldiner

Abstract

High-throughput methodologies have created new opportunities for studying biological phenomena in an unbiased manner. Using automated cell manipulations and microscopy platforms, it is now possible to easily screen entire genomes for genes that affect any cellular process that can be visualized. The onset of these methodologies promises that the near future will bring with it a more comprehensive and richly integrated understanding of complex and dynamic cellular structures and processes. In this review, we describe how to couple systematic genetic tools in the budding yeast *Saccharomyces cerevisiae* alongside robotic visualization systems to attack biological questions. The combination of high-throughput microscopy screens with the powerful, yet simple, yeast model system for studying the eukaryotic cell should pioneer new knowledge in all areas of cell biology.

Key words: Yeast, *Saccharomyces cerevisiae*, Systematic libraries, SGA, Genetic screen, High-throughput, Automated microscopy, Fluorescent marker, Cellular phenotypes

1. Introduction

The field of cell biology was established following the invention of the microscope and the publication of the book "Micrographia" by Robert Hooke in 1665. As microscopes became more advanced, they enabled descriptive works that uncovered constituents of cells and the basic processes that occur within them, such as the cell cycle, division, and death. Two examples for such important observations are the discovery of the Golgi apparatus through microscopic studies of Purkinje cells by Camillo Golgi in 1898 (1, 2) and the first description of chromosomes and their division by Walther Flemming in 1882 (3).

Gerard Cagney and Andrew Emili (eds.), *Network Biology: Methods and Applications*, Methods in Molecular Biology, vol. 781, DOI 10.1007/978-1-61779-276-2_8, © Springer Science+Business Media, LLC 2011

Over the years, modern cell biology emerged with the attempts to reach a mechanistic understanding of cellular processes using a plethora of molecular, genetic, and biochemical tools. In recent years, such studies have been aided by the emergence of systematic and automated tools for studying the secrets of the cell.

One powerful, yet simple, model system for studying the eukaryotic cell is the budding yeast, *Saccharomyces cerevisiae*. The ease of genetic manipulation in *S. cerevisiae* has enabled the creation of a large number of diverse systematic libraries of either mutated, tagged, or alternatively expressed genes that make up the majority, if not the entire, yeast genome (Table 1). Using sophisticated genetic tricks (4–6), it is now possible to easily create custom-made libraries for monitoring the effect of each gene on a phenomenon of interest. Finally, the combination of these "tailor-made" libraries with the development of automated microscopy and analysis platforms enables high-throughput visual assays to get insights into a broad range of basic processes in cell biology. Such questions range from the fate of a single protein (either its level or localization) through questions of structure and homeostasis of entire organelles and finally questions that address the variability that is found at the population level.

In this chapter, we describe how to easily use these genetic and robotic tools in yeast (Fig. 1) to attack, in an unbiased manner, questions in cell biology that were previously hard to tackle.

2. Materials

2.1. Defining a Biological Question

The updated information on the ever-growing availability of yeast libraries is available at: http://www.openbiosystems.com/GeneExpression/Yeast/.

2.2. Designing a Marker for Screening

2.2.1. Chemical Probes

Chemical probes enable us to detect particular components in living cells, with exquisite sensitivity and selectivity. Today, a large variety of fluorescent chemical probes are available for staining specific subcellular structures or following intracellular conditions, such as viability, redox status, nutrient levels, etc. (http://www.invitrogen.com/site/us/en/home/brands/Molecular-Probes.html).

2.2.2. Genetic Probes

Over the past years, an immense array of genetically encoded fluorescent proteins displaying a wide spectrum of emission and excitation peaks has been described (7, 8). Such a wide selection of fluorophores now enables simultaneous imaging of multiple fusion proteins in living cells. By the use of homologous recombination, these can easily be introduced into yeast to form either fusion proteins or to provide information on activation of promoters.

Table 1
Summary of the major commercially available systematic yeast libraries (in S288C background) important for microscopic screening procedures

Library name and reference	Description	Selection marker
Yeast Knockout Library (27)	This collection includes complete knockouts of the entire open reading frame (from start to stop codon) of all the nonessential yeast genes. Each deletion is also bar-coded. Available in MATa, MATα, and heterozygous diploids	G418 resistance
Yeast DAmP Library (28)	This collection includes hypomorphic alleles for ~82% of essential yeast genes using the decreased abundance by mRNA perturbation (DAmP) method. In short, the 3′ UTR of these genes is disrupted by insertion of an antibiotic resistance marker. The result is a full-length endogenous protein under its natural transcriptional regulation, but at reduced levels	MATa with G418 resistance MATα with nourseothricin resistance
Yeast TS Library (29)	This collection includes temperature-sensitive (TS) alleles of 250 essential yeast genes. Available as MATa	URA+
Yeast Tet-off Promoter Library (30)	This collection includes promoter shutoff strains for ~70% of the essential yeast genes. Created by replacement of the native promoter with a repressible tetracycline (Tet-off)-regulated promoter on the background of a strain containing the Tet inducer cassette at the Ura locus. Available as MATa	URA+ for the Tet inducer locus G418 resistance for the promoter locus
Yeast Gal-GST Library (31)	This collection includes high copy number (2 μ) plasmids that encode for ~80% of yeast ORFs under control of the GAL1/10 promoter (which is activated by growth in 2% galactose) and have an N-terminal GST tag that allows immunoprecipitation assays. Available as MATa	Amp for bacterial resistance URA+ for yeast
Molecular Bar-coded Yeast (MoBY) ORF Library (57)	This collection includes ~80% of yeast ORFs under control of their native promoters and terminator that were cloned into a low copy number (CEN/ARS) plasmid along with two unique DNA bar codes. These plasmids are maintained in bacterial cells	Kan for bacterial resistance URA+ for yeast
Yeast Genomic Tiling Library (58)	This collection contains the yeast genome in 1,588 high copy number (2 μ) plasmids representing a virtually complete overlapping clone collection. The genes are expressed untagged from their endogenous promoters. These plasmids are maintained in bacterial cells	Kan for bacterial resistance LEU+ for yeast

(continued)

Table 1 (continued)

Library name and reference	Description	Selection marker
Yeast GFP Library (32)	This collection includes ~75% of the yeast ORFs tagged at the carboxy terminus with the coding region of green fluorescent protein (*Aequorea victoria* GFP (S65T)). These ORFs are expressed under the control of their native promoter. Available as MATa	HIS+

During the last years, the variety of fluorophores has been enriched by protein molecules that have the ability to act as biological sensors of cellular conditions. This is done by changing the spectral properties of the fluorophores under defined conditions (9). To date, fluorescent-based biosensors have been created to measure cellular redox state, Ca^{2+} concentration, pH, and various trace elements (10–14). By fusing these molecules to specific proteins, it is also possible to target them to specific organelles, thus providing compartment-specific measurements.

2.2.3. Plasmids and Strains

Note: If during creation of a query strain multiple events of targeting are required, it is recommended to use different plasmid families to reduce the possibility of nonspecific recombination between cassettes.

2.3. Creating a Library for Screening

Manipulation of libraries can be performed either manually or robotically. Below are tools that allow either method:

2.3.1. Manual Replication Tools

In order to manually maintain, manipulate, and freeze down libraries, it is best to use a 96, 384, or 1,536 floating pin E-clip style manual replicator alongside grids that enable accurate pinning (all replicators and accessories can be purchased from V & P Scientific, Inc (http://www.vp-scientific.com)).

2.3.2. Robotic Replication Tools

There are a number of robotic systems available that can be programmed to manipulate yeast cell arrays, such as:

Singer RoTor bench top robot (Singer Instruments, http://www.singerinst.co.uk).

VersArray colony arrayer system (BioRad Laboratories, http://www.bio-rad.com).

QBot, QPixXT, MegaPix (Genetix, http://www.genetix.co.uk).

The RoTor is unique in that it uses disposable plastic replicator pads, whereas most other robotic systems use metal pinning tools,

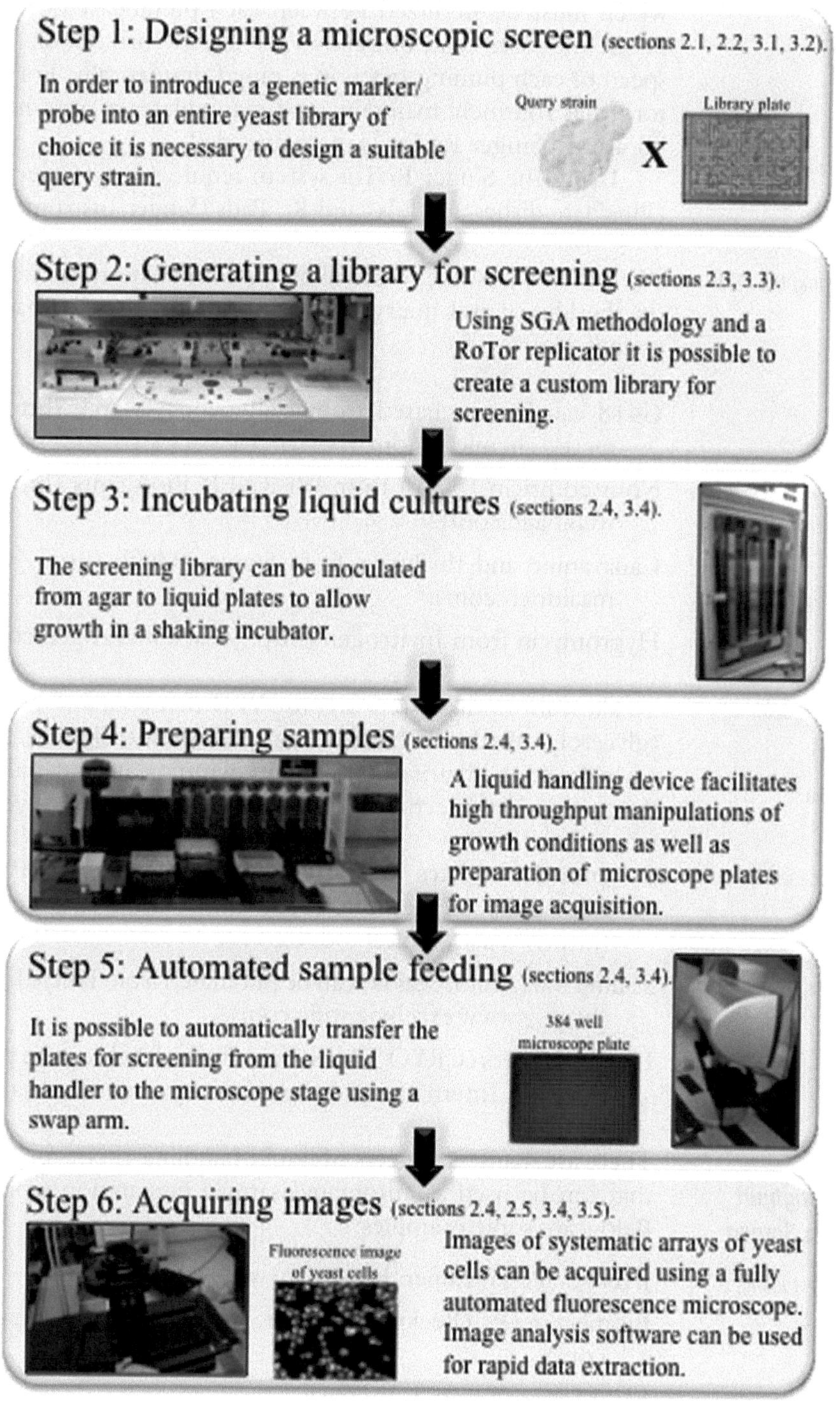

Fig. 1. Schematic representation of the steps required to set up and perform a whole genome microscopic screen in yeast.

which must be sterilized between each pinning step. Disposable plastic increases cost, but allows for higher throughput, since the speed of each pinning reaction is raised dramatically. For a laboratory that routinely maintains and manipulates a large number of libraries, a Singer RoTor is recommended.

Using the Singer RoTor system requires use of specific plates (PlusPlate dishes and lids) and Re-Pads (Singer Instruments).

2.3.3. Performing the SGA

Note: It is impossible to list all media combinations as they depend on the library and query strains used. However, for basic media recipes, see (6, 15).

Materials for Selections

G418 can be purchased from Calbiochem-Merck (http://www.merck-chemicals.com).

Nourseothricin (NAT) from WERNER BioAgents (http://www.webioage.com).

Canavanine and thialysine from Sigma-Aldrich (http://www.sigmaaldrich.com).

Hygromycin from Invitrogen (http://www.invitrogen.com).

2.3.4. Handling Yeast Libraries

Freezing Media

To allow for optimal freezing, use YPD that contains 15% glycerol (glycerol, Anhydrous, J.T. Baker, http://www.jtbaker.nl). However, when freezing libraries that contain plasmids, it is advisable to use the appropriate selection media (which contains 15% glycerol).

Plates and Accessories

Polypropylene plates (PP-MICROPLATE) and lids (PS Lid for Microplate Sterile) can be purchased from Greiner bio-one (http://www.greinerbioone.com).

Sealing foil (AlumaSeal II) can be purchased from EXCEL Scientific (http://www.excelscientific.com).

Library stickers (CRYO TAGS, deep freeze label) can be purchased from GA International (http://www.ga-international.com).

2.4. Running a High-Throughput Microscopic Screen

There are many producers of liquid-handling robots in the market that can be used as automated sample preparation workstations. Below are some examples:

2.4.1. Liquid Handling Platforms

JANUS (PerkinElmer, http://www.perkinelmer.com).

Biomek FXP (BeckmanCoulter, http://www.beckmancoulter.com).

EVO Series (Tecan, http://www.tecan.com).

EpMotion (Eppendorf, http://www.eppendorf.com).

Biorobot Universal System (Qiagen, http://www.qiagen.com).

2.4.2. Imaging Platforms

There are many producers of automated microscopes and high-content screening (HCS) systems in the market that can be used for imaging. Below are some examples:

ScanR (Olympus, http://www.microscopy.olympus.eu).

Cell/Axio Observer (Zeiss, http://www.zeiss.com).

ECLIPSE Ti (Nikon, http://www.nikon.com).

Opera (PerkinElmer, http://www.perkinelmer.com).

Cellomics ArrayScan (Thermo, http://www.thermoscientific.com).

IN Cell Analyzer (GE Healtcare, http://www.gelifesciences.com).

2.4.3. Sample Preparation for the High-Content Microscopy Screen

Growth plates (Polystyrene, PS-MICROPLATE) can be purchased from Greiner bio-one.

Microscope plates (Glass Bottom MicroWell Plate) can be purchased from matrical bioscience (http://www.matrical.com).

Concanavalin A (conA) (C2010-25MG) can be purchased from Sigma-Aldrich.

Paraformaldehyde (PFA) 16% solution can be purchased from Electron Microscopy Sciences (http://www.emsdiasum.com).

KH_2PO_4 – Potassium Phosphate, Monobasic, crystal can be purchased from J.T. Baker.

2.4.4. Image Acquisition

The starting point for any high-throughput microscopic screen is the collection of images for all the strains in the chosen libraries. These images are later used for either manual or automated analysis. Automated microscopes or HCS systems must be purchased/assembled so that they can acquire images with sufficient contrast, resolution, and signal-to-noise ratio to allow image algorithms to extract features of interest. Therefore, when purchasing such a system, focus on several parameters that can significantly impact the quality of the acquired data (16, 17).

1. *Speed*: Speed of the apparatus is important since when scanning large numbers of strains, even small differences in speed may result in large differences in throughput. In addition, for live cell imaging, the time difference between acquisitions of two different fluorophores can cause colocalized structures to move from each other. Parameters that affect the speed are the types of focus that the system employs (hardware or software – see below); motorized items, such as stage, filter wheels, or shutters; communication between the camera and the computer; and camera sensitivity that influences acquisition time.

   Once the system has been purchased, try to maximize the speed for each screen by reducing the number of channels captured, reducing exposure time, and limiting the number of images that are acquired per well (using 60× magnification may allow acquisition of hundreds of cells per single field. Depending on the type of later analysis (qualitative or quantitative), it may be sufficient to acquire even one position within the well). Another option for increasing the system speed is to use only the software autofocus or the hardware autofocus every tenth well or at the beginning of each row. The better the quality of the bottom of the plate, the less frequently software autofocus must be utilized.

2. *Type of focus*: Most microscopic systems in the market have two integrated focusing options: laser-based (hardware) or image-based (software). In laser-based focusing, the position of a reference point at the interface between the sample and the plate is measured by an external light source. Such autofocus is very rapid and can increase the throughput; however, changes in the thickness of the glass can highly impact focusing making any single reference point insufficient. Image-based focusing relies on acquisition of image stacks followed by selection of the plane for which parameters, such as contrast, resolution, and intensity, decrease on either side monotonically and symmetrically. This process enhances image quality, but increases the scan time dramatically and, if done on a fluorescent channel, can cause bleaching of the respective fluorophore.

3. *Magnification*: To distinguish intracellular compartments and events in yeast cells, a minimum of 60× magnification is required. Most scanning microscopes cannot take images with magnification higher than 60×. This is because most lenses with magnification higher than 60× require a non-air interface to function and this can be hard to achieve with automated screening throughput. To this end, it is most likely that your screen will utilize the 60× lens, so make sure to buy an optimal lens with high numerical aperture (NA).

   Increasing the NA reduces the working distance to the range of 0.17–0.22 mm, making it necessary to work with thin-bottomed plates. Although plastic plates with 0.17 mm thickness could potentially be used, the ones that we tested using our ScanR system could not give images with the same resolution as glass bottom plates.

4. *Compatibility with additional robotic components*: For increasing the throughput, it is recommended to integrate the automated microscopic system with other robotic components, such as liquid handler and automated incubator. This allows automated preparation of the microscopic plates, including incubation, and automated loading of the plates onto the microscopic system.

5. *Camera quality*: In order to acquire high-resolution images while capturing small cellular details, a high-end camera is necessary. Most automated screening microscopes to date already have high-quality, cooled CCD cameras for image acquisition. However, several additional parameters should be considered, such as number of pixels, size of pixels, and quantum efficiency. The number of pixels determines the size of your field, whereas image resolution is determined by the pixel size. The smaller the pixel size, the higher the resolution with optical resolution setting the limit. Thus, pixel size in the range of 6–9 μm is the optimum for 60× magnification. Increasing quantum efficiency increases sensitivity and optimizes the signal-to-noise ratio.

2.5. Analyzing the Data

A critical component of any automated imaging system is the software that serves to control image capture and analysis. Since high-throughput fluorescence microscopy quickly generates data at a much higher rate than can be annotated and evaluated manually, it is advisable to have image analysis tools which couple acquisition with automatic, computerized image processing methods to score the assay and annotate the data in a truly quantitative and unbiased fashion (18).

Regardless of which analysis tool you plan to use, the challenges posed by image analysis are at least as difficult as those faced in image capture. There are multiple levels of image processing: at the pixel level, the object level, the semantic concept level, and the pattern and knowledge level. Images comprise pixels, but it is generally the objects (e.g., cells) represented by the pixels that are of interest. It is, therefore, going from pixel-based representation of data to an object-based representation of data that is the principal challenge in analyzing images, especially in yeast which are small, have buds, and tend to clump together. However, once yeast cells are recognized as objects, it is straightforward for a computer to quantitate their properties, such as shape, size, and fluorescence. The image analysis process involves four basic steps (16, 19).

1. *Preprocessing*: Provides background subtraction and flat-field correction to filter out noise and normalize intensities. Flat-field correction removes artifacts that are caused by small changes in the relative light sensitivity of each pixel in the array of the detector or by nonuniformities in the focal plane. This is performed by compensating computationally for different gains and dark currents in the image.
2. *Segmentation*: Identifies all cells or objects in each image. Depending on the cellular assay, the object of interest could be the whole cell or subcellular structures that are labeled with an appropriate fluorescence marker or captured through the bright-field image. The difficulty of performing segmentation and the algorithms that should be used varies considerably

depending on the cells/objects being segmented and how close they are to each other.

3. *Classification*: Places objects identified as regions of interest into subpopulations of the relevant phenotypic categories. A classification decision can be simple, for example, if it is based on the size or the fluorescence intensity of the object, but it can also be more complex if many parameters or only subtle differences have to be recognized.
4. *Morphological measurements*: Provide quantitative data on important features, enabling production of a unique morphometric profile for the cells being examined. The basic approach is to extract a long list of numerical parameters that describe the shape, texture, and other derived characteristics, such as the pixel intensity statistics, of each object.

Most automated microscope systems come with a commercial software package that allows sophisticated image analysis. Such systems typically provide a number of algorithms to address specific biological applications – it is important to note the range of options provided in each package. It is important to ascertain that the automated image analysis software is intuitive and easy to navigate so that setting and validating the parameters for image segmentation and feature calculations are user-friendly even to biologists with no background in programming. However, for groups that do have expertise in programming, there are packages that allow programming within the main files to allow customization of analysis and increase the options for working with the program. As users become more experienced or require more advanced analysis methods, it is important to ensure that they can easily export files out into more powerful image analysis software packages (16, 17).

3. Methods

3.1. Defining a Biological Question

The availability of fluorescent markers and dyes, coupled with automated microscopy platforms, now allows us to ask an enormous variety of biological questions in a systematic manner (Fig. 1). Literally, anything that can be visualized can be used as a phenotype for genetic screens. Since a large number of systematic libraries exist in yeast, fluorescent markers can be coupled with libraries to create "custom-made" libraries for screening. By and large, three options for combining fluorescent markers with yeast libraries exist.

1. *Combining a fluorescent marker into a mutant library*: Such a process could aid in defining the functional requirements for subcellular processes. More specifically, in order to delineate

the proteins required (directly or indirectly) for the occurrence of a cellular process of choice, it is necessary to create a marker for that process. Processes could be the localization of a single protein (20, 21), physical interactions between proteins (22), the structure of an entire organelle (23, 24), the activation of a single promoter (25), or overall changes in cellular conditions (26). To study the proteins that are required for such cellular processes, these markers should be studied on the background of systematic libraries of mutants, such as the deletion library for nonessential proteins (27), several different libraries allowing downregulation or destabilization of essential proteins (such as the DAmP library (28); temperature-sensitive alleles (29); or the Tet-off collection (30)), as well as the protein overexpression library (31). By screening for the event of choice on the background of all library mutants, we can identify strains displaying an altered phenotype, such as a difference in the expression or localization of our marker, uncovering its biogenesis requirements or regulatory pathways.

2. *Combining a fluorescent marker into a fluorescent library*: This could aid in uncovering novel components of compartments. Although a systematic GFP library has already been created several years ago (32), we still do not know the subcellular localization of hundreds of proteins. These include proteins that could not be tagged or that are not expressed during mid-logarithmic growth in SD medium (the condition in which the GFP library was originally imaged) as well as strains whose tagged proteins exhibit a punctate pattern that could not be colocalized with known compartments of that time. As new subcellular compartments are being discovered (such as P-bodies (33–37); JUNK and IPOD (38); and autophagosomes (39)), their resident proteins can be found by screening the GFP library for colocalization with a query protein.

   For example, to uncover all resident proteins of a new punctate subcellular compartment, it is possible to use a query strain that contains a known marker for this compartment fused to RFP and photograph it in the background of the entire GFP library. A fusion protein with GFP that would colocalize with the RFP marker of the organelle reveals a new protein in this compartment.

3. *Combining a fluorescent marker and a mutation into a mutant library*: This would serve the purpose of understanding the functional relationships and hierarchy of processes within a cell. Specifically, once we have uncovered the basic components of processes, we can also find their functional hierarchy by combining mutations in a single cell (double-mutant analysis) and assaying for their genetic interactions. For this type of question, both a mutation in a gene of choice (such as

a deletion of a specific protein) and a marker protein for our process are studied on the background of systematic mutant libraries. By visualizing these libraries for the functional outcome of this combination, we can identify the cases, where genes are suppressing/buffering or aggravating the phenotype. Such observations would teach us about the structure of the cellular pathways controlling this process (15, 40–47).

Note: Since yeast is infamous for its ability to grow under an endless variety of conditions, it is advisable to perform any such screens under a variety of environments and growth conditions, as the signal may be altered.

3.2. Designing a Marker for Screening

Regardless of which of the three above options you choose to undertake, you first need to consider a way to visualize your process of choice. To do this for automated microscopy, in most cases you would need a fluorescent label (although it is also possible to screen for effects that can be seen by regular bright-field images, such as cell size, shape, cell cycle progress, etc.). There are two ways to create a fluorescent label for your process of choice.

1. *Using a chemical marker/probe*: If a chemical dye exists (Subheading 2.2) to report on the process of choice, then you can decide which library you would like to screen (Table 1) and no genetic manipulations are necessary (if this is the case, skip to Subheading 3.4). Such an approach would save you time; however, it is limited by the number of potential chemical markers and probes available and the high cost of reagents for use on entire libraries.
2. *Using a genetic marker/probe*: The ease of genetic manipulations in yeast alongside the vast array of genetically encoded fluorophores (Table 2 and Subheading 2.2) allows creation of genetic markers for literally any process of choice. For example, by fusing a fluorophore with a protein of choice, it is possible to follow its levels and localization in a living cell under various conditions. By fusing a fluorophore to a compartment-specific protein or by attaching a compartment-localization signal to the fluorophore, any cellular compartment can be labeled. By placing the fluorophore downstream of a transcription regulatory element, it is easy to measure promoter/enhancer activities. Using condition-sensitive fluorophores, readout of cellular/organellar conditions can be achieved. Finally, by using split fluorophores or FRET, it is possible to measure proximity/physical interactions between proteins.

An enormous breakthrough in our ability to use these fluorophores to study cell biological questions in a systematic manner came through the invention of Synthetic Genetic Array (SGA) technology (4–6). This technology allows integration of the marker of choice into only a single query strain (with specific genetic loci

Table 2
Examples of plasmid families for genetically manipulating query strains

Plasmid family and reference	Description
pCG (59)	These plasmids contain auxotrophic selectable markers whose source is from *Candida glabrata* (*Cg*). They can be used for gene deletions in all strains, but are especially useful for manipulating strains that do not have complete deletion of the auxotrophic markers. Include KO plasmids with HIS, LEU, TRP, MET, and URA markers
pFA6 (60, 61)	These plasmids allow for a large number of different genetic manipulations all on the same plasmid backbones: gene deletion, promoter swap, C- or N-terminal protein tagging, and partial N- or C-terminal deletions. The modular nature of these types of plasmids allows efficient and economical use of a small number of PCR primers for a wide variety of gene manipulations. Originally created for use with either antibiotic markers, NAT, Hygro, G418, or auxotrophic markers, HIS, TRP, URA
pYM (62)	These plasmids contain a broad variety of deletion cassettes, C-terminal epitope tags, nine different promoter substitutions (of variable expression levels), and N-terminal tags. The modular nature of these types of plasmids allows efficient and economical use of a small number of PCR primers for a wide variety of gene manipulations. Originally created for use with either antibiotic markers, NAT, Hygro, G418, or auxotrophic markers, HIS, TRP

outlined below), which is easily crossed into any yeast library by the use of pinning tools. This simple method, therefore, allows rapid insertion of any marker into an entire library of choice creating "custom-made" libraries for screening.

3.2.1. The SGA Query Strain

The SGA procedure (outlined in Subheading 3.3) is the method of choice to insert any marker quickly and systematically into entire yeast libraries. In contrast to mass transformation, it does not require liquid handling or affirmation of all strains at the end of the procedure and is fast and robust. In short, this method works by allowing the mating of the query strain to mutant libraries, generating diploids, inducing meiosis to retrieve haploid cells, and finally selecting for only haploid strains that contain the original genetic determinants from both the query strain and the library of choice. To enable this process, an SGA query strain must contain the following genetic features:

Cassette for selection of haploids of a specific mating type: Since mating is an extremely efficient process in yeast, the haploids that are generated at the end of the sporulation process could mate with their "siblings," thereby generating diploids that are heterozygote for both mutation alleles (of the query strain and the library strains). This would lead to false negatives, as the genetic marker would be imaged on the background of a functional genome. To overcome this problem, the selection of a single mating type following sporulation is required. Therefore, the query strain contains a haploid mating type-specific promoter linked to a selectable marker. Most strains contain a MATa-specific promoter (such as Ste2 or Mfa1) conjugated to the HIS selectable marker and a MATα-specific promoter (such as Ste3) driving the LEU marker.

Loci for selection against diploids: The common library strain (S288C (48)) undergoes sporulation at very low rates (usually, up to 10% of diploids in nitrogen starvation). This leads to very poor chances of finding the desired haploid harboring all the mutations of interest simply by selecting for the markers that are also present in the diploid strain. Moreover, because mitotic recombination can occur between homologous chromosomes in MATa/α diploids, a crossover event between the MAT locus and the centromere on chromosome III can result in MATa/a or MATα/α diploids (25, 49). These types of diploids actually behave like the haploids that were selected for expressing the selectable marker under a mating type-specific promoter, and therefore survive the haploid selection step. To avoid this complication, the SGA strains have two selection cassettes for removal of the diploids following the sporulation step: *can1Δ* and *lyp1Δ*. The CAN1 gene encodes an arginine permease that is permissive also to the toxic analog canavanine. Similarly, the LYP1 gene encodes a lysine permease that enables entry of its toxic analog thialysine (also called S-AEC). Since all library strains are CAN1+ and LYP1+, the heterozygous diploids formed by crossing them against the query strain are all sensitive to these toxic analogs. However, the haploid spores that have the deletions of these two permeases (*can1Δ* and *lyp1Δ*) survive on selection plates that include canavanine and thialysine.

To enable rapid crossing of markers into entire yeast libraries, a query strain should be created. When starting to plan your query strain, consider the following issues:

1. *Suitable genetic background for SGA* (Table 3): Since the first step of the SGA method is mating between a query strain to the library of interest, make sure to choose a strain that has the opposite mating type from the intended library and the right genetic background that is suitable to the specific selections that are going to be used (Tables 1 and 3). For example, a query strain with only a MATa promoter driving LEU (Table 3) should be used for crossing against a library that already harbors a HIS selectable marker (such as the GFP library (Table 1)).
2. *The fluorophore of choice*: A variety of genetically encoded fluorophores are now available (Subheading 2.2). When

Table 3
Examples of existing query strains (in S288C background) for systematically introducing markers into yeast libraries (4–6)

Genotype	Best used to cross against
his3Δ1, leu2Δ0, LYS2+, met15Δ0, ura3Δ0, can1Δ::STE2pr-spHIS5 lyp1Δ::STE3pr-LEU2 **MATα**	This is the strain of choice for most SGA crosses. It contains two selection alleles against the diploids (Δcan, Δlyp) as well as two options for selecting haploids: a MATa-specific promoter (STE2pr) driving the *Schizosaccharomyces pombe* HIS5 (which is the functional homologue of the *S. cerevisiae* HIS3) and a MATα promoter (STE3pr) driving LEU2
his3Δ1, leu2Δ0, LYS2+, met15Δ0, ura3Δ0, can1Δ::MFA1pr-LEU2 lyp1Δ **MATα**	This strain should be used to cross against libraries, which harbor the HIS selectable marker (such as the GFP library) since there is no internal HIS selection
his3Δ1, leu2Δ0, LYS2+, met15Δ0, ura3Δ0, cyh2, can1Δ::STE2pr-spHIS5 lyp1Δ::STE3pr-LEU2 **MATa**	This MATa strain can be used to cross against MATα libraries

choosing a fluorophore, it is necessary to consider its basic properties: Is it a monomer or a dimer? How quickly does it fold? Is it very stable? Is its codon optimized for the organism of choice? Does it aggregate at high expression levels? etc. (7, 8, 50). If the screen requires two or more different fluorescent markers, the wavelength overlap of both excitation and emission between these fluorophores must be taken into consideration (Subheading 3.4).

3. *The selection marker*: When transforming the genetic probe into the query strain (whether using a plasmid-based expression system or integration into the genome), a selectable marker must be chosen (Table 2). This selectable marker should be chosen so as not to overlap with the markers for mating type-specific selection (HIS+ for selecting MATa cells or LEU+ for selecting MATα cells) nor with the markers of the destination library. For SGA techniques, the use of positive selection markers (such as Kan^r, Nat^r, or $Hygro^r$) is advisable as they give better selection specificity during the manipulation of the strains and assist in preventing contamination of library plates.
4. *The second marker for an internal control*: The most accurate control is a second fluorophore that reports on the baseline in each cell. This helps assessing variability between cells, wells, plates, and days.
5. *Expression levels of the marker*: It is important that the protein that is chosen as a marker has a high expression level for easy

visualization. This requires either using a highly expressed protein from its endogenous promoter tagged at its C-terminus, N-terminally tagging a protein while changing its promoter to a constitutive and highly expressing one, or utilizing a high copy number plasmid-based expression system (Table 2). High copy number plasmids have variable expression in each cell, so when possible it is preferable not to use them.

6. *The function of tagged proteins*: To ensure that a protein has not lost its function following tagging, compare the tagged strain to a deletion/mutant strain under conditions, where the mutant exhibits a phenotype. If the tag does damage the functionality of the protein and if this is detrimental to the screen, you can insert the marker as a second copy either on a plasmid or at a nonendogenous locus (such as the URA or HO locus), thereby retaining one functional copy.

3.3. Creating a Library for Screening

Once you have created the query strain, you must use SGA technology to introduce your marker of choice into one of the available yeast libraries. Many protocols for performing SGA have been published (for example, (4–6, 15)). To avoid redundancy, we give only some of the basic procedures required to start working with this methodology.

In order to manipulate libraries, it is possible to work with either manual or robotic tools (Subheading 2.3).

3.3.1. Manual Replication Tools

If working with manual tools, it is essential to sterilize them efficiently between each pinning step, as pinning from dense colonies on agar plates causes many cells to remain on the pins, increasing the risk of cross-contamination between plates. Below is a simple, yet efficient, protocol to enable sterilization.

Sterilizing

Wash the pinner sequentially in the following solutions:

Tray 1: Sterile DDW – 1 min or until most of the yeast cells have dropped from the pins.

Tray 2: 10% bleach – 30 s

Trays 3–4: Sterile DDW – rinse

Tray 5: 70% ethanol – 10 s

Tray 6: 95% ethanol – 5 s

Allow excess ethanol to drip off the pins and then place the pinner carefully into the flame of a Bunsen burner (torch) and let it cool before use.

Note: To ensure that the pins are cleaned properly and to avoid contamination in the wash procedure, the volume of wash liquids in the cleaning reservoirs should rise in small increments along the wash steps. However, make sure that all of the pinheads are completely immersed in each solution.

Note: For containers you can use lids of tip box.

Note: The water in tray 1 should be replaced frequently when large amounts of yeast cells can be seen at the bottom of the basin.

Note: If you would like to control for cross-contaminations, press the pinners onto an agar YPD plate following the last daily round of sterilization. If no colonies grow after 2 days, then your sterilization technique is working correctly.

3.3.2. Robotic Replication Tools

Although manual pinning is possible, it is most time-efficient and accurate to use robotic pinning devices (Subheading 2.3). We use the RoToR robotic replicator (Singer Instruments) that utilizes disposable pins. However, all steps are identical for manual replicators or other pinner robots, except for the need to sterilize them between steps. Furthermore, all of the following steps are explained in 1,536 format, but 384 and 96 formats are effective as well.

3.3.3. Performing an SGA (Table 4)

The first steps of the SGA are to prepare a query strain and libraries in a fresh copy of 1,536 format to boost the effectivity of consequent mating.

Arraying the Query Strain and Mutant Library

Day 1: Grow an overnight (ON) culture of the query strain in 20 ml of the appropriate selection medium at 30°C.

Day 2:

1. Pour the query strain culture into a sterile container and use a replicator in order to create a 1,536 array of this culture onto an appropriate selection agar plate. Incubate the plate at 30°C for 1 day.
2. Replicate the relevant libraries in 1,536 format to fresh agar plates with the appropriate selections. Incubate the plate at 30°C for 1 day.

Mating the Query Strain with the Library Array

The purpose of this step is to create heterozygous diploid cells containing the query strain genotype in addition to the library genotype. It is extremely important to mate freshly plated cells. Cells that are incubated for more than 2 days or have been kept at 4°C will mate at a lower efficiency.

Day 3:

1. Replicate the 1,536 array of the query strain onto agar YPD plates.
2. Replicate the library array on top of the query cells.
3. Incubate the mating plates at room temperature (RT) for 1 day.

Note: Following this step, all 1,536 colonies should grow, as there has been no selection pressure.

Table 4
An outline of the SGA routine

Day	Step	Media	Time	Temperature
1	Query strain liquid starter	Liquid media with the specific selection. For example, liquid YPD + NAT for query strain that has nourseothricin resistance	ON	30°C
2	Query strain array	Agar plates with the specific selection. For example, YPD + NAT plates for query strain that has nourseothricin resistance	1 day	30°C
2	Library array	Agar plates with the specific selection. For example, YPD + G418 plates for the deletion library	1 day	30°C
3	Mating	YPD plates	1 day	RT
4	Diploid selection	Plates that contain specific selection to both the query strain markers and the library markers. For example, YPD + G418 + NAT is used for diploid selection of mating between the deletion library and a query strain with nourseothricin resistance	2 days	30°C
5	Sporulation	Nitrogen starvation plates	5 days	22°C/RT
10	Haploid selection	SD-His-Arg-Lys + canavanine + thialysine plates to select for MATa haploids or SD-Leu-Arg-Lys + canavanine + thialysine plates to select for MATα haploids	2 days	30°C
12	Final mutant selection	SD-His/-Leu and -Arg-Lys + canavanine + thialysine + selecting for all markers from both the query strain and the library	2 days	30°C

Diploid Selection

Although mating occurs at high efficiency, haploid cells remain on the YPD mating plates. The goal of this step is, therefore, to select for only the diploid cells that now contain the selectable markers of the query strain in addition to one of the relevant libraries.

Day 4: Replicate the cells from the mating plates onto diploid selection plates that contain the specific selections. For example, if the query strain had a mutation linked to a URA selection marker and it is crossed against the GFP library that has the HIS selection marker, then plate the cells on SD-HIS-URA medium for diploid selection. Incubate the diploid selection plates at 30°C for 2 days.

Note: G418 (for selecting KAN resistance) or hygromycin do not work well on regular SD medium due to the low pH; therefore, if such selection is required alongside an auxotrophic marker, then use synthetic media prepared with monosodium glutamic acid (SD (MSG)) instead of ammonium sulfate (6, 15) to alleviate the acidity of the medium.

Note: Following the diploid selection step, the plates should regain the pattern of the library plate as any "query" colony that had no library strains to mate with would die in this selection. If diploid selection has not eliminated all haploids (as can be seen by the appearance of colonies in places, where no colonies existed in the original library), then it is necessary to go through one additional round of diploid selection. Note that when using two auxotrophic markers, two diploid selection steps should always be performed.

Sporulation

Sporulation is required in order to produce haploid progeny from the selected diploids. Since sporulation is inefficient in S288C cells, it is essential to transfer a large number of diploid cells to the sporulation plates to enlarge the number of spores that are achieved at the end of this process. To increase the percent of sporulation, transfer fresh diploid cells to the sporulation plate (grown for no more than 2 days).

Day 5: Replicate the diploid cells to sporulation plates (nitrogen starvation plates). To transfer a large number of cells, use the "source mix" and "target mix" option in the RoTor or replicate the colonies three times (back and forth) when using manual pinning. Incubate the sporulation plates at 22°C (RT is usually OK) for 5 days.

Note: Make sure to keep the plates humidified and well-aerated. This can be done, for example, by placing them in a cardboard box with a trough of DDW in it. (DO NOT wrap in saran wrap so as to allow maximal airflow.)

Haploid Selection

Due to low sporulation levels, this step eliminates the diploid cells that did not undergo sporulation and selects for haploid spores of a single mating type. These plates are usually SD-HIS-based, if one wishes to select for MATa cells, or SD-LEU-based, if the MATα mating type is required (however, some SGA query strains have different markers, so always make sure to create plates that are tailored to your specific SGA query strain). In addition, to select against the diploids, plates contain canavanine and thialysine and lack arginine and lysine (to avoid allosteric competition on the transporters).

Day 10: Replicate spores onto SD-His-Arg-Lys + canavanine + thialysine plates (Subheading 2.3) to select for MATa haploids or SD-Leu-Arg-Lys + canavanine + thialysine plates to select for MATα haploids. Incubate the haploid selection plates at 30°C for 2 days.

Note: As for the sporulation step also in this haploid selection step, it is essential to transfer a large number of cells from the sporulation plates. For this purpose, use the "source/target mix" options in the RoTor or replicate the colonies three times (back and forth) when using manual pinning.

Note: If haploid spores have a growth retardation phenotype, they may be outgrown by other spores. In this case, consider adding a selection for the mutant cells at the stage of the haploid selection. However, it is advisable not to add antibiotics to this stage as spores may be sensitive to them while germinating even when they contain the resistance cassette.

Final Mutant Selection

The purpose of this step is to finalize the SGA by selecting for the haploid cells of specified mating type that also contains all markers of choice (from the query and the library).

Day 12: Replicate the haploid cells onto SD-His/-Leu and -Arg-Lys + canavanine + thialysine in addition to selecting for all markers from both the query and the library strains. Incubate the haploid selection plates at 30°C for 2 days.

Note: If performing SGA with the TS library, all incubation steps must be done at RT instead of 30°C.

Note: To ensure that the library is completely free of diploids and haploids of the wrong genotype, it is possible to repeat this step.

Quality Control

Once the SGA has been completed, you can ensure that it was successful by selecting several colonies and manually checking that they are indeed haploids carrying both query and library markers. This can be performed by either imaging the cells if two fluorescent markers were selected for or in combination with PCR if one marker is a deletion. In addition, mating-type assays can be performed to verify that the cells are indeed haploids (51). Finally, if your screen includes a positive control, before screening the full library ascertain the validity of the SGA procedure by ensuring visualization of the phenotype in the control strains.

3.3.4. Handling Yeast Libraries

Once a library has been created or purchased, it is very important to handle it carefully in order to keep it in mint condition for long-term use. The agar "working copy" must be refreshed every month by replicating onto fresh agar plates with the appropriate selection. Libraries that have passed the SGA should be kept on the final mutant selection plates to eliminate the chances of remnant diploids overtaking the culture.

It is highly advisable to freeze down the libraries directly after they have been obtained so as to have a pristine stock for future experiments.

Freezing Libraries

1. Replicate the library destined for freezing onto fresh agar plates with the correct selection.
2. Incubate the plates at 30°C for 1 day or until colonies are of intermediate size.

3. Dispense liquid freezing medium (15% glycerol in YPD) to the preferable format of polypropylene plates (usually, 150 μl for 96-well, 50 μl for 384-well, and 10 μl for 1,536-well plates, but may depend on the plate dimensions). This step can be done by employing a liquid-handling robot or manually by using a multipipette. It is important to label plates carefully and descriptively by either writing on the plate or by using special library stickers (Subheading 2.3).
4. Replicate the library strains from the agar plates directly into the polypropylene plates and thoroughly mix the cells to ensure that yeasts are properly suspended. This step can be done using a handheld pinner or the RoTor. Make sure that the plate pattern can be seen at the bottom of the wells.
5. Immediately seal the plates with sterile sealing foil and secure the seal by placing a plastic lid above it. Do not let the yeast sit at RT in the freezing medium for over an hour. Store the plates in a −80°C freezer.

Thawing Libraries

1. Take out the library plates from the −80°C freezer and incubate them at RT or 30°C until completely thawed.
2. Centrifuge the plates for 1 min at 3,000 RCF. The purpose of this step is to spin down drops remaining on the sealing foil in order to avoid cross contamination while opening the seal.
3. Remove the seal and replicate the library strains from the polypropylene plates onto the appropriate agar plates with a handheld pinner or the RoTor. Because the yeast cells tend to sink, it is very important to mix the culture well in the polypropylene plates (with the pinner/RoTor) in order to transfer enough cells to the target agar plate. If this proves difficult, aspirate some of the top liquid following centrifugation to increase cell density.
4. Incubate the agar plates at 30°C for at least 2 days or until all colonies have grown to a significant size.
5. It is possible to reseal the thawed polypropylene plates immediately after use with sealing foil and a plastic lid above it and restore them in the −80°C freezer. Yeast can be frozen/thawed for up to three times. However, efficiency is reduced in every cycle and therefore it is recommended to have one frozen copy that is completely untouched to eliminate the chances of strain loss.

3.4. Running a High-Throughput Microscopic Screen

Once you have created a library for screening (either by use of a genetic or chemical marker), you can proceed to acquire microscopic images. High-throughput microscopic screens enable exploring cell biological changes on a large scale in response to either genetic or environmental perturbations. An

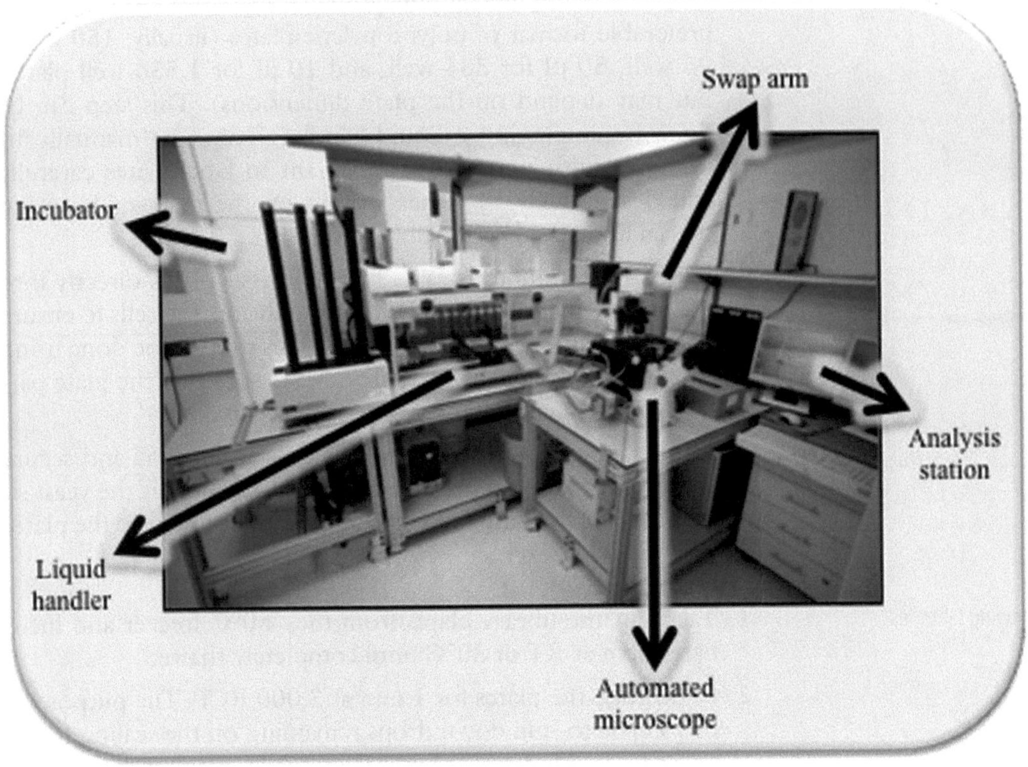

Fig. 2. Example of a setup for high-throughout microscopic screening. Shown is an example of a complete system for microscopic screening. Such a setup can be built in a single room and includes a shaking incubator, a liquid-handing device, a swamp arm, and an automated microscopy platform.

important key for the success of such screens is to use the most suitable automated system. Such a system may include automated sample preparation (liquid handling), automated image acquisition, and automated image analysis for identification of relevant “hits” (16).

Many liquid-handling platforms and automated stage microscopes exist in the market and can be used for such purposes (Subheading 2.4). We have created a fully automated system (Fig. 2) by connecting a shaking incubator (LiCONiC Instruments) through a conveyer belt to a JANUS liquid handler (PerkinElmer), which is connected via a swap robot (Hamilton) to an inverted fluorescent microscopic ScanR system (Olympus) (which brings together both the image acquisition and analysis software). All four stations are controlled via the JANUS Project Manager (JPM) program (PerkinElmer) enabling complete automation of screening procedures allowing us to screen up to 2,000 wells a day. However, good throughput does not require every part of this setup. Below are protocols allowing the screening stages once SGA has been completed.

3.4.1. Sample Preparation for High-Throughput Microscopic Screens

Once the SGA has been completed, the newly formed library is arrayed on agar plates. Before imaging, cells are transferred to liquid media to allow for easy manipulations and for optimal fluorescent signal.

Before transferring your library to liquid format would be a good time to introduce controls. Such controls can be placed within the library in coordinates, where no colonies are growing either before the SGA procedure or onto the final agar or liquid copy. If possible, positive and negative controls (including wild-type cells to measure the baseline or "normal" phenotype) should be placed in EACH plate so that plate-to-plate variability is accounted for. When possible, it is recommended to include an internal control within each cell.

Preparing Liquid Cultures

To ensure optimal image acquisition, prepare liquid cultures in a low-fluorescence medium. Since YPD has very high emission at the wavelengths of fluorophores, such as GFP, it is best to use either SD or the superior specifically formulated screening medium (yeast nitrogen base without riboflavin and folic acid; 5 g/l $(NH_4)_2SO_4$, 1 g/l KH_2 PO_4, 0.5 g/l $MgSO_4$, 0.1 g/l NaCl, 0.1 g/l $CaCl_2$, 0.5 mg/l H_3BO_4, 0.04 mg/l $CuSO_4$, 0.1 mg/l KI, 0.2 mg/l $FeCl_3$, 0.4 mg/l $MnSO_4$, 0.2 mg/l Na_2MoO_4, 0.4 mg/l $ZnSO_4$, 2 μg/l biotin, 0.4 mg/l calcium pantothenate, 2 mg/l inositol, 0.4 mg/l niacin, 0.2 mg/l PABA, 0.4 mg/l pyridoxine HCl, 0.4 mg/l thiamine. This medium has negligible autofluorescence – within 10% of water (52)). The microscopic screen can be performed in 96, 384, or 1,536 microscopic plates.

1. Dispense liquid medium to the preferable format of growth plates (150 μl for 96-well, 50 μl for 384-well, and 10 μl for 1,536-well plates; however, the volume can vary depending upon plate dimensions). This step can be done employing a liquid-handling robot or manually by using a multipipette.
2. Replicate the plates for screening from the agar plates into the growth plates. This step can be done using a handheld pinner or the RoTor. Make sure to transfer a small amount of cells (less than 6 OD/ml) so that cells can divide in the liquid medium.
3. Incubate the plates at 30°C ON until cells are in stationary phase.
4. For acquiring images of cells during logarithmic growth, dilute the ON cultures 1:25 into fresh liquid medium.
5. Incubate the plates at 30°C for 3 h.

Note: If any other manipulations are needed (staining, incubation with specific reagents, etc.), they should be done at this stage before transferring the cells onto the microscopic plates.

Note: Library strains differ in growth rates. To ensure that all strains are in mid-logarithmic growth at the time of image acquisition, it is possible to work with plate readers to read ODs and software for calculating and automatically diluting each strain accordingly (53). However, such hardware and software can be complicated to set up and therefore our method gives similar results with less dependence on such tools. The method is based on the assumption that following transfer of cells to media all strains (even ones with long division times) can reach stationary phase if given enough time. To this end, since yeast cultures slow growth dramatically when reaching a certain OD, culturing them ON "synchronizes" the OD of the cultures. The following day, cultures are diluted 1:25 (for example, a WT cell reaches 6 OD/ml which is diluted into 0.24 OD/ml), which ensures that cultures are sparse enough to enable at least two cell divisions after exit from stationary (assuming no culture grows faster than WT) until the diauxic shift (which occurs at 1 OD/ml in WT cells). This ensures that if grown for less than 6 h before acquisition that most cultures will be logarithmically growing.

3.4.2. Preparing Microscope Plates

A monolayer of well-separated cells is best for automated image acquisition and analysis as it easily allows for border detection during segmentation procedures. To achieve optimal adherence, plates should be coated with concanavalin A (conA) (or a similar adherence substance, such as poly-lysine).

Concanavalin A Coating of Plates and Transfer of Cells

1. Wipe the bottom of each microscopy plate with microscopy paper dipped in 100% ethanol to eliminate dust and ensure optimal autofocusing. From hereafter, refrain from touching the glass bottom during manipulations.
2. Dispense conA (0.25 mg/ml in DDW) to the microscopic plates so that the liquid covers the entire base (50 μl for 96-well, 35 μl for 384-well, and 2 μl for 1,536-well plates, but this can vary depending upon the plate dimensions).
3. Incubate the plates for 15 min at RT.
4. Remove the conA by aspiration and leave to completely dry for 30 min.

   Note: Placing plates in a chemical hood reduces drying time to several minutes.

   Note: Plates must be very dry for optimal adherence.

   Note: The residual conA collected from the plates can be recycled for several additional plates.
5. After the plates have completely dried, dispense premixed logarithmic cell cultures (at OD ~0.5/ml) onto the conA-coated microscope plates (150 μl for 96-well, 50 μl for 384-well, and 10 μl for 1,536-well plates, but this can vary depending upon the plate dimensions).

Note: If the cell density is higher than the above OD (for example, if working with stationary-phase cells), then the culture must first be diluted in a separate growth plate. Do not try to dilute cells in the conA-treated plates as this results in multilayers of cells that inhibit the autofocus and reduce image quality following automated acquisition.

6. Incubate at RT for 15 min.
7. Remove the liquid (that contains the cells that did not adhere to the conA) from the plates.
8. Wash the plates with SD medium. Although once is enough, washing three times is recommended for optimal monolayers.
9. Dispense SD (or low-fluorescence medium) to the plates (150 μl for 96-well, 50 μl for 384-well, and 10 μl for 1,536-well plates, but this can vary depending upon the plate dimensions) and visualize.

Fixing Cells

Scanning a multiwell plate at the microscope takes several hours. Exact time depends on the autofocus system used, number of fluorophores imaged, exposure times, and number of pictures that are taken from each well. For two fluorophores with a 1 second exposure time and three images per well, we find that 2 hours are usually required for a 384-well plate. This wide time frame creates a situation, where there might be differences between the acquisition of the first and last wells. Such differences can be detrimental if you are looking at a specific time point after induction of a process or if your process is cell cycle dependant. Moreover, if the conditions used include stress, then it may be alleviated by the time the last wells are scanned.

The solution for such cases is to fix the cells on the microscopic plates before visualization. The fixation of the cells is performed after cells have adhered to the conA plate since fixation reduces the ability of the cells to adhere to conA.

1. Perform the protocol above to adhere cells onto conA-treated plates until step 8.
2. Following one SD wash of the cells (step 8 above), aspirate the medium from the microscopic plates.
3. Add 4% PFA in 4.25% sucrose solution to the cells.
4. Incubate for 15 min at RT.
5. Aspirate the PFA solution and wash the plates with a solution of 0.1 M KH_2PO_4 in 1.2 M sorbitol (KPi solution).
6. Add KPi solution to the plates so as to cover samples (150 μl for 96-well, 50 μl for 384-well, and 10 μl for 1,536-well plates, but can vary depending upon the plate dimensions) and visualize.

3.4.3. Image Acquisition

In order to ensure optimal visualization of cells on the glass-bottomed microscopy plates, several parameters should be considered.

1. *Autofocus*: Most systems have both hardware and software autofocus, which ensure very high focusing accuracy (Subheading 2.4). Software autofocus can be done on bright field or fluorescent images. Use of bright field can result in shifts in focal plane due to differences in diffraction during various growth phases or of cells of various sizes; however, no photobleaching is incurred and the acquisition speed is higher due to lower exposure times. Use of a fluorescent channel requires adding a fluorophore that is distributed evenly in the cells (such as a cytosolic marker) at a high intensity that is different than the fluorophore for quantification of the phenotype since autofocus causes photobleaching; however, it can often be more accurate and also allows for easy image recognition with automated software.
2. *Offset from autofocus*: When looking at intracellular events, the optimal focus may not be the one achieved by autofocusing on the bright field (which usually catches the stack representing either the top or bottom of the cells as they are the most diffractive) or the cytosolic fluorophore. In these cases, in order to optimize image acquisition, calibrating a specific offset from the autofocus is required.
3. *Emission and excitation overlap*: In screens with several fluorescent markers, ensure that emission and excitation spectrums do not overlap in practice. To check if emission overlap exists, simply acquire images of strains expressing only one of the fluorophores in both channels. If no signal is acquired using the second filter, then there is no overlap in emission. Excitation overlap can also be problematic as it may cause bleaching. Since most fluorophores do have some degree of overlap, it is best to utilize narrow excitation filters.
4. *Exposure time*: Most fusion proteins in yeast require between 0.01- and 2 seconds exposure time to be well-visualized. Longer exposure times on living cells may cause blurry signal. This parameter needs to be optimized for each fluorophore before starting large-scale screening.
5. *Number of pictures per well*: This parameter depends on the screening phenotype and the statistics that is needed for ensuring that this phenotype can be measured. It also depends on the density of the cells in the wells. Higher cell density may inhibit the ability of analysis programs to define the edges of cells; however, low-density fields may require multiple pictures which increase the time of the screen and the data load.

Following calibration of the above parameters, automated screening can commence.

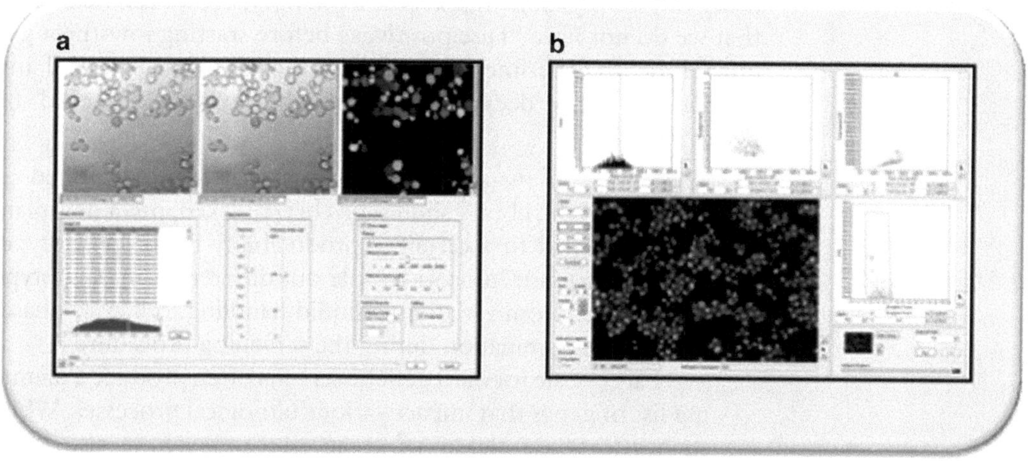

Fig. 3. Example of an image analysis tool. (**a**) Shown is a sample screen from the ScanR program (part of the Olympus ScanR automated microscopy platform). As in any analysis program, the first step of image analysis is segmentation, in which objects are detected within each image. The ScanR edge-based particle detector requires three steps: (1) clipping of unwanted noise or artifacts; (2) edge extraction of the objects; (3) edge closing by combining open edges into single objects. (**b**) Segmentation allows measuring various parameters for each object. Shown are examples of such cutouts for data analysis as either scatter plots or histograms. Gating of specific populations in these graphs can reduce noise and eliminate false measurements. By visualizing the parameters, it is then possible to uncover samples/cells that carry the sought-after phenotype.

3.5. Analyzing the Data

Once images have been acquired, the data must be analyzed to allow identification of strains ("hits") with phenotypes that differ from the control strain. If the screening phenotype is a qualitative one, it is possible to manually assess the images of each strain and score them for the phenotype of choice. However, if a quantitative phenotype is measured, then automation is required. Many screening microscopes come with analysis programs included. We use the ScanR analysis system which enables us to quantify the data automatically in thousands of samples with hundreds of cells in each sample, giving different parameter measurements both at the single cell level and for the whole population (Subheading 2.5 and Fig. 3).

3.6. Confirming Hit Phenotypes

High-throughput screens of systematic libraries are never saturated. This may result from strains that are not present in the libraries, cross contamination of neighboring strains, strains that could not be made during the SGA protocol (for example, two markers that are tightly linked on the chromosome can be hard to create using SGA as they require meiotic recombination to appear together in the final haploid strain), mistakes in the acquisition or analysis, and mutations whose phenotype is covered by a paralogue or a homeostatic response.

Although there is little we can do about such "false negatives" (except for using libraries of different sources and screen twice using different genetic markers as a phenotype), we can make sure that we do not have "false positives" before starting low-throughput follow-up experiments. To ensure that the "hits" are real and specific to the process of interest, the following should be performed:

1. *Verification of the phenotype*: Strains that were identified as "hits" from the library should be cherry picked into a new plate and rescreened to ascertain reproducibility of the phenotype. This is important in order to rule out differences in phenotype that resulted from variation in liquid handling, cell cycle phase, or strain contamination during the screening procedure.

   Large-scale forward genetic screens often produce a daunting list of genes that impact various biological processes. While such efforts are clearly of great value, they have also overwhelmed our ability to systematically follow up specific biological stories, often resulting in screen outputs comprising long lists of poorly characterized genes. One major problem with such screens is the inability to determine how direct the effect of each mutation is on the phenotype of choice. It is often impossible to determine a direct effect based on the intensity of the phenotype alone since essential genes from hypomorphic libraries have varying levels of downregulation and unessential components often have backup pathways that dampen the observed phenotype.

   Since follow-up experiments often take more time and energy than the initial screening phase, it is important to find ways to focus on the interesting "hits" and weed away uninteresting strains. This can be done by performing secondary screens tailored to the phenotype of interest. Generally, there are two types of possible secondary screens, both to be performed only on the "hit" strains.

2. *Ensuring generalizability of the phenotype*: For this, an additional marker carrying similar properties is assayed in the background of the mutations to assure that the phenotype is reproducible and general.

3. *Verifying specificity*: This screen tests whether other similar processes are not perturbed. For this kind of secondary screen, a marker that can report on the status of additional pathways/routes is integrated and screened.

   Strains that were verified as being both general and specific are good candidates for follow-up experiments. Before proceeding to hypothesis-based experiments, it is best to verify the genotype and its causality.

4. *Verification of the genotype*: In order to ensure that the "hits" harbor the specific mutation that they are thought to contain based on their coordinates in the library, it is essential to check their genotype by check PCR procedure. Another option is to

recreate the specific mutant strains manually and then rescreen. If remaking the strain, it is possible to also use an additional laboratory strain of yeast (for example, W303) to ensure that the phenotype found is general.

Note: It has been shown that many of the deletion library strains contain the selection cassette in the correct integration spot (making them "correct" by most check PCR techniques); however, due to aneuploidy, a normal copy of the deleted gene may still be present in the genome (54). To ensure that the strain with which you are working is indeed deleted for the gene of choice, it is recommended to follow up on the above PCR by using primers from within the ORF. By using such primers, a correct strain would show lack of PCR signal, whereas a product should always be found in a WT strain. To control for lack of DNA or the PCR not working, use control primers, such as actin (for more protocols on check PCR, see (15)).

5. *Verification of causality*: To verify that indeed the mutation that is in the gene of interest is responsible for the phenotype observed (and not a secondary change to promoters/terminators of nearby genes, which is often the case, or a mutation in an overlapping gene on the opposite strand), rescue experiments are essential. This can be done by creating expression plasmids for each hit or more systematically by using either the MoBY library or Gal-GST library (Table 1) as both contain yeast genes on expression plasmids. The plasmids are transformed into the appropriate deletion strains, and this should revert the phenotype back to the control status.

3.7. Creating Genetic Interaction Maps

An alternative to a specific secondary screen is to create a genetic interaction map of the "hits." This is done by systematically combining mutations to form all double-mutant strains carrying the fluorescent marker of choice. To easily perform this, all "hit" strains from the screen (which include both a deletion and the visualization marker) should be crossed against the same deletion strains but of the opposite mating type and harboring a different selection marker. One way to create such a library easily is by taking the deletion library (in either MATa or MATα to have the right mating type) and use a "marker swap" cassette (6, 55) to create a new selection version for these cells.

Previously, such genetic interaction maps have provided concrete information on the function of individual genes and the organization of such genes into pathways and complexes. This suggests that such an effort can dramatically increase the amount and quality of information that can be obtained from a screen while only modestly increasing the amount of time needed to complete it (15, 25, 41, 44–47, 56).

Designing and performing a good genetic screen can pinpoint to novel genes that take part in a cellular process of choice. Performing such screens on entire genomes in a systematic manner has recently become feasible with the development of a large number of yeast libraries, protocols for crossing in markers of choice, and tools to enable rapid microscopic screening. With the increasing ease to generate such data, it is important to use the screening procedure as a stepping-stone rather than a goal. A well-planned and executed screen (and secondary screens) allow for high-quality hypotheses to be generated. In depth follow up on these ideas should allow true detailed and mechanistic understanding of cell biology.

The protocols brought here demonstrate the basic principles of how state-of-the-art tools can be used to tackle fundamental problems in cell biology and as such provide a paradigm for tackling complex biological questions at a variety of organizational levels in eukaryotic cells. By complementing traditional genetic screening approaches with tools for generating custom libraries and systematically visualizing these libraries, this approach can dramatically impact the way cell biology is done today.

4. Notes

All notes can be found under the respective protocols.

Acknowledgments

Yifat Cohen and Maya Schuldiner were funded by the Legacy Heritage Biomedical Science Partnership Program of the Israel Science Foundation (grant No. 1995/08). Yifat Cohen is a recipient of the Karen Siem Fellowship for Women in Science. We would like to thank Michal Breker and Anat Shemesh for help with setting up the protocols and robotics to enable high-throughput microscopy in our laboratory. We would like to thank Moshe Gabso, Ayelet Rahat, and all the members of the Schuldiner laboratory for critical reading of the manuscript.

References

1. Farquhar, M. G., and Palade, G. E. (1998) The Golgi apparatus: 100 years of progress and controversy, *Trends Cell Biol 8*, 2–10.
2. Droscher, A. (1998) Camillo Golgi and the discovery of the Golgi apparatus, *Histochem Cell Biol 109*, 425–430.
3. Rieder, C. L., and Khodjakov, A. (2003) Mitosis through the microscope: advances in seeing inside live dividing cells, *Science 300*, 91–96.
4. Tong, A. H. Y., and Boone, C. (2007) High-Throughput Strain Construction and Systematic

Synthetic Lethal Screening in Saccharomyces cerevisiae, in *Methods in Microbiology* (Stansfield, I., and Stark, M. J., Eds.) Second ed., pp 369–386, Elsevier Academic Press.
5. Tong, A. H., Evangelista, M., Parsons, A. B., Xu, H., Bader, G. D., Page, N., Robinson, M., Raghibizadeh, S., Hogue, C. W., Bussey, H., Andrews, B., Tyers, M., and Boone, C. (2001) Systematic genetic analysis with ordered arrays of yeast deletion mutants, *Science 294*, 2364–2368.
6. Tong, A. H., and Boone, C. (2006) Synthetic genetic array analysis in Saccharomyces cerevisiae, *Methods Mol Biol 313*, 171–192.
7. Shaner, N. C., Steinbach, P. A., and Tsien, R. Y. (2005) A guide to choosing fluorescent proteins, *Nat Methods 2*, 905–909.
8. Shaner, N. C., Patterson, G. H., and Davidson, M. W. (2007) Advances in fluorescent protein technology, *J Cell Sci 120*, 4247–4260.
9. Frommer, W. B., Davidson, M. W., and Campbell, R. E. (2009) Genetically encoded biosensors based on engineered fluorescent proteins, *Chem Soc Rev 38*, 2833–2841.
10. Morris, M. C. (2010) Fluorescent biosensors of intracellular targets from genetically encoded reporters to modular polypeptide probes, *Cell Biochem Biophys 56*, 19–37.
11. Ibraheem, A., and Campbell, R. E. (2010) Designs and applications of fluorescent protein-based biosensors, *Curr Opin Chem Biol 14*, 30–36.
12. Wang, H., Nakata, E., and Hamachi, I. (2009) Recent progress in strategies for the creation of protein-based fluorescent biosensors, *Chembiochem 10*, 2560–2577.
13. Miyawaki, A., Llopis, J., Heim, R., McCaffery, J. M., Adams, J. A., Ikura, M., and Tsien, R. Y. (1997) Fluorescent indicators for Ca^{2+} based on green fluorescent proteins and calmodulin, *Nature 388*, 882–887.
14. Paroutis, P., Touret, N., and Grinstein, S. (2004) The pH of the secretory pathway: measurement, determinants, and regulation, *Physiology (Bethesda) 19*, 207–215.
15. Schuldiner, M., Collins, S. R., Weissman, J. S., and Krogan, N. J. (2006) Quantitative genetic analysis in Saccharomyces cerevisiae using epistatic miniarray profiles (E-MAPs) and its application to chromatin functions, *Methods 40*, 344–352.
16. Vizeacoumar, F. J., Chong, Y., Boone, C., and Andrews, B. J. (2009) A picture is worth a thousand words: genomics to phenomics in the yeast Saccharomyces cerevisiae, *FEBS Lett 583*, 1656–1661.
17. Gough, A. H., and Johnston, P. A. (2007) Requirements, features, and performance of high content screening platforms, *Methods Mol Biol 356*, 41–61.
18. Pepperkok, R., and Ellenberg, J. (2006) High-throughput fluorescence microscopy for systems biology, *Nat Rev Mol Cell Biol 7*, 690–696.
19. Megason, S. G., and Fraser, S. E. (2007) Imaging in systems biology, *Cell 130*, 784–795.
20. Araki, Y., Wu, H., Kitagaki, H., Akao, T., Takagi, H., and Shimoi, H. (2009) Ethanol stress stimulates the Ca^{2+}-mediated calcineurin/Crz1 pathway in Saccharomyces cerevisiae, *J Biosci Bioeng 107*, 1–6.
21. Cai, L., Dalal, C. K., and Elowitz, M. B. (2008) Frequency-modulated nuclear localization bursts coordinate gene regulation, *Nature 455*, 485–490.
22. Lim, R. P., Misra, A., Wu, Z., and Thanabalu, T. (2007) Analysis of conformational changes in WASP using a split YFP, *Biochem Biophys Res Commun 362*, 1085–1089.
23. Anand, V. C., Daboussi, L., Lorenz, T. C., and Payne, G. S. (2009) Genome-wide analysis of AP-3-dependent protein transport in yeast, *Mol Biol Cell 20*, 1592–1604.
24. Kanki, T., Wang, K., Baba, M., Bartholomew, C. R., Lynch-Day, M. A., Du, Z., Geng, J., Mao, K., Yang, Z., Yen, W. L., and Klionsky, D. J. (2009) A genomic screen for yeast mutants defective in selective mitochondria autophagy, *Mol Biol Cell 20*, 4730–4738.
25. Jonikas, M. C., Collins, S. R., Denic, V., Oh, E., Quan, E. M., Schmid, V., Weibezahn, J., Schwappach, B., Walter, P., Weissman, J. S., and Schuldiner, M. (2009) Comprehensive characterization of genes required for protein folding in the endoplasmic reticulum, *Science 323*, 1693–1697.
26. Merksamer, P. I., Trusina, A., and Papa, F. R. (2008) Real-time redox measurements during endoplasmic reticulum stress reveal interlinked protein folding functions, *Cell 135*, 933–947.
27. Giaever, G., Chu, A. M., Ni, L., Connelly, C., Riles, L., Veronneau, S., Dow, S., Lucau-Danila, A., Anderson, K., Andre, B., Arkin, A. P., Astromoff, A., El-Bakkoury, M., Bangham, R., Benito, R., Brachat, S., Campanaro, S., Curtiss, M., Davis, K., Deutschbauer, A., Entian, K. D., Flaherty, P., Foury, F., Garfinkel, D. J., Gerstein, M., Gotte, D., Guldener, U., Hegemann, J. H., Hempel, S., Herman, Z., Jaramillo, D. F., Kelly, D. E., Kelly, S. L., Kotter, P., LaBonte, D., Lamb, D. C., Lan, N., Liang, H., Liao, H., Liu, L., Luo, C., Lussier, M., Mao, R., Menard, P., Ooi, S. L., Revuelta, J. L., Roberts, C. J., Rose, M., Ross-Macdonald, P., Scherens, B., Schimmack, G., Shafer, B., Shoemaker, D. D., Sookhai-Mahadeo, S.,

Storms, R. K., Strathern, J. N., Valle, G., Voet, M., Volckaert, G., Wang, C. Y., Ward, T. R., Wilhelmy, J., Winzeler, E. A., Yang, Y., Yen, G., Youngman, E., Yu, K., Bussey, H., Boeke, J. D., Snyder, M., Philippsen, P., Davis, R. W., and Johnston, M. (2002) Functional profiling of the Saccharomyces cerevisiae genome, *Nature 418*, 387–391.

28. Breslow, D. K., Cameron, D. M., Collins, S. R., Schuldiner, M., Stewart-Ornstein, J., Newman, H. W., Braun, S., Madhani, H. D., Krogan, N. J., and Weissman, J. S. (2008) A comprehensive strategy enabling high-resolution functional analysis of the yeast genome, *Nat Methods 5*, 711–718.
29. Ben-Aroya, S., Coombes, C., Kwok, T., O'Donnell, K. A., Boeke, J. D., and Hieter, P. (2008) Toward a comprehensive temperature-sensitive mutant repository of the essential genes of Saccharomyces cerevisiae, *Mol Cell 30*, 248–258.
30. Mnaimneh, S., Davierwala, A. P., Haynes, J., Moffat, J., Peng, W. T., Zhang, W., Yang, X., Pootoolal, J., Chua, G., Lopez, A., Trochesset, M., Morse, D., Krogan, N. J., Hiley, S. L., Li, Z., Morris, Q., Grigull, J., Mitsakakis, N., Roberts, C. J., Greenblatt, J. F., Boone, C., Kaiser, C. A., Andrews, B. J., and Hughes, T. R. (2004) Exploration of essential gene functions via titratable promoter alleles, *Cell 118*, 31–44.
31. Sopko, R., Huang, D., Preston, N., Chua, G., Papp, B., Kafadar, K., Snyder, M., Oliver, S. G., Cyert, M., Hughes, T. R., Boone, C., and Andrews, B. (2006) Mapping pathways and phenotypes by systematic gene overexpression, *Mol Cell 21*, 319–330.
32. Huh, W. K., Falvo, J. V., Gerke, L. C., Carroll, A. S., Howson, R. W., Weissman, J. S., and O'Shea, E. K. (2003) Global analysis of protein localization in budding yeast, *Nature 425*, 686–691.
33. Tudisca, V., Recouvreux, V., Moreno, S., Boy-Marcotte, E., Jacquet, M., and Portela, P. (2010) Differential localization to cytoplasm, nucleus or P-bodies of yeast PKA subunits under different growth conditions, *Eur J Cell Biol 89*, 339–348.
34. Sheth, U., and Parker, R. (2003) Decapping and decay of messenger RNA occur in cytoplasmic processing bodies, *Science 300*, 805–808.
35. Brengues, M., and Parker, R. (2007) Accumulation of polyadenylated mRNA, Pab1p, eIF4E, and eIF4G with P-bodies in Saccharomyces cerevisiae, *Mol Biol Cell 18*, 2592–2602.
36. Kulkarni, M., Ozgur, S., and Stoecklin, G. (2010) On track with P-bodies, *Biochem Soc Trans 38*, 242–251.
37. Buchan, J. R., Muhlrad, D., and Parker, R. (2008) P bodies promote stress granule assembly in Saccharomyces cerevisiae, *J Cell Biol 183*, 441–455.
38. Kaganovich, D., Kopito, R., and Frydman, J. (2008) Misfolded proteins partition between two distinct quality control compartments, *Nature 454*, 1088–1095.
39. Xie, Z., and Klionsky, D. J. (2007) Autophagosome formation: core machinery and adaptations, *Nat Cell Biol 9*, 1102–1109.
40. Ryan, C., Greene, D., Cagney, G., and Cunningham, P. (2010) Missing value imputation for epistatic MAPs, *BMC Bioinformatics 11*, 197.
41. Fiedler, D., Braberg, H., Mehta, M., Chechik, G., Cagney, G., Mukherjee, P., Silva, A. C., Shales, M., Collins, S. R., van Wageningen, S., Kemmeren, P., Holstege, F. C., Weissman, J. S., Keogh, M. C., Koller, D., Shokat, K. M., and Krogan, N. J. (2009) Functional organization of the S. cerevisiae phosphorylation network, *Cell 136*, 952–963.
42. Dixon, S. J., Costanzo, M., Baryshnikova, A., Andrews, B., and Boone, C. (2009) Systematic mapping of genetic interaction networks, *Annu Rev Genet 43*, 601–625.
43. Boone, C., Bussey, H., and Andrews, B. J. (2007) Exploring genetic interactions and networks with yeast, *Nat Rev Genet 8*, 437–449.
44. Beltrao, P., Cagney, G., and Krogan, N. J. (2010) Quantitative genetic interactions reveal biological modularity, *Cell 141*, 739–745.
45. Schuldiner, M., Collins, S. R., Thompson, N. J., Denic, V., Bhamidipati, A., Punna, T., Ihmels, J., Andrews, B., Boone, C., Greenblatt, J. F., Weissman, J. S., and Krogan, N. J. (2005) Exploration of the function and organization of the yeast early secretory pathway through an epistatic miniarray profile, *Cell 123*, 507–519.
46. Collins, S. R., Miller, K. M., Maas, N. L., Roguev, A., Fillingham, J., Chu, C. S., Schuldiner, M., Gebbia, M., Recht, J., Shales, M., Ding, H., Xu, H., Han, J., Ingvarsdottir, K., Cheng, B., Andrews, B., Boone, C., Berger, S. L., Hieter, P., Zhang, Z., Brown, G. W., Ingles, C. J., Emili, A., Allis, C. D., Toczyski, D. P., Weissman, J. S., Greenblatt, J. F., and Krogan, N. J. (2007) Functional dissection of protein complexes involved in yeast chromosome biology using a genetic interaction map, *Nature 446*, 806–810.
47. Breker, M., and Schuldiner, M. (2009) Explorations in topology-delving underneath the surface of genetic interaction maps, *Mol Biosyst 5*, 1473–1481.
48. Brachmann, C. B., Davies, A., Cost, G. J., Caputo, E., Li, J., Hieter, P., and Boeke, J. D. (1998) Designer deletion strains derived from

Saccharomyces cerevisiae S288C: a useful set of strains and plasmids for PCR-mediated gene disruption and other applications, *Yeast 14*, 115–132.

49. Sychrova, H., and Chevallier, M. R. (1993) Cloning and sequencing of the Saccharomyces cerevisiae gene LYP1 coding for a lysine-specific permease, *Yeast 9*, 771–782.
50. Day, R. N., and Davidson, M. W. (2009) The fluorescent protein palette: tools for cellular imaging, *Chem Soc Rev 38*, 2887–2921.
51. Sprague, G. F., Jr. (2004) Assay of Yeast Mating Reaction, in *Guide to Yeast Genetics and Molecular and Cell Biology* (Guthrie, C., and Fink, G. R., Eds.), pp 77–93, Elsevier Academic Press, San Diego.
52. Sheff, M. A., and Thorn, K. S. (2004) Optimized cassettes for fluorescent protein tagging in Saccharomyces cerevisiae, *Yeast 21*, 661–670.
53. DeLuna, A., Vetsigian, K., Shoresh, N., Hegreness, M., Colon-Gonzalez, M., Chao, S., and Kishony, R. (2008) Exposing the fitness contribution of duplicated genes, *Nat Genet 40*, 676–681.
54. Hughes, T. R., Roberts, C. J., Dai, H., Jones, A. R., Meyer, M. R., Slade, D., Burchard, J., Dow, S., Ward, T. R., Kidd, M. J., Friend, S. H., and Marton, M. J. (2000) Widespread aneuploidy revealed by DNA microarray expression profiling, *Nat Genet 25*, 333–337.
55. Voth, W. P., Jiang, Y. W., and Stillman, D. J. (2003) New 'marker swap' plasmids for converting selectable markers on budding yeast gene disruptions and plasmids, *Yeast 20*, 985–993.
56. Collins, S. R., Schuldiner, M., Krogan, N. J., and Weissman, J. S. (2006) A strategy for extracting and analyzing large-scale quantitative epistatic interaction data, *Genome Biol in press.*
57. Ho, C. H., Magtanong, L., Barker, S. L., Gresham, D., Nishimura, S., Natarajan, P., Koh, J. L., Porter, J., Gray, C. A., Andersen, R. J., Giaever, G., Nislow, C., Andrews, B., Botstein, D., Graham, T. R., Yoshida, M., and Boone, C. (2009) A molecular barcoded yeast ORF library enables mode-of-action analysis of bioactive compounds, *Nat Biotechnol 27*, 369–377.
58. Jones, G. M., Stalker, J., Humphray, S., West, A., Cox, T., Rogers, J., Dunham, I., and Prelich, G. (2008) A systematic library for comprehensive overexpression screens in Saccharomyces cerevisiae, *Nat Methods 5*, 239–241.
59. Kitada, K., Yamaguchi, E., and Arisawa, M. (1995) Cloning of the Candida glabrata TRP1 and HIS3 genes, and construction of their disruptant strains by sequential integrative transformation, *Gene 165*, 203–206.
60. Longtine, M. S., McKenzie, A., 3rd, Demarini, D. J., Shah, N. G., Wach, A., Brachat, A., Philippsen, P., and Pringle, J. R. (1998) Additional modules for versatile and economical PCR-based gene deletion and modification in Saccharomyces cerevisiae, *Yeast 14*, 953–961.
61. Goldstein, A. L., and McCusker, J. H. (1999) Three new dominant drug resistance cassettes for gene disruption in Saccharomyces cerevisiae, *Yeast 15*, 1541–1553.
62. Janke, C., Magiera, M. M., Rathfelder, N., Taxis, C., Reber, S., Maekawa, H., Moreno-Borchart, A., Doenges, G., Schwob, E., Schiebel, E., and Knop, M. (2004) A versatile toolbox for PCR-based tagging of yeast genes: new fluorescent proteins, more markers and promoter substitution cassettes, *Yeast 21*, 947–962.

Chapter 9

Pooled Lentiviral shRNA Screening for Functional Genomics in Mammalian Cells

Kim Blakely, Troy Ketela, and Jason Moffat

Abstract

Genome sequencing efforts have reformed the nature of biological inquiry, prompting the development of technologies for the functional annotation of mammalian genes. Based on methodologies originally discovered in plants and *Caenorhabditis elegans*, RNA interference has offered cell biologists an effective and reproducible approach to perturb gene function in mammalian cells and whole organisms. Initial application of RNA interference libraries targeting the human and mouse genomes relied on arrayed screening approaches, whereby each unique RNA interference reagent is arrayed into individual wells of a microtiter plate. These screens are not trivial to perform, requiring a substantial investment in infrastructure. In the past decade, many technological advances have been made that make genome-wide RNA interference screening more accessible to researchers and more feasible to perform in nonspecialized laboratories. Here, we describe a comprehensive protocol for pooled short-hairpin RNA screening, including methodologies for pooled lentivirus production, cell infection, genome-wide negative selection screening and resources for pooled screen deconvolution, and data analysis. As a technique, pooled shRNA screening is still in its infancy, but the methodology has already been successfully applied to probe diverse signaling pathways, as a means of drug target identification, and to identify essential genes in normal and cancer cell lines.

Key words: Gene Modulation Array Platform microarray, High-throughput screening, Lentivirus, Mammalian functional genomics, Pooled-screening, RNA interference, Short-hairpin RNA

1. Introduction

Over the past decade, RNA interference (RNAi) has revolutionized scientific research, not only as an active field of study leading to the discovery of microRNAs and other small noncoding RNAs (reviewed in ref. 1), but also as a powerful research tool to interrogate gene function and develop novel therapeutic agents (2–5). Continuing research is leading to a deeper understanding of the basic mechanics of RNAi and the development of improved RNAi reagents.

Gerard Cagney and Andrew Emili (eds.), *Network Biology: Methods and Applications*, Methods in Molecular Biology, vol. 781, DOI 10.1007/978-1-61779-276-2_9, © Springer Science+Business Media, LLC 2011

RNAi is an evolutionarily conserved post-transcriptional gene silencing program mediated by double-stranded RNA (dsRNA). Endogenous or exogenously introduced dsRNA will initiate the cellular RNAi pathway, whereby specific RNAses will cleave the dsRNA into 21–23 nt small interfering RNAs (siRNAs). These mature siRNAs are incorporated into the RNA-induced silencing complex (RISC), which aids in binding siRNA to the target mRNA. The RISC–siRNA complex can then repress expression of specific target mRNAs through either translational inhibition or mRNA degradation ((6, 7) and reviewed in (8)). Since RNAi is unlikely to completely eliminate the mRNA of interest, RNAi studies are generally considered a reduction-of-function genetic approach, generating hypomorphic alleles.

In the post-genomic era, essentially complete sequences of the human and mouse genomes have made it possible to generate RNAi reagents for systematic, functional analyses of the entire human and mouse genomes. Recent advances in RNAi library design have focused on improving the quality of RNAi reagents, including better sequence-selection algorithms, decreasing off-target and interferon responses, and improving the ease and cost associated with performing high-throughput gene interrogation studies. Together, this has led to the generation of several versions of RNAi libraries including siRNA (9–11), endoribonuclease-prepared siRNA (esiRNA) (12, 13), and short-hairpin RNA (shRNA) (14–17). In general, each of these reagents can be used for short-term perturbation of gene function by simple transfection approaches. However, long-term experiments, or analyses of nontransfectable cell types rely on the transduction of cells with a virus that integrates a stably inherited shRNA construct into the genome of a target cell. Libraries that rely on the introduction of RNAi constructs by either retrovirus or lentivirus have been constructed. For the purposes of this chapter, we will be focusing on the latter approach.

The RNAi Consortium (TRC) is a collaborative group of academic and corporate life science researchers that have designed and optimized lentiviral-encoded shRNA libraries against both the human and mouse genomes (15, 18). The first version of the TRC libraries contain >160,000 unique shRNA constructs targeting ~16,000 genes in each of the human and mouse genomes. TRC shRNAs are contained in the pLKO.1 vector, allowing for production of lentiviral particles, stable transduction and gene silencing, and selection of infected clones through use of the puromycin selectable marker. Use of this vector with a second or third generation packaging system ensures the production of self-inactivating, replication-incompetent viral particles. Unlike adenoviral or retroviral systems, lentiviral-based shRNA delivery permits efficient transduction of the specific shRNA construct into slow and nondividing cell types, such as stem cells and terminally differentiated cells. To reduce the cost and labor associated with genome-wide shRNA screening, these libraries have been adapted to a pooled format (14).

Single-copy integration, or low multiplicity of infection (MOI), is of critical importance for pooled screening applications to avoid synthetic-genetic effects. This is achieved through limiting dilution where only a fraction of cells become infected upon exposure to the viral pool, and most infected cells integrate only a single shRNA construct. Uninfected cells are subsequently eliminated by puromycin selection. The pooled lentiviral shRNA screening strategy relies upon readout of the changes in abundance of each shRNA clone in a pool of cells over a defined period of time. A decline in representation of specific clones infers that their target mRNA is essential to cell viability (14, 16, 17). Genes and pathways responsible for cellular reaction to perturbagens (e.g., compounds, radiation, nutritional starvation, hypoxia, etc.) may be identified by comparison of shRNA abundance profiles to those derived from an untreated, control population (19–22). Alternatively, shRNA libraries can be used to identify novel components of cellular pathways, or genes synthetic-lethal with known signaling defects in cancer (23–26). Deconvolution of TRC-based shRNA libraries may be accomplished by PCR amplification of the integrated shRNA sequences from the genomic DNA of cell pools. The complex mixture of shRNA sequences may then be applied to a DNA microarray or subject to next generation sequencing so that each shRNA species may be quantitated (Fig. 1). In this chapter, we focus on the application of pooled shRNA screening for identification of essential genes in mammalian cell lines. These protocols may be adapted for screening and analysis of pooled lentiviral shRNA-libraries in a wide variety of contexts.

2. Materials

2.1. Lentivirus Production

1. Human embryonic kidney (HEK) 293 T cells (below passage 15).
2. Transfection Medium: Dulbecco's Modified Eagle's Medium 1× (DMEM) (Wisent Inc., cat. no. 319-005-CL) supplemented with 10% heat-inactivated fetal bovine serum (iFBS) (Wisent Inc., cat. no. 080-450) and 0.1× penicillin–streptomycin (P/S) (Wisent Inc., cat. no. 450-201-EL).
3. 78 k pLKO.1 TRC shRNA plasmid pool (Sigma-Aldrich).
4. Second generation packaging plasmid (e.g., psPAX2).
5. Envelope plasmid (e.g., pMD2.G).
6. FuGENE®6 Transfection Reagent (Roche, cat. no. 11 814 443 001).

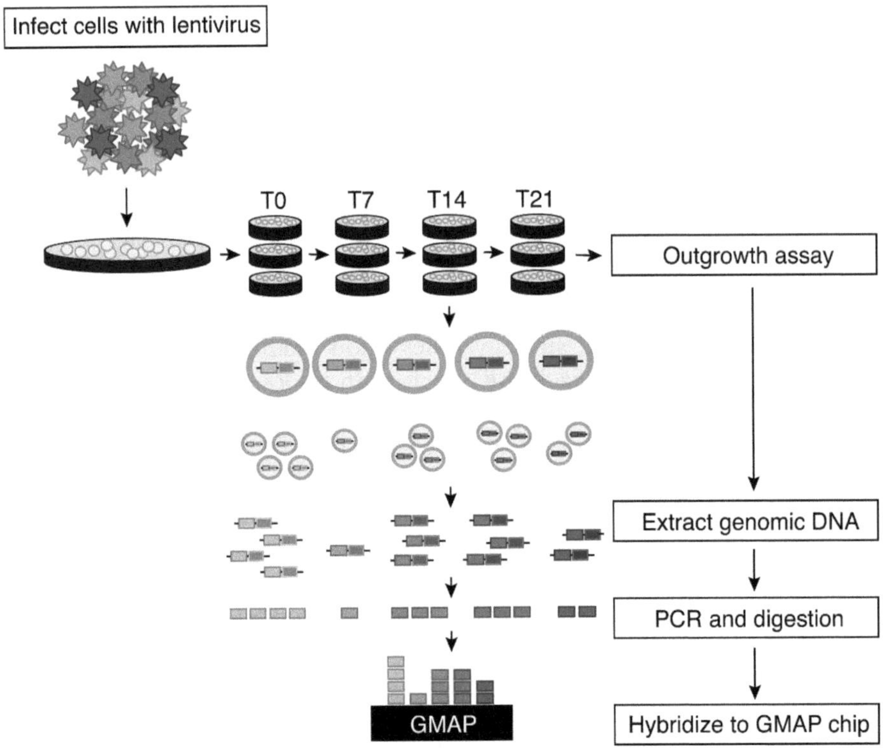

Fig. 1. Schematic of the pooled shRNA screening procedure. Cells are infected with the 78 k TRC lentivirus-encoded shRNA library at an MOI ~ 0.3–0.4. Cells are grown in culture, during which time certain shRNAs will become enriched or depleted from the pool depending on whether knockdown of their complementary mRNA imparts a growth advantage or disadvantage to the cells. At each passage, genomic DNA is isolated from a sub-pool of the cells and the shRNA inserts are PCR amplified using a 5′-biotinylated forward primer. PCR amplicons are digested to reveal the half-hairpin barcodes, which are hybridized to the GMAP microarray to determine the relative abundance of each shRNA in the pool. T7, T14 and T21 are example passage numbers.

7. Opti-MEM® I Reduced Serum Medium (1×) (Invitrogen, cat. no. 31985-062).
8. Viral Harvest Medium: DMEM supplemented with 1.1% bovine serum albumin (BSA) (Wisent, cat. no. 800-095-EG) and 1× P/S.

2.2. Dose–Response Tests for Puromycin, Polybrene, and Protamine Sulfate

1. Cell line to be screened, and appropriate growth medium.
2. Clear 96-well flat-bottom, tissue culture-treated microtiter plate (BD Falcon, cat. no. 353072).
3. Puromycin (Sigma-Aldrich, cat. no. P8833) is dissolved at 2 mg/mL in ddH_2O and filter sterilized.

4. Hexadimethrine bromide (polybrene) (Sigma-Aldrich, cat. no. H9268) is dissolved at 2 mg/mL in ddH_2O and filter sterilized.
5. Protamine sulfate (Sigma-Aldrich, cat. no. P3369) is dissolved at 5 mg/mL in ddH_2O and filter sterilized.
6. AlamarBlue® (Invitrogen, cat. no. DAL1100).
7. 96-well fluorescence-capable plate reader.

2.3. MOI Determination

1. 10 mL aliquot of 78 k virus pool (see Note 1).
2. Cell line to be screened, and appropriate growth medium.
3. If desired, appropriate transduction enhancer (e.g., polybrene or protamine sulfate).
4. Puromycin (2 mg/mL in ddH_2O, filter sterilized).

2.4. Primary Screen Infection and Cell Passaging

1. Aliquot of 78 k virus pool (see Note 1).
2. Cell line to be screened, and appropriate growth medium.
3. If desired, appropriate transduction enhancer (e.g., polybrene or protamine sulfate).
4. Puromycin (2 mg/mL in ddH_2O, filter sterilized).

2.5. Genomic DNA Isolation

1. QIAamp DNA Blood Maxi Kit (Qiagen, cat. no. 51194). Store QIAamp Maxi spin columns at 2–8°C. All other QIAamp Maxi components can be stored at room temperature (15–25°C).
2. Phosphate-buffered saline (PBS) (Wisent Inc., cat. no. 311-010-CL).
3. Aerosol filter pipette tips (VWR, cat. no. 89003-420).
4. 96–100% ethanol.

2.6. Genomic DNA Precipitation

1. Sodium chloride solution (5 M) (Sigma-Aldrich, cat. no. S5150).
2. 96–100% ethanol (−20°C).
3. 70% ethanol (−20°C).

2.7. shRNA Barcode Amplification

1. Platinum® Pfx DNA Polymerase (Invitrogen, cat. no. 11708-021).
2. 10 mM dNTP mix (Invitrogen, cat. no. 18427-088).
3. PCR strip tubes (VWR, cat. no. 93001-118).
4. 20 μM primer mix (Table 1).

2.8. shRNA Barcode Purification and Restriction Digestion

1. QIAquick PCR Purification Kit (Qiagen, cat. no. 28106).
2. 96–100% ethanol.
3. *Xho*I restriction enzyme (20,000 U/mL) (New England BioLabs, R0146S).

2.9. Gel Purification of Digested shRNA Barcodes

1. 100 bp DNA ladder and 6× loading dye (Fermentas, cat. no. SM1143).
2. QIAquick Gel Extraction Kit (Qiagen, cat. no. 28706).
3. 100% isopropanol.
4. 96–100% ethanol.
5. Heating block or water bath at 50°C.

2.10. PCR Purification of Digested shRNA Barcodes

1. QIAquick PCR Purification Kit.
2. 96–100% ethanol.

2.11. GMAP Microarray Pre-conditioning

1. GeneChip® Gene Modulation Array Platform (GMAP) microarrays (one per sample) (Affymetrix, part no. 520601).
2. Sodium hydroxide solution (10 mM) (Sigma-Aldrich, cat. no. S2770) at 40°C.
3. Wash Buffer A: 300 mL 20× SSPE (Sigma-Aldrich, cat. no. MS2015), 1 mL 10% Tween 20 (Sigma-Aldrich, cat. no. P1379), 699 mL ddH_2O.
4. 0.0005% Triton®X-100 (Sigma-Aldrich, cat. no. T8787).

2.12. Hybridization Mix Preparation

1. 12× MES: 7.04 g MES-free acid monohydrate (Sigma-Aldrich, cat. no. M5287), 19.3 g MES sodium salt (Sigma-Aldrich, cat. no. M5057), 80 mL ddH_2O. Mix well for >10 min, pH to 6.5–6.7. Adjust volume to 100 mL with ddH_2O and sterilize by passage through 0.2-μm filter. Store in dark at 4°C.
2. 2× Hybridization Buffer: 8.3 mL 12× MES, 17.7 mL 5 M NaCl, 4.0 mL 0.5 M EDTA (Sigma-Aldrich, cat. no. E7889), 0.1 mL 10% Tween 20, 19.9 mL ddH_2O. Sterilize by passage through 0.2-μm filter. Store in the dark at 4°C. Discard if the solution turns yellow.
3. BSA (50 mg/mL in ddH_2O).
4. Herring sperm DNA solution (10 mg/mL) (Invitrogen, cat. no. 15634-017).
5. B213 biotinylated control oligonucelotide and blocking oligonucleotide mix (Table 1).
6. DMSO (Sigma-Aldrich, cat. no. D8418).

2.13. GMAP Microarray Staining and Washing

1. Wash Buffer A (see Subheading 2.11).
2. Wash Buffer B: 83.3 mL 12× MES stock, 5.2 mL 5 M NaCl, 1.0 mL 10% Tween 20, 910.5 mL ddH_2O. Sterilize by passage through 0.2-μm filter. Store in dark at 4°C.
3. 2× MES Stain Buffer: 8.3 mL 12× MES stock, 18.5 mL 5 M NaCl, 1.0 mL 10% Tween 20, 50.0 mL ddH_2O. Sterilize by passage through 0.2-μm filter. Store in dark at 4°C.

Table 1
Oligonucleotides used in this protocol

Name	5′ Modification	Sequence (5′–3′)
shRNA amplification		
PCR-B-F	Biotin	AATGGACTATCATATGCTTACCGTAACTTGAA
PCR-R	N/A	TGTGGATGAATACTGCCATTTGTCTCGAGGTC
Prepare a primer mix (20 μM each) in sterile 10 mM Tris–HCl pH 7.5, aliquot and store at −20°C		
B213 biotinylated control oligonucleotide		
B213	Biotin	CTGAACGGTAGCATCTTGAC
Prepare in sterile 10 mM Tris pH 7.5. Using serial dilutions, prepare a 5 nM stock and store at −20°C		
Blocking oligonucleotides		
Block-1	N/A	AATGGACTATCATATGCTTACCGTAACTTGAA
Block-2	N/A	TTCAAGTTACGGTAAGCATATGATAGTCCATT
Block-3	N/A	GTATTTCGATTTCTTGGCTTTATATATCTTGTGGAAAGG ACGAAACACCG
Block-4	N/A	CGGTGTTTCGTCCTTTCCACAAGATATATAA AGCCAAGAAATCGAAATAC
Prepare oligonucleotides individually in sterile 10 mM Tris pH 7.5. Prepare a mix of all four blocking oligonucleotides, each with a final concentration of 20 μM. Individual oligonucleotides and oligonucleotide mixes should be stored at −20°C		

4. BSA (50 mg/mL in ddH_2O).
5. Streptavidin, R-phycoerythrin conjugate (SAPE) (1 mg/mL) (Invitrogen, cat. no. S866).
6. Goat IgG (Sigma-Aldrich, cat. no. I5256) is dissolved at 30 mg/mL in 150 mM NaCl.
7. Biotinylated antistreptavidin–phycoerythrin (BAS) (Vector Labs, cat. no. BA-0500) is dissolved at 0.5 mg/mL in ddH_2O.
8. Laser Tough-Spots® (Sigma-Aldrich, cat. no. T1316).
9. GeneChip® Fluidics Station 450 (Affymetrix, Inc.).
10. GeneChip® Scanner 3000 (Affymetrix, Inc.).

3. Methods

If you have purchased the 78 k pLKO.1 TRC shRNA plasmid pool, begin from Subheading 3.1. If you have purchased lentiviral stocks, begin from Subheading 3.2.

3.1. Lentivirus Production

See Note 2 prior to beginning protocol.

1. Harvest HEK293T packaging cells in Transfection Medium and measure cell concentration. Seed 10E6 cells per 150-mm plate in 30 mL Transfection Medium. Let cells adhere to plate at room temperature (~1 h) to ensure even distribution, then transfer to incubator (37°C, 5% CO_2).
2. Twenty-four hours post-seeding, cells should be ~70% confluent, and are ready to transfect for lentivirus production.
3. For each 150-mm plate to be transfected, combine 16.2 μg psPAX2, 1.8 μg pMD2.G, and 18.0 μg 78 k library plasmid pool in a 2.0-mL microcentrifuge tube.
4. For each plate to be transfected, add 1,620 μL Opti-MEM® to a new 2.0-mL microcentrifuge tube. Add 108 μL FuGENE®6 Transfection Reagent directly into Opti-MEM® and mix by gently flicking the tube. Do not mix by pipetting or vortexing. Incubate 5 min at room temperature. If preparing several plates of virus, prepare a master mix ($n+1$) of Opti-MEM® and FuGENE®6 Transfection Reagent.
5. Following the 5-min incubation, swirl pipette tip in Opti-MEM®/FuGENE®6 Transfection Reagent mixture and transfer 1,728 μL to the microcentrifuge tube containing DNA. Mix by gently flicking the tube.
6. Incubate the transfection mix for 30-min to 1 h at room temperature.

7. Carefully transfer the transfection mix to the HEK293T packaging cells by adding drop-wise. Tilt plate a few times to ensure mixture is evenly distributed over cells. Place cells in incubator.
8. Eighteen hours post-transfection, remove media from cells using a 30-mL serological pipette, and dispense into a container of 20% bleach. Add 30 mL Viral Harvest Media to plate and return plates to incubator.
9. Twenty-four hours post-media change, perform the first virus harvest. Collect the media off the HEK293T cells using a 30-mL serological pipette and transfer to a polypropylene storage tube. Add 30 mL of fresh Viral Harvest Media to cells.
10. Repeat viral harvest again after 24 h and combine with the first harvest. Following this harvest, discard the viral production plate in 20% bleach.
11. Centrifuge the virus media at 335 × *g* for 5-min to pellet any HEK293T cells that were collected during harvesting. Transfer the supernatant to a sterile polypropylene storage tube.
12. Virus may be stored at 4°C for short periods (hours to days), but should be frozen at −80°C for long-term storage. To reduce the number of freeze/thaw cycles, aliquot large-scale virus preps to smaller storage tubes prior to long-term storage.

3.2. Dose–Response Tests for Puromycin, Polybrene, and Protamine Sulfate

See Note 3 prior to beginning protocol.

1. Harvest cells to be screened and measure cell concentration per mL.
2. Prepare cell suspension and seed cells in rows A–F of a 96-well plate. Seed cells in 95 μL total volume at a density that will allow for 3 days growth without becoming overly confluent (~80–90% confluent). Usually 4,000–10,000 cells per well are appropriate.
3. Add 100 μL media without cells to rows G and H.
4. Place plate in incubator and allow cells to adhere/recover from handling (~3–6 h).
5. In a 12-well strip tube, add 25 μL of 2 mg/mL puromycin stock to tube 1. Add 12.5 μL ddH$_2$O to remaining 11 tubes. Take 12.5 μL from the first tube and add to tube 2. Mix thoroughly, and take 12.5 μL from the second tube and add to tube 3. Continue doing twofold serial dilutions until tube 11, leaving tube 12 as a "water only" control.
6. Using a 12-well multichannel, take 5 μL from the strip tubes, and add to row A of the 96-well plate containing cells. Repeat with another 5 μL for row B. This will generate a dilution curve for polybrene from 100 to 0 μg/mL, in duplicate.

7. Using new strip tubes, repeat for polybrene and protamine sulfate solutions, using rows C/D and E/F of the 96-well plate. Place plate in incubator.
8. Forty-five hours post-addition of compounds, add 10 μL alamarBlue® solution to each well of the 96-well plate. The rows without cells (G and H) will become the "background fluorescence" wells. Replace plate in incubator for 3 h.
9. Measure fluorescence intensity using a 96-well fluorescence plate reader with an excitation filter of 535 nm and an emission filter of 590 nm. Calculate percentage reduction of alamarBlue® using the equation (mean fluorescence of compound-treated wells - background fluorescence/mean fluorescence of water-treated wells - background fluorescence) × 100.
10. The concentration of puromycin to be used for MOI determination and the screen should be 1 μg/mL higher than that required to kill 100% of uninfected cells in 48 h in this assay (usually 1–4 μg/mL). Polybrene and protamine sulfate nontoxic doses are variable, usually between 4–8 μg/mL and 2–5 μg/mL, respectively. Most cells tolerate polybrene, and we recommend the use of polybrene over protamine sulfate to improve viral infection. However, if polybrene proves to be toxic to cells at ≤4 μg/mL, opt for protamine sulfate.

3.3. MOI Determination

1. Harvest cells to be infected and measure cell concentration per mL.
2. Prepare ten 150-mm tissue culture treated plates at a cell density that will allow for 3 days growth, without becoming overly confluent (~80–90% confluent). Usually two to five million cells per plate is suitable in a 30-mL volume. This seeding density is required since the puromycin selection step will kill between 60 and 70% of the cells at the desired MOI.
3. If nontoxic to your cells, include polybrene or protamine sulfate as a transduction enhancer at the appropriate dose determined in Subheading 3.2.
4. Ensure virus prep is mixed thoroughly by inversion five times. Add virus to plates to generate a five-point dilution series (e.g., 0, 0.031, 0.125, 0.5, and 2.0 mL). Prepare plates in duplicate (same volume of virus for two vessels at each dilution point).
5. Mix plates thoroughly by tilting back and forth.
6. For adherent lines, let plates sit at room temperature on a level surface to ensure even distribution after settling.
7. Transfer plates to incubator. Ensure that they sit level.

8a. *Adherent cells.* Twenty-four hours post-infection, remove media using serological pipettes and dispose of media and pipettes in 20% bleach solution.

8b. *Suspension cells.* Twenty-four hours post-infection, transfer cells to 50-mL conical bottom centrifuge tubes, and pellet cells at 335 ×*g* for 5 min. Remove media using serological pipettes and dispose of media and pipettes in 20% bleach solution.

9a. *Adherent cells.* Add 30 mL fresh media containing puromycin at the required concentration to vessels of one virus dilution series and fresh media without puromycin to the other virus dilution series vessels. Return plates to incubator.

9b. *Suspension cells.* Resuspend cell pellets of one virus dilution series with fresh media containing puromycin at the required concentration, and fresh media without puromycin to the other virus dilution series. Seed cells in new 150-mm plates and return to incubator.

10. Forty-eight hours post-selection, dispose of any remaining media in 20% bleach. Wash cells once with PBS, and dispose of PBS in 20% bleach.

11. Count cells from all plates and plot results for the two series (±puromycin), percent survival on the *Y*-axis and virus volume on the *X*-axis.

12. Determine virus volume that gives 30–40% survival with puromycin selection versus the same virus volume without puromycin. This is the volume of pooled virus that gives a MOI of 0.3–0.4 with the tissue culture conditions that you used, and the volume that should be used for the primary screen (see Note 4).

3.4. Primary Screen Infection and Cell Passaging

1. Expand cell line to be screened such that you have ~200 million cells for the primary infection. This will result in a ~1,000-fold representation of the 78 k shRNA library post-infection when using a MOI ~ 0.4, e.g., (200E6x0.4)/7.8E4.
2. Harvest cells to be infected and measure cell concentration per mL.
3. Prepare a mix of cells, virus, and transduction enhancer using the same proportions established in Subheading 3.3. For example, if it was determined that you require 250 μL virus per five million cells in a 30-mL volume, using 8 μg/mL polybrene for a MOI 0.3–0.4, prepare a mixture of ~200 million cells, 4.8 mL polybrene (2 mg/mL), and 10 mL virus in a total volume of 1,200 mL. Mix thoroughly by inversion.
4. Seed 30 mL of the mixture per 150-mm plate. Add 20% bleach to original vessel to decontaminate any remaining virus.
5. For adherent lines, let plates sit at room temperature on a level surface to ensure even adherence of cells.
6. Transfer plates to incubator. Ensure that they sit level.

7a. *Adherent cells.* Twenty-four hours post-infection, remove media using serological pipettes and dispose of media and pipettes in 20% bleach solution.

7b. *Suspension cells.* Twenty-four hours post-infection, transfer cells to 50-mL conical bottom centrifuge tubes, and pellet cells at 335 × *g* for 5 min. Remove media using serological pipettes and dispose of media and pipettes in 20% bleach solution.

8a. *Adherent cells.* Add 30 mL fresh media containing puromycin at the required concentration. Return plates to incubator.

8b. *Suspension cells.* Resuspend cell pellets in fresh media containing puromycin at the required concentration. Seed cells in new 150-mm plates and return to incubator.

9. Forty-eight hours later, remove media using serological pipettes and dispose in 20% bleach. Wash cells with warm PBS, dislodging any dead cells.

10. Collect cells from all plates into a single sterile container. Gently triturate cells with a sterile pipette to break up any remaining clumps. Mix cell suspension well and then measure the cell concentration.

    Note: From this point onward, there will be three parallel replicates maintained per screen (A, B, and C). Each replicate will maintain a minimum of 250-fold representation throughout the screen.

11. Prepare three separate, sterile containers, one for each new replicate. Prepare a mix of cells and media at the required density for normal maintenance of the cells – do not include puromycin at this, or any subsequent passaging.

    For example, if you would normally seed two million cells in a 150-mm plate for 3 days growth, prepare three separate containers, each containing 20 million cells in a total volume of 300 mL. Seed ten 150-mm plates, each with 30 mL of the suspension. This cell density will be based on the normal density and passaging requirement of the cell line.

12. Place plates in incubator.

13. Collect remaining cells in labeled 50-mL conical bottom centrifuge tubes. Collect at least two individual tubes, each containing 20 million cells. Pellet cells, remove media, and freeze the cell pellets at −80°C. These are the time zero (T0) samples.

14. Passage cells as normal – the infected cells may show a transient growth delay post-infection, and will return to a normal growth rate 5–10 days post-infection. The length and severity of this effect is cell line-dependent. Most lines require passaging every 3–7 days. Each time you passage, collect cells from all plates for

each individual replicate together in a single vessel. This serves to minimize stochastic growth effects in individual vessels within a replicate group. Do not mix cells from different replicates together.

15. Each time you passage, plate 20 million cells per replicate exactly as was done in steps 11–13 for T0.
16. Each time you passage, collect at least two pellets per replicate, each containing 20 million cells. Each pellet gets a time (*T*) number and a replicate designation. This number corresponds to the number of days post T0 that it was collected (e.g., T3_A, T21_B, etc.).
17. Continue screen for ~15–20 cell doublings. For most cells lines, this corresponds to 3–5 weeks, so screen samples will range from T0 to T21–T35.

3.5. Genomic DNA Isolation

See Note 5 prior to beginning protocol.

1. Thoroughly resuspend one cell pellet in 4.5 mL PBS by vortexing. Keep duplicate cell pellet at −80°C in case a mistake or contamination occurs with the first.
2. Add 500 μL Qiagen protease solution to cell suspension and vortex thoroughly to mix.
3. Add 6 mL Buffer AL, and mix thoroughly by inverting the tube 15 times, followed by additional vigorous shaking for at least 1 min.
4. Incubate in a 70°C water bath for 15 min.
5. Add 5 mL ethanol (96–100%) to the sample, and mix by inverting the tube ten times, followed by additional vigorous shaking. Do not use denatured alcohol, which contains other substances such as methanol or methylethylketone.
6. Carefully transfer the entire solution from step 5 onto the QIAamp Maxi column placed in a 50-mL centrifuge tube (provided), taking care not to moisten the rim. Close the cap and centrifuge at 1,850 × *g* for 3 min.
7. Remove the QIAamp Maxi column, remove the filtrate by aspiration, and place the QIAamp Maxi column back into the 50-mL centrifuge tube.
8. Carefully, without moistening the rim, add 5 mL Buffer AW1 to the QIAamp Maxi column. Close the cap and centrifuge at 4,500 × *g* for 2 min.
9. Carefully, without moistening the rim, add 5 mL Buffer AW2 to the QIAamp Maxi column. Close the cap and centrifuge at 4,500 × *g* for 15 min.

10. Place the QIAamp Maxi column in a clean 50-mL centrifuge tube (provided), and discard the collection tube containing the filtrate.
11. Dispense 600 μL of Buffer AE directly onto the membrane of the QIAamp Maxi column and close the cap. Incubate at room temperature for 5 min, then centrifuge at 4,500 × *g* for 3 min.
12. Reload the eluate (~300–600 μL) containing the DNA onto the membrane of the QIAamp Maxi column. Add an additional 200 μL Buffer AE to the column and close the cap. Incubate at room temperature for 5 min, then centrifuge at 4,500 × *g* for 5 min.
13. Collect flow through and transfer to 1.5-mL microcentrifuge tubes. Store samples at −20°C or proceed directly to Subheading 3.6.

3.6. Genomic DNA Precipitation

See Note 6 prior to beginning protocol.

1. Transfer gDNA to be precipitated to a 1.5-mL microcentrifuge tube. Determine the total quantity of gDNA to be precipitated based on volume and measured concentration.
2. Add 5 M NaCl to a final concentration of 0.2 M. Invert tubes five times to mix.
3. Add two volumes of −20°C 96–100% ethanol. Invert tubes five times to mix.
4. Centrifuge at 15,000 × *g* for 15 min at 4°C.
5. Aspirate supernatant, being careful to avoid gDNA pellet.
6. Add 500 μL −20°C 70% EtOH.
7. Invert tube five times to mix, then centrifuge at 15,000 × *g* for 15 min at 4°C.
8. Aspirate supernatant. Pulse-spin down any remaining liquid and aspirate it, being careful to avoid the gDNA pellet.
9. Air dry pellet ~5–10 min and resuspend with repeated pipetting in 10 mM Tris–HCl pH 7.5 to a final estimated concentration of 500 ng/μL.
10. Heat samples at 50°C for 1 h. Vortex repeatedly, followed by passaging sample four to five times through a sterile 30-gauge needle to fully solubilize and shear the gDNA.
11. Check concentration of sample and adjust to 400 ng/μL.

3.7. shRNA Barcode Amplification

See Note 6 prior to beginning protocol.

1. Combine the following reagents in a 1.5-mL microcentrifuge tube in the order indicated: 240 μL 10× PCR Amplification Buffer, 240 μL 10× Enhancer Buffer, 36 μL 10 mM dNTPs,

54 μL 20 μM Primer Mix, 24 μL 50 mM $MgSO_4$, 588 μL–X μL DNAase-free ddH_2O, 18 μL Pfx DNA Polymerase, X μL (30 μg) genomic DNA template.

2. Mix thoroughly by pipetting three-fourths volume ten times. Centrifuge briefly and aliquot 100 μL into each tube of a 12-well strip tube.
3. Place 12-well strip tube in a PCR cycler and run the following thermal cycler program: Step 1, 94°C for 5:00 min; Step 2, 94°C for 0:15 min; Step 3, 55°C for 0:15 min; Step 4, 68°C for 0:20 min; Step 5, Go to Step 2 (×30); Step 6, 68°C for 5:00 min; Step 7, 4°C hold.
4. Once complete, immediately remove samples from cycler and continue to Subheading 3.8. Do not store product for extended periods of time at 4°C or −20°C as this may enable the conversion of the double-stranded product to a restriction-digest resistant cruciform DNA structure (Fig. 2a).

3.8. shRNA Barcode Purification and Restriction Digestion

See Note 7 prior to beginning protocol.

1. Combine the twelve 100 μL PCR reactions into one 15-mL screw-cap tube. Take 10 μL and set aside in a 1.5-mL microcentrifuge tube to run on a gel later.
2. Add 6 mL Qiagen Buffer PB to the sample and vortex 10 s to mix thoroughly.
3. Prepare three QIAquick spin columns per sample by placing the columns in the provided 2-mL collection tubes. Load 750 μL sample onto each of the three columns. Do not decrease column number per sample otherwise columns will become saturated, resulting in significant loss of PCR product.
4. Centrifuge for 30 s at 15,000 × *g*.
5. Discard flow-through and place the QIAquick columns back into the same tubes. Repeat steps 3 and 4 until the entire sample has been passed though the columns.
6. To wash, add 750 μL Buffer PE to each column. Centrifuge for 30 s at 15,000 × *g*.
7. Aspirate flow-through from collection tube and replace column. Centrifuge for an additional 1 min at 15,000 × *g* to remove any residual wash buffer from the column.
8. Place each column in a new 1.5-mL microcentrifuge tube.
9. Add 40 μL Buffer EB to the center of each QIAquick membrane (three per sample). Let sit for 1 min, then centrifuge for 1 min at 15,000 × *g*.
10. Combine three eluates in one tube. Take 1 μL and set aside in a new 1.5-mL microcentrifuge tube to run on a gel later.

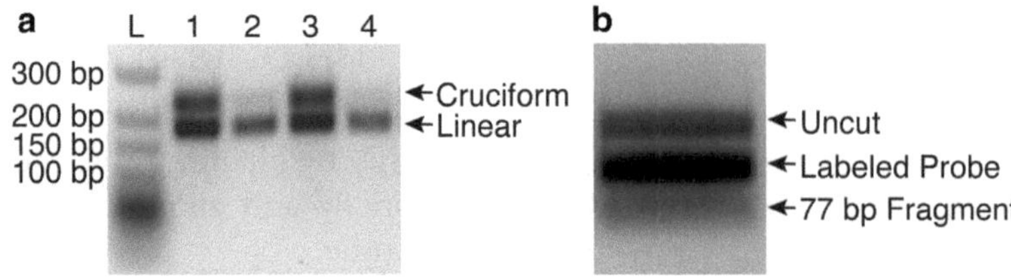

Fig. 2. Agarose gel images of PCR amplification and digestion of shRNA barcodes. (**a**) Agarose gel image of PCR amplicons pre- and post-PCR purification. The agarose gel should show a predominance of the ~177 bp linear PCR product. The ~225 bp product contains the same sequence of bases, in an alternative cruciform secondary structure. If this product is predominant on the agarose gel, the PCR reaction should be repeated using fewer cycles since this structure is resistant to restriction digestion. L, 100 bp ladder; 1, PCR reaction 1 pre-purification; 2, PCR reaction 1 post-purification; 3, PCR reaction 2 pre-purification; 4, PCR reaction 2 post-purification. (**b**) Agarose gel image of the digested PCR product. Post-digest, three bands will be evident on the agarose gel. The *middle band* of ~100 bp is the probe to be excised. The *upper band* is undigested product, and the *lower band* is the unlabeled half of the shRNA sequence.

11. Prepare the restriction digestion reaction: 105 μL purified PCR product, 12 μL 10× NEB Buffer 2, 1.2 μL 100× BSA (10 mg/mL), 2 μL *Xho*I (20,000 U/mL).
12. Mix contents thoroughly by pipetting. Briefly centrifuge to collect contents.
13. Add 60 μL into each of two PCR tubes and run the following thermal cycler program: Step 1, 37°C for 2:00 h; Step 2, 65°C for 20:00 min; Step 3, 4°C hold.
14. During the digest, run a 2% agarose TAE gel using fractions set aside in steps 1 and 10. The pre- and post-column purification samples should contain product bands of approximately equal intensity, and the primers should be absent in the post-column purification samples (Fig. 2a).
15. Following the digest, continue to Subheading 3.9. Alternatively, digested product may be stored at 4°C overnight, or for extended periods at −20°C.

3.9. Gel Purification of Digested shRNA Barcode

1. Prepare a 2% agarose TAE gel with wells capable of holding 150 μL. Thin, wide (~1.5 cm) lanes are preferred to avoid band smearing.
2. Combine the two digest reactions per sample in a 1.5-mL microcentrifuge tube. Add 20 μL 6× loading dye for a total volume of 140 μL.
3. Run gel for 45 min at 120 V to fully resolve bands.
4. Using a UV transilluminator, carefully excise the middle DNA fragment from the agarose gel with a clean, sharp scalpel. Minimize excess agarose in the excised fragment. The (middle) band of ~100 bp is the desired probe DNA, and should be the

brightest band of the three. Upper band is uncut PCR product, and lower band is the unbiotinylated half-hairpin fragment (~77 bp) released by the restriction digest (Fig. 2b).

5. Place a 15-mL screw cap tube on a tabletop balance. Tare balance. Transfer the gel slice to the tube and measure gel slice weight (should be ~300–400 mg).
6. Add three volumes of Buffer QG to the gel slice (e.g. 300 μL per 100 mg).
7. Incubate at 50°C for 15 min (or until the gel slice has completely dissolved). To help dissolve gel, mix by vortexing the tube every 2–3 min during the incubation.
8. Add 1 gel volume of isopropanol to the sample and mix.
9. Place a QIAquick spin column in a provided 2-mL collection tube.
10. Apply 750 μL of the sample to the QIAquick column and centrifuge for 1 min at 15,000 × *g*.
11. Discard flow-through and place QIAquick column back in the same collection tube. Repeat steps 10 and 11 until the entire sample has been passed though the column.
12. Add 750 μL of Buffer PE to the column and centrifuge for 1 min at 15,000 × *g*.
13. Discard the flow-through and centrifuge the column for an additional 1 min at 15,000 × *g*.
14. Place column into a clean 1.5-mL microcentrifuge tube.
15. Add 30 μL Buffer EB to column and let sit for 1 min. Centrifuge for 1 min at 15,000 × *g*.
16. Add an additional 30 μL Buffer EB to column and let sit for 1 min. Centrifuge for 1 min at 15,000 × *g*.

3.10. PCR Purification of Digested shRNA Barcode

See Notes 7 and 8 prior to beginning protocol.

1. Add 300 μL Buffer PB to the 60 μL sample generated in Subheading 3.9 and vortex 10 s to mix thoroughly.
2. Load entire sample (~360 μL) onto a QIAquick column and centrifuge for 30 s at 15,000 × *g*.
3. Discard flow-through and place the QIAquick column back into the same collection tube.
4. To wash, add 750 μL Buffer PE to each column. Let sit for 5 min then centrifuge for 30 s at 15,000 × *g*.
5. Aspirate flow-through from collection tube and replace column. Centrifuge for an additional minute at 15,000 × *g* to remove any residual wash buffer from the column.
6. Place each column in a new 1.5-mL microcentrifuge tube.

7. Add 30 μL Buffer EB to the center of the QIAquick membrane. Let sit for 1 min, then centrifuge for 1 min at 15,000 ×*g*.
8. Add an additional 30 μL Buffer EB to column and let sit for 1 min. Centrifuge for 1 min at 15,000 ×*g*.
9. Measure DNA concentration and purity using a Nanodrop (ThermoFisher) or equivalent instrument. The 260/280 and 260/230 ratios should be between 1.8–2.0 and 1.7–2.1, respectively. Total yields range from 3 to 6 μg in a ~55 μL volume. You require 2 μg to continue with the hybridization protocol.

3.11. GMAP Microarray Pre-conditioning

1. Let GMAP microarrays warm to room temperature for ~1 h prior to beginning protocol.
2. Fill the microarray slowly with 40°C 10 mM NaOH, then remove the liquid.
3. Fill the microarray again with 40°C 10 mM NaOH. Incubate with rotation in hybridization oven for 10 min at 40°C, 40 rpm.
4. Cut the head off of a 200-μL pipette tip and fit the tip onto a 5-mL syringe.
5. Rinse microarray slowly with 3–5 mL Wash Buffer A using the 5-mL syringe, collecting the flow-through in a beaker.
6. Remove Wash Buffer A, then fill the array with 0.0005% Triton®X-100 and incubate with rotation in hybridization oven for 10 min at 40°C, 40 rpm.
7. Empty the microarray, then rinse it twice with Wash Buffer A by filling and emptying the microarray.
8. Fill up microarray with Wash Buffer A.
9. Incubate with rotation in hybridization oven for 2–4 h at 40°C, 40–60 rpm.

3.12. Hybridization Mix Preparation

1. In a 1.5-mL microcentrifuge tube, prepare the probe mixture: 150 μL 2× hybridization buffer, 3 μL 50 mg/mL BSA, 3 μL 10 mg/mL herring sperm DNA, 2.9 μL 5 nM B213, 25 μL blocking oligonucleotide mix, 30 μL DMSO, X μL (2 μg) sample DNA, 100 μL–X μL ddH_2O.
2. Incubate the tube in a boiling water bath for 10 min.
3. Transfer sample to a 40°C water bath for 5 min.
4. Centrifuge tubes at 15,000 ×*g* for 5 min.
5. Remove Wash Buffer A from the microarrays and fill them with the sample solution, avoiding any debris in the bottom of the tube.
6. Cover the ports on the microarrays with Laser Tough-Spots® seals.

7. Incubate with rotation in hybridization oven for 15–16 h at 40°C, 60 rpm.

3.13. GMAP Microarray Staining and Washing

See Note 9 prior to beginning protocol.

1. Prime the fluidics station by putting all tubing in place, emptying the waste bottle, filling wash A and B bottles, and filling the deionized water bottle. Select GeneChip® Fluidics Station 450 protocol "Prime_450". Select the correct number of modules and click "Run".
2. Prepare SAPE labeling mix. Make two aliquots of 590 μL each per microarray in microcentrifuge tubes labeled (1) and (3): 600 μL 2× MES stain buffer, 48 μL 50 mg/mL BSA, 12 μL 1 mg/mL SAPE, 540 μL ddH_2O.
3. Prepare antibody labeling mix. Make 1 aliquot of 590 μL per microarray in a microcentrifuge tube labeled (2): 300 μL 2× MES stain buffer, 24 μL 50 mg/mL BSA, 2 μL 30 mg/mL IgG, 3.6 μL 500 μg/mL BAS, 270.4 μL ddH_2O.
4. Remove hybridization mix from microarray (~2 × 150 μL) and store at −20°C for potential future use.
5. Fill microarray with Wash Buffer A.
6. On scanner computer, under "Experiments", scan the barcode and type in the sample information.
7. Run the protocol "FlexGE-WS2v5_450" with the exception that the BWashTemp should be set to 30, not 50.
8. When the microarrays are ready, check for air bubbles and wash again if necessary.
9. Put Laser Tough-Spots® seals over the two microarray ports, then scan microarrays using the GeneChip® Scanner 3000.
10. When done with the fluidics, put all tubing in deionized water and run Shutdown_450 program.

3.14. GMAP Data Extraction and Analysis

1. Most GMAP users may elect to use publicly available data extraction and analysis tools. A full toolset and instructions for use are available online at: http://chemogenomics.med.utoronto.ca/supplemental/gmap/(27).

4. Notes

1. Virus aliquots used for titer and primary screen infections should be previously unthawed. Repeated freeze/thaw cycles will reduce the quality/titer of the virus preparations, resulting in inconsistent infection efficiency. Always use unthawed

aliquots of virus for infections, and freeze viral harvests in convenient (5 or 10 mL) aliquots post-harvesting.

2. Stringent safety procedures should be enforced when working with lentivirus. All lentivirus work should be done in accordance with your facility's biosafety regulations. All reagents that contact (or potentially have come in contact) with lentivirus should be immediately submerged into containers of freshly diluted 20% bleach inside of the biological safety cabinet. After 30 min in bleach, pipettes or other reagents may be transferred to the proper disposal container.
3. Polybrene and protamine sulfate are cationic polymers used to neutralize the charge repulsion between cells and viral particles. This effectively increases viral transduction, but can also have toxic effects on cells. One may forgo the use of polybrene or protamine sulfate if cells infect well in their absence, e.g. <2 mL virus per 150-mm plate of cells to be screened (for an MOI ~ 0.3–04).
4. Do not decrease the number of virus dilution points assessed when doing the MOI determination. High virus concentrations are toxic due to multiple integrations per cell. It is therefore necessary to test both high and low virus concentrations to distinguish cell survival/death based on appropriate MOI (0.3–0.4) versus death due to infection.
5. For all centrifuge steps, use a centrifuge and swinging bucket rotor combination that is capable of generating at least 4,500 × *g*. Use aerosol filter pipette tips for all pipetting steps to prevent contamination of genomic DNA. Do not use pipettes or other instruments that are used in preparation of PCR product and GMAP probe for preparation of genomic DNA samples as they can transfer contaminating DNA into the genomic DNA preparations.
6. Use aerosol filter pipette tips for all pipetting steps to prevent contamination of samples.
7. Do not add pH Indicator I to Buffer PB as it can adversely affect microarray hybridization.
8. The gel extraction process leaves contaminating chaotropic salts in the eluted probe that can be inhibitory to the hybridization process. Subsequent PCR column purification of the resulting product from Subheading 3.9 is highly recommended to remove these salts.
9. This protocol outlines programs and solutions to be used for array washing and staining, but assumes that user is familiar with the general use of the GeneChip® Fluidics Station 450 and GeneChip® Scanner 3000. For more detailed information, please refer to the Affymetrix website.

References

1. Ghildiyal, M., and Zamore, P. D. (2009) Small silencing RNAs: an expanding universe, *Nat Rev Genet* ***10***, 94–108.
2. Gondi, C. S., and Rao, J. S. (2009) Concepts in in vivo siRNA delivery for cancer therapy, *J Cell Physiol* ***220***, 285–291.
3. Hajeri, P. B., and Singh, S. K. (2009) siRNAs: their potential as therapeutic agents – Part I. Designing of siRNAs, *Drug Discov Today* ***14***, 851–858.
4. Lee, S. K., and Kumar, P. (2009) Conditional RNAi: towards a silent gene therapy, *Adv Drug Deliv Rev* ***61***, 650–664.
5. Singh, S. K., and Hajeri, P. B. (2009) siRNAs: their potential as therapeutic agents – Part II. Methods of delivery, *Drug Discov Today* ***14***, 859–865.
6. Elbashir, S. M., Harborth, J., Lendeckel, W., Yalcin, A., Weber, K., and Tuschl, T. (2001) Duplexes of 21-nucleotide RNAs mediate RNA interference in cultured mammalian cells, *Nature* ***411***, 494–498.
7. Fire, A., Xu, S., Montgomery, M. K., Kostas, S. A., Driver, S. E., and Mello, C. C. (1998) Potent and specific genetic interference by double-stranded RNA in *Caenorhabditis elegans*, *Nature* ***391***, 806–811.
8. Filipowicz, W., Jaskiewicz, L., Kolb, F. A., and Pillai, R. S. (2005) Post-transcriptional gene silencing by siRNAs and miRNAs, *Curr Opin Struct Biol* ***15***, 331–341.
9. Aza-Blanc, P., Cooper, C. L., Wagner, K., Batalov, S., Deveraux, Q. L., and Cooke, M. P. (2003) Identification of modulators of TRAIL-induced apoptosis via RNAi-based phenotypic screening, *Mol Cell* ***12***, 627–637.
10. Huesken, D., Lange, J., Mickanin, C., Weiler, J., Asselbergs, F., Warner, J., Meloon, B., Engel, S., Rosenberg, A., Cohen, D., Labow, M., Reinhardt, M., Natt, F., and Hall, J. (2005) Design of a genome-wide siRNA library using an artificial neural network, *Nat Biotechnol* ***23***, 995–1001.
11. Zheng, L., Liu, J., Batalov, S., Zhou, D., Orth, A., Ding, S., and Schultz, P. G. (2004) An approach to genomewide screens of expressed small interfering RNAs in mammalian cells, *Proc Natl Acad Sci USA* ***101***, 135–140.
12. Buchholz, F., Kittler, R., Slabicki, M., and Theis, M. (2006) Enzymatically prepared RNAi libraries, *Nat Methods* ***3***, 696–700.
13. Theis, M., and Buchholz, F. MISSION esiRNA for RNAi screening in mammalian cells, *J Vis Exp*.
14. Luo, B., Cheung, H. W., Subramanian, A., Sharifnia, T., Okamoto, M., Yang, X., Hinkle, G., Boehm, J. S., Beroukhim, R., Weir, B. A., Mermel, C., Barbie, D. A., Awad, T., Zhou, X., Nguyen, T., Piqani, B., Li, C., Golub, T. R., Meyerson, M., Hacohen, N., Hahn, W. C., Lander, E. S., Sabatini, D. M., and Root, D. E. (2008) Highly parallel identification of essential genes in cancer cells, *Proc Natl Acad Sci USA* ***105***, 20380–20385.
15. Moffat, J., Grueneberg, D. A., Yang, X., Kim, S. Y., Kloepfer, A. M., Hinkle, G., Piqani, B., Eisenhaure, T. M., Luo, B., Grenier, J. K., Carpenter, A. E., Foo, S. Y., Stewart, S. A., Stockwell, B. R., Hacohen, N., Hahn, W. C., Lander, E. S., Sabatini, D. M., and Root, D. E. (2006) A lentiviral RNAi library for human and mouse genes applied to an arrayed viral high-content screen, *Cell* ***124***, 1283–1298.
16. Schlabach, M. R., Luo, J., Solimini, N. L., Hu, G., Xu, Q., Li, M. Z., Zhao, Z., Smogorzewska, A., Sowa, M. E., Ang, X. L., Westbrook, T. F., Liang, A. C., Chang, K., Hackett, J. A., Harper, J. W., Hannon, G. J., and Elledge, S. J. (2008) Cancer proliferation gene discovery through functional genomics, *Science* ***319***, 620–624.
17. Silva, J. M., Marran, K., Parker, J. S., Silva, J., Golding, M., Schlabach, M. R., Elledge, S. J., Hannon, G. J., and Chang, K. (2008) Profiling essential genes in human mammary cells by multiplex RNAi screening, *Science* ***319***, 617–620.
18. Root, D. E., Hacohen, N., Hahn, W. C., Lander, E. S., and Sabatini, D. M. (2006) Genome-scale loss-of-function screening with a lentiviral RNAi library, *Nat Methods* ***3***, 715–719.
19. Bauer, J. A., Ye, F., Marshall, C. B., Lehmann, B. D., Pendleton, C. S., Shyr, Y., Arteaga, C. L., and Pietenpol, J. A. RNA interference (RNAi) screening approach identifies agents that enhance paclitaxel activity in breast cancer cells, *Breast Cancer Res* ***12***, R41.
20. Duan, Z., Weinstein, E. J., Ji, D., Ames, R. Y., Choy, E., Mankin, H., and Hornicek, F. J. (2008) Lentiviral short hairpin RNA screen of genes associated with multidrug resistance identifies PRP-4 as a new regulator of chemoresistance in human ovarian cancer, *Mol Cancer Ther* ***7***, 2377–2385.
21. Hurov, K. E., Cotta-Ramusino, C., and Elledge, S. J. A genetic screen identifies the Triple T complex required for DNA damage

signaling and ATM and ATR stability, *Genes Dev 24*, 1939–1950.

22. Brummelkamp, T. R., Fabius, A. W., Mullenders, J., Madiredjo, M., Velds, A., Kerkhoven, R. M., Bernards, R., and Beijersbergen, R. L. (2006) An shRNA barcode screen provides insight into cancer cell vulnerability to MDM2 inhibitors, *Nat Chem Biol 2*, 202–206.
23. Barbie, D. A., Tamayo, P., Boehm, J. S., Kim, S. Y., Moody, S. E., Dunn, I. F., Schinzel, A. C., Sandy, P., Meylan, E., Scholl, C., Frohling, S., Chan, E. M., Sos, M. L., Michel, K., Mermel, C., Silver, S. J., Weir, B. A., Reiling, J. H., Sheng, Q., Gupta, P. B., Wadlow, R. C., Le, H., Hoersch, S., Wittner, B. S., Ramaswamy, S., Livingston, D. M., Sabatini, D. M., Meyerson, M., Thomas, R. K., Lander, E. S., Mesirov, J. P., Root, D. E., Gilliland, D. G., Jacks, T., and Hahn, W. C. (2009) Systematic RNA interference reveals that oncogenic KRAS-driven cancers require TBK1, *Nature* ***462***, 108–112.
24. Klinghoffer, R. A., Frazier, J., Annis, J., Berndt, J. D., Roberts, B. S., Arthur, W. T., Lacson, R., Zhang, X. D., Ferrer, M., Moon, R. T., and Cleary, M. A. (2009) A lentivirus-mediated genetic screen identifies dihydrofolate reductase (DHFR) as a modulator of beta-catenin/GSK3 signaling, *PLoS One 4*, e6892.
25. Luo, J., Emanuele, M. J., Li, D., Creighton, C. J., Schlabach, M. R., Westbrook, T. F., Wong, K. K., and Elledge, S. J. (2009) A genome-wide RNAi screen identifies multiple synthetic lethal interactions with the Ras oncogene, *Cell 137*, 835–848.
26. Moffat, J., and Sabatini, D. M. (2006) Building mammalian signalling pathways with RNAi screens, *Nat Rev Mol Cell Biol 7*, 177–187.
27. Ketela pmid: 21548937.

Chapter 10

Plant DNA Sequencing for Phylogenetic Analyses: From Plants to Sequences

Susana S. Neves and Laura L. Forrest

Abstract

DNA sequences are important sources of data for phylogenetic analysis. Nowadays, DNA sequencing is a routine technique in molecular biology laboratories. However, there are specific questions associated with project design and sequencing of plant samples for phylogenetic analysis, which may not be familiar to researchers starting in the field. This chapter gives an overview of methods and protocols involved in the sequencing of plant samples, including general recommendations on the selection of species/taxa and DNA regions to be sequenced, and field collection of plant samples. Protocols of plant sample preparation, DNA extraction, PCR and cloning, which are critical to the success of molecular phylogenetic projects, are described in detail. Common problems of sequencing (using the Sanger method) are also addressed. Possible applications of second-generation sequencing techniques in plant phylogenetics are briefly discussed. Finally, orientation on the preparation of sequence data for phylogenetic analyses and submission to public databases is also given.

Key words: DNA extraction, DNA marker selection, DNA sequencing, Herbarium, PCR, Cloning, Plant systematics, Phylogenetics, Taxon sampling, Vouchers

1. Introduction

1.1. Plant Systematics and Phylogenetics

Systematics (or Taxonomy, in a broad sense) is the science of biological diversity; it is the study of relationships among organisms based on their evolutionary history (i.e. their phylogeny), leading to classifications of organisms and tools (including keys) that facilitate identification of species and other taxonomic groups (see e.g. ref. 1).

Phylogenetic trees are useful in the establishment of natural classifications and deepen the understanding of many biological processes, including character evolution, hybridization, polyploidy,

Gerard Cagney and Andrew Emili (eds.), *Network Biology: Methods and Applications*, Methods in Molecular Biology, vol. 781, DOI 10.1007/978-1-61779-276-2_10,

speciation and rates, and timing of species and lineages diversification. Phylogenies also provide the information needed to develop comparative systems in genomics and developmental biology (2). Phylogenetic studies are useful in multiple research fields, such as agriculture (to trace origin of crops and in breeding programmes), biotechnology and pharmaceutical sciences (e.g. for comparative analyses of metabolic pathways of different organisms), and medicine (e.g. to track the history of infections).

In the last few decades, in virtue of the increasing availability of DNA sequence data, there has been remarkable progress in our understanding of the phylogeny of major groups of plants, particularly among angiosperm lineages (e.g. (3–5)) and bryophytes (e.g. (6)). However, much work is still required at lower taxonomic levels, particularly regarding species-level phylogenies (2). There is still a long way to go until the "Tree of Life" for the plant species of the world is complete. The DNA barcoding of plants (7, 8), a methodology that uses DNA sequences for species identification, is also at the most preliminary stages. Thus, DNA sequencing and molecular phylogenetic methods continue to be fundamental research tools for the investigation of plant diversity and evolution.

DNA sequencing for plant phylogenetics is more than just a sequencing method: it involves specific problems related to project design, selection of species/taxa and DNA regions to be analyzed, gathering of plant material (often across the world), and sequencing samples that show a wide range of chemical components and various levels of genetic divergence. In this chapter, several of these important issues are summarized and discussed.

There are a number of useful, accessible references to introduce students or other researchers to phylogenetic analysis or plant systematics, including several books (e.g. (1, 9, 10)) and articles (e.g. (11, 12)). These provide sound foundations to the more specialized literature of molecular phylogenetics.

Other valuable molecular methods (polymerase chain reaction (PCR)-based techniques, molecular cytogenetics, etc.) for the study of evolution and diversity of plants are not discussed herein. For more information on these, see e.g. refs. 13, 14.

1.2. Designing the Project

When designing a project in molecular phylogenetics, consider: (1) past and current research on the systematics and phylogeny of the group of interest; (2) taxon sampling and biological material (plant samples) needed for analyses; (3) DNA regions to be analyzed; and, of course, available funds. An interesting project in plant molecular phylogenetics need not be very expensive, but it does need to be well-planned to be successful.

1.2.1. Review of Taxonomic and Phylogenetic Work

Prior knowledge on the systematics of a group of organisms is the starting point for taxon selection in any phylogenetic project.

The review of previous work should cover the entire taxonomic history of the group of interest: early studies often represent the most comprehensive work on taxonomy, particularly with respect to morphology and knowledge of overall species diversity. Rare or historical publications may be restricted to a few botanical libraries worldwide and so be difficult to obtain. Fortunately, multiple online databases facilitate literature searches and, in some cases, provide free online access to publications, including historical books and journals (see Note 1). Previous research on the plant group of interest may include molecular studies and DNA sequences (see Note 2). This prior sequence data can provide valuable information on the genetic variability of species/taxa, which will be useful for the selection of DNA regions or markers for further study (see Subheading 1.2.3). Previous molecular work can also provide valuable information on protocols specific to the plants of interest – for example, if they have a high mucilage or polysaccharide content, which DNA extraction protocol was optimal?

1.2.2. Taxon Sampling

The problem of taxon sampling has been extensively studied (e.g. (15–19)); results suggest that the greater and denser the sampling, the better for phylogenetic accuracy (16, 19). Therefore, it is important to sample as many species/taxa as possible from the group of interest ("ingroup"), and include multiple samples from related taxa as an "outgroup". Outgroups are used to root phylogenetic trees (see e.g. refs. 9, 20). Whenever possible, include more than one sample per ingroup species or taxon. Multiple samples per taxon can be useful to detect several problems, namely: (a) misidentification of plant material, (b) problems in taxon delimitation, and (c) sample contamination. If all species in the group of interest cannot be sampled, a representative sample of the group can be made by selecting species that represent the (1) morphological diversity within the group, (2) taxonomic divisions proposed for the group in current and previous classifications, and (3) geographical distribution of the taxa.

When studying plant groups that have a wide geographical distribution or include species from distant or inaccessible regions, there are often difficulties obtaining material for molecular work. Therefore, make an early, though preliminary, selection of taxa to be analyzed, and start searching immediately for the necessary samples. Optimally, these are from living plants (in the wild or cultivation), although herbarium specimens are also potential sources of DNA (see Subheading 1.3).

1.2.3. Selection of DNA Regions

Different genes or DNA regions in the three plant genomes (nuclear, plastid, and mitochondrial) vary considerably in their characteristics (nucleotide composition, secondary structure, functional constraints), and therefore evolve at different rates.

Because of this, different DNA regions contain different levels of phylogenetic information and can be useful at diverse taxonomic levels (see e.g. ref. 21). Genes that are highly conserved, in nucleotide sequence and structure, are suitable for the examination of evolutionary relationships among lineages that diverged a very long time ago, i.e. they allow the study of relationships at higher levels of the classification of organisms. DNA regions that have a high-mutation rate ("fast genes"), but that still have some constrains on size variation (allowing sequence homology assessments to be made, in the form of an alignment) are ideal for analyses of relationships at lower taxonomic levels, among closely related genera and species.

However, finding appropriate DNA regions for comparative sequencing is not simple. Mutation rates of genes may vary considerably across different lineages or even within the same plant lineage. In addition, levels of genetic diversity and divergence within a given taxonomic rank may vary considerably; it is important to remember that these ranks are artificial constructs that have been applied in an arbitrary and inconsistent manner over the tree of life. For instance, there are plant genera that have thousands of species while others have only one; some genera represent lineages that have been evolving for many millions of years while species in other genera diverged from each other in relatively recent times; even genera that contain similar numbers of species may have very different levels of genetic divergence.

Extensive molecular comparative analyses in plants have clearly shown that no DNA regions work equally well across different plant groups, and that the best choice of DNA markers for phylogenetics is lineage-specific (2, 22). For that reason, before a set of DNA regions is selected for analyses across a particular group, pilot, or preliminary, sequence studies should be conducted (2, 22, 23). In a pilot study, a small number of taxa from the group of interest (perhaps 2–3 closely related taxa, plus 2–3 others that are more distantly related) are selected and sequenced for several DNA regions that have proved useful in other studies at a similar taxonomic level. The amount and type of sequence variation seen for each DNA region in the exemplar taxa is compared and used to select markers that prove most informative for the taxon of interest. This sampling strategy should include a minimum of four exemplar taxa to be effective (22).

If molecular data are available in public sequence databases for the group of interest (see Note 2), they should be accessed and used for preliminary analyses to assess levels of genetic variation; such analyses provide further clues about DNA regions that could be useful. In addition, when planning to complement a prior study by adding sequences from other gene regions, a similar set of species or taxa should be used. It has been shown that combining gene data sets, even if they are incomplete, can provide valuable results (24); obtaining fully complete (squared) matrices, although preferable, is not essential. As discussed above, to increase phylogenetic accuracy, it is useful to increase the number of taxa sequenced, not just the number of genes ((19), and references therein).

A great number of DNA regions from the three plant genomes have had their potential utility as phylogenetic markers evaluated. For a comprehensive earlier review of the most widely used DNA regions in plant phylogenetics, see ref. 21. More recent studies have compared the relative utility of many plastid regions, including coding versus non-coding, not only for phylogenetic analyses (e.g. (23, 25)), but also for their potential utility as DNA barcodes (e.g. (7, 26–28)). Plastid genes that have been particularly useful at suprageneric and higher taxonomic levels include *rbc*L, *atp*B, *mat*K, and *ndh*F (e.g. (4, 21, 24, 29, 30)). Loci within the genes *mat*K and *rbc*L have been chosen as the DNA barcode for land plants (7).

For many years, nuclear ribosomal DNA (nrDNA), including the 18S and 26S genes, but in particular the ITS region (at lower taxonomic levels), was the dominant source of plant nuclear sequence data (31–33). However, interest in resolving phylogenies at lower taxonomic levels has increased the search for other potentially useful nuclear regions, in particular single-copy nuclear genes and low-copy nuclear genes (LCNG) (34–37). For discussion on strategies and markers that can be used at low-taxonomic levels, particularly for species-level phylogenies, see refs. 2, 22, 38.

Due to limitations in time or funding, researchers may have to choose between two sampling strategies: (1) sequencing a large number of taxa for a limited number of DNA regions; or (2) sequencing far fewer accessions and taxa for many more genes. The results of simulation analyses and some experimental studies suggest that, facing time and cost constrains, it is better to sequence a moderate number of DNA regions but many more taxa, than to sequence a very large number of genes for far fewer taxa (39). However, this can be project-specific – if there is little phylogenetic resolution with low sampling, adding taxa is unlikely to help while adding loci might. To improve the accuracy and support of phylogenetic analysis, it is also wise to sequence genes or DNA regions that have different characteristics and that are dispersed across the genome. This avoids biases that may be created by analyzing DNA regions that evolved under similar or identical constrains, or are somewhat linked (e.g. falsely increasing support due to non-independent evolution of sites) (39). It is also important to remember that different genomes have different evolutionary histories due to their modes of inheritance; the nuclear genome has biparental inheritance while in most angiosperms, for example, inheritance of the plastid genome is maternal (e.g. (40, 41)). This can be used when considering, for example, the effects of hybridization in a plant's history.

1.3. Sources of Plant Material

1.3.1. Living Plants: Botanical Gardens and Germplasm Collections

Living plants are the best sources of plant material for molecular analyses because they provide the highest quality DNA. Some species are available as living accessions in botanical gardens, or can be obtained from germplasm collections as seeds, bulbs, corms, or rhizomes (see Note 3). A limited number of species (usually crops or garden plants) can be obtained from commercial sources. In all cases, it is essential to verify the identification of plant material, as misidentification and mislabelling exist even in the best collections. Weedy plants are quite capable of "jumping" pots in a greenhouse and outcompeting the original inhabitant. It is also essential that, concurrent with the removal of DNA samples from living plants, a voucher (preferably a reproductive sample, e.g. a flowering or fruiting specimen) is collected for every plant sampled (see Subheading 1.4) – if this is done, the identity of accessions with phylogenetically surprising placements can be verified prior to publication.

Germination of seeds or spores may require special treatments, so check literature or inquire among colleagues for the best germination protocol for the plants of interest. When growing plants from seeds, be aware that mixed seeds can be found in a single package (accession): before sowing, examine seed packages carefully to check for mixed samples. When growing plants whose identification has not yet been rechecked, it is better to allow them to reach fertility before DNA samples and vouchers are collected. If samples are needed at an earlier stage, label the plants used for DNA extraction and allow them to develop fully, preferably until fertile; then, collect the voucher. Be careful not to let plants die before vouchers have been collected; it is better to have a non-fertile voucher than no voucher at all (see Subheading 1.4).

It is also possible to obtain plant DNA samples from various plant science organizations; for a list of the most important Plant DNA Banks, see ref. 42. However, when requesting DNA samples from other laboratories, make sure to receive complete voucher information (and specimen images, if possible) together with the DNA samples (see Subheading 1.4).

1.3.2. Living Plants in the Wild: Fieldwork

For most plant species, living material is only available from the wild, often distant from research centres. For that reason, when designing a project in plant phylogenetics, one should plan for fieldwork whenever possible, and consider the time and funds required for organizing and completing one or more collection trips. Fieldwork provides valuable plant material for laboratory analyses, helps enlarge and diversify herbarium and germplasm collections, and increases knowledge of the taxon's biology and ecology.

Planning a field collection trip requires many months of preparation. First, information on the geographical distribution of the species of interest should be gathered, to target field work. Floras and other floristic publications contain very useful information on species distribution in regions or countries. However, often more detailed information on localities is required to find particular species in the wild. Such locality information can be obtained from specimen labels in existing herbarium collections. It is also important to be able to recognize the species in the field. Therefore, study of herbarium collections and familiarization with the morphology of the species is a prerequisite to fieldwork. If the local herbarium is not particularly rich in the relevant species, visit other herbaria and/or request loans of specimens from other collections. Information on existing herbaria and collections is available in *Index Herbariorum* (see Note 4).

After choosing fieldwork target areas, contact natural resource agencies and local botanical organizations to obtain information on collecting regulations and permits. Plant collecting is restricted within natural parks and other protected areas, but is usually allowed when requested for scientific purposes. When planning fieldwork in other countries, collaboration with local botanists facilitates organization and eases the process of obtaining collecting permits. Additional permits may be required to export/import plant material from/to other countries. Apply for any permits several months in advance of fieldwork. Also, consider obtaining funding for overseas colleagues who provide assistance in the field.

Advice on fieldwork techniques, plant collecting and herbarium specimen preparation is available in various publications (e.g. (1, 42, 43); see Note 5). For DNA analyses, silica gel drying is normally the recommended method of field preservation of plant samples ((44); see Subheading 3.2). This has been successfully used in a wide variety of plants, generally with very good results in terms of DNA quality. However, in some plants (e.g. ferns, parasitic plants) extraction from silica gel dried material has resulted in low DNA yields. For those plants, preservation (pickling) in saturated NaCl/CTAB solution has given good results (42, 45–48). Ethanol treatment (prior to air drying) is also used in the field to preserve plant material and avoid fungal growth. The use of ethanol should be avoided in material intended for DNA sampling as it accelerates DNA degradation (46, 49), although it can still be used for the accompanying voucher specimen. Live samples are ideal, but can only be brought back from international travel with relevant permits.

1.3.3. Herbarium Collections

Herbarium specimens are valuable sources of DNA for molecular analysis (49, 50), especially for sequencing short highly repeated regions in the plant genomes (plastid, mitochondrial, nuclear ribosomal), which can usually be easily amplified even from degraded DNA. The use of herbarium specimens is also critical for DNA barcoding – for most of the 300,000 species of land plants, the only specimens available are those in herbarium collections, and obtaining fresh material for all plant species in the form of new field collections is unrealistic in the short or medium term (51, 52). There is considerable interest in extracting DNA from old and historical specimens (53).

DNA extracts from herbarium specimens are often degraded and low in concentration, and some may have PCR inhibiting activities (50, 53). However, the use of plant DNA extraction kits, such as the DNeasy Plant Mini Kit (Qiagen), considerably increase extraction success, even from old herbarium specimens (53, 54).

In order to complete taxon sampling, many plant studies need to use DNA obtained from herbarium specimens. In spite of its low yields, this can be used successfully in DNA sequencing.

Herbarium collections are very valuable and, in many cases, irreplaceable. The specimens sampled for DNA analysis must be carefully selected. This involves careful study of the morphology of plant material to be certain of what is being sampled. In poorly studied plant groups or collections, misidentification is common. Species names on the labels may be right, but double check the identification of specimens to be certain.

Limit the amount of material removed from specimens, as this affects their quality and value for other studies, for example, morphology or anatomy. In addition, DNA extraction protocols can be optimized with fresh samples of related species prior to use of herbarium samples – and with common herbarium collections before proceeding to rare ones.

Usually, more recently collected specimens are better for DNA extraction. However, in general experience ((54), SSN and LLF personal experience), specimens that look "greener" (rather than brown) provide DNA of better quality no matter the age. The green colour in specimens often means that the drying process was fast, but, of course, the colour of the specimen also depends on the species. Herbarium samples that appear similar to material from related plants that have been rapidly dried (e.g. in silica gel) are optimal. Sometimes, dark discoloured leaves are indicative that the specimen has been dried with ethanol (see Subheading 1.3.2). Unfortunately, the herbarium labels often do not reflect this, although for recently collected specimens it may be possible to make enquiries from the collector themselves. DNA extraction from ethanol-dried material is usually unsuccessful.

Once the specimens have been selected, permission must be requested from herbarium staff before removing any samples, for molecular or any other analyses. If allowed to remove the samples personally, place on each specimen a small label with your name, date, type of sample removed (leaf, fruit, wood section or other), an accession number, and the type of study (molecular, anatomical or other). In some herbaria, the staff prefers to remove the samples themselves and place the "sample" labels in the specimens.

The herbarium specimen that was the source of the sample is now the DNA voucher (see Subheading 1.4), so record full information about the specimen. Whenever possible, digital images (photos) should be obtained from every specimen sampled. This is particularly useful if the specimens are deposited in collections that, for logistic reasons, cannot easily be re-examined.

1.4. The Importance of Vouchers

A voucher is a herbarium specimen that needs to be prepared for every plant specimen used for molecular (or other) analyses. Vouchers are fundamental for verification of species identity; they are also sources of morphological or other characters, which are useful for interpreting character evolution. Vouchers are also a requirement for publication; most journals do not accept molecular phylogenetic work without voucher information.

All vouchers should be deposited in a public herbarium, preferably in an institution that can guarantee long-term preservation of specimens. The voucher specimens must be collected at the same time as the plant samples for DNA (or other) analyses. If samples (plant material or DNA) are obtained from other researchers or organizations, either: (a) get the vouchers together with the samples, or (b) get complete voucher information (collection details, herbarium location) and, if possible, images (photos) of the vouchers. Do not trust that this can easily be obtained at a later date. In addition, keep well-organized records about all accessions and samples, and the corresponding vouchers. These records are most useful when maintained in spreadsheet or database format.

Preparing herbarium specimens (vouchers) is a simple and inexpensive procedure. Information on how to prepare herbarium specimens is available in several manuals or books (e.g. ref. 1) and Web sites (see Note 5). Advice may also be obtained from staff of the herbarium, where the vouchers will be deposited.

2. Materials

In the following list of materials, general equipment and consumables that are typically available in molecular biology laboratories, such as benchtop centrifuge, water bath, heating block, orbital shaker, vortexer, thermocycler, gel tank and trays, UV or blue-light transilluminator, laminar flow hood, fume hood, water purification system, autoclave, micropipettes, tips, microcentrifuge tubes, tube racks, are omitted.

2.1. Preparation of Plant Samples and Vouchers

1. Silica gel desiccant (28–200 mesh, Grade 12), available from Fisher Scientific or Sigma-Aldrich.

2. Silica gel self-indicating type (blue or orange) from various companies (orange silica gel is considered to be safer).
3. Liquid nitrogen (optional).
4. Ziplock plastic bags, paper labels, and permanent markers.
5. Fieldwork equipment and tools (see e.g. ref. 43).
6. Voucher preparation: Plant press, drying paper, newspaper, paper packages, mounting paper, labels (archival paper), glue, and others (see Note 5).

2.2. Plant DNA Extraction

2.2.1. CTAB Method

1. Plant tissue disruption:
   (a) Plant samples: Freshly collected or dry (silica gel dried or herbarium material).
   (b) Sterile sand (acid washed sand).
   (c) Liquid nitrogen (not required for dry plant material).
   (d) Sterile mini-pestle (or ceramic pestle and mortar, particularly for grinding larger samples), OR Mixer Mill (Retsch) or TissueLyser (Qiagen), adapter racks for microcentrifuge tubes, stainless steel cone balls (5 mm; Retsch), or tungsten carbide beads (3 mm; Retsch or Qiagen).
2. Plant DNA extraction:
   (a) CTAB buffer (2×): 2% (w/v) CTAB, 1.4 M NaCl, 20 mM EDTA, 100 mM Tris–HCl (pH 8.0).
   (b) β-mercaptoethanol.
   (c) Polyvinylpolypyrrolidone (PVPP).
   (d) Proteinase K (Invitrogen), 10 mg/mL (optional).
   (e) "Wet chloroform": 24:1 (v/v) chloroform/isoamyl alcohol (see Note 6).
   (f) Wash buffer: 76% (v/v) ethanol, 10 mM ammonium acetate.
   (g) TE buffer: 10 mM Tris–HCl (pH 7.4), 1 mM EDTA.
   (h) RNase A (optional).

2.2.2. Plant DNA Extraction Using Kits

DNeasy® Plant kits (Qiagen), NucleoSpin® Plant II kit (Clontech), or other.

2.3. Agarose Gel Electrophoresis and DNA Quantification

1. SeaKem® LE Agarose (Lonza), Fisher Molecular Biology Grade Low EEO/multipurpose Agarose, or similar.
2. TBE or TAE buffer (10× stock solution; 1× or 0.5× working solution).
3. DNA gel stain: GelRed™ or GelGreen™ (Biotium), SYBR® Safe (Invitrogen), or ethidium bromide.
4. Loading buffer (e.g. 30% glycerol, 0.25% bromophenol blue).

5. DNA ladders (e.g. 1 Kb Plus DNA and Low DNA Mass ladders, from Invitrogen).
6. DNA quantification: Thermo-Scientific's NanoDrop® Spectrophotometer, Invitrogen's Qubit® Fluorometer.

2.4. DNA Amplification (PCR)

1. Genomic DNA.
2. Sterile double-distilled water (from inhouse purification, or available e.g. from Fisher Scientific).
3. Oligonucleotide primers for the relevant DNA regions (usually a 10 μM working solution).
4. *Taq* polymerase and reaction buffer.
5. $MgCl_2$ (if not included in the reaction buffer).
6. Deoxynucleotide triphosphates (dNTPs) mix (usually a 10 mM working solution).
7. Dimethyl sulfoxide (DMSO) (optional PCR additive).
8. Bovine serum albumin (BSA) (optional PCR additive).
9. Betaine (optional PCR additive).
10. PCR tubes, strips or plates (96 wells), and caps or sealing mats.

2.5. PCR Product Purification

1. PCR product purification kit: e.g. Illustra™ GFX™ PCR DNA and Gel Band Purification Kit (GE Healthcare), NucleoSpin® Extract II Kit or NucleoTraP®CR PCR purification kit (Clontech), QIAquick® PCR Purification kit and QIAquick® Gel Extraction Kit (Qiagen), or Montage® PCR (Millipore, for PCR products larger than 300 bp).
2. For high-throughput PCR product purification: MultiScreen$_{HTS}$® PCR$_{96}$ Filter Plate, with MultiScreen$_{HTS}$®™ Vacuum Manifold (Millipore), OR enzymes (e.g. ExoSap-IT™ (GE Healthcare), or Exonuclease-I (ExoI) and shrimp alkaline phosphatase (SAP)).
3. Enzymatic clean-up: ExoI and SAP, or ExoSap-IT (GE Healthcare).

2.6. Cloning of PCR Products

1. LB (Luria–Bertani) broth base (Sigma-Aldrich, No. L-3022).
2. Micro agar (Duchefa, Cat. No. M1002) or agarose.
3. Distilled water, and double-distilled water (ddH_2O).
4. Glycerol (80%).
5. Ampicillin or kanamycin (50 mg/mL).
6. Toothpicks (or other tool to pick colonies).
7. Petri plates (plastic) or autoclaved glass plates.
8. PCR plates (96 wells) and sealing mats or lids.

9. Microtiter plates (96 wells).
10. TOPO® TA Cloning® Kit for Sequencing, with TOP10 chemically competent *Escherichia coli* cells (Invitrogen), or other cloning kit.
11. *Taq* polymerase.
12. PCR product.
13. Other PCR reagents: Reaction buffer, primers, dNTP mix, $MgCl_2$.
14. Agarose gel electrophoresis reagents (see Subheading 2.3).
15. PCR product purification kit (see Subheading 2.5) or, for high-throughput, $\text{MultiScreen}_{\text{HTS}}$® PCR_{96} Filter Plate, with $\text{MultiScreen}_{\text{HTS}}$®™ Vacuum Manifold.

2.7. DNA Sequencing: Sanger Method

1. Dye terminator cycle sequencing kit appropriate to the automated sequencing machine: e.g. BigDye® Terminator v1.1 and v3.1 Cycle Sequencing Kits (Applied Biosystems), GenomeLab™ DTCS with Quick Start Kit (Beckman Coulter), or other.
2. Automated sequencer and supplies: e.g. ABI PRISM® 377 (Applied Biosystems, Life Technologies), CEQ™ 8000 Genetic Analysis System (Beckman Coulter).
3. Reagents for ethanol precipitation (as recommended for the particular sequencing chemistry), OR column-based cleaning of sequencing reactions, e.g. Sephadex™ G-50 fine and Centri-Sep™ columns (Applied Biosystems) or plates (optional).

3. Methods

3.1. General Notes on Laboratory Work

One of the most important things in laboratory work is maintaining a good laboratory book. Often, this book remains the property of the laboratory, and the notes in it should allow co-workers to understand what has been done, and why. Each experiment should be dated and self-explanatory. In the face of criticism or disbelief, a laboratory book is evidence to prove the experiment was done and back up results. When publishing a study, the laboratory book is the primary resource for the methods. It is also invaluable in explaining unexpected results: returning to the original notes may help track down potential labelling or sampling errors (for example, if a sequence that should be from one genus comes out unexpectedly in a phylogeny, what other samples were extracted that day – could there have been tube inversions or DNA contamination rather than horizontal gene transfer or remarkable morphological convergences?).

A second general rule is: Aliquot. Freeze–thaw cycles are harmful to many reagents, so it is better to keep reagents and samples in multiple small easily thawed tubes. Using aliquots also reduces incidences of stock contamination. Reagents may be accidentally left sitting on the bench over the weekend, or thrown away with the ice: doing this with an aliquot is far less painful than doing it with the stock solution. It is best to aliquot the entire tube or bottle when you first open it (see Note 7).

3.2. Preparation of Plant Samples

Fresh leaf material immediately frozen in liquid nitrogen is the ideal choice for DNA extraction. However, freezing plant samples in the field is not usually possible or advisable, for logistic or safety reasons. When freezing is not possible, the best choice is to quickly dry the plant in silica gel (44). Samples from succulent or other xerophytic plants (cactus, bromeliads, pine needles, etc.), as well as bryophytes, can be kept in plastic bags for several hours, up to a day, at ambient temperature without significant DNA degradation. This facilitates bulk processing of samples (freezing or drying in silica gel) at the end of the day (see Note 8). Most plant species can remain "fresh" and without significant DNA degradation for 1–2 days when kept in the refrigerator (4°C) after collection.

Drying samples in silica gel:

1. Prepare ziplock bags with silica gel (mostly non-indicating type, with a small portion of self-indicating silica).
2. Collect fresh leaves or thallus (preferably, although other plant materials, such as herbaceous stems or inflorescences, can be used) and place them in silica gel bags. Large leaves should be torn into smaller pieces. The ratio of plant material to silica gel should be 1:10 or greater (44), but if plants are fleshy or slightly damp because of weather conditions, use more silica; if plants are wet, blot dry the surface first. Place a label (paper) with sample information in pencil (species name, locality, accession number) inside the bag as well as annotating the outside of the bag with permanent marker. Distribute (shake) silica gel evenly around the sample, and leave bags for several hours (in a vehicle, do not leave them exposed to sunlight, as high heat will cause DNA degradation).
3. After 24 h, check if the sample is dry ("crispy"). If it is not dry and silica gel seems saturated (crystals of the indicating silica have changed colour), add more silica or replace it (although this is not usually necessary).
4. After 2–3 days, remove used silica gel (recommended, but not essential), and place bags in a cold room or freezer (−20°C or −80°C) for long-term storage. After samples have been dried in silica gel, they can be kept at room temperature in dry conditions for many days or months without the effect on DNA quality. Silica gel can be removed prior to air travel, firstly because it can add considerably to baggage weight, and

secondarily because fine white powders may be viewed suspiciously by security and customs personnel.

Reusing silica gel:

5. The used (humid) silica gel can be reused; simply clean the silica of any plant remains with a pair of tweezers, and then dry it in an oven at 150–175°C for 30 min up to 1 h. If the silica is cleaned, there is minimal risk of cross-contamination of samples because the high heat destroys the DNA of any plant remains. Short exposures to ultraviolet light may also be used to destroy any remaining DNA. The silica gel can be reused and heated many times without losing its drying properties; only a change in colour is noticeable (the white, non-indicating silica will turn beige). More indicating silica can be added. Reuse may not be advisable for delicate plant material that easily breaks into powder when dry, as it will be difficult and time-consuming to clean the silica.

3.3. Plant DNA Extraction

Many laboratories regularly use plant DNA extraction kits in their molecular work. However, high-quality DNA can also be obtained with a modified CTAB method (55, 56), which is simple and inexpensive. The CTAB method has been successfully used in a wide range of plants from either fresh, silica-gel dried or herbarium material. However, in some plant groups, especially those where contents of polyphenols or polysaccharides are high, other protocols are recommended (e.g. (57, 58)). For a review of methods and common problems in plant DNA isolation, see ref. 59.

When extracting from multiple accessions, try to alternate taxa whenever possible. One extraction can quite easily be contaminated by another, particularly if filter tips (e.g. Molecular BioProducts ART® Aerosol Resistant Tips) are not being used, and the most likely way for this to happen is the contamination of a sample with the DNA that is physically closest to it. Alternating taxa will not help avoid cross-contamination, but can help subsequently identify it.

3.3.1. CTAB Method

The following procedure is based on a widely used modified CTAB method (55, 56, 60, 61), which has been proven effective in a wide range of plants:

1. Add 0.2% (v/v) of β-mercaptoethanol to 2× CTAB buffer just before use. Prepare ca. 1.2 mL of CTAB mixture per sample in a tube. Mercaptoethanol is toxic; work in a fume hood (see Note 9).
2. Preheat the total volume of 2× CTAB buffer to 65°C in a water bath or heating block.
3. *Grinding and homogenization of plant material.* Physical maceration is required to rupture plant cell walls. If working

with dry plant samples (which can be left at room temperature), prepare all tubes in succession. If working with fresh samples that need to be maintained frozen with liquid nitrogen, prepare (and grind, if using mini-pestle) one sample at a time. Some taxa can be ground directly, without prior drying or freezing. Be extremely careful to avoid cross-contamination. Between samples, clean utensils (forceps, tweezers, scissors) used to cut or handle plant material with water and 70% ethanol.

Option 1 (*Using the mixer mill – dry or liquid nitrogen frozen plant samples*): Cut a piece of leaf (1–2 cm^2 or 0.1–0.5 g) or other plant material (herbaceous stems, inflorescences, and even cambium can also be used), and place it inside a 2-mL microcentrifuge tube. Add one cone ball (or three 3-mm tungsten carbide beads) to each sample tube (a pinch of sterile sand can also be added to help macerate hard tissues). Place tubes in the mixer mill tube adapters; distribute tubes equally between the two adapters to balance. When dealing with fresh plant material, small holes should be punched in the microcentrifuge tubes to allow for pressure changes during freezing and thawing, and adapters and tubes should be frozen in liquid nitrogen. Grind samples in the mixer mill for 30 s to 1 min at a frequency of 20–30 cycles per second (see Note 10). Once plant material is completely ground (to powder), add 1 mL of preheated 2× CTAB (with β-mercaptoethanol) and a pinch of PVPP, and then mix contents gently. If a hole has been punched in the tube, transfer to a new tube now. If the tube is under any pressure, the grinding ball can shoot out – wear eye protection.

Option 2 (*Using a mini-pestle – for fresh or dry material*): Add a pinch of sterile sand to a 2-mL microcentrifuge tube. Cut a piece of leaf (1–2 cm^2 or 0.1–0.5 g) or other plant material, and place it inside the tube. With fresh material, freeze the sample by partially submerging the open tube in liquid nitrogen for 30 s; do not let liquid nitrogen enter the tube. Lift the tube and grind plant tissue with a sterile mini-pestle (either by hand or using a pestle inserted into an electric drill); keep the sample frozen (by dipping the tube repeatedly in liquid nitrogen) until tissue is completely ground. If you are working with unfrozen material, add a little buffer prior to grinding. When using dry plant material, it is not necessary to freeze the sample during grinding. Finally, add 200 μL of the preheated 2× CTAB (with β-mercaptoethanol) and a pinch of PVPP, and continue to macerate the tissue with the pestle. Add a further 800 μL of CTAB (or enough CTAB for the sample to be liquid rather than a paste; exact amounts will depend on the plant tissue used) and mix contents gently by inverting the tube or on a vortex machine.

4. Incubate samples at 65°C for 45 min to 1 h; mix tube contents gently every 10 min. Heat helps to lyse membranes and inactivate nucleases while chelating agents in the buffer also inhibit nucleases (62), thus protecting the extracted DNA.
5. Leave samples to cool to room temperature for a few minutes.
6. Add 700 μL of "wet chloroform" and mix gently to obtain a momentary single phase, for 5–10 min (or 10–20 min when using an orbital shaker). The chloroform causes surface denaturing of proteins while the alcohol helps to keep the phases apart in steps 7–8 (62). Keeping tubes horizontal while shaking increases contact between phases, allowing for better mixing.
7. Centrifuge for 10 min at 13,000 × *g* (see Note 11).
8. Two phases (layers of liquid) should be visible. The DNA is in the upper (aqueous) phase while many proteins and compounds such as chlorophyll are in the lower (chloroform) phase, which is often strongly coloured (usually green). A semi-solid band is often visible between the two phases. Most of the plant debris and any sand or metal beads will be at the bottom of the tube. With a micropipette, carefully transfer the aqueous upper phase (supernatant) to a new tube (1.5 mL volume), taking care not to pick up parts of the semi-solid band or the lower phase. If a small amount of chloroform is accidentally picked up, it will usually settle to the bottom of the pipette tip and can be expelled. If the phases get too mixed, the sample can simply be re-spun.
9. Repeat the chloroform wash step by adding 700 μL of "wet chloroform" to the recovered aqueous phase. If using the mixer mill and grinding beads, which can be reused, recover beads from the bottom of sample tubes (see Note 12).
10. Repeat steps 7–8. This time, little or no debris should be seen, and both phases should be almost colourless, so more careful pipetting is required.
11. Transfer aqueous supernatant to a new 1.5-mL microtube and add 700 μL of cold isopropanol (ca. 0.6–0.7 volumes relative to the recovered supernatant) (at −20°C; taken from the freezer just before use). Place tubes in a freezer (−20°C) for at least 2 h; this is the DNA precipitation stage. Longer precipitation times (overnight to several days) are recommended for samples obtained from herbarium material. Samples where very low recovery of DNA is expected may benefit from the addition of ca. 1 μL of glycogen, which can act as a DNA carrier and help improve pellet visibility. DNA precipitation occurs most effectively in a high salt concentration; this salt should already be present in the supernatant. Isopropanol precipitation is used

instead of ethanol precipitation as isopropanol precipitates DNA at lower concentrations.

12. Centrifuge at 13,000 × *g* for 10 min to pellet DNA (see Note 13).
13. Remove and discard all supernatant by tipping off supernatant and/or with a micropipette. The DNA should be visible as a small, thin and clear or whitish pellet (PCR from discoloured DNA can be successful, although post-extraction clean-up or using dilutions of the CTAB extraction usually increases success; see Note 14).
14. Add 1 mL wash buffer. Make sure that the pellet is released from the bottom of the tube. Then, leave tubes in the refrigerator (4°C) for at least 1 h, preferably for several hours (overnight, up to a maximum of 24 h). The wash buffer removes salts that co-precipitated with the DNA. Replacing the isopropanol that was used to precipitate the DNA with ethanol (in the wash buffer) also makes the DNA pellet easier to dry.
15. Centrifuge at 13,000 × *g* for 5 min. Remove the supernatant (again, by tipping and/or pipetting) and leave tubes to air dry for 1 h, or dry in a spin-vacuum. A whitish DNA pellet will often become transparent after drying.
16. Re-suspend DNA in 100–200 μL TE buffer. If the DNA pellet is very small (e.g. extracts from herbarium material), one can reduce the volume of added TE to 25 or 50 μL (see Note 15).
17. (Optional) Add 1–2 μL of RNAse (final concentration 50–200 μg/mL) and incubate at 37°C for 45 min to 1 h to remove any co-precipitated RNA from the extraction.
18. Store DNA extracts at 4°C for short-term use, or at −20°C, for long-term storage. Avoid repeated freeze–thaw cycles, which are damaging to DNA. Aliquoting a working solution from stock DNA, and keeping the working aliquot at 4°C, is advisable. Some laboratories avoid storing stock DNA at −80°C due to concerns about nicks developing in the DNA strands. However, for most phylogenetic purposes, sample storage at −80°C is acceptable.

The use of a CTAB protocol instead of a DNA isolation kit in general DNA extractions considerably reduces laboratory extraction costs (although as the resulting DNA is often of lower quality, use of the method can increase downstream costs with less successful PCRs and more time required for optimization). Using a DNA extraction kit for plant samples that have been difficult to obtain or cannot be replaced is recommended. When there is sufficient plant material (fresh or silica gel dried) and the DNA extraction fails at the first try, the extraction can always be repeated with another method. But failure at the first, and only, try when using rare or irreplaceable plant samples is to be avoided. Plant DNA extraction kits are optimized for recovery and purification of DNA in a great range of plants, so the chances of success considerably increase with their use.

Once again, the importance of adequate record keeping for all plant and DNA samples is emphasized; this includes proper labelling of DNA tubes, preferably with a numbering and database system that facilitates the finding of DNA tubes and corresponding plant information. Each research organization or laboratory should implement a record system that unambiguously links DNA tubes to the original plant accession information; these records should include complete voucher information (i.e. plant collection details and herbarium location – see Subheading 1.4).

3.3.2. Plant DNA Extraction Using Kits

The most commonly used plant DNA extraction kits utilize plastic columns with permeable membranes. Columns are used both to remove plant debris (in which case, the DNA is eluted through the column) and to trap the DNA (in which case elutes are discarded, until the final step when the DNA is removed from the column) – for example, Qiagen's Plant DNeasy kits and Macherey-Nagel's NucleoSpin Plant II kits (Clontech). Prior to using a kit, carefully read the manufacturer's instructions, including the technical notes and troubleshooting sections (see Note 16).

An alternative to DNA extraction is offered by Clontech, which sells Terra PCR polymerase optimized to work directly from plant tissue. This may be of use when only amplifying one or two regions per plant sample; in all other cases, extraction of high-quality DNA is the best strategy.

3.4. Agarose Gel Electrophoresis and DNA Quantification

DNA extracts and PCR products should be run on an agarose gel for verification of extraction or PCR reaction success, as well as for rough quantification and estimation of size (length) of the DNA molecules. Agarose gel electrophoresis is a well-known laboratory technique, and detailed protocols are given in many references (e.g. (63)).

In general, a 1% agarose gel (1 g agarose per 100 mL 1× or 0.5× TBE or TAE buffer) is adequate for electrophoresis of DNA extracts and PCR products, although some laboratories routinely use 0.7–0.8% agarose gels to reduce expenses. The same gel can be used several times, particularly if the DNA stain is in the gel tank buffer or in the gel loading solution. After visualizing the samples, the gel is returned to the tank (or rig) and electrophoresis continued in order to run the samples into the buffer. Costs can also be reduced by re-melting and reusing agarose gels two to three times; if the DNA stain is in the gel, a little more can be added each time the gel is melted. New gels should be prepared for excising and purifying DNA bands; but reused gels are suitable for most other purposes. Higher concentrations of agarose (2–4%) are used to facilitate separation of small DNA fragments (100–500 bp) that are close in size. There are various types of agarose, including standard (e.g. SeaKem LE) and low- and intermediate melting temperature agaroses (63). For example, Metaphor® agarose (Lonza) is a high-resolution agarose with intermediate melting temperature that provides a safer alternative to polyacrylamide for the separation of PCR products and small DNA fragments (20–800 bp) with similar sizes, although Metaphor agarose is more expensive and its gels are more laborious to prepare. Concerning buffers, both TBE and TAE are well-suited for gel electrophoresis; double-stranded DNA (dsDNA) migrates ca. 10% faster in TAE than in TBE, but TBE has a significantly higher buffering capacity than TAE (63). This means that TBE maintains its properties in long runs while TAE may need to be replaced (63). Some laboratories prefer to use TAE buffer when a PCR product band needs to be cut from a gel, although this is not usually necessary.

3.4.1. DNA Staining

The location of DNA on the gel is determined by the use of fluorescent dye that binds to DNA. For many years, the DNA gel stain of choice was ethidium bromide (EtBr), a proven mutagen that laboratories have started to replace. Among alternative DNA stains are the SYBR dyes, including SYBR Safe (Invitrogen), and GelRed and GelGreen (Biotium), which have low toxicity and are considered environmentally safe. These stains are considerably more expensive than EtBr, but this may even out when costs of decontamination or safe disposal of contaminated materials are considered. GelRed and EtBr have very similar emission spectra, different from the spectrum of SYBR Safe. For that reason, replacement of EtBr with GelRed does not require changes in the imaging system. SYBR Safe is best viewed with a blue-light transilluminator, but can also be viewed with a standard UV transilluminator that includes an emission filter. SYBR Safe can be stored for up to 6 months (see Note 17), but is light sensitive so that gel tanks should be covered while running gels stained with SYBR Safe to avoid degradation of the dye. In comparison, GelRed is very stable under light or heat, and is suitable for long-term storage. SYBR dyes are considered safe for normal use in laboratories, but in higher concentrations they are cytotoxic, as they can cross cell membranes and accumulate in living cells. In this respect, GelRed or GelGreen are considered safer than SYBR Safe, as they do not enter living cells.

Gel stains can be added to the actual gel after melting the agarose, just prior to pouring, i.e. when the flask is cool to touch (ca. 2.5 μL of SYBR Safe in a 50 mL gel). They can also be added to the loading buffer. This is the most cost-effective route, with only 2.5 μL SYBR Safe added to 500 μL loading buffer (then, use ca. 1 μL loading buffer to 2.5 μL of DNA sample) while for SYBR Green the recommended dilution is 1:1,000.

3.4.2. Loading and Running Gels

Before loading a DNA sample or PCR product, mix it with loading buffer (or loading dye). The loading buffer contains: (a) one or two dyes (e.g. bromophenol blue, xylene cyanol, orange G) that give colour and show the sample's approximate progress on the gel, and (b) a solute (glycerol, ficoll, or sucrose) that gives density to the samples, facilitating their loading (63). The dyes in the loading buffer migrate on gel at a predictable rate; for example, bromophenol blue migrates at a rate similar to that of 300 bp linear dsDNA (63). When running gels, a DNA ladder should be run in parallel with samples. The DNA ladder contains DNA fragments of known size and quantity that allows estimation of the size and concentration of DNA molecules in the samples. Various DNA ladders are commercially available; select one containing a range of fragment sizes appropriate for the samples.

3.4.3. Excision of DNA Bands on Gel

Prepare a deep gel. Low melting temperature agarose is recommended when extracting DNA bands from gels because it helps with product clean-up, but it is not essential and is often far more expensive than standard agarose. Load samples to be cut at least one lane apart on the gel; if possible, load the whole sample into a single well. Using TAE instead of TBE in the gel may lead to more amplifiable product; however, using a 0.5× TBE gel seems to work adequately in most situations. Ultraviolet light damages DNA, so when cutting the gel on a UV light-box it is imperative to be prepared and to work quickly – and also to minimize exposed skin, as the light will cause sunburn (see Note 18). If available, dark readers or blue-light transilluminators are ideal for excision of DNA bands from gel because the light used does not damage DNA. Bands can be cut using a scalpel, razor, or specialized gel cutters (e.g. USA Scientific's x-tracta gel extractor, which is sold as disposable, but can be cleaned with 0.1 M HCl and reused many times). The bands should then be cleaned up using a column-based kit (see Subheadings 2.5 and 3.6.1) rather than using enzymes, as the kit will remove agarose.

3.4.4. DNA Quantification

While it is possible to obtain approximate quantifications of DNA (or PCR products) by comparison of band strength to those in a known DNA ladder, other methods are also commonplace. The most straightforward and inexpensive is quantification by optical density (e.g. using Thermo-Scientific's NanoDrop system), which provides estimates of both DNA concentration and purity (using the ratios of absorbance at different wavelengths of UV light). However, concentration estimates from optical density can be wildly inaccurate, particularly for high-molecular weight DNA. The most accurate DNA quantifications are produced through fluorometry, wherein DNA molecules are labelled with a fluorescent dye, the levels of light emitted are recorded and compared to a standard curve. One of the simplest systems is Invitrogen's Qubit® fluorometer, with the appropriate dsDNA assay kit (High-Sensitivity (HS) for most small-scale DNA extractions and PCRs).

The best possible picture of DNA quality and quantity is obtained using a combination of all three methods: agarose gel electrophoresis, to assess levels of degradation (i.e. does the DNA sample produce a single high-molecular weight band, or a smear of different sized pieces); optical density, to give a measure of purity through the A260/A280 ratio; and fluorometry, to assess concentration. Alternatively, if available, DNA can be run in a microfluidics system on a DNA chip (e.g. Bio-Rad Experion™). However, this is by far the most expensive option, and only required for really critical samples.

3.5. DNA Amplification (PCR)

The PCR is a fundamental and versatile technique in molecular biology (63); it is also an essential tool in molecular systematics and phylogenetics (e.g. (64)). The PCR amplifies exponentially (over a million-fold) a specific DNA region, with the help of a pair of oligonucleotide primers (short stretches of single-stranded DNA that are complementary, on opposite DNA strands, to the ends of the target sequence) and a thermostable DNA polymerase (usually *Taq*, isolated from *Thermus aquaticus*, a thermophilic bacterium). The essential components of the reaction are the thermostable enzyme and corresponding buffer, nucleotides (dNTPs), Mg^{2+} (required by the enzyme), a pair of primers, and DNA template (e.g. genomic or plasmid DNA). Composition of a typical PCR reaction is given in Table 1. The PCR process involves multiple cycles of DNA denaturation, primer annealing, and sequence extension. First, in the denaturation phase, the two strands of DNA are separated into single strands by heating the reaction up to 94–95°C. This is followed by rapid lowering of the temperature to a predetermined annealing temperature (T_a), usually 48–65°C so that primers can bind to their complementary sites in the DNA template. Finally, the temperature is increased again, up to 72°C (extension phase) to allow the *Taq* polymerase to synthesize a new DNA strand, in the 5′→3′ direction, starting at the annealing sites of the primers (less commonly, extension temperatures down to 60°C can be used). These three phases constitute the PCR cycle, which is then repeated 20–40 or more times. The amplified DNA region is known as PCR product or amplicon, and can be visualized by gel electrophoresis (see Subheading 3.4). For detailed description of the PCR process, its components, primer design, reaction conditions, protocols, and troubleshooting, see e.g. refs. 63–65.

3.5.1. Primer Design

The design of primers is absolutely crucial to the efficiency and specificity of PCR (63). At present, there are many primer sequences available in the literature for well-characterized or widely used DNA regions in plant phylogenetics, such as plastid

Table 1
Components in a PCR reaction: Example of a 25 μL volume reaction

Chemical	Final concentration	Volume per reaction
Sterile, double-distilled water (ddH_2O)	–	17.65 μL
10× reaction buffer	1×	2.5 μL
$MgCl_2$ (50 mM)	2.5 mM	1.25 μL
dNTPs (10 mM)	0.2 mM	0.5 μL
Forward primer (10 μM)	0.4 μM	1 μL
Reverse primer (10 μM)	0.4 μM	1 μL
Taq polymerase (5 U/μL)	0.5 U/25 μL	0.1 μL
DNA (usually diluted from stock)[a]	2–20 ng/μL	1 μL

[a]DNA dilutions (e.g. 1:10, 1:50, 1:100, 1:1,000) depend on the concentration of DNA stocks. Dilutions are also used to decrease the concentration of PCR inhibitors that may be present in DNA extracts

regions and nrDNA (e.g. (22, 27–31)). However, most researchers will need to design new primers at some point. For instance, analyses of LCNG, which are highly desirable in studies at low taxonomic levels, often require design of new primers for every taxonomic group studied (2, 34).

Selection of primers that amplify a target region in a diverse set of uncharacterized species is generally a challenge (66). Important criteria for designing primers are the identification of conserved segments in the sequences of related organisms, and the kinetic properties of the primers (66). Primer properties (e.g. melting temperature (T_m), self-complementarity and intramolecular hairpin formation) can be calculated using free online software (see Note 19).

The optimal properties of primers can be summarized as follows: (a) Length: 18–25 nucleotides, with the two primers not differing in length by more than 3 bp; (b) GC content between 40–60%; (c) uniform distribution of all four bases along the primers; (d) no self-complementary region longer than 3 bp; (e) the 3′ end of one primer should not bind to any part of the other primer, to avoid the formation of primer–dimers; (f) the 3′ terminal base in the primer should be G or C, when possible, but avoiding NNCG or NNGC configurations that promote self-annealing; (g) melting temperatures (T_m: the temperature at which the primer dissociates from its complementary DNA template) of the pair of primers should not differ by more than 5°C (63). In coding regions, it is also desirable that the 3′ terminal base of the primer binds, whenever possible, to a second codon position nucleotide in the target sequence (66). Second codon positions are less likely to mutate, reducing the risk of mismatch at the critical 3′ end of the primer (34, 66).

Optimal primers for known sequences can be designed with the help of free online software, such as Primer3 (67), Primer3Plus (68), IDT's PrimerQuest℠ (69), or Oligo Perfect™ Designer (Invitrogen). These and other online tools design primers for one sequence at a time. However, in phylogenetic and other comparative studies, it is convenient to find primers that can successfully amplify target sequences across multiple related organisms. Primaclade (70) is a free, Web-based programme that designs primers for an alignment of multiple sequences (see Subheading 3.11), and greatly facilitates the work of finding PCR primers for comparative molecular studies.

It is difficult to design perfect primers, mainly because it may not be possible to construct primers that have similar T_m or to find sites of about 20 nucleotides that are invariant in related organisms (34, 66). A way to obviate these difficulties is to use degenerate primers and/or lower annealing temperatures in PCR (34). A degenerate primer is a complex mix of primers that differ by one to several nucleotides at positions that are known, or suspected, to be variable in related organisms and across the DNA samples under analysis. PCR products obtained by using degenerate primers can be purified and sequenced; in case of multiple PCR bands, these can be excised from gel and/or cloned, and then sequenced (see Subheadings 3.4.3 and 3.7). Once the sequences are obtained, it is possible to design specific primers for the taxa of interest if required (34). When ordering degenerate primers, use the International Union of Pure and Applied Chemistry (IUPAC) symbols for degenerate bases (W=A or T; S=C or G; M=A or C; K=G or T; R=A or G; Y=C or T; B=C, G or T; D=A, G or T; H=A, C or T; V=A, C or G; N=any base), but be aware that the more degenerate bases you add, the fewer primer copies with any specific combination of bases will be present in the oligonucleotide mixture. This can be partially compensated by increasing the amount of degenerate primers added to the PCR reaction.

3.5.2. PCR Components and Enhancers

Thermostable Polymerases, Buffers, and Magnesium

Taq polymerase has no 3′→5′ exonuclease (proofreading) activity and has been reported to have an error rate of 0.38–2.1 errors per 10 kb (reviewed in ref. 71). For routine DNA sequencing and molecular phylogenetics, this error rate is generally acceptable, but confirmation of critical mutations, or certain genetic or biotechnological applications, may require the use of polymerases with higher fidelity. Many other thermostable polymerases are commercially available, including proofreading enzymes, such as iProof™ High-Fidelity DNA polymerase (4.4 errors per 10,000 kb; BioRad), Advantage® HD DNA polymerase (12 errors per 250 kb; Clontech, Takara Bio Company), and *LA Taq*™ Polymerase (8.7 errors per 100 kb; Takara). iProof and *LA Taq* are specially recommended for accurate amplification of long fragments (>10 kb, up to 37 or 48 kb, respectively). HotMaster™ *Taq* DNA polymerase (5 Prime), HotStartTaq® (Qiagen), and TITANIUM™ *Taq* (Clontech) are modified *Taq* enzymes (nonproofreading) that are especially suitable for Hot Start PCR (see Subheading 3.5.3), a protocol that is used to reduce non-specific amplification, and increase yields of PCR products, even from very low-copy targets or complex templates.

Each polymerase has its own specific buffer, many of which already include $MgCl_2$ (e.g. TITANIUM *Taq* PCR Buffer and Promega's GoTaq® Reaction Buffer), but some of which do not (e.g. Promega's GoTaq® Flexi Reaction Buffer), so check this for each new enzyme before adding more Mg^{2+} to the reaction. Magnesium is an important cofactor for the polymerase, but it is also attracted to the negative phosphate groups of dNTPs and primers. For that reason, Mg^{2+} concentrations need to be higher than those of dNTPs plus primers (63). A concentration of 1.5 mM $MgCl_2$ is sufficient for most reactions (63, 64). However, for some polymerases variation in Mg^{2+} concentration can improve reaction efficiency, by increasing yields and specificity. Magnesium concentration needs to be optimized empirically for each set of primers and template because, in some cases, increasing Mg^{2+} (up to 6 mM) may reduce non-specific priming while increasing it in others (63).

Primers, dNTPs, and DNA Template

In general, dNTPs can be used at 0.2–0.25 mM in reactions using 1.5 mM $MgCl_2$. Concentrations of dNTPs higher than 4 mM are inhibitory, possibly because dNTPs sequester Mg^{2+} (63). Minimize freeze–thaw cycles for dNTPs; when doing a lot of lab work, it can be worth leaving a working aliquot at −4°C instead of repeatedly freezing it. During long-term freezing, small amounts of water evaporate and freeze on the walls of tubes containing dNTP solutions; so, after thawing, they should be centrifuged for a few seconds to minimize changes in concentration (63).

Standard reactions contain 0.1–0.5 μM of each primer; higher concentrations of primers promote mispriming and non-specific amplification (63).

Although theoretically only a single-DNA molecule is necessary for PCR, in practice up to around 500 ng of DNA is added (and reactions with less DNA may not be successful). When judging how much template DNA to add to a PCR reaction, consider both quantity and quality – for degraded DNA, it may be best to add more DNA to increase the chance of the primers annealing to an unbroken segment; to amplify plastid regions, less DNA is often required than when amplifying low-copy number nuclear regions (because many more copies of the plastid genome are present in each plant cell, compared to the nuclear genome).

For information on DNA quantification using agarose gel electrophoresis, spectrometry and fluorometry, see Subheading 3.4.4.

PCR Enhancers

Many additives have been used to increase yield and specificity of PCR, and reduce the effects of inhibitors. Examples of commonly used enhancers are BSA, gelatine, Tween-20, glycerol, formamide,

DMSO, and betaine (e.g. (63, 64, 72, 73)). BSA (0.2–0.6 mg/mL) is widely used to relieve the effects of inhibitors in PCR (63, 72). Betaine (1 M) and DMSO (1–10%) are often used to increase yield and specificity of reactions in PCR (73). Both DMSO and betaine promote strand separation and lower the melting temperature of DNA, thus facilitating the access of primers and polymerase to the DNA. Betaine is also known to improve amplification of GC-rich sequences (74). Any of these additives will inhibit the PCR if used in excess (63, 64).

3.5.3. PCR Programmes and Strategies

A typical PCR programme may consist of: (1) 1–5 min at 94–95°C (initial heating helps to completely denaturate DNA); (2) PCR cycle (to be repeated 30–40 times): 1 min at 94°C (denaturation); 1 min at the selected T_a (48–65°C) (annealing); 1–3 min at 72°C (extension); (3) 5–10 min at 72°C (final extension). Notice that T_a is usually 3–5°C lower than primers' T_m (63).

Hot Start PCR

In a basic hot start PCR (75), DNA template, primers, and other PCR components, with the exception of the thermostable polymerase, are mixed and heated to a temperature (e.g. 70–95°C) that inhibits non-specific binding of primers. The polymerase is then added to the preheated PCR mixture, and the thermal cycling starts. Hot start eliminates the warm-up phase of the first PCR cycle, reducing opportunities for non-specific binding of primers, and the absence of the polymerase in the initial heating of the mixture prevents the extension of mismatched primers (63). However, instead of having to stop the thermocycler and physically add polymerase to each individual tube, the "hot start" is usually achieved chemically (e.g. HotStartTaq, Qiagen) or using antibodies (e.g. Titanium taq, Clontech), or by physical separation of reagents that combine on reaching a certain temperature (e.g. *Taq*Bead™ Hot Start Polymerase, Promega). A hot start protocol is particularly useful to increase yields of PCR product and when non-specific amplification is a problem (63, 75). A combination of hot start and "touchdown PCR" (see below) is also recommended to further reduce non-specific amplification. More information about hot start and its variants is given in ref. 63.

Touchdown PCR

In "touchdown PCR" (76) the reaction starts above the expected annealing temperature, which is then decreased 1°C every second cycle until reaching the selected "touchdown" temperature that is then maintained for ten cycles (or more). The initial high annealing temperature will allow the amplification of very specific copies, and the lower annealing temperatures are used for more efficient replication of those copies. Touchdown PCR can be used as a way to circumvent time-consuming, empirical optimizations of reactions and to prevent primer–dimer formation (63, 76), wherein the primers anneal to each other and are then amplified in competition with the desired amplicon in the PCR reaction.

Touch-Up PCR

In touch-up (or "step-up") PCR, the annealing temperature is low in the first 3–5 cycles to ensure that some copies of the target region are produced, even if the primers are not a perfect match. The annealing temperature is then progressively increased (1°C every one or two cycles) to a "touch-up" temperature, which is maintained for ten or more cycles. With the increase of temperature, available PCR products will be preferentially amplified, because they have now incorporated the synthetic primers and are thus a perfect match to the primers that are free in the reaction (64). Touch-up PCR is also used to increase the yields of weak PCR products.

Nested PCR

This protocol involves two successive PCR amplifications. The second PCR uses amplicon (diluted) from the first amplification as template, and different primers ("nested primers") that specifically bind within the target DNA region. The second PCR amplifies a shorter amplicon that is more likely to be the region of interest because if the initial amplicon included non-specific products, these will not be re-amplified in the second reaction for lack of specific binding site for the nested primers. Even when there is no evident product from the initial PCR on an agarose gel, this nested approach can be successful in generating the desired amplicon.

A similar strategy that can also help increase yields involves including a cocktail of 3–4 primers in the initial PCR reaction, then sequencing any resulting product using the most nested of the PCR primers.

Long Range PCR

Under normal conditions, PCR easily amplifies fragments 1–2 kb long (or, in optimized reactions, up to 3–4 kb), but this range of sizes is insufficient to amplify many genes (63). The main reason for this length limitation is the relatively high error rate of *Taq* polymerase. However, the inclusion of a proofreading polymerase (e.g. Takara's *LA Taq* and iProof; see Subheading 3.5.2) in the reaction mix has greatly increased the length and yields of PCR products, up to more than 40 kb. Long range PCR is especially useful to survey for structural rearrangements or to amplify entire protein-coding genes. Long novel fragments can subsequently be sequenced using primer-walking techniques whereby one starts from the known ends (the PCR primers) and sequences in as far as possible in both directions, designing new primers each time the end of the readable sequence is reached, until the entire region is sequenced.

Multiplexing

It is possible to amplify product from several primer pairs in a single PCR tube. This can cut costs (less PCR reagent, fewer clean-ups required) and time. However, it is only worthwhile when the products of all reactions are known to be of different sizes and clearly identifiable on an agarose gel (so that the presence of each amplicon in the PCR product can be assessed). But getting sufficient quantities of all amplicons from a mixture may take excessive time and effort in comparison to amplifying the regions separately. When multiplex PCR does work well, all the products can be directly sequenced from a single cleaned-up PCR reaction, as each amplicon has specific primers. Multiplexing is not routinely used in PCR for Sanger sequencing, but is a common technique in microsatellite studies.

3.5.4. PCR Troubleshooting

For a list of common PCR problems, possible causes and solutions, see Table 2 (PCR troubleshooting).

The extreme sensitivity of the PCR means that contamination is a real issue; only minute amounts of extraneous DNA can lead to the wrong taxon amplifying. This can be obvious (e.g. a contaminant from a completely different plant family), but if all samples in a set of PCRs are from the same genus, or even species, contamination will be far less easy to identify.

In order to minimize the possibility of crossover, physically separate areas where DNA extraction and PCR occur (at least on separate benches, if not separate rooms). Maintain a clean and tidy laboratory – benches should be regularly wiped down with ethanol and bench-coat, if used, should be replaced frequently. Ultraviolet light can be used to degrade any DNA in the work area, although this is not usually necessary. Use new latex or neoprene gloves (and never reuse a pair that was previously worn during DNA extractions). Pipettes that are used for DNA extraction should not be used for PCR. Using pipette tips with filters in them (e.g. Molecular Bioproducts ART® Aerosol Resistant Tips), although considerably more expensive, can also be worthwhile, particularly in a shared laboratory, where each pipette has multiple users. Always include a negative control (a sample with all the reagents but no DNA added to it) in the PCR, to identify possible contamination.

However, some DNA contamination cannot be avoided in spite of all efforts because some plant tissues include endophytic fungi that are not removed by surface sterilization techniques (77). DNA extracts obtained from such plants include fungal DNA that can be amplified with "universal" primers (e.g. some of those used in ribosomal DNA), which target highly conserved regions of organisms. The problem of fungal DNA contamination may be minimized by using primers specific to plants. When this is not possible, DNA band gel excision and/or cloning of PCR products may be required (see Subheadings 3.4.3 and 3.7).

3.6. PCR Product Purification

Samples that have successfully undergone PCR amplification need to be purified to remove excess primers, dNTPs, and other impurities. It is essential to remove any remaining primers in the PCR products before sequencing. When using terminator-labelled sequencing reactions, if more than one type of primer is present in

Table 2
PCR troubleshooting

Problem	Possible causes	Potential solutions
No PCR product	Master mix error, e.g. pipetting error, missing reagent, wrong reagent concentrations	• Revise calculations of volumes and concentrations of PCR reagents. Repeat PCR with same samples and primers • Use a positive control (a sample that has amplified for this region in the past) when available
	Gel electrophoresis failure	• Always load a DNA ladder onto the gel • Rerun gel – or post-stain (soak in a container filled with buffer and gel stain)
	Low quantity of DNA (in stock tubes)	• Increase volume of DNA template • Attempt amplification of a locus that is usually successful (e.g. plastid *trn*L)
	Too much DNA (high amounts of high-molecular-weight DNA in stock tubes)	• Repeat reaction with diluted DNA template (e.g. 1:100; 1:1,000) • Attempt amplification of a locus that is usually successful (e.g. *trn*L)
	Low-quality DNA (contains PCR inhibitors)	• Repeat reaction with diluted DNA (1:50, 1:100; 1:1,000) • Add BSA (0.2–0.6 mg/mL) to the reaction (63, 72) • Use a column-based kit or Glassmilk to clean the DNA (see Note 14) • Attempt amplification of a locus that is usually successful (e.g. *trn*L)
	Degraded DNA (no high-molecular-weight DNA when run out on gel, only a smear of fragmented DNA)	• Use primers that amplify a shorter region
	Annealing temperature too high	• Try lower annealing temperatures. If possible, use a gradient thermocycler to test various temperatures simultaneously • Use touchdown PCR programme (see Subheading 3.5.3) • Use touch-up PCR programme (see Subheading 3.5.3)
	Inadequate primers (not a good match to the target in that particular sample)	• Try different primers or design new ones
	PCR machine failure	• Try a different machine; ramp times, etc. may need optimization • Check the volume of the final product – problems with heated lid contact lead to evaporation

Weak PCR product (expected size)	DNA in low quantity or degraded	• Increase number of PCR cycles (to 40–50) • Increase volume of DNA template • Use nested PCR (see Subheading 3.5.3) • Try a PCR cocktail, with three or more primers (and, if successful, use the most internal primers to sequence from)
	DNA may have PCR inhibitors	• Repeat reaction with diluted DNA (1:50, 1:100, 1:1,000) • Add BSA (0.2–0.6 mg/mL) to the reaction • Use a column-based kit or Glasssmilk to clean the DNA • Use nested PCR
	Annealing temperature too high	• Decrease annealing temperature • Use touchdown PCR
	Reaction too stringent	• Decrease annealing temperature • Increase Mg^{2+} concentration
	PCR enhancers required	• Add betaine and/or DMSO (73)
	Specialist polymerases required	• Try different enzymes, e.g. Advantage, Titanium, HotStartTaq (see Subheading 3.5.2) • Lower extension temperature from 72°C to 68°C
	More reagents required	• Increase the quantity of polymerase and primer in the reaction
Weak PCR product (wrong size)	Non-specific primer annealing	• Try different primers or design new ones
Multiple PCR bands (2 or more clearly visible DNA bands)	Non-specific primer annealing	• Increase specificity of primers: use higher annealing temperature and PCR additives (e.g. 5% DMSO and/or 1 M betaine; see ref. 73); decrease the amount of Mg^{2+} • Decrease the number of cycles in the PCR • To reduce long unspecific products: reduce extension times • To reduce short unspecific products: increase extension times • Add less polymerase and/or primer • Try different primers or design new ones • Use nested PCR

(continued)

Table 2 (continued)

Problem	Possible causes	Potential solutions
	Annealing temperature too low (less than 50°C)	• Increase annealing temperature 2–5°C • Use touchdown PCR
	DNA target multi-copy or with pseudogenes; contaminated DNA	• Excise and purify DNA bands from gel (see Subheading 3.4.3) or clone PCR products (see Subheading 3.7)
Some samples work, others fail	Taxonomic pattern Sample pattern	• Redesign primers • Check if failures occur from a single batch/extraction technique/collection technique, etc. Repeat extractions; try a different method and/or purify the DNA
	No pattern	• Check DNA concentrations; it may be that there is an optimal amount of DNA to add to the reaction
Smeared PCR products (no clear PCR bands)	Non-specific primer annealing	• Increase annealing temperature 2–5°C • Use touchdown PCR • Try different primers or design new ones • Use as template for a nested PCR (replacing either one or both primers with internal primers for the second PCR)
	Degraded primers	• Prepare new working solutions from stock primers
PCR product in negative control	DNA template may have been added to negative control by mistake	• Prepare new negative control and repeat PCR
	One or more PCR reagents (working solutions) contaminated with DNA	• Remake all PCR working solutions. Clean and decontaminate micropipettes (as recommended by manufacturer). Discard potentially contaminated aliquots

the reaction then more than one type of sequence may be produced, resulting in difficult to unreadable data. In the case of primer-labelled sequencing reactions, the sequence products of unlabelled primers will not be visible but these primers still compete with the labelled primer, affecting the efficiency of the sequence reaction. Therefore, independent of the type of sequencing reaction, cleaning of PCR products is required.

3.6.1. Column-Based PCR Product Purification

Purification of PCR products using column-based kits (see Subheading 2.5) is simple and fast. However, for optimal results, choose a type of column best suited to the size range of the PCR amplicons. With many kits, columns can be reused multiple times, and it is often possible to order the buffers separately. Simply rinse columns thoroughly with water, or boil in a microwave if preferred. In a laboratory that routinely amplifies several different loci, columns used for one locus can be rotated for reuse with another locus, minimizing any possibility of contamination (as sequencing primers will not anneal to any contaminating PCR product). However, reused columns are not suitable for products that will later be cloned.

For high-throughput purification, PCR products can be loaded into a MultiScreen PCR 96-well filter plate (Millipore), and the MultiScreen vacuum manifold is used for PCR product purification.

When using a column- or precipitation-based clean-up method, if the PCR product of a particular reaction is very faint when run on a gel (which often happens when using degraded DNA obtained from old herbarium material) repeat the same reaction several times or in a higher volume, and combine the volumes of the same reaction in the PCR purification.

3.6.2. Enzymatic PCR Product Purification

Enzymatic methods are highly suitable for use in a high- or low-throughput setting when the products will not be cloned later. These have some benefits over column-based clean-ups: they are comparatively cheap; no product is lost on the column membrane, so faint bands are often able to be sequenced; the concentration of cleaned PCR product is directly proportional to its band strength on a gel pre-cleanup (thus, there is no need to run another gel or re-quantify DNA); there is less possibility of muddling tubes (particularly if strips or plates are used) than when using individual columns; finally, it is easy to scale up with strips/plates/multichannel pipettes. There are also disadvantages, namely, that it is not possible to concentrate weak bands by eluting into smaller volumes; it requires a PCR machine; it is not possible to accurately quantify cleaned PCR product using optical densities (e.g. on a Nanodrop), and it cannot be used to clean excised bands from gel-cuts (see Note 20).

*Exo*I and SAP can be purchased premixed from GE Healthcare as ExoSap-IT; however, the more cost-effective option is to buy both enzymes separately.

Enzymatic clean-up:

1. Aliquot 7.5 μL PCR product into 200-μL PCR tube/strip.
2. Keep enzymes on ice (see Note 21). Make enzyme master mix as follows: for each PCR use 0.5 μL SAP (to remove the phosphate groups from excess dNTPs left over from the PCR reaction); 0.075 μL *Exo*I (to digest single-stranded PCR primers into dNTPs, then the phosphate groups are removed by the SAP); 0.675 μL SAP buffer.
3. Pipette 1.25 μL of this master mix into each tube, spin down briefly, and put into thermocycler. Programme: 37°C for 15–30 min (for enzymes to work); 80°C for 15 min (to kill enzymes); cool to 4–10°C.

Troubleshooting:

If the *Exo*I enzyme is not active or present in insufficient quantity, incomplete removal of PCR primers results; carryover of these primers can lead to messy chromatograms in sequencing, as both forward and reverse directions are read together. Too little SAP causes incomplete removal of dNTPs from the PCR reaction. Competition from this excess of unlabelled dNTPs in the sequencing reaction can cause low chromatogram signal.

3.7. Cloning of PCR Products

Cloning PCR products, although more costly and time-consuming, is recommended or required when sequencing multi- or low-copy genes, or other DNA regions, that are repeated along the genome – e.g. nrDNA. Cloning is also required when analyzing contaminated or mixed biological material, like forensic or ancient DNA samples (78). In general, nrDNA repeats are highly similar or identical, but in many plants divergent repeats have been detected see e.g. refs. 31–33. Divergent repeats in the nrDNA represent paralogous sequences (or paralogues – gene copies that derived from duplication of a single ancestral gene but that have since evolved independently from each other) or pseudogenes (genes or DNA regions that have become non-functional and that mutate at a higher rate). In the case of hybrids or plants that are suspected to have an allopolyploid origin, analysis of different nrDNA repeats present in a single genome may confirm the occurrence of hybridization or provide information on the origin of the polyploid genome. An allopolyploid contains the genomes of two ancestors, and thus there are two different sets of nrDNA in recent polyploids. If two different copies of the gene/DNA region of interest can be amplified, only a single PCR product may appear on a gel, or perhaps two or more bands that are too close in size to be efficiently cut and separated. A PCR product that contains more than one type of DNA can result in unreadable or bad sequence data, as the signal of the different sequences overlaps each other. In cases like this, cloning allows separation of different sequences present in the PCR product.

When cloning PCR products, several to many clones must be sequenced before all the sequences (gene copies, alleles, or variants) present in the sample are obtained. But how many clones should be sequenced before it is likely all existing DNAs have been sampled? For low-copy genes, if different alleles amplify equally well, sequencing a small number of clones (5–6) should be sufficient to detect all existing copies (34). However, with multicopy genes, there is no straightforward answer to the question of clone number because it is not possible to predict the number of copies for a particular gene or other DNA regions (e.g. (79, 80)). Therefore, for multicopy genes or highly repeated DNA, sequence as many clones as possible. A good strategy to sample multicopy genes is to perform various PCR reactions with different conditions (e.g. with and without DMSO, and with different annealing temperatures) for each DNA sample; then, sequence at least 6–10 clones per PCR reaction. Once different repeat sequences have been detected in one or various DNA samples, it may be possible to design specific primers to search for and specifically amplify those repeats in all other samples of the taxa under study (80), increasing the chance of creating a matrix of orthologous loci.

3.7.1. Preparation of Culture Media and Plates

1. LB medium (solid): Dissolve 20 g LB broth base in 1 L distilled water, then add 15 g agar/agarose and mix well. Divide LB-agar solution equally between three 500-mL bottles (ca. 333 mL each) and autoclave. Store for later use.
2. LB medium (liquid): Dissolve 20 g LB broth base in 1 L distilled water and autoclave.
3. Preparation of LB-ampicillin (or kanamycin) plates: Prepare plates in a laminar flow hood that has been on for at least 30 min (see Note 22); clean surfaces with 70% ethanol. Use new, sterile Petri plates or autoclaved, sterile glass plates. Melt solid LB medium in a microwave. Let the LB-agar solution cool until the bottle can be held with a bare hand. Then, add 1 μL ampicillin (50 mg/mL) per mL LB-agar solution (final concentration of ampicillin: 50 μg/mL). Stir briefly and pour 10–15 mL LB-agar ampicillin into each plate. Cover plates partially with their lids and allow the medium to solidify. Once the medium is firm, close the plates (avoid passing a hand over the medium, as shed skin may inadvertently contaminate plates). Place LB-ampicillin plates in a sealed bag and store at 4°C, in the dark, for up to 4 weeks.

3.7.2. Cloning of PCR Products Using Kits

A fast and efficient option for cloning PCR products is to use the TOPO TA Cloning Kit for Sequencing (with TOP10 chemically competent *E. coli* cells) from Invitrogen; but explore the range of cloning kits available from them and other companies. Other kits may be more suitable. For example, if competent cells can easily be produced in-house, another kit may provide the option of buying cloning reagents without the cells, reducing costs considerably.

The TOPO TA Cloning Kit for Sequencing uses the pCR4-TOPO vector (plasmid) that includes a lethal *E. coli* gene, *ccd*B, as well as ampicillin and kanamycin resistance genes. Ligation of a PCR product to the pCR4-TOPO vector disrupts expression of the lethal *ccd*B gene. Bacteria (TOP10 cells) transformed with plasmids that include the PCR product are able to grow on the plates; these are the positive recombinants. Cells that contain non-recombinant vector (plasmid without PCR product) are killed. Bacteria that contain no vector are also killed due the presence of antibiotic (ampicillin or kanamycin) in the plates.

1. Set up TOPO cloning reaction and transform TOP10 competent cells following manufacturer's instructions.
2. Work in a sterile situation; clean surfaces with 70% ethanol.
3. Spread transformed cells on LB-ampicillin plates (follow instructions in the kit manual).
4. Incubate plates overnight at 37°C.
5. Examine plates: Each colony that appears on the Petri plate corresponds to a positive clone, i.e. *E. coli* cells that have been transformed with plasmid containing the PCR product.

At this stage, there are two options for sequencing the PCR product that has been inserted into the plasmid:

Option 1: *Use purified "Colony PCR" product for sequencing* (see Subheading 3.7.3).

1. Perform a "colony PCR" reaction (using bacterial colony as the direct source of DNA template).
2. Prepare a glycerol stock for each colony (bacteria are frozen in culture medium that includes glycerol).
3. Run "colony PCR" product on agarose gel to confirm the presence of PCR insert.
4. Purify the whole volume of the "colony PCR" product.
5. Use purified "colony PCR" product (including PCR insert) as template for DNA sequencing.

Option 2: *Use purified plasmid for sequencing* (see Subheading 3.7.4).

1. Perform "colony PCR" reaction.
2. Run "colony PCR" product on agarose gel to confirm the presence of PCR insert.

3. Grow the same colony used for "colony PCR" overnight in LB liquid medium (with antibiotic); part of this culture is used to prepare a glycerol stock.
4. Isolate plasmid from the bacterial culture using miniprep kit.
5. Confirm success of plasmid purification by PCR.
6. Run PCR product on agarose gel.
7. Use purified plasmid (including PCR insert) as template for DNA sequencing.

Option 1 saves time and expense when sequencing many clones. Option 2, although more time-consuming and costly, avoids performing a double PCR reaction on the original DNA region before sequencing. If the possibility of errors introduced by multiple PCR reactions is a concern, then direct sequencing of the plasmid with the PCR insert (option 2) is preferred. Using a high-fidelity *Taq* polymerase in PCR is another possibility, though this also increases costs. However, as glycerol stocks are prepared in both these methods, a good strategy to save time and money is to use option 1 for general sequencing of clones. Then, if confirmation of a particular sequence is needed, recover the colony of interest from its glycerol stock, extract the plasmid, and sequence again, but this time the original PCR insert.

3.7.3. Colony PCR and PCR Product Purification

1. Prepare microtiter plate (96-well) for bacterial culture (work in a sterile environment): Add 75 μL of a solution of LB liquid medium with 50 μg/μL ampicillin and 15% glycerol to each well.
2. Prepare PCR reactions (in microcentrifuge tubes or a 96-well PCR plate); each reaction (50 μL) contains 39.8 μL ddH_2O, 5 μL 10× reaction buffer, 2 μL $MgCl_2$ (50 mM), 1 μL dNTP mix (10 mM), 1 μL of each M13 primer (10 μM), and 0.2 μL *Taq* polymerase (5 U/μL).
3. With a sterile toothpick (or other picking tool), touch one colony (from step 5 in Subheading 3.7.2); dip the toothpick first into a tube with PCR mix, and then into one of the wells of the microtiter plate (prepared in step 1, above). Repeat the process for other colonies.
4. Run "colony PCR" programme as follows: 5 min at 94°C; 30 cycles of (94°C for 30 s; 55°C for 30 s; 72°C for 1 min); 10 min at 72°C.
5. Run 2 μL "colony PCR" product on 1% agarose gel to confirm the presence of PCR insert.
6. Incubate the inoculated microtiter plate (from step 3, above) at 37°C overnight, with shaking set at 150 rpm. Then, freeze plate at –80°C for long-term storage – these are the glycerol stocks of the recombinant colonies.
7. Purify "colony PCR" products using PCR product purification kit (see Subheading 2.5). For high-throughput (96-well plates),

purify PCR products by loading samples into a 96-well MultiScreen PCR 96 Filter Plate. Use the MultiScreen vacuum manifold for PCR product purification (follow manufacturer's instructions).

8. Purified "colony PCR" product (M13 amplicon) can subsequently be sequenced using T3 and T7 primers; these primers are internal to the M13 primers, but external to the PCR insert (i.e. the original PCR product that was cloned).

3.7.4. Plasmid Purification

1. Prepare culture tubes (15 mL) with 3.5 mL of LB liquid medium with ampicillin (50 μg/mL).
2. Prepare PCR reactions ("colony PCR" test): Each reaction (25 μL) contains 19.9 μL ddH_2O, 2.5 μL 10× reaction buffer, 1 μL $MgCl_2$ (50 mM), 0.5 μL dNTP mix (10 mM), 0.5 μL of each M13 primer (10 μM), and 0.1 μL *Taq* polymerase (5 U/μL).
3. With a sterile toothpick, touch one colony (from step 5 in Subheading 3.7.2); dip toothpick first into a tube with PCR mix, and then inoculate the LB-ampicillin liquid medium in a culture tube. Repeat process for other colonies.
4. Run "colony PCR" programme (see step 4 in Subheading 3.7.3).
5. Run 2 μL of "colony PCR" products on 1% agarose gel to confirm the presence of PCR insert.
6. Grow clones in LB-ampicillin liquid medium overnight (up to 16 h) at 37°C, with shaking set at 175–200 rpm.
7. Use 1.2 mL of clone culture to prepare glycerol stocks in 2-mL tubes; add 0.4 mL 80% glycerol to the culture; shake well and freeze at −80°C.
8. Use 2 mL of clone culture to isolate the plasmid using QIAprep Spin Miniprep kit (Qiagen), following manufacturer's instructions.
9. Confirm success of plasmid purification by PCR: Each reaction (25 μL) contains 17.15 μL ddH_2O, 2.5 μL 10× reaction buffer, 1 μL $MgCl_2$ (50 mM), 0.25 μL dNTP mix (10 mM), 0.5 μL each M13 primer (10 μM), 0.1 μL *Taq* polymerase (5 U/μL), and 3 μL plasmid (diluted 1:100 in ddH_2O).
10. Run PCR programme as follows: 30 cycles of (95°C for 30 s; 55°C for 30 s; 72°C for 1 min); 10 min at 72°C.
11. Run 5 μL PCR product on 1% agarose gel to confirm the presence of plasmid and PCR insert.
12. Purified plasmid and PCR insert can subsequently be sequenced using T3 and T7 primers (as above).

3.8. DNA Sequencing: Sanger Method

3.8.1. Cycle Sequencing: General Issues

Automated DNA sequencing using the Sanger method (or chain termination) involves an initial sequencing PCR reaction in which PCR products are randomly terminated by fluorescent-labelled dideoxynucleotide triphosphates (ddNTPs). The ABI system is perhaps the most commonly used today, wherein each of the four terminator nucleotides is labelled with a different colour, the fragments are separated by polyacrylamide gel electrophoresis on an automated sequencer machine, and a laser/camera setup captures the order of the bases, based on sizes of the DNA fragment they terminate. The sequencing systems used by other manufacturers differ; for example, LI-COR uses a two-channel infrared detection system while Beckman Coulter also has a four colour system.

Older automated sequencers use slab polyacrylamide gels, which are poured between two glass plates and allowed to polymerize. The gel is mounted vertically into the machine, and samples are loaded from the top into wells. The advent of capillary sequencing, where samples are loaded into a gel matrix that has been injected into thin capillaries, improved the speed and efficiency of automated sequencing. However, the cost of capillary systems and of the gel matrix is high, and although the individual run time for a sequence is only in the order of 1–2 h, many of the low-end machines can only sequence 8–16 samples concurrently.

Because sequencing reactions are basically PCR reactions, some of the modifications that are used to improve PCR can also improve sequencing reactions (e.g. addition of DMSO when dealing with templates with strong secondary structure, like the nuclear small and large ribosomal subunits; altering thermocycle temperatures to improve denaturing; increasing the number of PCR cycles when the product is weak). Specific to the ABI BigDye chemistry, there are two versions, each best suited to different templates: v 1.1 is optimized for base-calling adjacent to the primer and for sequencing short PCR product templates; v 3.1 has more flexible chemistry, making it preferable for de novo sequencing of difficult or variable templates.

Precise quantification of PCR products prior to sequencing is not necessary. Although low-throughput labs may choose to quantify each product and optimize sequencing reactions individually, inaccuracies involved in pipetting minute volumes of reagents into individual tubes rather than constructing a master mix almost always outweigh any advantage conferred by adding in a precise quantity of PCR product. However, some automated sequencer systems seem more vulnerable than others to overloading the reaction; trial and error for an individual sequencer and sequencing chemistry is necessary. This should stay within a range that can be estimated by eye from comparison of the PCR product to a known ladder on an agarose gel.

Fluorescent dyes used in sequencing are light sensitive, so sequencing reactions should be kept out of direct sunlight when possible.

In most cases, "forward" and "reverse" sequencing reactions are made for a given amplicon so that each base is read from two directions, once from the 5′ end of the amplicon, and secondly from the 3′ end, to obtain the highest level of confidence in the final sequence data. This is particularly critical when sequence reads are suboptimal due to ambiguities, high levels of background signal, or polymerase slippage (for example, after short sequence repeat regions like mononucleotides or dinucleotides). Either the initial PCR primers, or a set of primers that sits inside the PCR amplicon, can be used for sequencing. Because Sanger sequence read lengths often do not extend beyond 1,000 bp, sequencing of long products (e.g. the complete *rbc*L gene, which is ca. 1,400 bases long) always requires internal primers.

3.8.2. Cleaning-Up of Sequencing Reactions

Ethanol precipitation. In order to remove unincorporated dNTPs and primers from the sequencing reaction, a clean-up should be performed. The cheapest, and still very effective, way to clean sequencing products is through ethanol precipitation (where salt and ethanol are added to the sequencing reaction product, causing the DNA to precipitate out of solution; the DNA is then separated out using centrifugation). Different sequencing chemistries have different optimized protocols, so it is best to refer to the literature for a specific system. A small amount of glycogen (ca. 1 μL) can be added to this reaction – this will act as a carrier for small DNA molecules, and also increases visibility of the DNA pellet on the side of the microcentrifuge tube. Although ethanol precipitation requires two to three spins of 5–15 min in a refrigerated centrifuge as well as final drying of the pellet so that clean-ups of individual samples are quite lengthy, it can easily be scaled up for use on plates in a high-throughput situation.

Sephadex clean-up. Sephadex methods of sequencing product clean-up are also commonly used. Sephadex™ G-50 fine (GE Healthcare) is obtained as a powder of dry beads, hydrated in water (see Note 23) and packed into Centri-Sep columns (Princeton Separations Inc.) by centrifugation, following the manufacturer's instructions. Packed columns that have been centrifuged should be used almost immediately as the surface will dry out. The sequencing product is passed through the Sephadex columns in a centrifuge, and the elute can be loaded directly onto an automated sequencer. Both individual columns and 96-well plates can be used. Although this method is faster than ethanol precipitation and does not require a refrigerated centrifuge, it is more costly. Individual columns and plates can be reused multiple times. However, the Sephadex beads themselves are quite expensive, leading some laboratories to collect, autoclave, and reuse the beads.

3.8.3. Reducing Sequencing Costs

Automatic sequencing costs represent a major part of any molecular project. However, it is possible to minimize these by reducing the size of the sequencing reaction. For example, very few laboratories use a "full" sized ABI BigDye reaction (8 μL of BigDye in a 20 μL reaction). Commonly, ca. 1/8 of the quantity of BigDye is used in a 10 μL reaction, with the addition of some form of buffer (most commonly that supplied with the BigDye). To improve efficiency even further, specialized buffers on the market allow up to 64-fold reductions in the amount of BigDye used per reaction as well as optimizing other parts of the sequencing reaction (e.g. BDX64, from MCLAB, which also reduces the required PCR cycling time).

Lane costs. These are hard to reduce, as they usually directly reflect the cost of the sequencer machine's capillaries and gel matrix. However, most institutes extend the use of capillaries well beyond their recommended shelf-life, until a noticeable decline in performance is observed or a capillary malfunctions. One temptation can be to reduce costs by not running the second-strand sequencing reaction; however, this is not a preferred option in most cases as it reduces the ability to find mix-ups in the sequences or errors associated with the polymerase. Also, reads for loci that have been unidirectionally sequenced are often of unequal quality along their lengths. Many smaller laboratories now send their sequences out to lower cost, high-throughput laboratories rather than maintaining their own machines.

3.8.4. General Sequencing Problems: Troubleshooting

Background signal. Failure to adequately reduce the quantity of PCR primers prior to a sequencing reaction can lead to strong background signal, as competing reactions take place in the sequencing reaction tube. Other problems can be caused by contaminants in the DNA extraction that are co-amplifying (re-extract the DNA; if this is not possible, clone the PCR products into a vector and sequence from the plasmid), or by non-specific products (if there were different sized products in the initial PCR, try a gel cut; alternatively clone, or increase the specificity of the initial PCR with a higher annealing temperature).

Overlapping peaks. Enzyme slippage in the PCR reaction can lead to observation of double peaks. This is particularly visible after short repetitive regions like mono- or dinucleotide repeats. Solutions involve trying a higher fidelity polymerase enzyme in the initial PCR, or looking for alternative sequencing primers that avoid the problem area. When forward and reverse sequence reads are combined, it is usually possible to read through regions of slippage (as each read will only be problematic on the 3′ side of the repetitive region).

Variable peak heights. Although sequencing chemistry is continually improving, variance in the relative heights of different peaks can occur. In many cases, these are not problematic, particularly when they are predictable (for example, the sequencing chemistry may tend to give a weak G after a row of Ts, or may tend to insert Gs between real peaks if the gel is overloaded).

Weak signal. This can sometimes be corrected by increasing the number of cycles in the sequencing PCR reaction. Also, try increasing the amount of the PCR product added to the sequencing reaction.

Failure. Sometimes, despite the presence of a clear single band in the initial PCR, sequencing reactions fail. In this case, try again, vary conditions, and if there is access to different primers that sit internal to the initial amplicon, try them. When an enzymatic digestion method like ExoSap-IT has been used to clean up PCR reactions, check carefully that the enzymes were fully denatured at the end – otherwise, it is possible that they have carried over and are interfering with the enzyme in the sequencing reaction. Occasionally, part of the amplicon sequences well, but signal then plummets or is lost completely. This often correlates with secondary structure of the molecule that is being sequenced; adding ca. 5% DMSO to relax stem-loop structures may fix the problem.

Mechanical problems. There is a wide spectrum of things that can go wrong with the automated sequencer itself. For example, if runs are consistently shorter than expected, this may be caused by low current, which, in turn, may be caused by a problem with the buffer or the polymer, or by a leak in the machine. Pay attention to raw data files. If reactions were sequenced in-house, consult the troubleshooting section of the sequencer manual. For reactions that were sent out for sequencing, it is necessary to be more proactive in identifying both problem and solution.

3.9. Second-Generation Sequencing Technologies

Next- or second-generation technology is still in a state of flux – constant advances are being made, with ever more impressive technologies rumoured to be just around the corner. Thus, any discussion of specific products or protocols here will be outdated even prior to publication.

The single factor that is least likely to alter is the prime importance of the starting DNA for any project (see Note 24), although even here advances have made the technology more accessible to those working on small or rare plants. In autumn 2009, the starting point for a 454 shotgun library was 3–5 μg total DNA; today, the 454 Rapid Library protocol requires a tenth of the amount, 300–500 ng. Quantification using a spectrometer is notoriously inaccurate, and it is also difficult to get a precise quantification using gel-based methods, so if at all possible DNA should be quantified using fluorometry (for example, with the Invitrogen Qubit system). The quality of starting DNA should be high, it should be non-degraded, with an optical density 260/280 of around 1.8, and should be dissolved in TE. Ideally, the DNA extract should be white and should contain no particulate matter – however, successful libraries can be generated from DNA that does not fully meet these criteria. Rolling circle amplification (e.g. using the Qiagen REPLI-g® UltraFast kit) or targeted long-range PCR can also be used to increase the quantity of DNA in the original extraction, or to increase representation of a particular region (e.g. the plastid genome). While these methods are becoming more accessible for a wide range of taxa, many plants are available only as herbarium specimens, with low-quality degraded DNA. In some respects, next-generation technology, with read lengths for many platforms well below those expected from Sanger sequencing, has no trouble with degraded DNA: however, DNA degradation is non-random throughout the genome, so using degraded DNA to construct a random library potentially leads to bias. Given that DNA quantity and quality remain critical, this technology is unlikely to be widely applied to DNA obtained from herbarium collections of plants in the near future.

Currently, long-read next-generation technology (454) is being used for microsatellite development (e.g. (81)), proving far more rapid and less labour intensive than enrichment or selective hybridization protocols (e.g. (82, 83)). Plastid genome assembly is another growing field, after plastid enrichment (either physically on a gradient, or using long range PCR) (e.g. (84–87)) or from total genomic DNA libraries (e.g. (88)). Transcriptomes also offer valuable data, including the identification of candidate genes for evolution and development studies and new phylogenetic markers. Transcriptome sequencing from multiple accessions using libraries that have been RNA depleted but not normalized maximizes the chance of obtaining sequences from an overlapping set of genes, with great potential for the development of novel phylogenetic markers.

3.10. Sequence Data Submission

Before publication of a study that uses DNA sequencing data, it is normally incumbent on the authors to make their data available to the scientific community. Indeed, sequence data generated using money from publicly funded agencies should rapidly be made publicly available. One of the most efficient tools is NCBI–GenBank's Sequin programme (downloadable from the internet; http://www.ncbi.nlm.nih.gov/Sequin/). In order to submit the data, it helps to have prepared a few simple things:

1. A voucher table in a spreadsheet programme like Microsoft Excel™ that can be used to generate FASTA-style headings for the Sequin file, using a concatenate command. Authorities (i.e. authors of taxon names) should be in a separate column to taxon names. Add columns for other information wanted in the final submission (e.g. collection locality and date; PCR primer names and sequences).
2. The final matrix, containing sequences that have been checked for premature stop-codons and other problems, compiled in a

consistent way (e.g. using "?" for missing data at the start and end of the sequence, and "-" for insertions or deletions), without any data-free columns (i.e. made up entirely of "-"). Accessions in both the voucher table and the aligned sequence matrix must be in the same order.

3. Information required for annotation of the sequences. For sequences generated from a single exon, this is straightforward. However, if the sequencing primers are located outside the gene of interest or when amplifying a region like *trn*L that includes genes and intergenic spacer regions, or if the region contains gene introns and exons, these all need separate annotation. When submitting a phylogenetic study rather than single sequences, always use an aligned matrix: only one annotation is required per matrix, as it can be propagated to all the sequences. One way to identify the boundaries of each region is to extract the corresponding region from a well-annotated genome in NCBI–GenBank. For example, to annotate a *trn*K-*pbs*A-*trn*H region, extract the three separate genes from the published plastid genome closest to the study group and align those genes to one of the sequences to be annotated. Select that sequence as the one to annotate in Sequin, and draw on relevant regions and CDs. Propagate these to all the other sequences in the submission.

Plant barcoding data has rather more stringent guidelines, including submission of chromatograms along with sequence files. Submission through the BOLD interface (*Barcode of Life Data Systems*: http://www.barcodinglife.com) will facilitate meeting these requirements.

TreeBase (http://www.treebase.org) is a tool for databasing the aligned data matrix, phylogenetic tree and details of published analyses. It is independent from sequence databases like NCBI–GenBank – one does not replace the other. Submission to TreeBase is a requisite for publication in some journals (e.g. Systematic Botany; The Bryologist). In order to maximize the usefulness of data to other researchers, the matrix should be tidy and well-annotated (e.g. with relevant data blocks, like exclusion sets), and taxon names in the matrix should be as full and accurate as possible; abbreviations should never be used.

3.11. Multiple Sequence Alignment

Sequence alignment is the procedure where two or more sequences are compared and lined up in a data matrix, to determine which positions (sites) in the various sequences can be considered homologous (e.g. (89, 90)) – establishing homology is fundamental in comparative biology. Homology can be defined as similarity (character identity) that is shared among organisms (or sequences) due to common descent (e.g. (91)). In DNA sequences, each nucleotide position is a character, with four possible character states

(A, C, G, and T). If there are length mutations in the sequences, spaces or gaps must be opened to align homologous bases to each other; each gap is an "indel" (insertion or deletion) and can also be scored as a character for phylogenetic analysis.

In sequence data, homology can be recognized at three different levels: (1) The sequences; (2) the positions within the sequences; (3) the character states at a homologous position in the sequences (90). Sequence alignment is concerned essentially with homology assessment of positions in the sequences because it is usually assumed that the sequences being aligned are homologous (orthologous) (89, 90). Homology of character states can only be inferred after phylogenetic analyses, i.e. once the relationships among the various sequences are known (10). For example, if the nucleotide G is in the same (homologous) position in two different sequences (A and B), the state "G" may be homologous or homoplasious (not homologous). It is homologous if the "G" of both sequences derived from a single mutation in the ancestor of A and B, but homoplasious if the Gs resulted from two distinct mutation events in separate lineages.

Sequence alignment is a critical step preceding phylogenetic analyses because all analyses of relationships derived from sequence data are primarily based on the alignment (89). An alignment is a hypothesis of homology and like other hypotheses may contain errors, some of which can have huge effects on subsequent analyses (90). Therefore, great care is necessary in the preparation of sequence alignments for phylogenetic analyses.

Alignment of coding sequences (i.e. genes) among related taxa is usually straightforward because, due to functional constrains, genes are conservative in both length and nucleotide variation. Indels are rare or absent in coding sequences of related taxa and usually do not affect the gene reading frame, as gaps are found as multiples of three nucleotides. Coding regions have been very useful in phylogenetic analyses at suprageneric levels, but less conserved regions are often required for species-level phylogenies. Non-coding regions (introns and intergenic spacers) are more variable, both in sequence composition and in length, and are thus potentially a good source of characters at low taxonomic levels. However, non-coding DNA is subjected to various mutation mechanisms that often produce regions of variable length, where alignment can be ambiguous or impossible (92). Examples of these mechanisms are slipped-strand mispairing and insertion or deletions that are linked to the sequences' secondary structure (hairpins and stem-loops). Such types of mutations are non-random and highly homoplasious (i.e. likely to occur repeatedly in distinct lineages), so may generate misleading results in phylogenetic analysis. It is important to recognize these mutations to optimize alignments and scoring of characters for phylogenetic analyses (92).

When preparing an alignment with sequences that have variable lengths, it is best to use a multiple sequence alignment programme to produce an initial alignment. This alignment should then be examined and manually edited, if necessary.

In computation terms, there are two modes of alignment: local and global. Local methods try to align segments of the sequences without considering overall sequence patterns (89, 90). These are mostly used in database searching (e.g. BLAST: "Basic Local Alignment Search Tool"; see ref. 93), since they facilitate alignment of similar regions while allowing highly divergent regions of the same sequences to remain unaligned (89, 90). However, the principal method of alignment for phylogenetic analysis is global alignment, where sequence comparisons are made over the entire length of sequences (89).

Most multiple sequence alignment programmes use progressive global alignment, e.g. the widely used Clustal W (94). Progressive alignment starts by constructing all possible pairwise alignments (two sequences at a time), which are used to calculate pairwise similarity scores and build a distance-based guide tree. Following this, the intermediate pairwise alignments are added to each other following this guide tree, to generate the complete alignment (89, 90). Other progressive alignment programmes are parsimony-based and use multiple guide trees before choosing an optimal alignment (89); (see Note 25).

Each pairwise alignment is chosen based on the minimum edit distance between two sequences (89, 90). The edit distance is the number of changes (substitutions or indels) required to convert one sequence into another (89). Each of these changes is given a cost or penalty. A high gap penalty (on indels) leads to increased nucleotide mismatches and decreased indels in the final alignment while a low gap penalty leads to increased indels and decreased mismatches (89, 90).

The main disadvantage of progressive alignment is that any errors that occur early in the procedure, i.e. during pairwise alignment, are not corrected when producing the complete alignment (90). But even when the alignment produced is computationally optimal, it could be biologically incorrect (90). Therefore, it is essential to carefully revise any alignments that are produced by sequence alignment software before proceeding with phylogenetic analysis.

After a sequence alignment has been produced, it may still contain regions where alignment is ambiguous and unreliable. Regions that are poorly aligned or highly divergent should be removed before phylogenetic analyses because they may not be homologous or may have been saturated by multiple substitutions (95). However, removing regions of problematic alignment also means loss of phylogenetic information. The programme Gblocks (95) can be used to eliminate poorly aligned and highly divergent regions while minimizing the loss of informative sites, to produce an alignment that is more suitable for phylogenetic analysis. One advantage in using a programme to select the regions of alignment to exclude from subsequence analyses is that it removes some of the subjectivity from the process (i.e. following the same steps, another person would be able to reach exactly the same conclusion from an alignment).

3.12. Phylogenetic Analyses

It is not the aim of this chapter to introduce or discuss methods of phylogenetic analysis, which is an extensive and complex subject. Readers are directed to some relevant literature in the field: for less experienced users, there are a number of accessible books (see e.g. refs. 1, 9, 10, 96, 97) and articles (e.g. (11, 12)) that introduce theory and practice of phylogenetic analysis. For introduction to Bayesian analysis and comparison with "traditional" (maximum parsimony and maximum likelihood) methods, see refs. 98, 99. For intermediate and advanced users, there is a vast literature on phylogenetic methods, but for comprehensive references, see "Inferring Phylogenies" by J. Felsenstein (100) and "The Phylogenetic Handbook" by Lemey et al. (97). The earlier book "Molecular Systematics" (101) is another fundamental reference, particularly the chapter dedicated to phylogenetic inference (102). Other useful papers are dedicated to "troubleshooting molecular phylogenetic analyses" (103), and "gene trees vs. species trees", referring to the problems of using gene trees when reconstructing phylogeny of organisms (104, 105). Finally, ref. 106 highlights and clarifies common errors in interpretation of phylogenetic trees.

4. Notes

1. At present, there are many Web resources that facilitate literature search and provide online access to publications in all fields of science (search online for "academic databases"). There are also Web resources that are specific for plant systematics, such as the Kew Bibliographic Databases (KBD: http://kbd.kew.org/), which give access to an extensive reference catalogue in botany and systematics of vascular plants; coverage of KBD is particularly good between 1971 and 2007. For literature searches on bryophytes, check, for example, the W^{3}MOST bibliography database (http://www.mobot.org/MOBOT/tropicos/most/welcome.shtml); for algae references, search AlgaeBase (http://www.algaebase.org/). There are also digital libraries that provide free access to historic or rare books and journals, such as the Biodiversity Heritage Library (BHL) (http://www.biodiversitylibrary.org/), Botanicus (http:www.botanicus.org/), the Digital Library of the Real Jardín Botánico de Madrid (http://bibdigital.rjb.csic.es/), and Gallica, Bibliothèque Nationale de France (http://gallica.bnf.fr/).

2. Published molecular phylogenetic studies usually include sequence accession numbers that facilitate finding of sequences on databases (GenBank, EMBL, DNA Data Bank of Japan). Also search for sequences available for the taxa of interest in the "Taxonomy" database of GenBank (http://www.ncbi.nlm.nih.gov/Taxonomy/).
3. Many botanical gardens around the world maintain a living seed collection, gathered from wild or cultivated plants; the list of seeds is published in a catalogue, often available online, known as "Index Seminum". Multiple living plant collections around the world can be accessed with an online search tool provided by the Royal Botanic Garden Edinburgh (http://rbg-web2.rbge.org.uk/multisite/multisite3.php). Many other important germplasm resources are available online, such as the USDA National Plant Germplasm System (http://www.ars-grin.gov/npgs/acc/acc_queries.html) and the GenBank Information System of the IPK Gatersleben (http://gbis.ipk-gatersleben.de/gbis_i/). For research purposes, it can be possible to obtain small amount of seeds from these collections by contacting the organizations.
4. Information on herbaria of the world is available at the *Index Herbariorum* Web site (http://sciweb.nybg.org/science2/IndexHerbariorum.asp).
5. Various Web sites include detailed information on how to prepare herbarium specimens/vouchers; see, for example, Web pages of the University of Florida Herbarium (107), and the University of Reading (http://www.herbarium.rdg.ac.uk/making_specimens/index.html).
6. "Wet chloroform" (24:1 chloroform/isoamyl alcohol) is light sensitive; after preparation, cover the bottle with foil and store at −20°C. In the "wet chloroform" solution, the isoamyl alcohol (3-methyl-1-butanol) can be replaced by 1-octanol, to be used at a proportion of 24:1 chloroform/1-octanol, with similar results in the DNA extraction procedure.
7. With large bottles, e.g. 1 L DNAse-free or RNAse-free water, aliquot some into 1.5-mL microcentrifuge tubes and pour the rest into a selection of 15 mL and 50 mL disposable centrifuge tubes; label and date all aliquots.
8. In the case of bryophytes, which are usually very small plants, it is advisable to process (separate) samples for DNA analysis with the help of a microscope, to guarantee that only one species is included in each sample bag.
9. β-mercaptoethanol can be increased to 0.4–1%. For hard tissues, proteinase K (final concentration: 1 mg/mL) may also be added to the CTAB buffer.

10. Before using the mixer mill with valuable samples, carry out tests to determine the best grinding frequency for the microcentrifuge tubes being used; some tubes may break at maximum cycle frequency.
11. The value 13,000 × *g* corresponds to 10,000–13,000 rpm in most benchtop centrifuges. For precise centrifugation conditions, relative centrifugal force (RCF) values, expressed as × *g* (times gravity), must be used because, for the same revolutions per minute (RPM), the centrifugal force varies depending on the radius of the rotor used.
12. To avoid cross-sample contamination, DNA must be thoroughly removed from grinding beads or cone balls before reuse. Remove tissue debris with running water (use a sieve or a small beaker) and incubate the beads in 0.4 M HCl for 1 min. Then, rinse thoroughly with water and let the beads dry before storage.
13. It is good laboratory procedure to always spin microcentrifuge tubes with the hinge facing towards the outside of the centrifuge. This means that any time a pellet is due to form (e.g. when precipitating DNA), it will be directly below the tube hinge.
14. DNA extracts can be further cleaned using a column-based DNA extraction kit, like the DNeasy Plant Mini kit (Qiagen) or the NucleoSpin Plant II kit (Clontech); only omit steps that correspond to plant tissue disruption. Alternatively, the Geneclean Kit (bio101) uses GLASSMILK to clean DNA extractions. Most simply, however, run PCR using a dilution series made from the original CTAB extraction – try 1:10, 1:50, 1:100, and 1:1,000 dilutions; this can help if the original sample contains PCR inhibitors.
15. Some researchers prefer to suspend and store DNA in double-distilled water because TE buffer contains a chelator (EDTA) that may interfere with PCR reactions (EDTA chelates Mg^{2+}, which is required for the activity of *Taq* polymerases). However, many laboratories use TE to re-suspend DNA without experiencing any PCR problems attributable to the presence of TE (EDTA) in the DNA template. This may be because the final concentration of EDTA in the PCR reaction is usually too low to interfere with the activity of the *Taq*. If there are still concerns that TE may affect PCR reactions, re-suspend DNA in water; in which case samples should be stored at −20°C as DNA degrades much faster in water than in TE buffer. Alternatively, when column-based kits for DNA extraction or PCR purification are in use, left-over elution buffers from old kits can be used to suspend the sample; these

are optimized for DNA storage and often consist of 5 mM Tris–HCl, pH 8.5.

16. When centrifuging hinged columns to remove ethanol, a "spin-flip-spin" is more effective than leaving the columns the same orientation. First spin with the hinges facing outwards, then stop the centrifuge, rotate all the columns 180°, and spin again.
17. SYBR Safe should be protected from light and stored and used at room temperature; it can freeze in a refrigerator or even near a cool window.
18. If there are general problems when sequencing PCR products excised from a gel, UV damage is a likely cause. Repeat the PCR and gel extraction protocols, reducing DNA exposure to UV light as much as possible.
19. For calculation of primer properties, see, for example, "OligoCalc" (108) and tools provided by companies that synthesize primers, e.g. IDT's OligoAnalyzer (69) and Promega's T_m calculator (http://www.promega.com/biomath/).
20. If using a product like 5× Green GoTaq® Buffer (Promega) in PCR reactions, enzymatic clean-ups will not remove the dyes, which may interfere with subsequent sequencing of the PCR product. This can be avoided by instead using the 5× Colorless GoTaq® buffer (also supplied with GoTaq polymerase).
21. Working aliquots of enzymes should be stored at −20°C, but all other aliquots should be stored at −80°C to best preserve enzymatic activity.
22. When a laminar hood is not available, culture plates can be set up on a clean bench. Putting a naked flame, e.g. a lit Bunsen burner, next to the plates provides an up-current of air at that spot, reducing the chance of contamination.
23. Hydration of Sephadex beads: Slowly add ddH_2O to a tablespoonful or so of beads, repeatedly stirring and allowing to settle, until a thin layer of clear water forms on top, at which point the Sephadex is saturated; the saturated Sephadex can be autoclaved (if desired) and stored at 4°C for later use.
24. When dealing with very small plants, an issue to be aware of is how many individuals have gone into each DNA extraction, as differences between contigs assembled from the reads may represent errors, misalignments or heterozygosity, or real differences between individuals.
25. See information on MALIGN and POY (parsimony-based alignment) and other tree-based sequence alignment software in J. Felsenstein's "Phylogeny Programs" Web site: http://evolution.genetics.washington.edu/phylip/software.html.

Acknowledgments

We are grateful to numerous colleagues from various organizations and laboratories, who over the years gave us training or advice in molecular biology techniques and/or plant phylogenetics. In particular, we thank current and former colleagues from the Royal Botanic Garden Edinburgh (UK); ITQB, Universidade Nova de Lisboa (Portugal); and EEB and MCB, University of Connecticut; Plant Biology Southern Illinois University; Department of Biological Sciences of Virginia Tech and University of Wisconsin-Milwaukee (USA). Jessica Budke provided helpful comments on a draft of this paper. Susana Neves is currently supported by a post-doctoral fellowship (SFRH/BPD/26669/2006) from Fundação para a Ciência e a Tecnologia, Portugal, and Laura Forrest by National Science Foundation grant EF-0531557 (Assembling the Liverwort Tree of Life).

References

1. Judd WS, Campbell CS, Kellogg EA, Stevens PF, Donoghue MJ (2007) Plant systematics: a phylogenetic approach, 2nd edn. Sinauer Associates, Sunderland, MA
2. Hughes C, Eastwood RJ, Bailey CD (2006) From famine to feast? Selecting nuclear DNA sequence loci for plant species-level phylogeny reconstruction. Phil Trans R Soc B 361:211–225
3. Savolainen V, Chase MW (2003) A decade of progress in plant molecular phylogenetics. Trends Genet 19:717–724
4. Soltis DE, Gitzendanner MA, Soltis PS (2007) A 567-taxon data set for angiosperms: the challenges posed by Bayesian analyses of large data sets. Int J Plant Sci 168:137–157
5. The Angiosperm Phylogeny Group (2009) An update of the Angiosperm Phylogeny Group classification for the orders and families of flowering plants: APG III. Bot J Linn Soc 161:105–121
6. Renzaglia KS, Schuette S, Duff RJ, Ligrone R, Shaw AJ, Mishler BD, Duckett JG (2007) Bryophyte phylogeny: advancing the molecular and morphological frontiers. Bryologist 110:179–213
7. CBOL Plant Working Group (2009) A DNA barcode for land plants. Proc Nat Acad Sci USA 106:12794–12797
8. Chase MW, Fay MF (2009) Barcoding of plants and fungi. Science 325:682–683
9. Wiley EO, Siegel-Causey D, Brooks DR, Funk VA (1991) The compleat cladist. Museum of Natural History, Special Publication No 19. The University of Kansas, Lawrence. http://www.archive.org/details/compleatcladistp00wile. Accessed 25 October 2010
10. Page RDM, Holmes EC (1998) Molecular evolution: a phylogenetic approach. Blackwell Science Ltd, Oxford
11. Baldauf SL (2003) Phylogeny for the faint of heart: a tutorial. Trends Genet 19:345–351
12. Harrison CH, Langdale JA (2006) A step by step guide to phylogeny reconstruction. Plant J 45:561–572
13. Soltis DE, Soltis PS, Doyle JJ (1998) Molecular systematics of plants II: DNA sequencing. Kluwer Academic Publishers, Boston/Dordrecht/London
14. Hollingsworth PM, Bateman RM, Gornall RJ (1999) Molecular systematics and plant evolution. Systematics Association Special Vol Ser 57. Taylor & Francis, London and New York
15. Rosenberg MS, Kumar S (2001) Incomplete taxon sampling is not a problem for phylogenetic inference. Proc Natl Acad Sci USA 98:10751–10756
16. Zwickl DJ, Hillis DM (2002) Increased taxon sampling greatly reduces phylogenetic error. Syst Biol 51:588–598
17. Hillis DM, Pollock DD, McGuire JA, Zwickl DJ (2003) Is sparse taxon sampling a problem for phylogenetic inference? Syst Biol 52:124–126

18. Hedtke SM, Townsend TM, Hillis DM (2006) Resolution of phylogenetic conflict in large data sets by increased taxon sampling. Syst Biol 55:522–529
19. Heath TA, Hedte SM, Hillis DM (2008) Taxon sampling and the accuracy of phylogenetic analyses. J Syst Evol 46:239–257
20. Maddison WP, Donoghue MJ, Maddison DR (1984) Outgroup analysis and parsimony. Syst Zool 33:83–103
21. Soltis DE, Soltis PS (1998) Choosing an approach and an appropriate gene for phylogenetic analysis. In: Soltis DE, Soltis PS, Doyle JJ (eds) Molecular systematics of plants II: DNA sequencing. Kluwer Academic Publishers, Boston/Dordrecht/London, pp 1–42
22. Mort ME, Archibald JK, Randle CP, Levsen ND, O'Leary TR, Topalov K, Wiegand CM, Crawford DJ (2007) Inferring phylogeny at low taxonomic levels: utility of rapidly evolving cpDNA and nuclear ITS loci. Amer J Bot 94:173–183
23. Shaw J, Lickey EB, Beck JT, Farmer SB, Liu W, Miller J, Siripun KC, Winder CT, Schilling EE, Small RL (2005) The tortoise and the hare II: relative utility of 21 noncoding chloroplast DNA sequences for phylogenetic analysis. Amer J Bot 92:142–166
24. Burleigh JG, Hilu KW, Soltis DE (2009) Inferring phylogenies with incomplete data sets: a 5-gene, 567-taxon analysis of angiosperms. BMC Evol Biol 9:61. doi:10.1186/1471-2148-9-61
25. Graham SW and Olmstead RG (2000) Utility of 17 chloroplast genes for inferring the phylogeny of the basal angiosperms. Amer J Bot 87:1712–1730
26. Kress WJ, Wurdack KJ, Zimmer EA, Weigt LA, Janzen DH (2005) Use of DNA barcodes to identify flowering plants. Proc Natl Acad Sci USA 102:8369–8374
27. Fazekas AJ, Burgess KS, Kesanakurti PR, Graham SW, Newmaster SG, Husband BC, Percy DM, Hajibabaei M, Barrett SCH (2008) Multiple multilocus DNA barcodes from the plastid genome discriminate plant species equally well. PLoS ONE 3(7):e2802. doi:10.1371/journal.pone.0002802
28. Hollingsworth ML, Clark AA, Forrest LL, Richardson J, Pennington RT, Long DG, Cowan R, Chase MW, Gaudeul M, Hollingsworth PM (2009) Selecting barcoding loci for plants: evaluation of seven candidate loci with species-level sampling in three divergent groups of land plants. Mol Ecol Res 9:439–457
29. Hilu KW, Borsch T, Müller K, Soltis DE, Soltis PS, Savolainen V, Chase MW, Powell MP, Alice LA, Evans R, Sauquet H, Neinhuis C, Slotta TAB, Rohwer JG, Campbell CS, Chatrou LW (2003) Angiosperm phylogeny based on matK sequence information. Amer J Bot 90:1758–1776
30. Qiu YL, Dombrovska O, Lee JH, Li LB, Whitlock BA, Bernasconi-Quadroni F, Rest JS, Davis CC, Borsch T, Hilu KW, Renner SS, Soltis DE, Soltis PS, Zanis MJ, Cannone JJ, Gutell RR, Powell M, Savolainen V, Chatrou LW, Chase MW (2005) Phylogenetic analyses of basal angiosperms based on nine plastid, mitochondrial, and nuclear genes. Int J Plant Sci 166:815–842
31. Álvarez I., Wendel JF (2003) Ribosomal ITS sequences and plant phylogenetic inference. Mol Phylogenet Evol 29:417–434
32. Bailey CD, Carr TG, Harris SA, Hughes CE (2003) Characterization of angiosperm nrDNA polymorphism, paralogy, and pseudogenes. Mol Phylogenet Evol 29:435–455
33. Nieto Feliner G, Rosselló JA (2007) Better the devil you know? Guidelines for insightful utilization of nrDNA ITS in species-level evolutionary studies in plants. Mol Phylogenet Evol 44:911–919
34. Mort ME, Crawford DJ (2004) The continuing search: low-copy nuclear sequences for lower-level plant molecular phylogenetic studies. Taxon 53:257–261
35. Small RL, Cronn RC, Wendel JF (2004) Use of nuclear genes for phylogeny reconstruction in plants. Aust Syst Bot 17:145–170
36. Álvarez I, Costa A, Nieto Feliner G (2008) Selecting single-copy nuclear genes for plant phylogenetics: a preliminary analysis for the Senecioneae (Asteraceae). J Mol Evol 66:276–291
37. Steele PR, Guisinger-Bellian M, Linder CR, Jansen RK (2008) Phylogenetic utility of 144 low-copy nuclear regions in taxa at different taxonomic levels in two distantly related families of rosids. Mol Phylogenet Evol 48:1013–1026
38. Chatrou LW, Escribano MP, Viruel MA, Maas JW, Richardson JE, Hormaza JI (2009) Flanking regions of monomorphic microsatellite loci provide a new source of data for plant species-level phylogenetics. Mol Phylogenet Evol 53:726–733
39. Cummings MP, Meyer A (2005) Magic bullets and golden rules: data sampling in molecular phylogenetics. Zoology 108:329–336
40. Birky CW (1995) Uniparental inheritance of mitochondrial and chloroplast genes:

mechanisms and evolution. Proc Natl Acad Sci 92:11331–11338
41. Nagata N (2010) Mechanisms for independent cytoplasmic inheritance of mitochondria and plastids in angiosperms. J Plant Res 123:193–199
42. Hodkinson TR, Waldren S, Parnell JAN, Kelleher CT, Salamin K, Salamin N (2007) DNA banking for plant breeding, biotechnology and biodiversity evaluation. J Plant Res 120:17–29
43. Liesner R (1995) Field techniques used by Missouri Botanical Garden (English, French and Spanish versions). Missouri Botanical Garden, St Louis, MO. http://www.mobot.org/MOBOT/molib/fieldtechbook/. Accessed 25 October 2010
44. Chase MW, Hills HH (1991) Silica gel: an ideal material for field preservation of leaf samples for DNA studies. Taxon 40:215–220
45. Rogstad SH (1992) Saturated NaCl-CTAB solution as a means of field preservation of leaves for DNA analyses. Taxon 41:701–708
46. Nickrent DL (1994) From field to film: rapid sequencing methods for field-collected plant species. BioTechniques 16:470–475
47. Štorchová H, Hrdličková R, Chrtek J Jr, Tetera M, Fitze D, Fehrer J. (2000) An improved method of DNA isolation from plants collected in the field and conserved in saturated NaC1/CTAB solution. Taxon 49:79–84
48. Thomson JA (2002) An improved non-cryogenic transport and storage preservative facilitating DNA extraction from 'difficult' plants collected at remote sites. Telopea 9:755–760
49. Doyle JJ, Dickson EE (1987) Preservation of plant samples for DNA restriction endonuclease analysis. Taxon 36:715–722
50. Savolainen V, Cuénoud P, Spichiger R, Martinez MDP, Crèvecoeur M, Manen JF (1995) The use of herbarium specimens in DNA phylogenetics: evaluation and improvement. Plant Syst Evol 197:87–98
51. Chase MW, Salamin N, Wilkinson M, Dunwell JM, Kesanakurthi RP, Haidar N, Savolainen V (2005) Land plants and DNA barcodes: short-term and long-term goals. Phil Trans R Soc B 360:1889–1895
52. Cowan RS, Chase MW, Kress WJ, Savolainen V (2006) 300,000 species to identify: problems, progress, and prospects in DNA barcoding of land plants. Taxon 55:611–616
53. Telle S, Thines M (2008) Amplification of cox2 (~620 bp) from 2 mg of up to 129 years old herbarium specimens, comparing 19 extraction methods and 15 polymerases. PLoS One 10:e3584. doi:10.1371/journal.pone.0003584
54. Jankowiak K, Buczkowska K, Szweykowska-Kulinska Z (2005) Successful extraction of DNA from 100-year-old herbarium specimens of the liverwort *Bazzania trilobata*. Taxon 54:335–336
55. Doyle JJ, Doyle, JL (1987) A rapid DNA isolation procedure for small quantities of fresh leaf tissue. Phytochemical Bull 19:11–15
56. Doyle JJ, Doyle JL (1990) Isolation of plant DNA from fresh tissue. Focus 12: 13–15
57. Porebski S, Bailey LG, Baum BR (1997) Modification of a CTAB DNA extraction protocol for plants containing high polysaccharide and polyphenol components. Plant Mol Biol Rep 15:8–15
58. Allen GC, Flores-Vergara MA, Krasnyanski S, Kumar S, Thompson WF (2006) A modified protocol for rapid DNA isolation from plant tissues using cetyltrimethylammonium bromide. Nature Protocols 1:2320–2325
59. Varma A, Padh H, Shrivastana (2007) Plant genomic DNA isolation: an art or a science. Biotechnol J 2:386–392
60. Saghai-Maroof MA, Soliman KM, Jorgensen RA, Allard RW (1984) Ribosomal spacer-length polymorphisms in barley: Mendelian inheritance, chromosomal location, and population dynamics. Proc Natl Acad Sci USA 81: 8014–8019
61. Stewart CN, Via LE (1993) A rapid CTAB DNA isolation technique useful for RAPD fingerprinting and other PCR applications. Biotechniques 14:748–749
62. Marmur J (1961) A procedure for the isolation of deoxyribonucleic acid from microorganisms. J Mol Biol 3: 208–218
63. Sambrook J, Russell DW (2001) Molecular cloning: a laboratory manual (3 vols), 3rd edn. Cold Spring Harbor Laboratory Press, New York
64. Palumbi SR (1996) Nucleic acids II: the polymerase chain reaction. In: Hillis DM, Moritz C, Mable BK. Molecular Systematics, 2nd edn. Sinauer Associates, Sunderland, MA, pp 205–247
65. Dieffenbach CW, Dveksler GS (2003) PCR primer: a laboratory manual, 2nd edn. Cold Spring Harbor Laboratory Press, Cold Spring Harbor, NY
66. Strand AE, Leebens-Mack J, Milligan BG (1997) Nuclear DNA-based markers for plant evolutionary biology. Mol Ecol 6:113–118
67. Rozen S, Skaletsky H (2000) Primer3 on the WWW for general users and for biologist programmers. In: Misener S, Krawetz SA (eds.)

Bioinformatics methods and protocols. Methods in Molecular Biology, vol 132. Humana Press, Totowa, NJ, pp 365–386 (*Primer 3* interface: http://frodo.wi.mit.edu/primer3/. Accessed 25 October 2010)

68. Untergasser A, Nijveen H, Rao X, Bisseling T, Geurts R, Leunissen JAM (2007) Primer3Plus, an enhanced web interface to Primer3. Nucleic Acids Res 35:W71–W74. doi:10.1093/nar/gkm306 (*Primer3Plus* interface: http://www.bioinformatics.nl/primer3plus. Accessed 25 October 2010)
69. Owczarzy R, Tataurov AV, Wu Y, Manthey JA, McQuisten KA, Almabrazi HG, Pedersen KF, Lin Y, Garretson J, McEntaggart NO, Sailor CA, Dawson RB, Peek AS (2008) IDT SciTools: a suite for analysis and design of nucleic acid oligomers. Nucleic Acids Res 36:W163–169 (*IDT SciTools*: http://www.idtdna.com/scitools/. Accessed 25 October 2010)
70. Gadberry MD, Malcomber ST, Doust AN, Kellogg EA (2005) Primaclade – a flexible tool to find conserved PCR primers across multiple species. Bioinformatics 21:1263–1264 (*Primaclade* website: http://www.umsl.edu/services/kellogg/primaclade.html. Accessed 25 October 2010)
71. Pavlov AR, Pavlova NV, Kozyavkin SA, Slesarev AI (2004) Recent developments in the optimization of thermostable DNA polymerases for efficient applications. Trends Biotechnol 22:253–260
72. Kreader CA (1996) Relief of amplification inhibition in PCR with bovine serum albumin or T4 gene 32 protein. Appl Environ Microbiol 62:1102–1106
73. Frackman S, Kobs G, Simpson D, Storts D (1998) Betaine and DMSO: enhancing agents for PCR. Promega Notes 65:27
74. Henke W, Herdel K, Jung K, Schnorr D, Loening SA (1997) Betaine improves the PCR amplification of GC-rich DNA sequences. Nucleic Acids Res 25:3957–3958
75. D'Aquila RT, Bechtel LJ, Videler JA, Eron JJ, Gorczyca P, Kaplan JC (1991) Maximizing sensitivity and specificity of PCR by preamplification heating. Nucleic Acids Res 19:3749
76. Don RH, Cox PT, Wainwright BJ, Baker K, Mattick JS (1991) 'Touchdown' PCR to circumvent spurious priming during gene amplification. Nucleic Acids Res 19:4008
77. Saar DE, Polans NO, Sørensen PD, Duvall MR (2001) Angiosperm DNA contamination by endophytic fungi: detection and methods of avoidance. Plant Mol Biol Rep 19:249–260
78. Bower MA, Spencer M, Matsumura S, Nisbet RER, Howe C (2005) How many clones need to be sequenced from a single forensic or ancient DNA sample in order to determine a reliable consensus sequence? Nucleic Acids Res 33:2549–2556
79. Baum BR, Johnson DA, Bailey LG (2001) Defining orthologous groups among multicopy genes prior to inferring phylogeny, with special emphasis on the Triticeae (Poaceae). Hereditas 135:123–138
80. Rauscher JT, Doyle JJ, Brown HD (2002) Internal transcribed spacer repeat-specific primers and the analysis of hybridization in the *Glycine tomentella* (Leguminosae) polyploid complex. Mol Ecol 11:2691–2702
81. Magain N, Forrest LL, Sérusiaux E, Goffinet B (2010) Microsatellite primers in the *Peltigera dolichorhiza* complex (lichenized ascomycete, Peltigerales). Amer J Bot 97:e102–e104. doi:10.3732/ajb.1000283
82. Zane L, Bargelloni L, Patarnello T (2002). Strategies for microsatellite isolation: a review. *Mol Ecol* 11:1–16
83. Squirrel J, Hollingsworth PM, Woodhead M, Russell J, Lowe AJ, Gibby M, Powell W (2003) How much effort is required to isolate nuclear microsatellites from plants? Mol Ecol 12:1339–1348
84. Moore MJ, Dhingra A, Soltis PS, Shaw R, Farmerie WG, Folta KM, Soltis DE (2006) Rapid and accurate pyrosequencing of angiosperm plastid genomes. BMC Plant Biol 6:17. doi:10.1186/1471-2229-6-17
85. Cronn R, Liston A, Parks M., Gernandt DS, Shen R, Mockler T (2008) Multiplex sequencing of plant chloroplast genomes using Solexa sequencing-by-synthesis technology. Nucleic Acids Res 36:e122. doi: 10.1093/nar/gkn502
86. Donaher N, Tanifuji G, Onodera NT, Malfatti SA, Chain PSG, Hara Y, Archibald JM (2009) The complete plastid genome sequence of the secondarily nonphotosynthetic alga *Cryptomonas paramecium*: reduction, compaction, and accelerated evolutionary rate. Genome Biol Evol 1:439–448
87. Atherton RA, McComish BJ, Shepherd LD, Berry LA, Albert NW, Lockhart PJ (2010) Whole genome sequencing of enriched chloroplast DNA using the Illumina GAII platform. Plant Methods 6:22. doi:10.1186/1746-4811-6-22
88. Tangphatsornruang S, Sangsrakru D, Chanprasert J, Uthaipaisanwong P, Yoocha T, Jomchai N, Tragoonrung S (2010) The chloroplast genome sequence of mungbean

(*Vinga radiata*) determined by high-throughput pyrosequencing: structural organization and phylogenetic relationships. DNA Research 17:11–22
89. Phillips A, Janies D, Wheeler W (2000) Multiple sequence alignment in phylogenetic analysis. Mol Phylogenet Evol 16:317–330
90. Rosenberg MS (2009) Sequence alignment: concepts and history. In: Rosenberg MS (ed) Sequence alignment: methods, models, concepts, and strategies. University of California Press, Berkeley, pp 1–22
91. Doyle JJ, Davis JI (1998) Homology in molecular phylogenetics: a parsimony perspective. In: Soltis DE, Soltis PS, Doyle JJ (eds) Molecular systematics of plants II: DNA sequencing. Kluwer Academic Publishers, Boston/Dordrecht/London, pp 101–131
92. Kelchner S.A. (2000) The evolution of noncoding chloroplast DNA and its application in plant systematics. Ann Missouri Bot Gard 87:482–498
93. Altschul SF, Gish W, Miller W, Myers EW, Lipman DJ (1990) Basic local alignment search tool. J Mol Biol 215:403–410
94. Thompson JD, Higgins DG, Gibson TJ (1994) CLUSTAL W: improving the sensitivity of progressive multiple sequence alignment through sequence weighting, position specific gap penalties and weight matrix choice. Nucleic Acids Res 22:4673–4680
95. Castresana J (2000) Selection of conserved blocks from multiple alignments for their use in phylogenetic analysis. Mol Biol Evol 17:540–552 (*Gblocks* website: http://molevol.cmima.csic.es/castresana/Gblocks.html. Accessed 25 October 2010)
96. Hall BG (2007) Phylogenetic trees made easy: a how-to manual, 3rd edn. Sinauer Associates, Sunderland, MA
97. Lemey P, Salemi M, Vandamme AM (2009) The phylogenetic handbook: a practical approach to phylogenetic analysis and hypothesis testing, 2nd edn. Cambridge University Press, Cambridge
98. Lewis PO (2001) Phylogenetic systematics turns over a new leaf. Trend Ecol Evol 16:30–37
99. Holder M, Lewis PO (2003) Phylogeny estimation: traditional and Bayesian approaches. Nature Rev Genet 4:275–284
100. Felsenstein J (2004) Inferring phylogenies. Sinauer Associates, Sunderland, MA
101. Hillis DM, Moritz C, Mable BK (1996) Molecular Systematics, 2nd edn. Sinauer Associates, Sunderland, MA
102. Swofford DL, Olsen GJ, Waddell PJ, Hillis DM (1996) Phylogenetic inference. In: Hillis DM, Moritz C, Mable BK. Molecular Systematics, 2nd edn. Sinauer Associates, Sunderland, MA, pp. 407–514
103. Sanderson MJ, Shaffer HB (2002) Troubleshooting molecular phylogenetic analyses. Annu Rev Ecol Syst 33:49–72
104. Page RDM, Charleston MA (1997) From gene to organismal phylogeny: reconciled trees and the gene tree/species tree problem. Mol Phylogenet Evol 7:231–240
105. Slowinski JB, Page RDM (1999) How should species phylogenies be inferred from sequence data? Syst Biol 48:814–825
106. Crisp MD, Cook LG (2005) Do early branching lineages signify ancestral traits? Trends Ecol Evol 20:122–128
107. Frank MS, Perkins KD (2007) Preparation of plant specimens for deposit as herbarium vouchers. University of Florida Herbarium. http://www.flmnh.ufl.edu/herbarium/voucher.htm. Accessed 25 October 2010
108. Kibbe WA (2007) OligoCalc: an online oligonucleotide properties calculator. Nucleic Acids Res 35:W43–W46. doi:10.1093/nar/gkm234 (*OligoCalc* interface: http://www.basic.northwestern.edu/biotools/OligoCalc.html. Accessed 25 October 2010)

Chapter 11

Using Coevolution to Predict Protein–Protein Interactions

Gregory W. Clark, Vaqaar-un-Nisa Dar, Alexandr Bezginov, Jinghao M. Yang, Robert L. Charlebois, and Elisabeth R.M. Tillier

Abstract

Bioinformatic methods to predict protein–protein interactions (PPI) via coevolutionary analysis have positioned themselves to compete alongside established in vitro methods, despite a lack of understanding for the underlying molecular mechanisms of the coevolutionary process. Investigating the alignment of coevolutionary predictions of PPI with experimental data can focus the effective scope of prediction and lead to better accuracies. A new rate-based coevolutionary method, MMM, preferentially finds obligate interacting proteins that form complexes, conforming to results from studies based on coimmunoprecipitation coupled with mass spectrometry. Using gold-standard databases as a benchmark for accuracy, MMM surpasses methods based on abundance ratios, suggesting that correlated evolutionary rates may yet be better than coexpression at predicting interacting proteins. At the level of protein domains, coevolution is difficult to detect, even with MMM, except when considering small-scale experimental data involving proteins with multiple domains. Overall, these findings confirm that coevolutionary methods can be confidently used in predicting PPI, either independently or as drivers of coimmunoprecipitation experiments.

Key words: Protein coevolution, Bioinformatics, Protein–protein interactions, Protein interaction network, Proteomics, Phylogenetic tree, Domain–domain interactions

1. Introduction

Proteins mediate the control and function of cellular processes through their interactions with one another, with nucleic acids and with many other cellular components. Emerging computational methodologies promise to overcome some of the restrictions that limit the discovery of protein–protein interactions (PPIs) using in vitro methods alone. The use of multiple tools to detect PPI is,

Gerard Cagney and Andrew Emili (eds.), *Network Biology: Methods and Applications*, Methods in Molecular Biology, vol. 781, DOI 10.1007/978-1-61779-276-2_11, © Springer Science+Business Media, LLC 2011

however, necessary in constructing high confidence protein networks due to the inherent limitations that are specific to each method. Recent methods make use of protein sequences, understanding that these sequences ultimately direct a protein's higher-order structure and hence function. These sequences, collected from widely divergent species, permit the analysis of evolutionary rates and relationships that are known to influence, or to be influenced by, PPI (1, 2). The basis of methods for the computational prediction of PPI thus relies on the premise that proteins that interact tend to coevolve. Within this framework, identifying a pairwise interaction amounts to finding well-correlated rates of evolution between proteins. Significantly, such coevolutionary methods have already been successful in predicting PPIs in both bacteria and humans (3, 4). They are now included alongside other high-throughput technologies for investigating PPIs, rapidly increasing the number of putative protein interactions, thus helping to fulfill the goal of defining the interactome – all known PPIs in a given organism.

1.1. Coevolution and Protein–Protein Interactions

Coevolution is a process whereby modification in one biological entity drives a compensatory modification in another. Although the concept of coevolution originated in the examination of inter-species interactions at the anatomical level, it applies equally well to macromolecules (5). Specifically, coevolutionary theory has been adapted both to determine contact sites within proteins, as well as to identify between-protein interactions. Two sequence-based coevolutionary analyses can be performed. The first of these is correlated mutation analysis.

1.1.1. Correlated Mutation Analysis

CMA relies on finding correlated mutations of amino-acid residues to predict protein contact. Residues in close contact contribute to an energetic optimum that maintains structure and stability in proteins. Mutation of a residue may adversely affect local protein stability through suboptimal interactions with nearby residues, unless those other residues also mutate in compensation to preserve protein stability and function (6, 7). Such residue-based methods have been successful in identifying intraprotein and sometimes even interprotein contacts, and can find important motifs within protein secondary structures (8–15).

1.1.2. Correlated Evolutionary Rates

The observation that interacting proteins tend to coevolve (16) has spawned many investigations into the use of coevolutionary analysis to predict PPI (3, 4, 17–22). Rate-based strategies aim to find proteins that evolve at similar rates such that they present similar phylogenetic trees in terms of both branching topology and branch lengths. (However, rather than matching phylogenetic trees directly, these approaches typically make use of the tree's underlying distance matrix, thus avoiding the pitfalls of phylogenetic reconstruction). A protein family's (see Subheading 4.1) distance matrix contains pairwise information about the evolutionary divergence of its included sequences. Since each pairwise entry is a relative evolutionary distance between two proteins, the matrix itself can be interpreted as a collection of relative evolutionary rates. Comparing these corresponding rates between two distance matrices lies at the heart of several distinct methods.

1.2. Coevolution Methods

1.2.1. MirrorTree

The *MirrorTree* method for comparing phylogenetic trees, by correlating their underlying distance matrices, was developed by Pazos and Valencia (4). It is commonly used in coevolutionary analysis (18–20, 22). This method uses Pearson's correlation coefficient for each taxon-matched distance entry between two paired distance matrices as a proxy measure of tree similarity. A high correlation between two matrices indicates that the two matrices are similar and, therefore, that the proteins may interact. In general, a correlation between two matrices of >0.8 is a reliable cut-off for predicting an interaction. An inconvenience of *MirrorTree* is that distance matrices must be trimmed to the same size and species composition prior to calculating the correlation coefficient. However, this method has proved effective in generating the *Escherichia coli* protein interaction network (4). In its original implementation, using 14 complete bacterial genomes, *MirrorTree* was able to reliably predict interaction between structural domains (cut-off $r > 0.77$), and between individual proteins (cut-off $r > 0.8$) with a high true positive rate.

Recently, *MirrorTree* has been extended to account for complex evolutionary factors that influence the coevolution of two proteins. The *ContextMirror* method (22) adapts *MirrorTree* in a manner that considers other possible PPIs that could influence an individual protein's evolutionary signal, thereby accounting for coevolution of complexes altogether, and reducing the inherent phylogenetic signal stemming from speciation events. The algorithm uses partial correlation coefficients to reduce the effect of speciation, with increasing levels of partial correlation relating to how many other proteins are considered to influence the coevolution of a single protein of interest. Using *ContextMirror* (10th level partial correlation), the accuracy of PPI prediction was improved, particularly when considering interactions among protein complexes, and proteins in the same pathway (22). This indicates that correcting for phylogenetic interference can improve the prediction accuracy of PPIs.

1.2.2. Matrix Match Maker

A more recent coevolutionary algorithm, MMM (3) avoids some of the difficulties associated with *MirrorTree*. The main advantage of MMM lies in its flexibility. While the principle of matching corresponding rates of evolution between two protein families remains, MMM is implemented so that coevolution is not required between all entries of the distance matrices, allowing for lineage-specific evolution (Fig. 1). The algorithm begins by selecting a triplet of distances within each matrix. For example, consider three corresponding (species-matched) sequences in each of the matrices under analysis: Human_CCT1 *(A)*, Yeast_CCT1 *(B)*, and Drosophila_CCT1 *(C)* in matrix 1 and Human_CCT2 (α), Yeast_CCT2 (β), and Drosophila_CCT2 (χ) in matrix 2. Pairings of distances in each matrix [human-yeast $(AB/\alpha\beta)$, human-*Drosophila* $(AC/\alpha\chi)$, yeast-*Drosophila* $(BC/\beta\chi)$) are compared to obtain corresponding match ratios. If each ratio satisfies a tolerance threshold and, therefore, if $AB/\alpha\beta \approx AC/\alpha\chi \approx BC/\beta\chi$, the program recursively tries to build up the match by adding other matrix entries whose ratios remain consistent with prior entries. Using this method, not all sequences are required to become a part of the solution; the largest common submatrix will comprise only those sequences that coevolve at the same relative rates.

MMM also offers flexibility by restricting matches to a user-specified tolerance argument. The tolerance threshold limits the differences in the ratio of evolutionary distance (by, e.g., 10%) within which the two protein families must match, but can be adjusted to accommodate for varying degrees of coevolution depending on the question at hand. In contrast to an approach such as *MirrorTree* that provides a single correlation score for a pair of matrices, MMM instead finds the largest common submatrix (of size ≥3, if any) that respects the chosen tolerance threshold.

The submatrix approach to detecting coevolution between two protein families makes MMM resilient to poorly categorized orthologs so that erroneous entries in a given matrix can automatically be ignored without impacting the score. This same feature allows MMM to handle paralogs; in cases where paralogs are included in distance matrices, they can be considered as a part of the solution, provided they fit the match criteria.

MMM has already been used in the construction of the human coevolution network (3). Like other protein networks, the human coevolutionary network is scale-free and small world (23), and includes a number of known interactions. Importantly, as the

Fig. 1. Illustration of steps setting up for MMM, and function of the MMM algorithm. (1) Having obtained protein clusters from a reliable database, multiple sequence alignments are created using MAFFT (or another suitably accurate alignment method). (2) Corrected distance matrices from all multiple sequence alignments are generated using programs such as protdist from PHYLIP. These distance matrices give pairwise information about the evolutionary relationship between all sequences and are representative of the phylogenetic tree. (3) The coevolution strategy of MMM attempts to find matching submatrices (within the tolerance threshold) between two distance matrices, which is equivalent to finding the matching phylogenetic subtree. If three branches are matched, MMM recursively attempts to match all other branches, until no more can be added. The solution with the highest number of branches matched – the largest common submatrix – provides the MMM score.

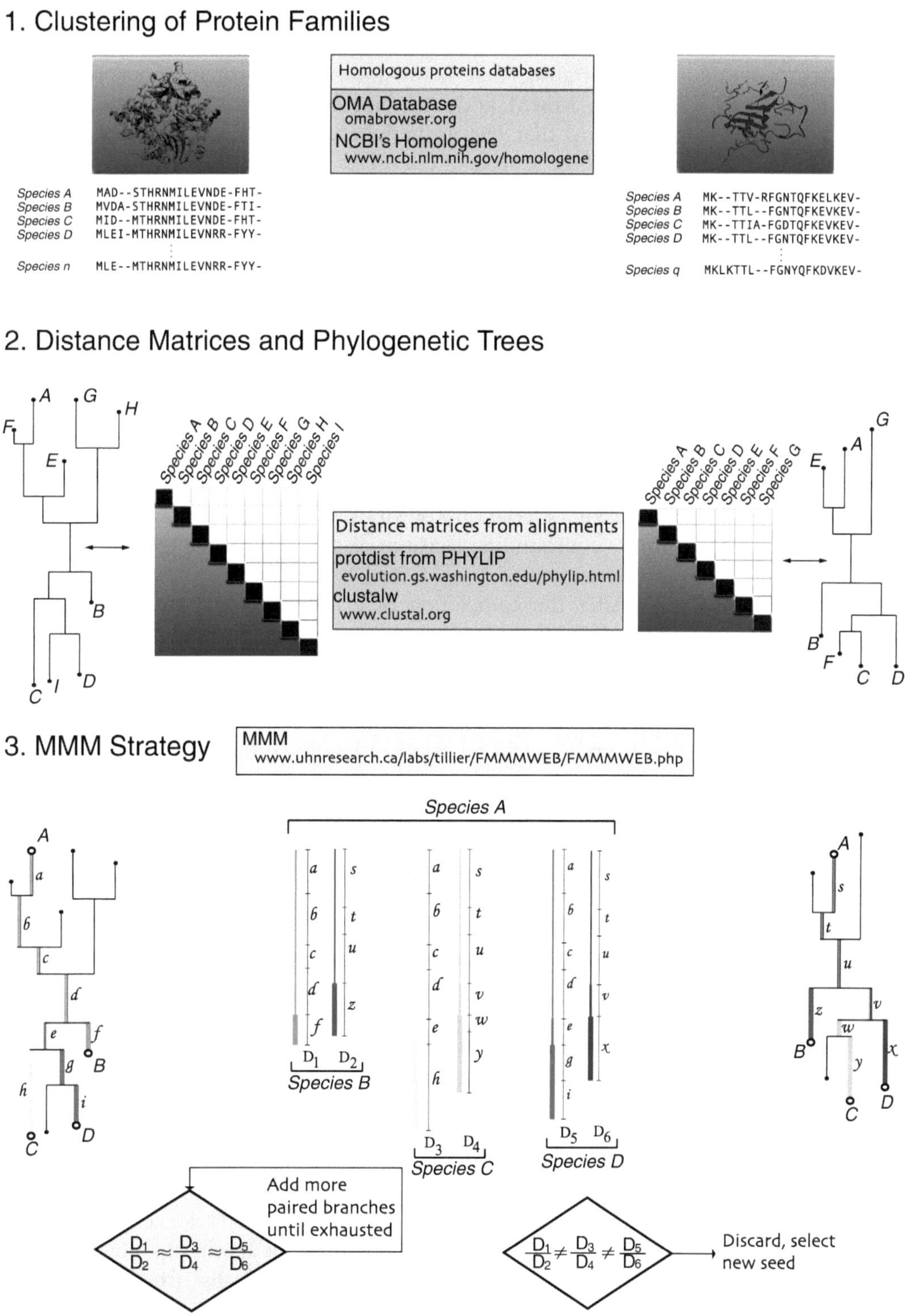
1. Clustering of Protein Families
Homologous proteins databases
OMA Database
omabrowser.org
NCBI's Homologene
www.ncbi.nlm.nih.gov/homologene
Species A MAD--STHRNMILEVNDE-FHT-
Species B MVDA-STHRNMILEVNDE-FTI-
Species C MID--MTHRNMILEVNDE-FHT-
Species D MLEI-MTHRNMILEVNRR-FYY-
Species n MLE--MTHRNMILEVNRR-FYY-
Species A MK--TTV-RFGNTQFKELKEV-
Species B MK--TTL--FGNTQFKEVKEV-
Species C MK--TTIA-FGDTQFKEVKEV-
Species D MK--TTL--FGNTQFKEVKEV-
Species q MKLKTTL--FGNYQFKDVKEV-
2. Distance Matrices and Phylogenetic Trees
Species A Species B Species C Species D Species E Species F Species G Species H Species I
Distance matrices from alignments
protdist from PHYLIP
evolution.gs.washington.edu/phylip.html
clustalw
www.clustal.org
Species A Species B Species C Species D Species E Species F Species G
3. MMM Strategy
MMM
www.uhnresearch.ca/labs/tillier/FMMMWEB/FMMMWEB.php
Species A
Species B
Species C
Species D
Add more paired branches until exhausted
D1/D2 ≈ D3/D4 ≈ D5/D6
D1/D2 ≠ D3/D4 ≠ D5/D6
Discard, select new seed

constraint on the ratios between evolutionary rates of proteins is increased, the accuracy of MMM also increases, once again demonstrating that many interacting proteins coevolve. Additionally, MMM finds proteins belonging to the same pathway, indicating that factors peripheral to the actual interaction may define some aspects of coevolutionary signal.

Although still computationally demanding, we have converted MMM's original algorithm to make use of a maximum compatible cliques approach to improve MMM's speed without sacrificing accuracy (23a). Despite being slower than some other coevolutionary analyses, MMM's approach provides distinct advantages that can mitigate this cost.

1.3. Effect of the Underlying Species Phylogeny

One of the difficulties with any coevolutionary analysis is in correcting for the underlying phylogenetic signal, comprising a large part of the evolutionary rate of a protein. Whereas compensatory mutations provide a distinctive coevolutionary signal to protein evolution, speciation contributes a common signal. Phylogeny is thus a confounding factor in discovering coadaptation among two proteins.

After the completion of a coevolutionary rate analysis (e.g., MMM or *MirrorTree*) on a given dataset, two post-processing methods can be used to correct this problem. The first method requires an average distance matrix, which can be obtained by averaging matched entries either from the entire set of input distance matrices, or from a careful selection of highly conserved proteins. To apply this correction method, MMM runs the average matrix against each of the input matrices. An MMM score between two protein families of interest can then be corrected for their individual match to the average matrix, thus removing the background signal. A second type of correction for the phylogenetic signal can be achieved through methods similar to *ContextMirror* as described above. (For MMM, we accomplish this using a program we call MMMC).

1.4. Does Coevolution Work?

Although coevolutionary approaches have been successful as measured against experimental data, they are currently treated as a "black-box" for PPI prediction owing to our lack of understanding regarding the fundamental processes that guide coevolution between proteins (24). The incumbent theory regarding the roots of coevolution contends that functional coadaptation through the correlated mutation of residues is the cause of coevolution between two proteins (16, 21, 22, 24, 25). Recently, Hakes et al. (2007) have suggested that interacting proteins are simply subject to coincidental evolutionary pressures, specifically from coordinated expression and its effect on evolutionary rates, resulting in coevolution (19). Their analysis raises two troubling issues regarding the understanding and application of current coevolutionary analysis: (1) coadaptation is not apparent at interfacial/interacting residues where it should be prominent and detectable and (2) correlation of mRNA abundance ratios performs better than *MirrorTree* at finding protein–protein interactions. The implication is that the coadaptive component of two interacting proteins is indistinguishable from the general coevolutionary signal, and that coexpression (via abundance ratios in this case) is largely responsible for coevolution. This correlation is expected, given the link between coexpression and evolutionary rate (26, 27) so that methods using either correlated rates or coexpression are finding elements of the same coevolutionary signal.

Pressure to maintain stoichiometry in protein complexes (28) is believed to regulate the expression of their constituent proteins (19, 29). This reasoning is fine for proteins that act within a larger complex or when considering limiting abundances creating bottlenecks in signaling systems. However, the value of this argument weakens when reflecting that coregulated genes (also implying gene neighborhood methods for PPI prediction) are also more likely to be physically interacting (30, 31). Implicitly, coregulated genes are coexpressed; but another way to match expression levels independent of transcription initiation is through codon usage bias at the gene level (32). In this scenario, changes in nucleotide sequence affect the fidelity of transcription, thereby influencing individual expression levels. Indeed, codon usage bias alone has been successful at predicting PPIs (33). However, these nucleotide changes need not effect changes at the protein level to exert their influence so that pressure to maintain the stoichiometry of an interaction may occur independently of changes in protein sequence. Yet, strongly correlated rates between proteins at the sequence level finds interacting proteins (3, 4), arguing that while coexpression and protein coadaptation may drift together, each should provide independent information regarding PPIs.

We used MMM to explore the types of interactions coevolutionary arguments find, and to question the relationship between coadaptation and coevolution.

2. Results

2.1. Comparison of MMM Predictions and Yeast Gold-Standard PPIs

Coevolutionary analysis using MMM is an easy and effective way to complement currently available data on PPI (see Subheading 4.1). Currently, two experimental PPI methods, yeast two-hybrid (Y2H) and affinity purification coupled with mass spectrometry (AP-MS), dominate the knowledge-base of “known” PPIs. However, the inconsistency that exists even between experiments using the same technology (34, 35) has prompted the creation of “gold-standard” datasets. The use of gold-standard datasets aims to increase accuracy in known, true PPIs by only using interactions that have been

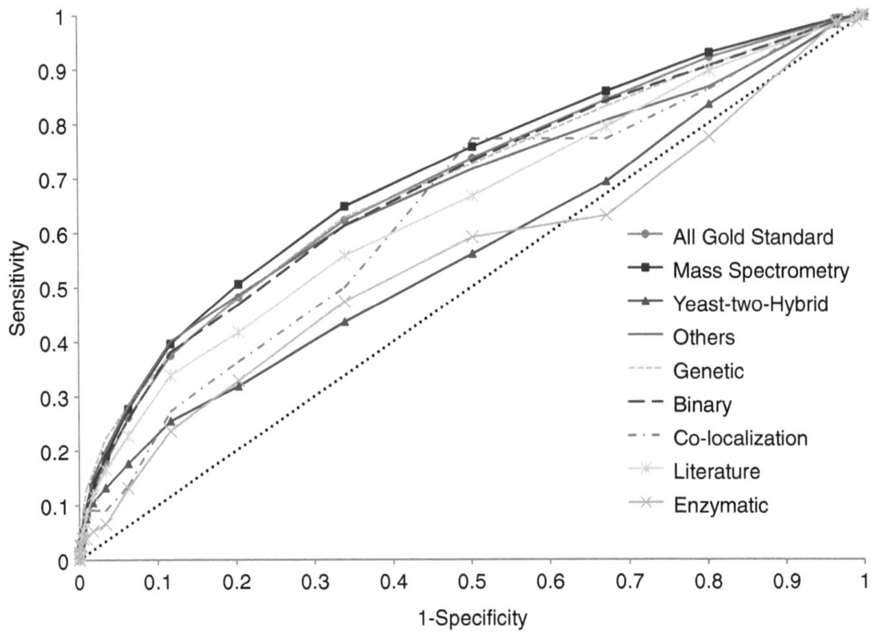

Fig. 2. Accuracy of MMM on subsets of the gold-standard datasets. ROC curve showing that MMM predicts mass spectrometry (*squares*) and genetic (*short-dashed line*) interactions best, while poorly predicting yeast two-hybrid (*triangles*) and enzymatic interactions ("x" markers). This plot suggests that MMM, and coevolutionary methods in general, find complexes with obligate protein members as opposed to transient interactions.

identified by multiple methods. MMM PPI prediction, projected onto these other datasets, can reveal idiosyncrasies underlying each methodology, while the datasets can in turn be used to gauge the accuracy of MMM predictions. Yeast (*Saccharomyces cerevisiae*) has the most complete coverage of PPIs of any interactome gold-standard dataset (36) (see Subheading 4.2), so we compared the results of MMM for yeast proteins with this gold-standard interactome. MMM preferentially identifies AP-MS over Y2H interactions (Fig. 2), while poorly finding enzymatic and colocalization interactions.

These findings not only provide information regarding the types of interactions that MMM finds but also reveal differences between the Y2H and AP-MS methods. In general, AP-MS methods predominantly predict protein complexes where individual proteins are strongly interacting with each other (obligate interactions), while Y2H methods specialize in finding in vivo binary interactions that are weak or transient. Another important factor that distinguishes AP-MS from Y2H is that acquisition of AP-MS data requires highly abundant protein complexes for detection. This bias may favor detection by coevolutionary methods due to the well-known link between protein expression and conservation (26, 27). Consideration should be given to the post-runtime assignment of interactions by AP-MS, where nondirect protein interactions can be included. As expected, MMM detects AP-MS, binary, and genetic

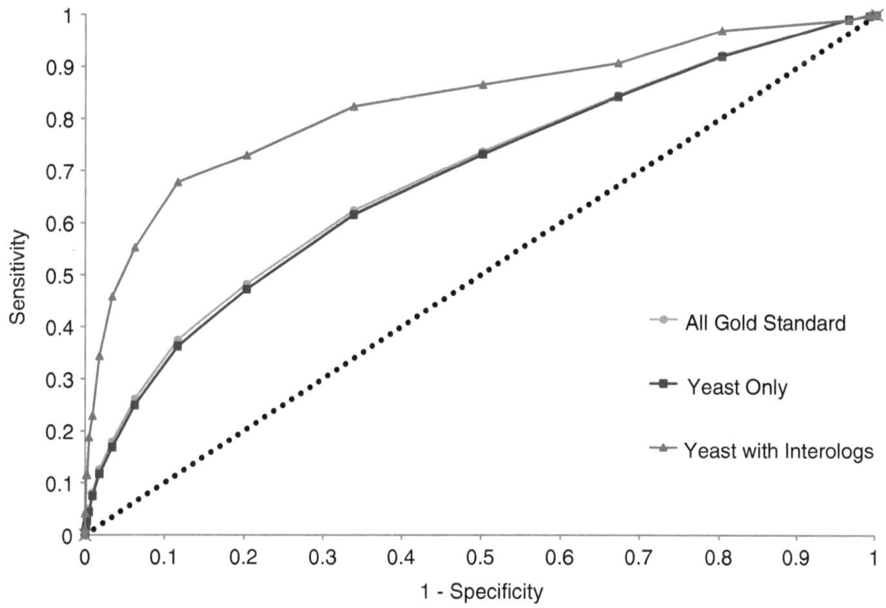

Fig. 3. Interactions confirmed in multiple species are more likely to coevolve. MMM finds interactions that are only known in yeast (*squares*), but is most accurate when considering interactions that have been observed in at least one other species as well (interologs; *triangles*).

interactions better than it does Y2H, since MMM (and coevolutionary methods in general) necessarily requires that proteins have intimate contact to facilitate the coevolutionary process. As such, obligate interactions should more likely be detected by coevolutionary methods (25). Furthermore, obligate interactions tend to have larger interaction interfaces than transient interactions (37) providing a greater number of relevant residues for detection of coadaptation among obligate interactions and thus a stronger signal that can more likely be detected by coevolutionary methods. MMM's bias is also shown by its poor performance on colocalization or enzymatic interactions that are transient in nature (Fig. 2).

Fittingly, the addition of interologs (interactions transferred from other organisms) improved MMM prediction accuracy (Fig. 3). Inclusion of these data is appropriate since MMM (and other coevolutionary approaches) requires that protein groups be preserved across numerous species to score well. These obligate interactions/protein complexes are likely conserved throughout evolution and thus may have a greater tendency to maintain their interactions (38, 39); corroboration of the protein interaction datasets increases the overall coverage of the network, with limited risk of false assignment.

2.2. Comparison of MMM to MirrorTree and Coabundance

As a challenge to current thinking regarding the basis of coevolution, Hakes et al. (2007) suggested that coexpression (via mRNA abundance ratios) is the dominant factor driving

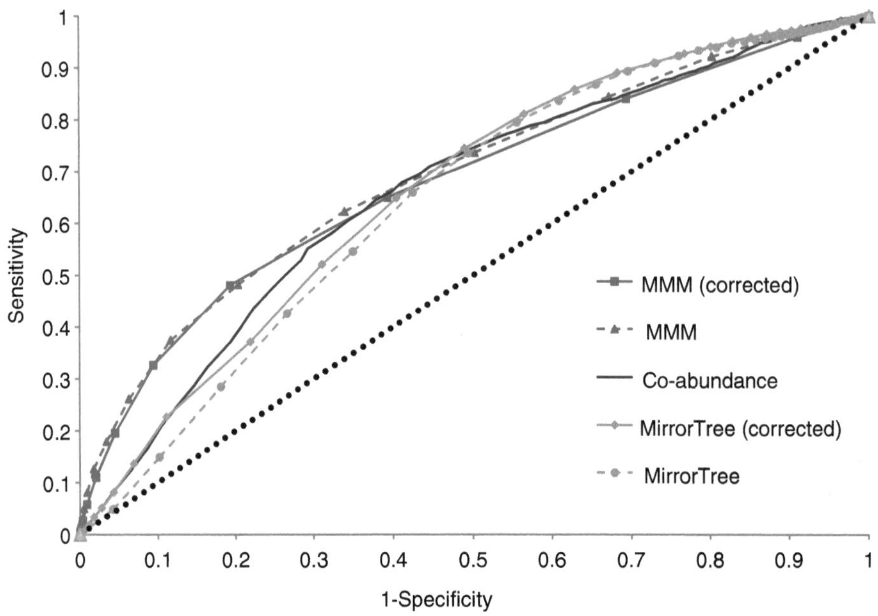

Fig. 4. Comparison of methods for predicting interactions. ROC curves for MMM and the standard *MirrorTree* approach are compared with mRNA abundance ratios to predict gold-standard interactions. This analysis replicates the analysis of Hakes et al. (2007), albeit on a much larger set of protein pairs. It confirms that similar expression levels are more predictive than *MirrorTree*. However, we additionally show that the performance of *MirrorTree* is improved by the phylogeny correction, and that MMM is a much more specific predictor.

coevolution. Since this analysis used *MirrorTree* on a small subset of yeast interacting proteins, we carried out the same analysis as Hakes et al. (2007) for all gold-standard interaction proteins for which abundance levels were available from Holstege et al. (1998) (40). The difference in mRNA abundance was measured as the absolute value of the difference in the log of each protein's abundance. Small values indicate similar abundance levels. Our analysis confirms the finding that low abundance ratios (indicative of coabundance) are a better predictor of interactions than *MirrorTree*. However, we also found that MMM performs better than both *MirrorTree* and coabundance (Fig. 4) on gold-standard PPIs in yeast. Interestingly, *MirrorTree*'s performance is greatly augmented by phylogeny correction, while MMM is only moderated improved. Importantly, MMM performs better than coabundance alone, demonstrating that factors other than coexpression are important components of the coevolutionary signal between proteins. This is consistent with the suggestion that MMM predicts obligate interactions better than transient interactions. Although MMM predicts interactions better than do coabundances on gold-standard datasets, MMM cannot specifically distinguish coadaptation from other factors that may influence coevolution.

2.3. Domain Coevolution

In eukaryotes, proteins are composed of modular, independent domains that often play direct roles in mediating PPIs (41–43). If an interaction is accommodated through domains that mediate an interaction, the coevolutionary signal between these domains should be higher than that for full-length sequences owing to a higher localized concentration of residues that are eligible for coadaptation between the interacting domains. In order to more directly address how coadaptation contributes to the coevolutionary signal, we used MMM to detect coevolution between protein domains (domains between protein pairs) to extract information on potential PPIs (see Subheading 4.3). As opposed to other approaches that use coevolutionary information derived at the residue level of domains – either directly (44, 45) or in preprocessing (21) – examination with MMM shows that individual domains discover interactions more poorly than full-length proteins (Fig. 5a). By contrast, interaction predictions using domain-based profiling have worked better than profiling based on full-length proteins (46). This suggests that while domains are relevant informational units in detection of protein interaction, the evolutionary signal is difficult to detect from individual domains alone through the use of rate-based coevolution methods. These results seem to confirm results of the Hakes et al. (2007) study that showed interface residues do not select for interacting proteins. However, when domains from multiple-domain proteins are considered, MMM's specificity increases compared to its full-length binary protein predictions (Fig. 5b), showing that MMM has analytical value at the biologically relevant level of domains. However, the results of MMM on multidomain proteins relative to individual domains could still be influenced by improper annotation. Error could arise in two ways: proteins that are annotated with only one domain may be incompletely annotated and missing additional domains, or single-domain annotations may incorrectly occupy a large proportion of the protein. In either case, valuable coevolutionary information will be lost by these definitions. While essential information regarding links between coexpression, coadaptation, and coevolution remain elusive, it is clear that coevolution of proteins based on sequence-level analysis has a predictive value that rivals current high-throughput in vitro methods.

2.4. The Yeast Coevolution Network

We used MMM to create the first coevolutionary network of the yeast *S. cerevisiae* (Fig. 6). The presented network shows some of the most highly coevolving (highest MMM scores) interactions, and features several known interactions among highly abundant proteins linked together, including all subunit components of cytosolic chaperonin, TRiC/CCT. This finding shows similarities to the human coevolution network (3), which also finds these strongly interacting proteins together. Interestingly, MMM finds currently unknown links between the cytosolic chaperonin to other proteins that are necessary for

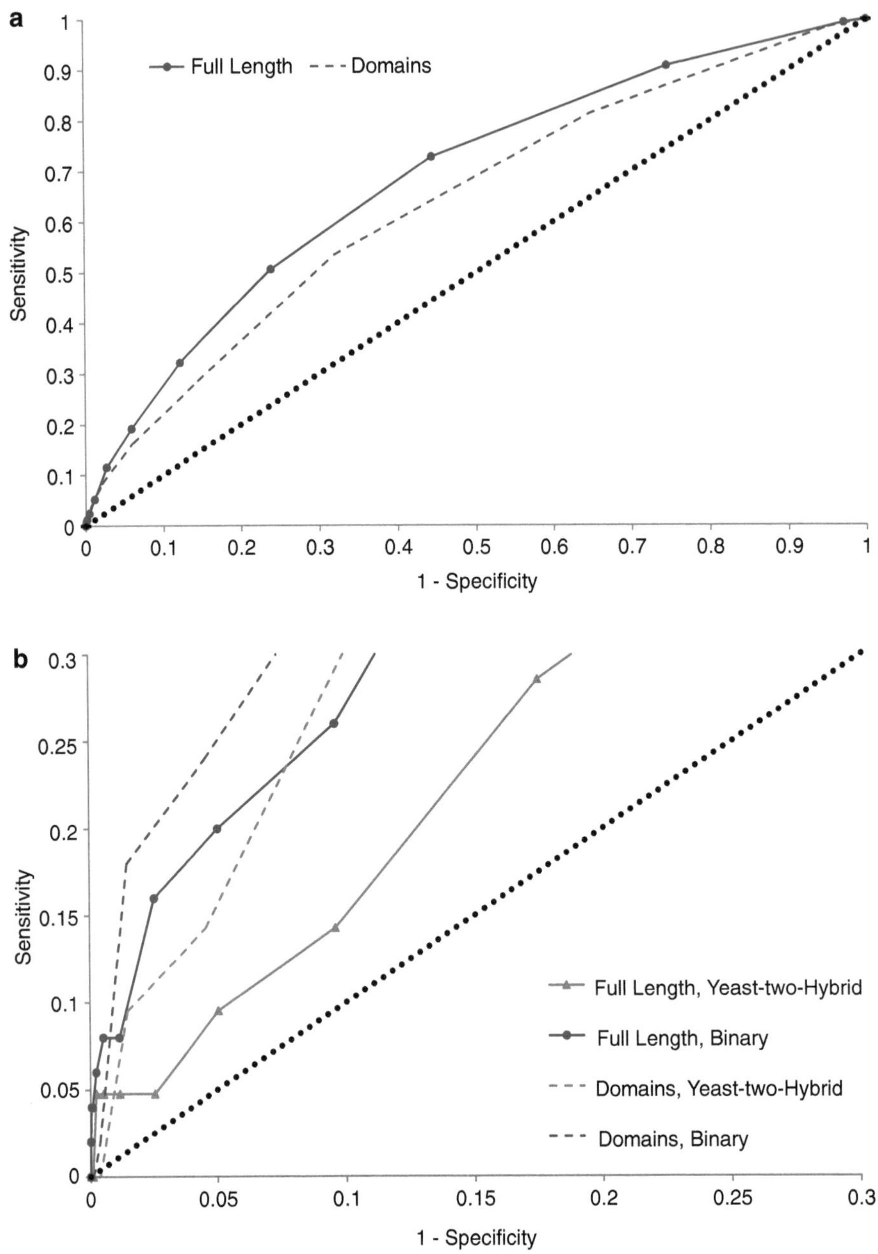

Fig. 5. Coevolution of domains. In (**a**) we see that the coevolution of full-length proteins (*circle marker*) is a generally better predictor than domains (*dashed line*) if only considering the portion of sequences with annotated domains from gold-standard data sets. However, (**b**) we see that for gold-standard yeast 2-hybrid and binary interactions where only well-annotated proteins with *multiple domains* are considered, the specificity of predicting interactions is improved over predictions based on full-length proteins.

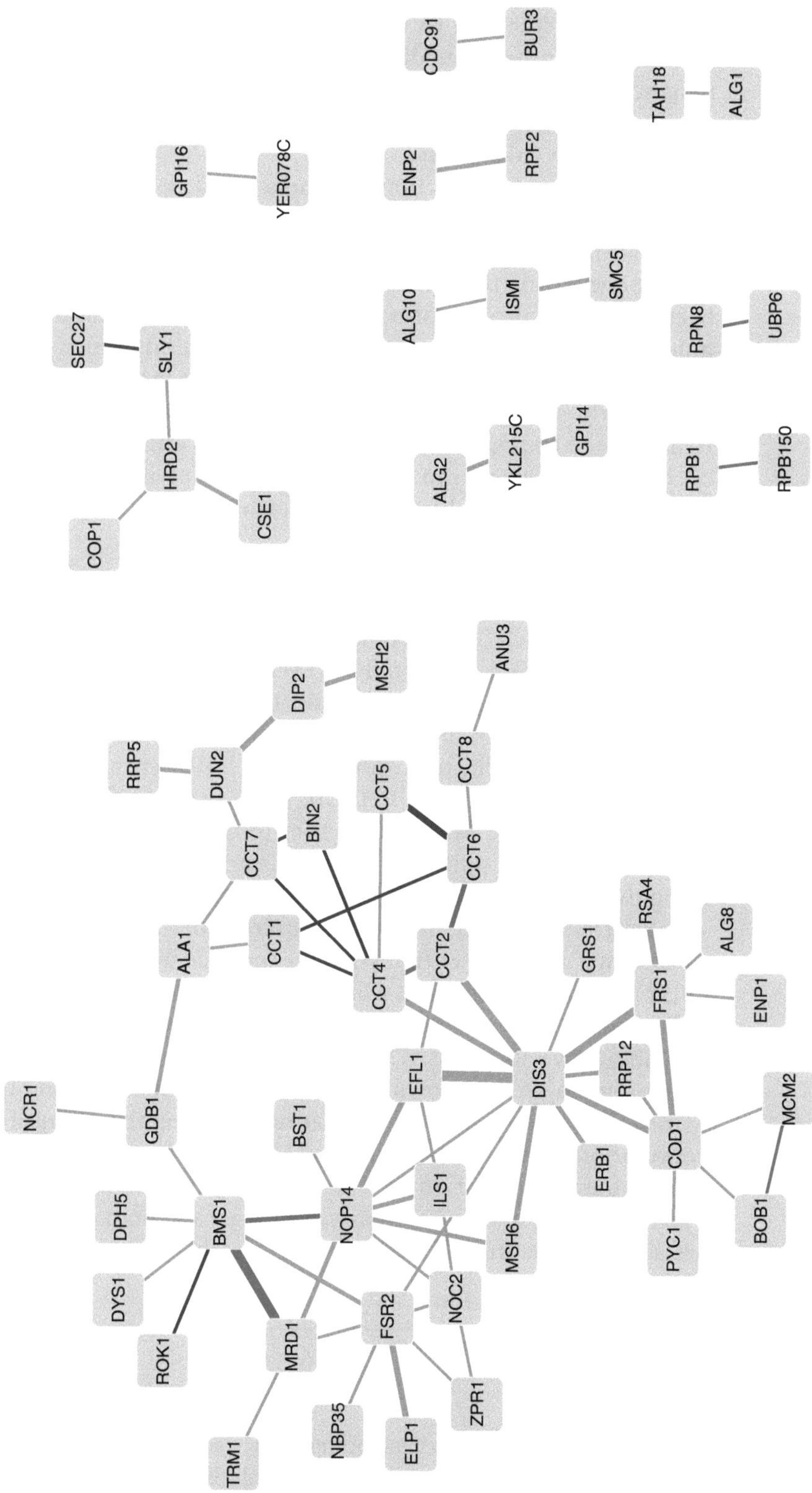

Fig. 6. The yeast coevolution network generated using MMM. In this illustration of the most highly coevolving yeast proteins, gold-standard interactions are represented by *red lines* while known, non-gold-standard interactions are represented by *blue lines*. In *gray* are interactions predicted by MMM that are not currently known. The width of the edge represents the strength of coevolution, as determined by MMM. At the center of the network, components of the essential eukaryotic cytosolic chaperonin TRiC/CCT are grouped together.

protein translation and transport of proteins from the rough endoplasmic reticulum to the Golgi apparatus, suggesting that the chaperonin helps mediate the trafficking of unfolded proteins to the Golgi. These interactions converge on eukaryotic-specific organelles, highlighting MMM's ability to find coevolution in eukaryotes.

While protein interaction networks based on physical interaction represent a "snapshot" of the current biological state, coevolution may also reveal ancestral features of the network, finding coevolving proteins where ancient interactions have since been lost in the process of network remodeling. Specifically, the quickly evolving yeast network loses interactions much more rapidly than it gains them (47) so that ancient interactions may significantly subsidize the coevolutionary signal between proteins, despite the current absence of a physical interaction. Thus, coevolution networks can show the remnants of ancient interactions, providing insight into how protein networks may be dynamically rewired (48).

2.5. MMM Web Server and Database

In order to provide an easy to use portal for coevolutionary analysis, the newly designed MMM Web server (http://www.uhnresearch.ca/labs/tillier/FMMMWEB/FMMMWEB.php) permits the testing of individual pairs of protein families by simply providing FASTA files of protein sequences. The Web server aligns the protein clusters using MAFFT (49), creates distance matrices with protdist from PHYLIP (50) (using the PMB matrix (51)), and runs the MMM algorithm to detect coevolving proteins. Files can be uploaded directly from the user's computer using two input boxes. Two options are available to the user: the scalable tolerance and the taxon requirement. If specified, the taxon requirement necessitates that the specified taxon be included in the MMM solution – this feature is necessary if a user wants to build a species-specific protein interaction network, such as we describe above for yeast. Users have access to all output from each program used, as downloadable files. Output from using the MMM Web server will give information regarding the strength of coevolution between two proteins (MMM score) which can be used toward the creation of a coevolutionary protein network. Precomputed results for our analysis of yeast and human proteins are also available from a Web-accessible database.

3. Conclusion

Despite an incomplete understanding of the causal molecular mechanisms of coevolution, coevolutionary analysis of proteins remains useful in the accurate prediction of PPIs (19, 24). In particular, coevolutionary methods preferentially predict obligate interacting proteins, as well as proteins in the same pathway. While obligate interactions provide a greater interfacial area for studying coadaptation, these types of interactions are also more likely to be coexpressed (37), adding little insight into the fundamental workings of coevolution. Proteins within the same pathway must necessarily be detected by means other than coadaptation, since they lack direct contact. However, coadaptation functions prominently in intraprotein interactions (10–15), indicating that the general biochemical principles are sound and should transitively apply to PPIs, and in particular to obligate interactions. Unraveling the collaborative forces that impact coevolution will ultimately lead to better prediction of PPIs.

4. Methods

4.1. From Sequences to Matrices to MMM

1. *Select orthologous proteins.* There are several important elements that are prerequisites to coevolutionary analysis. The most crucial of these is the selection of orthologous sequences (homologs across species) since coevolutionary methods implicitly require that a sufficient number of species have maintained a copy of each protein under analysis. Appropriate classification into protein families requires meticulous effort. Ideally, each protein family contains proteins from various species that all have the same function. In practice, orthologous sequences are determined on the basis of sequence similarity, often determined by the BLAST algorithm (52), which depending on the selected cut-off for sequence similarity can either be too strict and exclude true orthologs or too lenient leading to inclusion of spurious orthologs. Additionally, the choice of a homology cut-off score for the proper inclusion into orthologous families varies depending on evolutionary conservation of the family so that using a single cut-off value can be unsatisfactory for both slowly evolving and rapidly evolving proteins.

   The difficulty in correctly assigning protein families has led to the creation of many precomputed databases. Currently, there are several well-maintained databases that are both searchable (via keywords, protein identifiers or even sequence) and that permit downloading access to all information within the database (some examples are listed in Fig. 1). Large, accessible and curated datasets are vital for the study of coevolutionary sequence analysis since the number of sequences included in these analyses affects the amount of coevolutionary information. Biological relevance from orthologous sequences is extracted by the proper creation of a multiple sequence alignment for each protein family.

For the creation of the yeast coevolution network, we used the OMA Database (53) to obtain 204, 689 eukaryotic protein clusters that span 96 species. This database is in many ways ideal because of the large number of species for which coevolution can be detected.

2. *Align protein clusters.* Using the alignment program MAFFT (49), we were able obtain fast and accurate multiple sequence alignments (54) by utilizing its capabilities for automation. The alignment is another important step in coevolution analysis, especially for correlated mutation analysis where the result directly depends on the alignment.
3. *Obtain distance matrices.* Distance matrices were obtained by using protdist (from PHYLIP (50)) with the PMB distance matrix (51) to correct for multiple substitutions. The new, faster version of MMM must use corrected distances because distances converge to 100% for highly dissimilar sequences and will, therefore, give false-positive coevolution identification: identical branch lengths from saturated evolution do not represent coevolution. Similarly, MMM will ignore distances of zero, as the lack of evolution of a pair of proteins cannot be taken as evidence for their coevolution.
4. In using MMM, we selected a tolerance of 0.1 (10%) and chose to use taxon information such that only sequences from the same species could be matched. Previous analyses (not shown) have shown that this tolerance threshold provides adequate sensitivity and specificity (see Subheading 4.2). Our analysis of PPIs using MMM was filtered for all OMA clusters of protein orthologs where the match potential (the number of species in common) was greater than 10, below which coevolutionary signal would be difficult to detect.

4.2. Yeast Gold-Standard PPI Datasets

In order to determine the accuracy of interactions obtained using MMM, *MirrorTree* and abundance ratios (or any such protein interaction network), "gold-standard" and technology-specific datasets were used. We used carefully selected yeast interactions from the Yeast Interactome Database (36), which includes a "gold-standard" binary interaction dataset that attempts to represent the constitution of the yeast interactome.

To annotate these gold-standard interactions and the yeast coevolution network, we also obtained interactions from Biogrid (55), ChaperoneDB (56), Phosphogrid (57), DIP (58), HPRD (59), IntAct (60), MINT (61), and NCBI gene (62).

For the comparison of MMM's performance on various experimental PPI methods, interactions were broken down into several categories: cocomplex (mass spectrometry, etc.), yeast two-hybrid, binary (mostly small-scale interactions), genetic, literature-curated, reaction, colocalization, and "other."

To compare the various methods to predict PPIs, we used Receiver Operating Characteristic (ROC) curves, which act as a discriminating classification measure. For the purpose of PPIs, binary classification as either false positive or true positive is based on known protein interactions as described from gold-standard data sets. Although these gold-standard data sets are thought to include only a subset of true interactions, and possibly some false-positive interactions, they can still be used to gauge the performance of our predictions. In ROC curves (Figs. 2–5), the true positive rate is equivalent to *sensitivity* and the false-positive rate is equivalent to 1 – *specificity*. Through comparing the fraction of true positives to the fraction of false positives across threshold criteria, a ROC curve can survey the accuracy for a PPI prediction algorithm. The closer the ROC curve is to the upper left corner, the higher the overall accuracy of the measure, while the diagonal represents zero predictive value.

4.3. Domain–Domain Analysis by MMM

In order to carry out the domain-based coevolutionary analysis, protein clusters from NCBI's HomoloGene were obtained and annotated using the Conserved Domains Database (CDD) from NCBI (62). Briefly, Position Specific Scoring Matices (PSSM) were obtained for all domains in CDD and applied to all protein clusters containing a yeast sequence using RPS-BLAST (52). In cases where domain annotations overlapped along the sequence, the solution that had the better *E*-value was chosen. Additionally, sequences (protein families) were discarded in cases where the domain represented more than 70% of the entire sequence.

Multiple sequence alignments (MSA) of all protein families were carried out using MAFFT. If an acceptable domain from a yeast sequence was found, the domain boundaries were determined within the MSA, and the sequences of yeast and all orthologous species were extracted from the MSA. Domains were aligned using MAFFT, and distance matrices were obtained using protdist.

MMM was run on all possible domain–domain pairings for all permutations of proteins in which a domain was found. In cases where multidomain proteins were compared, the highest score between domains was taken as the MMM score.

Acknowledgements

This research was funded in part by the Ontario Ministry of Health and Long Term Care. The views expressed do not necessarily reflect those of the Ministry. Funding was also provided by a fellowship from the CIHR Training Program in Protein Folding: Principles and Diseases to GWC, an NSERC USRA to JMY and an NSERC discovery grant to ERMT. ERMT holds a Canada Research Chair in Analytical Genomics.

References

1. Fraser HB, Wall DP, and Hirsh AE (2003) A simple dependence between protein evolution rate and the number of protein–protein interactions, BMC Evol Biol 3, 11.
2. Krylov DM, Wolf YI, Rogozin IB, and Koonin EV (2003) Gene loss, protein sequence divergence, gene dispensability, expression level, and interactivity are correlated in eukaryotic evolution, Genome Res 13, 2229–2235.
3. Tillier ERM, and Charlebois RL (2009) The human protein coevolution network, Genome Res 19, 1861–1871.
4. Pazos F, and Valencia A (2001) Similarity of phylogenetic trees as indicator of protein–protein interaction, Protein Eng 14, 609–614.
5. Atchley WR, Wollenberg KR, Fitch WM, Terhalle W, and Dress AW (2000) Correlations among amino acid sites in bHLH protein domains: an information theoretic analysis, Mol Biol Evol 17, 164–178.
6. Bloom JD, Lu Z, Chen D, Raval A, Venturelli OS, and Arnold FH (2007) Evolution favors protein mutational robustness in sufficiently large populations, BMC Biol 5, 29.
7. Worth CL, Gong S, and Blundell TL (2009) Structural and functional constraints in the evolution of protein families, Nat Rev Mol Cell Biol 10, 709–720.
8. Singer MS, Vriend G, and Bywater RP (2002) Prediction of protein residue contacts with a PDB-derived likelihood matrix, Protein Eng 15, 721–725.
9. Saraf MC, Moore GL, and Maranas CD (2003) Using multiple sequence correlation analysis to characterize functionally important protein regions, Protein Eng 16, 397–406.
10. Tillier ERM, Biro L, Li G, and Tillo D (2006) Codep: maximizing coevolutionary interdependencies to discover interacting proteins, Proteins 63, 822–831.
11. Choi SS, Li W, and Lahn BT (2005) Robust signals of coevolution of interacting residues in mammalian proteomes identified by phylogeny-aided structural analysis, Nat Genet 37, 1367–1371.
12. Tress ML, and Valencia A (2010) Predicted residue-residue contacts can help the scoring of 3D models, Proteins 78, 1980–1991.
13. Horner DS, Pirovano W, and Pesole G (2008) Correlated substitution analysis and the prediction of amino acid structural contacts, Brief Bioinform 9, 46–56.
14. Pazos F, and Valencia A (2002) In silico two-hybrid system for the selection of physically interacting protein pairs, Proteins 47, 219–227.
15. Xu Y, and Tillier ERM (2010) Regional covariation and its application for predicting protein contact patches, Proteins 78, 548–558.
16. Goh CS, Bogan AA, Joachimiak M, Walther D, and Cohen FE (2000) Coevolution of proteins with their interaction partners, J Mol Biol 299, 283–293.
17. Aytuna AS, Gursoy A, and Keskin O (2005) Prediction of protein–protein interactions by combining structure and sequence conservation in protein interfaces, Bioinformatics 21, 2850–2855.
18. Sato T, Yamanishi Y, Kanehisa M, and Toh H (2005) The inference of protein–protein interactions by coevolutionary analysis is improved by excluding the information about the phylogenetic relationships, Bioinformatics 21, 3482–3489.
19. Hakes L, Lovell SC, Oliver SG, and Robertson DL (2007) Specificity in protein interactions and its relationship with sequence diversity and coevolution, Proc Natl Acad Sci USA 104, 7999–8004.
20. Craig RA, and Liao L (2007) Phylogenetic tree information aids supervised learning for predicting protein–protein interaction based on distance matrices, BMC Bioinformatics 8, 6.
21. Kann MG, Jothi R, Cherukuri PF, and Przytycka TM (2007) Predicting protein domain interactions from coevolution of conserved regions, Proteins 67, 811–820.
22. Juan D, Pazos F, and Valencia A (2008) Coevolution and co-adaptation in protein networks, FEBS Lett 582, 1225–1230.
23. Barabasi AL, and Albert R (1999) Emergence of scaling in random networks, Science 286, 509–512.

23a. Rodionov A, Bezginov A, Rose J, and Tillier ERM (2011) A new, fast algorithm for detecting protein coevolution using maximum compatible cliques, Algorithms Mol Biol 6, 17.

24. Pazos F, and Valencia A (2008) Protein coevolution, co-adaptation and interactions, EMBO J 27, 2648–2655.
25. Mintseris J, and Weng Z (2005) Structure, function, and evolution of transient and obligate protein–protein interactions, Proc Natl Acad Sci USA 102, 10930–10935.
26. Drummond DA, Bloom JD, Adami C, Wilke CO, and Arnold FH (2005) Why highly expressed proteins evolve slowly, Proc Natl Acad Sci USA 102, 14338–14343.
27. Rocha EP, and Danchin A (2004) An analysis of determinants of amino acids substitution rates in bacterial proteins, Mol Biol Evol 21, 108–116.

28. Papp B, Pál C, and Hurst LD (2003) Dosage sensitivity and the evolution of gene families in yeast, Nature 424, 194–197.
29. Kann MG, Shoemaker BA, Panchenko AR, and Przytycka TM (2009) Correlated evolution of interacting proteins: looking behind the mirrortree, J Mol Biol 385, 91–98.
30. Dandekar T, Snel B, Huynen M, and Bork P (1998) Conservation of gene order: a fingerprint of proteins that physically interact, Trends Biochem Sci 23, 324–328.
31. Huynen M, Snel B, Lathe W, and Bork P (2000) Predicting protein function by genomic context: quantitative evaluation and qualitative inferences, Genome Res 10, 1204–1210.
32. Sharp PM, and Li WH (1987) The codon Adaptation Index–a measure of directional synonymous codon usage bias, and its potential applications, Nucleic Acids Res 15, 1281–1295.
33. Fraser HB, Hirsh AE, Wall DP, and Eisen MB (2004) Coevolution of gene expression among interacting proteins, Proc Natl Acad Sci USA 101, 9033–9038.
34. Matthews LR, Vaglio P, Reboul J, Ge H, Davis BP, Garrels J, Vincent S, and Vidal M (2001) Identification of potential interaction networks using sequence-based searches for conserved protein–protein interactions or "interologs," Genome Res 11, 2120–2126.
35. von Mering C, Krause R, Snel B, Cornell M, Oliver SG, Fields S, and Bork P (2002) Comparative assessment of large-scale data sets of protein–protein interactions, Nature 417, 399–403.
36. Yu H, Braun P, Yildirim MA, Lemmens I, Venkatesan K, Sahalie J, Hirozane-Kishikawa T, Gebreab F, Li N, Simonis N, Hao T, Rual JF, Dricot A, Vazquez A, Murray RR, Simon C, Tardivo L, Tam S, Svrzikapa N, Fan C, de Smet AS, Motyl A, Hudson ME. Park J, Xin X, Cusick ME, Moore T, Boone C, Snyder M, Roth FP, Barabási AL, Tavernier J, Hill DE, and Vidal M (2008) High-quality binary protein interaction map of the yeast interactome network, Science 322, 104–110.
37. Nooren IMA, and Thorton JM (2003) Diversity of protein–protein interactions, The EMBO Journal 22.
38. Wuchty S, Barabási AL, and Ferdig MT (2006) Stable evolutionary signal in a yeast protein interaction network, BMC Evol Biol 6, 8.
39. Yellaboina S, Dudekula DB, and Ko MSh (2008) Prediction of evolutionarily conserved interologs in Mus musculus, BMC Genomics 9, 465.
40. Holstege FC, Jennings EG, Wyrick JJ, Lee TI, Hengartner CJ, Green MR, Golub TR, Lander ES, Young RA (1998) Cell 95, 717–728.
41. Pawson T, and Nash P (2003) Assembly of cell regulatory systems through protein interaction domains, Science 300, 445–452.
42. del Sol A, and Carbonell P (2007) The modular organization of domain structures: insights into protein–protein binding, PLoS Comput Biol 3, e239.
43. Itzhaki Z, Akiva E, Altuvia Y, and Margalit H (2006) Evolutionary conservation of domain–domain interactions, Genome Biol 7, R125.
44. Kim Y, Koyutürk M, Topkara U, Grama A, and Subramaniam S (2006) Inferring functional information from domain coevolution, Bioinformatics 22, 40–49.
45. Jothi R, Cherukuri, PF, Tasneem A, and Przytycka TM (2006) Coevolutionary analysis of domains in interacting proteins reveals insights into domain–domain interactions mediating protein–protein interactions, J Mol Biol 362, 861–875.
46. Wojcik J, and Schächter V (2001) Protein–protein interaction map inference using interacting domain profile pairs, Bioinformatics 17 Suppl 1, S296–S305.
47. Wagner A (2001) The Yeast Protein Interaction Network Evolves Rapidly and Contains Few Redundant Duplicate Genes, Mol Biol Evol 18, 1283–1292.
48. Berg J, Lässig M, and Wagner A (2004) Structure and evolution of protein interaction networks: a statistical model for link dynamics and gene duplications, BMC Evol Biol 4, 51.
49. Katoh K, Kuma K, Miyata T, and Toh H (2005) Improvement in the accuracy of multiple sequence alignment program MAFFT, Genome Inform 16, 22–33.
50. Felsenstein J. PHYLIP (Phylogeny Inference Package) version 3.6. Distributed by the author. University of Washington, Seattle.
51. Veerassamy S, Smith A, and Tillier ERM (2003) A transition probability model for amino acid substitutions from blocks, J Comput Biol 10, 997–1010.
52. Altschul SF, Gish W, Miller W, Myers EW, and Lipman DJ (1990) Basic Local Alignment Search Tool, Journal of Molecular Biology 215.
53. Schneider A, Dessimoz C, and Gonnet GH (2007) OMA Browser–exploring orthologous relations across 352 complete genomes, Bioinformatics 23, 2180–2182.
54. Nuin PA, Wang Z, and Tillier ERM (2006) The accuracy of several multiple sequence alignment programs for proteins, BMC Bioinformatics 7, 471.
55. Stark C, Breitkreutz BJ, Reguly T, Boucher L, Breitkreutz A, and Tyers M (2006) BioGRID:

a general repository for interaction datasets, Nucleic Acids Res 34, D535–D539.

56. Gong Y, Kakihara Y, Krogan N, Greenblatt J, Emili A, Zhang Z, and Houry W (2009) An atlas of chaperone-protein interactions in *Saccharomyces cerevisiae*: implications to protein folding pathways in the cell, Mol Sys Biol 5, 275.

57. Breitkreutz A, Choi H, Sharom JR, Boucher L, Neduva V, Larsen B, Lin ZY, Breitkreutz BJ, Stark C, Liu G, Ahn J, Dewar-Darch D, Reguly T, Tang X, Almeida R, Qin ZS, Pawson T, Gingras AC, Nesvizhskii AI, and Tyers M (2010) A global protein kinase and phosphatase interaction network in yeast, Science 328, 1043–1046.

58. Xenarios I, Rice DW, Salwinski L, Baron MK, Marcotte EM, and Eisenberg D (2000) DIP: the Database of Interacting Proteins, Nucleic Acids Reseach 28, 289–291.

59. Keshava Prasad TS, Goel R, Kandasamy K, Keerthikumar S, Kumar S, Mathivanan S, Telikicherla D, Raju R, Shafreen B, Venugopal A, Balakrishnan L, Marimuthu A, Banerjee S, Somanathan DS, Sebastian A, Rani S, Ray S, Harrys Kishore CJ, Kanth S, Ahmed M, Kashyap MK, Mohmood R, Ramachandra YL, Krishna V, Rahiman BA, Mohan S, Ranganathan P, Ramabadran S, Chaerkady R, and Pandey A (2009) Human Protein Reference Database–2009 update, Nucleic Acids Res 37, D767–D772.

60. Kerrien S, Alam-Faruque Y, Aranda B, Bancarz I, Bridge A, Derow C, Dimmer E, Feuermann M, Friedrichsen A, Huntley R, Kohler C, Khadake J, Leroy C, Liban A, Lieftink C, Montecchi-Palazzi L, Orchard S, Risse J, Robbe K, Roechert B, Thorneycroft D, Zhang Y, Apweiler R, and Hermjakob H (2007) IntAct–open source resource for molecular interaction data, Nucleic Acids Res 35, D561–D565.

61. Ceol A, Chatr Aryamontri A, Licata L, Peluso D, Briganti L, Perfetto L, Castagnoli L, and Cesareni G (2010) MINT, the molecular interaction database: 2009 update, Nucleic Acids Res 38, D532–D539.

62. Wheeler DL, Barrett T, Benson DA, Bryant SH, Canese K, Chetvernin V, Church DM, Dicuccio M, Edgar R, Federhen S, Feolo M, Geer LY, Helmberg W, Kapustin Y, Khovayko O, Landsman D, Lipman DJ, Madden TL, Maglott DR, Miller V, Ostell J, Pruitt KD, Schuler GD, Shumway M, Sequeira E, Sherry ST, Sirotkin K, Souvorov A, Starchenko G, Tatusov RL, Tatusova TA, Wagner L, and Yaschenko E (2008) Database resources of the National Center for Biotechnology Information, Nucleic Acids Res 36, D13–D21.

Chapter 12

Visualizing Gene-Set Enrichment Results Using the Cytoscape Plug-in Enrichment Map

Daniele Merico, Ruth Isserlin, and Gary D. Bader

Abstract

Gene-set enrichment analysis finds functionally coherent gene-sets, such as pathways, that are statistically overrepresented in a given gene list. Ideally, the number of resulting sets is smaller than the number of genes in the list, thus simplifying interpretation. However, the increasing number and redundancy of gene-sets used by many current enrichment analysis resources work against this ideal.

"Enrichment Map" is a Cytoscape plug-in that helps overcome gene-set redundancy and aids in the interpretation of enrichment results. Gene-sets are organized in a network, where each set is a node and links represent gene overlap between sets. Automated network layout groups related gene-sets into network clusters, enabling the user to quickly identify the major enriched functional themes and more easily interpret enrichment results.

Key words: Gene-set enrichment, Functional enrichment, Microarray data analysis, Gene ontology, Pathways

1. Introduction

High-throughput genomic experiments often lead to the identification of large gene lists (1). Gene lists are typically defined using statistical methods appropriate to the experimental design. For instance, a frequently applied method is to score genes by their differential expression between two biological states (such as healthy vs. disease). However, these methods for finding interesting genes often do not help the interpretation of the resulting gene lists.

Enrichment analysis is an automated and statistically rigorous technique to analyze and interpret large gene lists using a priori knowledge (2–4). Enrichment analysis assesses the over (or under-) representation of a known set of genes (e.g., a biological pathway)

Gerard Cagney and Andrew Emili (eds.), *Network Biology: Methods and Applications*, Methods in Molecular Biology, vol. 781, DOI 10.1007/978-1-61779-276-2_12, © Springer Science+Business Media, LLC 2011

within the input gene list. If a statistically significant number of genes from the known set are present in the gene list, it may indicate that the biological pathway plays a role in the biological condition under study. This analysis is repeated for all available known gene-sets, which could number in the thousands.

The growing number of available gene-sets, due to the increased availability of functional annotations, makes enrichment analysis a powerful tool to help researchers gain interesting insights from their high-throughput data. However, this comes at a cost: as gene-set collections get larger and more complex, there may be longer lists of results and increased redundancy between sets. Redundancy is particularly problematic with gene-sets derived from hierarchical functional annotation systems, such as Gene Ontology (GO), as children terms are partially redundant with their parents by definition. Gene-set redundancy constitutes a major barrier for the interpretability of enrichment results, limiting the full exploitation of its analytic power.

Enrichment Map is a visualization tool that organizes gene-sets into a similarity network, where nodes represent gene-sets, links represent the overlap of member genes, and node color encodes the enrichment score (5). Nodes are automatically arranged so that highly similar gene-sets are placed close together; these "clusters" can be easily identified manually and related to biological functions. Enrichment Map is implemented as a freely available and open-source plug-in for the Cytoscape network visualization and analysis software (6, 7).

In the methods section, we clarify fundamental concepts of enrichment analysis and describe how to use Enrichment Map for enrichment visualization. *Protocol 1* describes the basic visualization of enrichment results. *Protocol 2* shows how to visualize and compare two different enrichment results. *Protocol 3* describes how to analyze the relationships between enriched gene-sets and an additional query gene-set.

2. Materials

2.1. Software Installation

Install Cytoscape, the Enrichment Map and the WordCloud plug-ins according to the following instructions. Cytoscape 2.7.0, Enrichment Map 1.0, and WordCloud 1.0 were used for the protocols; later versions are supposed to work as well, but minor changes in the instructions may be required. Please refer to online manuals (http://baderlab.org/Software/EnrichmentMap/UserManual, http://baderlab.org/Software/WordCloudPlugin/UserManual) in case of discrepancies.

2.1.1. Cytoscape Installation

Before proceeding to install Cytoscape, check what version of Java is installed.

Windows (XP or Vista):

1. From *Start button/All programs/Accessories* run: *Command Prompt*
2. Type: java -version

Mac (Mac OS X):

1. From *Applications/Utilities* run: *Terminal*
2. Type: java -version

If a version older than 5.0 is found, please install Java SE 5 (http://www.oracle.com/technetwork/java/javase/downloads/index-jdk5-jsp-142662.html), Java SE 6 (http://www.oracle.com/technetwork/java/javase/downloads/index.html) or later.

To install Cytoscape, download the appropriate installer package for your operating system (Windows, Mac, or Linux) from: http://www.cytoscape.org/download.php.

Make sure the Cytoscape icon is added to the *Start Button* program list (Windows) or to the *Applications* folder (Mac).

2.1.2. Plug-in Installation

To install the Enrichment Map plug-in, first download the plug-in package at: http://baderlab.org/Software/EnrichmentMap#Plugin_Download.

The plug-in file (.jar) will have to be moved into the Cytoscape/plug-in directory. To locate the Cytoscape directory, follow these instructions.

- Windows (XP or Vista): right-click on the Cytoscape icon and select *Properties*.
- Mac (Mac OS X): CTRL-click on the Cytoscape icon and select *Show in Finder*.

To install the WordCloud plug-in, follow the same instructions as above after downloading the .jar from: http://baderlab.org/Software/WordCloudPlugin.

2.2. Sample Data Download

The analysis presented in the protocols can be reproduced downloading the sample data at: http://baderlab.org/Software/EnrichmentMap#Sample_Data_Download.

2.3. Data Formats

Enrichment results can be in two formats:

- The one adopted by *GSEA* (Gene Set Enrichment Analysis)
- The *Generic* format, which is suitable when enrichment results are generated using other software tools.

Table 1
Summary of enrichment map input files

File type	File extension	Separator	Header	# Columns	Requirement
Gene-set	.GMT	Tab	No	Variable	Mandatory
Expression matrix	.txt	Tab	Yes	Constant (any≥3)	Mandatory
Expression matrix	.gct	Tab	Yes	Constant (any≥3)	Mandatory
Enrichment table (GSEA)	.xls	Tab	Yes	Fixed (11)	Mandatory
Enrichment table (Generic)	.txt	Tab	Yes	Fixed (5)	Mandatory
Rank	.RNK	Tab	No	Fixed (2)	Optional
Query gene-set	.GMT	Tab	No	Variable	Optional

The number of columns has been categorized as *Fixed (N)* when every row must have exactly the number of elements in brackets *(N)*, *Constant* when every row must have the same number of elements but this can be different for different files, *Variable* when every row can have a different length

The gene-set file and expression matrix formats are the same for the GSEA and Generic option, only the enrichment table format differs. Data formats are summarized in Table 1.

The Subheading 2.3.8. *Formatting DAVID Data for Enrichment Map* explains step-by-step how to load data from DAVID, a broadly used and cited enrichment tool for *subset gene lists.*

2.3.1. Gene-Set File (.GMT)

The gene-set file contains the gene-set identifiers (IDs), the gene-set names and the member gene IDs. The gene-set .GMT file is tab-separated; each line can have a different length, but must follow the syntax:

<gene-set id><tab><gene-set name><tab><gene-id><tab> <gene-id> <…>

No header should be present.

Gene-sets can be derived from gene annotation databases, pathway databases and other gene-centered resources. Gene-set files following the GMT format and compiled from public annotation databases can be downloaded from Pathway Commons (http://www.pathwaycommons.org/pc-snapshot/current-release/gsea/), WhichGenes (http://www.whichgenes.org/), MSigDB (http://www.broadinstitute.org/gsea/downloads.jsp) or from the Enrichment Map download page (http://baderlab.org/Software/EnrichmentMap Gene-sets_for_Enrichment_Analysis).

When downloading or generating the gene-set file, it is important to check ID consistency:

(a) Gene IDs must be consistent in the gene-set file and in the expression matrix.

(b) Gene-set IDs must be consistent in the gene-set file and in the enrichment table(s).

Gene-sets should be filtered by size, removing the ones larger or smaller than user-selected thresholds. In particular:

- Large gene-sets, especially when derived from Gene Ontology (GO), have little use when interpreting gene lists because they are often very general. A reasonable maximum size for human or mouse is between 400 and 900 genes. In addition, the accurate exploration of the map can reveal gene-sets connecting very different biological functions (e.g., *GO Protein complex assembly*); these should be removed on the user's discretion, and the layout should be recomputed.
- Small gene-sets (e.g., with less than 5–10 genes) may be more sensitive to small fluctuations in the query gene list, hence their presence may cause fewer gene-sets to be significant according to the FDR.

See Notes 3–4 for additional formatting tips.

2.3.2. Expression Matrix (.txt)

The .txt expression matrix file contains the gene IDs, the gene names (or any other textual annotation) and gene expression signals, or any other quantitative value characterizing genes (e.g., spectral counts, number of mutations). The .txt expression file is tab-separated; every line must have equal length following the syntax:

<gene-id> <tab> <gene-name> <tab> <value> <tab> <value> <...>

The header must be present at the first line. The labels of the first and second columns must be set to "*NAME*" and "*DESCRIPTION*," whereas the value columns can have any label, but repetitions must be avoided.

2.3.3. Expression Matrix (.gct)

The .gct expression matrix file must have two initial lines with this content:

#1.2

<number-of-genes> <tab> <number-of-samples>

followed by the same content as the .txt file (including the header at the first line).

Specifically, *<number-of-genes>* consists of the number of rows having a gene ID, name, and values; *<number-of-samples>* consists of the number of columns having an associated *<value>* (*NAME* and *DESCRIPTION* should not be counted).

2.3.4. Enrichment Table: GSEA (.xls)

GSEA follows the two-condition enrichment logic described in the methods section. Two separate files are generated for enrichment in condition A and B.

The file names have the prefix "*gsea_report_for_*" and extension *.xls* (although they are tab-separated files and not Excel files). Each line must have the same length, and the first line must be the header, with the following labels: "*NAME*", "*GS
follow link to MSigDB*", "*GS DETAILS*", "*SIZE*", "*ES*", "*NES*", "*NOM p-val*", "*FDR q-val*", "*FWER p-val*", "*RANK AT MAX*", and "*LEADING EDGE*". Every column must be present, but only "*NAME*" (the gene-set ID) and "*NES*", "*NOM p-val*", "*FDR q-val*" (gene-set enrichment statistics) are utilized by Enrichment Map.

These files are automatically generated by GSEA and they follow the syntax described above (as of October 2010).

2.3.5. Enrichment Table: Generic (.txt)

The Generic format enrichment table is a tab-separated file (.txt). Each line must have the same length and the first line must be the header. The file must have 5 columns:

1. Gene-set ID
2. Gene-set name
3. Enrichment *p*-value
4. Enrichment FDR (false discovery rate)
5. Phenotype

The header labels can be freely defined by the user, but the columns must be kept in the order defined above. If data are not available for any of the columns, the column should not be skipped, as the file needs to have exactly 5 columns. Any conventional value representing missing data (e.g., "NA") should be used instead.

For one-condition enrichment, all values in the phenotype column must be set to 1. For two-condition enrichment, phenotype values can be set to 1 or −1 to indicate the enrichment in either condition (see Note 5 for a detailed explanation of the numbering conventions).

2.3.6. Rank File (.rnk)

This file is optional. It is tab-separated, without a header, and each line must have the same length. The first column must be gene ID and the second column must be gene scores (e.g., *t statistic* or *log ratio*).

2.3.7. Query Set Gene-Set File (.GMT)

Same format as the main gene-set file.

2.3.8. Formatting DAVID Data for Enrichment Map

DAVID is a Web-based enrichment tool that accepts *subset gene lists* as input. Follow these instructions to import DAVID results using the *Generic* option. An additional format option dedicated to DAVID is currently under development, and it will be available in future releases of the Enrichment Map plug-in.

1. Obtain the gene-sets from DAVID. This currently requires registering at http://david.abcc.ncifcrf.gov/knowledgebase/register.htm. The DAVID Forum (http://david.abcc.ncifcrf.gov/forum) can be consulted for updated information on this topic, using "Download knowledgebase" as search keyword.
2. Format DAVID gene-sets as GMT. This has to be done programmatically. This R script can be easily used by readers who are not confident with scripting: http://baderlab.org/Software/EnrichmentMap/DAVIDgs_R.
3. Generate enrichment results using DAVID and select *Functional Annotation Chart* as output type.
4. Export DAVID results as tab-separated text file by following the *Download File* link of the *Functional Annotation Chart* page. Save the file from the Web browser as a plain text file.
5. Format DAVID enrichment results (tab-separated text) following the *Generic* enrichment table format. This can be done programmatically or using Excel:
   (a) Column 1: label "ID," copy and paste from "Term" column.
   (b) Column 2: label "Name," copy and paste from "Term" column.
   (c) Column 3: copy and paste from "Pvalue" column, original label can be kept.
   (d) Column 4: label "FDR," copy and paste from "Benjamini" column.
   (e) Column 5: label "Phenotype" and add phenotype values accordingly (see note below).

Note: for *one-condition* enrichment, you will generate only one enrichment table from DAVID; therefore, all phenotype values will have to be set to 1. For *two-condition* enrichment, you will generate two enrichment tables from DAVID, each associated to a specific condition and to a different *subset gene list*; phenotype values will have to be set to +1 for one table and −1 for the other table; then the two tables will have to be merged. If any gene-set appears twice with two different phenotype values, enrichment map will neglect the phenotype with higher *p*-value.

3. Methods

3.1. Types of Enrichment Analysis

Before describing how to use Enrichment Map, it is important to clarify a few concepts of enrichment analysis.

Enrichment analysis typically requires two inputs: the gene-sets based on a-priori knowledge (*known gene-sets*) and the gene list to be functionally characterized using the known gene-sets (*query gene list*). Known gene-sets usually consist of functional annotations based on controlled vocabularies (e.g., *Gene Ontology*), curated pathways (e.g., *KEGG*), protein families (e.g., proteins carrying the same *PFAM* domain) and other gene-sets based on published experimental data (e.g., tissue-specific gene expression, physical interactors of given disease genes, predicted targets of transcription factors or regulatory RNAs). The type of gene-set used for enrichment analysis should be chosen based on the biological question of interest; for instance, functional annotations and pathways are particularly appropriate to elucidate which biological processes are modulated in relation to cell state changes. The *Gene-set File* Subheading 2 provides more details on how to obtain gene-sets from public resources.

The query gene list is often derived from a high-throughput experiment (e.g., a microarray gene expression study), but it can also be derived from a meta-analysis of the literature or any other source. The query gene list can be a simple set of genes, without any associated quantitative attribute, in which case it will always be a *subset* of a larger gene-set, usually called *universe-set*, composed of all genes in the genome or all genes that could be possibly detected using a given experimental method. The definition of the *universe-set* is highly dependent on how the *subset gene list* was generated. If genes were quantitatively scored for some biological property, the subset gene list can be generated by selecting a score threshold; this is common for microarray analysis, when genes are scored for differentiality following a two-class design (e.g., estrogen-treated vs. untreated cell lines). In these cases, it is also possible to use the scores for all genes (i.e., a *scored gene list*) as the input of enrichment analysis. Unlike the *subset* gene list, all genes of the universe-set will be present in the *scored* gene list. Figure 1 summarizes the relations between query and known gene-sets.

Each known gene-set is tested for enrichment in the query list genes. Different statistical tests are available and have different input requirements. For instance, the broadly used *Fisher's Exact Test* (often referred to as *hypergeometric distribution test*) determines the significance of the overlap between a *subset* gene list and a known gene-set. *GSEA (Gene-Set Enrichment Analysis)* is a popular method, developed by the Broad Institute, that works on *scored* gene lists by testing if known gene-sets are enriched in top-scoring genes from the query list (4).

Depending on the experimental design underlying gene scoring or selection, enrichment analysis can be *one-condition*, *two-condition*, or *multi-condition*. The term "condition" here is synonymous with "class" or "phenotype" in the way these terms are used in enrichment analysis literature and software tools.

In *one-condition* enrichment, the query list genes are associated to a single biological condition. For instance, a ChIP-chip or

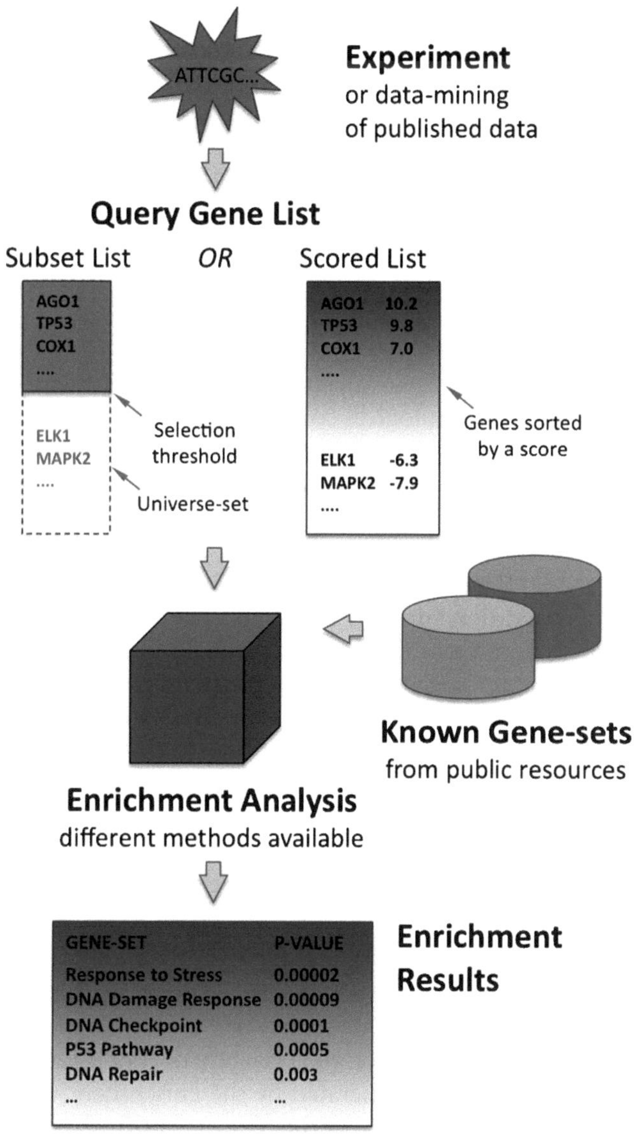

Fig. 1. The information flow from a high-throughput experiment to enrichment results. A high-throughput experiment, or data-mining of published data, typically produces a *query gene list* (of *subset* or *scored* type), which is tested for enrichment in *known gene-sets* utilizing a specific enrichment analysis method.

ChIP-seq experiment can be used to identify which genes are bound by the transcription factor *NF-Y*. Depending on the experimental platform used and the statistical model used to analyze the results, it may be more convenient to select a subset of genes that have a reliable signal for NF-Y binding (*subset gene list*), or to quantitatively score all genes for NF-Y binding (*scored gene list*). The query gene list will then be tested for enrichment in known gene-sets. Assuming that only NF-Y bound genes should be functionally characterized, the enrichment will be *one-condition*.

In *two-condition* enrichment, the query list genes are selected or scored by comparing *two* biological conditions; healthy vs. diseased and treated vs. untreated designs are common examples. Each gene can be more strongly associated to either of the two conditions, thus gene-sets are typically tested for enrichment in condition A or in condition B. For example, in gene expression studies, if condition *A* corresponds to *estrogen-treated cells* and condition *B* corresponds to *not-treated cells*, then the enrichment in condition A and B will be typically referred as enrichment in *upregulation* and *downregulation* after estrogen treatment, respectively.

In *multi-condition* enrichment, gene-sets can be enriched in multiple biological conditions. For instance, a time course experiment can be analyzed using clustering analysis. Each cluster can then be queried for enrichment in known gene-sets. In the end, each known gene-set can be enriched in no cluster, one or more clusters.

Enrichment Map currently supports one-condition and two-condition enrichment results, whereas multi-condition enrichment is an area of future development. The protocols presented in this chapter specifically focus on two-condition enrichment, which is the most common. Readers interested in one-condition enrichment would find instructions in Subheading 2 to appropriately structure their input files. Example enrichment results used in the protocols were derived from microarray gene expression data, as this is a common case; nonetheless, the methods presented here can be applied to any type of genomic data.

3.2. Files Required by Enrichment Map

Enrichment Map loads the following files:

1. The *gene-set file (.GMT)* defines available gene sets as a list of gene-set IDs, names or descriptions and member genes. Depending on the enrichment tool used, this file may be automatically generated by the enrichment tool by querying existing databases, or the user may be in charge of providing it. The Subheading 2 includes instruction to generate or retrieve this type of file.
2. The *expression matrix (.txt or .gct)* provides the expression values, or any other quantitative value at the gene level, that were used to score or select genes for gene-set enrichment. Within Enrichment Map, the expression matrix data are visualized using a heat map. These data should be directly provided by the user.
3. The *enrichment tables* consist of the gene-set enrichment results, i.e., the gene-set identifiers and enrichment statistics. These files are generated by the enrichment tool utilized for the analysis. Depending on the type of enrichment tool used, enrichment results can be loaded in the *GSEA* or *Generic*

format. Since *GSEA* is a broadly used tool with very clearly defined input and output files, its data format is directly supported by Enrichment Map. The *Generic* format was designed to be flexible enough to support any type of enrichment method with minimal user effort. The Subheading 2 includes a data reformatting protocol to load enrichment results from *DAVID* (2, 3), another popular enrichment tool developed by the NIAID/NIH.

4. The *gene rank file (.rnk)* is optional and consists of the scored gene list used for the enrichment analysis. It can be used to sort the expression matrix heat map. This file can be generated by the user, or by the enrichment tool.

The specifics of the data formats and the links to download the example enrichment data are in Subheading 2.

3.3. The Gene-Set Overlap Network

Enrichment Map arranges enriched gene-sets as a weighted similarity network. Nodes represent gene-sets passing a user-selected threshold of enrichment significance. Weighted links (i.e., edges) between the nodes represent an "overlap" score depending on the number of genes two gene-sets share. Nodes are automatically arranged so that highly similar gene-sets are placed close together; these clusters can be easily identified manually and related to biological functions. Gene-set enrichment results are graphically mapped to the Enrichment Map: node size represents the number of genes in the gene-set (see Note 1); edge thickness is proportional to the overlap between gene-sets, calculated using the Jaccard or overlap coefficients (see Protocol 1). The enrichment score (specifically, the enrichment p-value) is mapped to the node color as a color gradient. For one-condition enrichment results, node color ranges from white (no enrichment) to red (high enrichment). For two-condition enrichment results, node color ranges from red (high enrichment in the first condition, e.g., case) to white (no enrichment) to blue (high enrichment in the second condition, e.g., control). Figure 2 summarizes the Enrichment Map visualization of one-condition and two-condition enrichments.

3.4. Protocol 1: Analysis of One Enrichment

This protocol describes how to load results from a single enrichment analysis (one-condition or two-condition). Enrichment results of the transcriptional response to estrogen stimulation are used as an example in this protocol and in the following one. Microarray data profiling the response to estrogen treatment of MCF7 breast cancer cell lines were downloaded from Gene Expression Omnibus (GSE11352); genes were scored using the t statistic, comparing estrogen-treated vs. untreated cells; enrichment results were generated using GSEA. The example data in this protocol refer to the comparison of treated vs. untreated cells at 24 h of culture, thus a *two-condition* enrichment.

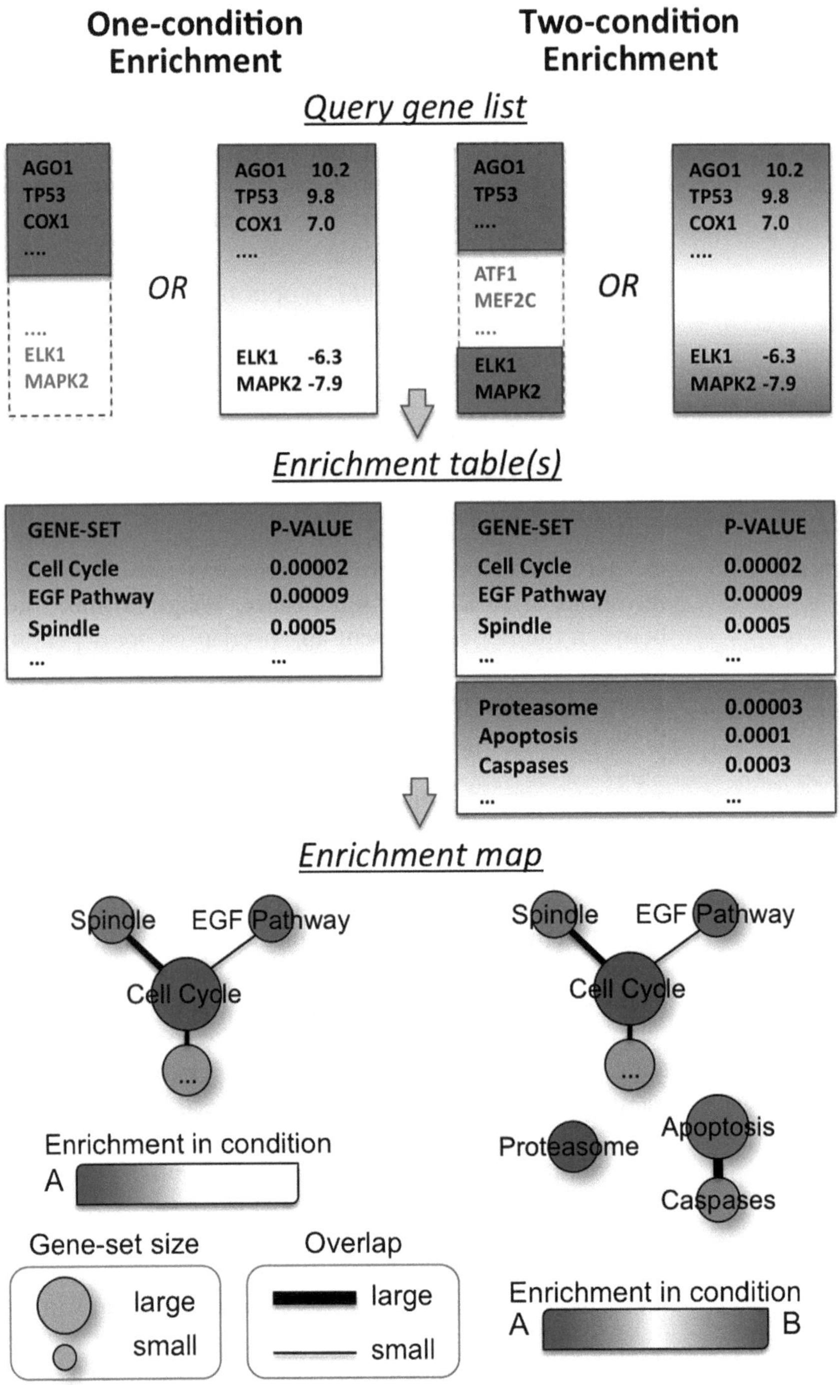

Fig. 2. One-condition and two-condition enrichment can be visualized using Enrichment Map. In one-condition enrichment, enriched gene-sets are associated to a single biological condition. In two-condition enrichment, enriched gene-sets are associated to one of two conditions. Enrichment Map arranges enriched gene-sets as a similarity network, where nodes correspond to gene-sets and links correspond to overlap of member genes.

3.4.1. Load Data and Set Parameters

1. Open Cytoscape.
2. From the top menu, select *Plugins/Enrichment Map/Load Enrichment Results.*
3. Select the format of enrichment analysis results, *Analysis Type*: *GSEA* or *Generic*. Example data: select *GSEA*.
4. If the selected analysis type is *GSEA* and the enrichment results generated by GSEA were not moved from the original directory, the *report file* (*.rpt*) can be used to automatically load all the other required files; load the report file in any of the *Dataset 1* boxes. Otherwise, follow instructions 5–8.
5. Load the *gene-set file (.GMT)*. Example data: load *GO_Hs_EG_f_hgu133p2_v2.gmt.*
6. Load the *expression matrix (.txt or .gct)*. Example data: load *MCF7_ExprMx_v2.txt.*
7. Load the *enrichment tables.* If you have selected the *GSEA* analysis type, you will have to load two files, each associated to a condition (*Enrichments 1*, *Enrichments 2* boxes); if you have otherwise selected the *Generic* analysis type, you will have to select only one file. Example data: load *EnrTable_24h_E2_v2.xls* and *EnrTable_24h_NT_v2.xls.*
8. If available, load rank data. Click on *Advanced* to display the loading text box and button. Example data: load *tTest_24h.rnk* and set *Phenotypes* to *ES_24h* and *NT_24h.*
9. Set the *p*-value and FDR (False Discovery Rate) parameters. The *P-value Cutoff* and the *FDR Q-value Cutoff* can be used to control the stringency of the analysis: only gene-sets with enrichment statistics satisfying these thresholds will be displayed by Enrichment Map. Example data: set the *p*-value cutoff to 0.001 and FDR cutoff to 0.05 (more stringent than defaults).
10. Select the similarity coefficient and its cutoff. Select the *Jaccard Coefficient* only if the gene-sets have comparable sizes (e.g., 200–300 genes). If Gene Ontology derived sets are present, select the *Overlap Coefficient.* The similarity coefficient cutoff should be tuned to optimize network connectivity. If many gene-sets are disconnected, try decreasing the cutoff value. If biologically unrelated gene-sets are connected, or if the network is close to being fully connected, increase the cutoff value. Example data: select the *Overlap Coefficient*, keep the default cutoff value of 0.5.
11. Generate the enrichment map by clicking on the *Build* button. A view of the network will be generated.

3.4.2. Globally Explore the Network

1. To explore different areas of the network, move to the *Network* tab in the Cytoscape *Control Panel* (left) and move the blue selection area in the network minimap. Use zoom icons on the Cytoscape icon bar (top) to zoom in and zoom out.
2. Automatic layout can be recalculated from the menu: *Layout/Cytoscape Layouts/Force Directed Layout/EM1_similarity_coefficient.*
3. Node color mapping and other graphical features can be tuned using the *VizMapper* tab in the Cytoscape *Control Panel* (left). For instance, the nodes can be colored according to the *GSEA NES score* (normalized enrichment score) instead of the *p*-value. From the *Visual Mapping Browser*, select *Node Color* and change its value from *EM1_Colouring_dataset1* to *EM1_NES_dataset1*; click on the color gradient, then click on the *Min/Max* button in the dialog box; set the NES min to −3 and the NES max to 3 (see Note 2 for details); finally, arrange the color markers in the dialogue box (top triangles) to obtain the preferred color mapping intensity. This is useful to better distinguish the differences in enrichment strength.
4. The slider bars in the *Results Panel* (right) can be used to interactively visualize how different *p*-value and FDR cutoff values affect the network.

3.4.3. Explore Specific Gene-Sets

1. By clicking on a gene-set, the *Data Panel* will display a heat map that visualizes gene expression signals (or any other quantitative value in the expression matrix file) on a magenta-green gradient. If signals have not been already normalized by-row, i.e., if different rows can have signals at different orders of magnitude, change *Normalization* from *Data As Is* to *Row Normalize Data.* This normalization works best with gene expression from one-dye arrays (e.g., Affymetrix), whereas for absolute count data (e.g., proteomics) *Log Transform Data* may work better. The heat map can be sorted by hierarchical clustering, using rank files or following the order of a single column. The first two options are available in the *Sorting* drop-down menu. To sort according to a column, just click on a column header in the heat map.
2. Gene-set attributes can be visualized by selecting the *Node Attribute Browser* tab (*Data Panel*, bottom); if certain attributes do not display, click on the first icon from the left of the *Data Panel* and modify the attribute selection.
3. By default, clicking on a node will use the Data Panel to display the heat map. If you want to change this setting, modify the *Advanced Preferences* in the *Results Panel* (bottom). If you do so, you will have to select the *EM Geneset Expression viewer* tab (*Data Panel*) to display the heat map.

3.4.4. Explore Gene-Set Clusters

1. Gene-set clusters can be usually spotted by eye throughout the network. They consist of highly overlapping gene-sets with slightly different biological meanings. A cluster typically corresponds to a biological function or "theme."
2. To investigate the biological meaning of a cluster, select all nodes and look through their name using the *Node Attribute Browser* as explained before.
3. The *WordCloud* Cytoscape plug-in can also be used to produce summaries of the gene-set names. Select the nodes of interest (their color will turn to yellow), then select from the menu *Plugins/WordCloud/Create Cloud.* The word cloud will be visualized in the *Data Panel.* In the *Cloud Parameters/ Attribute Choice* frame, click on the *Edit* button to set the *Current Values* to *EM1_GS_DESCR* and then click on the *update* button in the bottom; in this way, a summary of gene-set names will be displayed.

Figure 3 displays the enrichment map for estrogen treatment at 24 h, and shows how the WordCloud plug-in can help summarizing the content of a gene-set cluster.

3.5. Protocol 2: Analysis of Two Enrichments

Enrichment results of the transcriptional response to estrogen stimulation are used as an example in this protocol. Microarray data were obtained and analyzed as described previously. In this protocol, two enrichments are loaded in the same map to compare the estrogen response at different time points (at 12 and 24 h of culture). Note that each enrichment follows a two-condition logic, as it results from the comparison of treated vs. untreated cells at a specific time point.

3.5.1. Load Data and Set Parameters

1. Follow the instructions in Protocol 1. In addition, load the enrichment tables for *Dataset 2* (click on the frame name or the triangle to display the boxes and buttons). Example data: load *EnrTable_12h_E2_v2.xls* and *EnrTable_12h_NT_v2.xls* as *Dataset 1*, whereas *EnrTable_24h_E2_v2.xls* and *EnrTable_24h_NT_v2.xls* as *Dataset 2.*
2. The node center will be colored according to *Dataset 1* and the node border according to *Dataset 2.*

3.5.2. Interpretation (Example Data)

1. The majority of nodes have the same color, meaning that the estrogen response is similar at 12 and 24 h. Several nodes have more intense border colors, suggesting they are more induced at 24 h, whereas a few have more intense center color, suggesting they are more induced at 12 h.

Figure 4 shows the 12–24 h estrogen response enrichment map after some manual node repositioning to improve clarity;

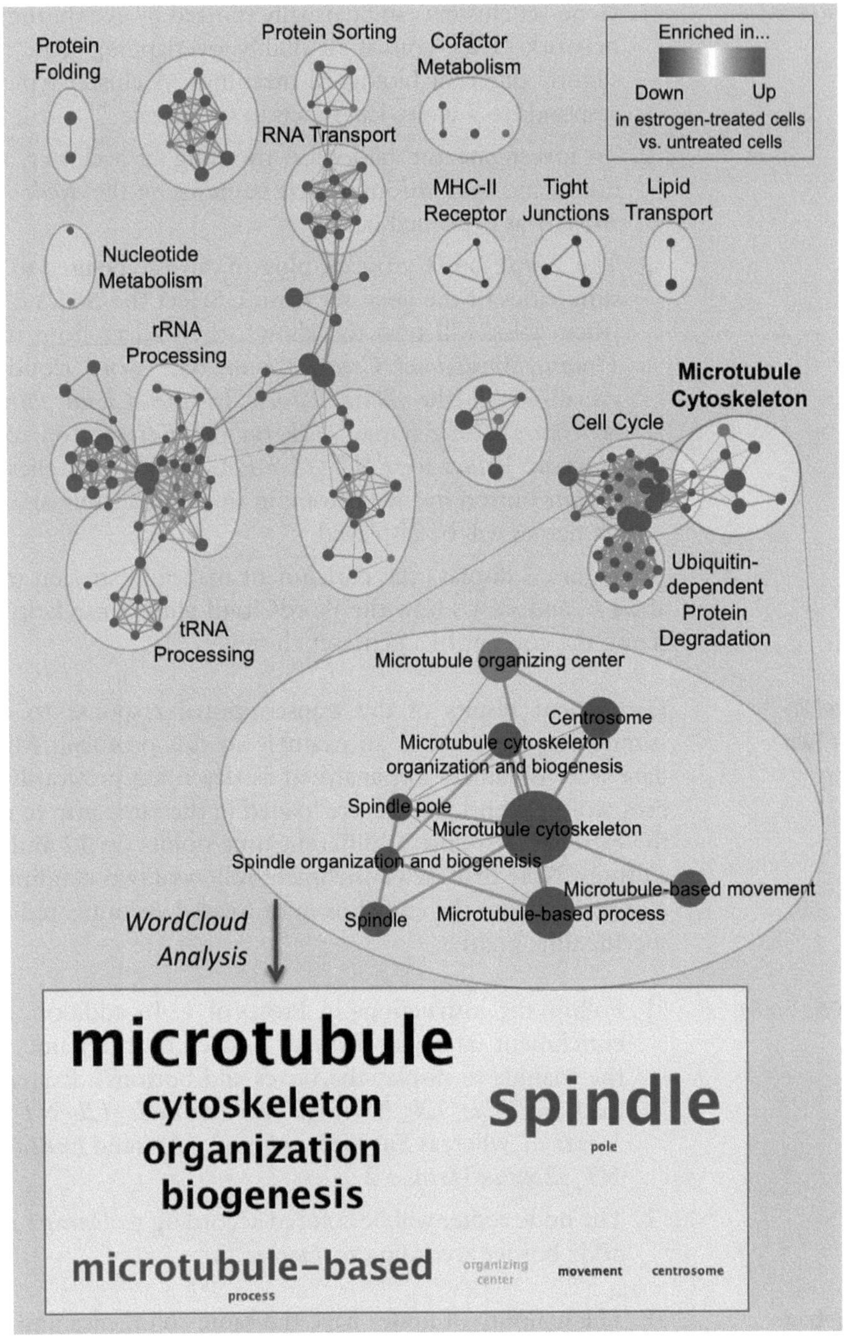

Fig. 3. Enrichment Map of the 24-h estrogen response. The network was manually rearranged to improve layout, and major clusters were manually labeled. The WordCloud plug-in can be used to automatically process gene-set names and help with cluster labeling. This is exemplified for the *Microtubule Cytoskeleton* cluster; the clustered word cloud summarizing its gene-set names is displayed in the bottom.

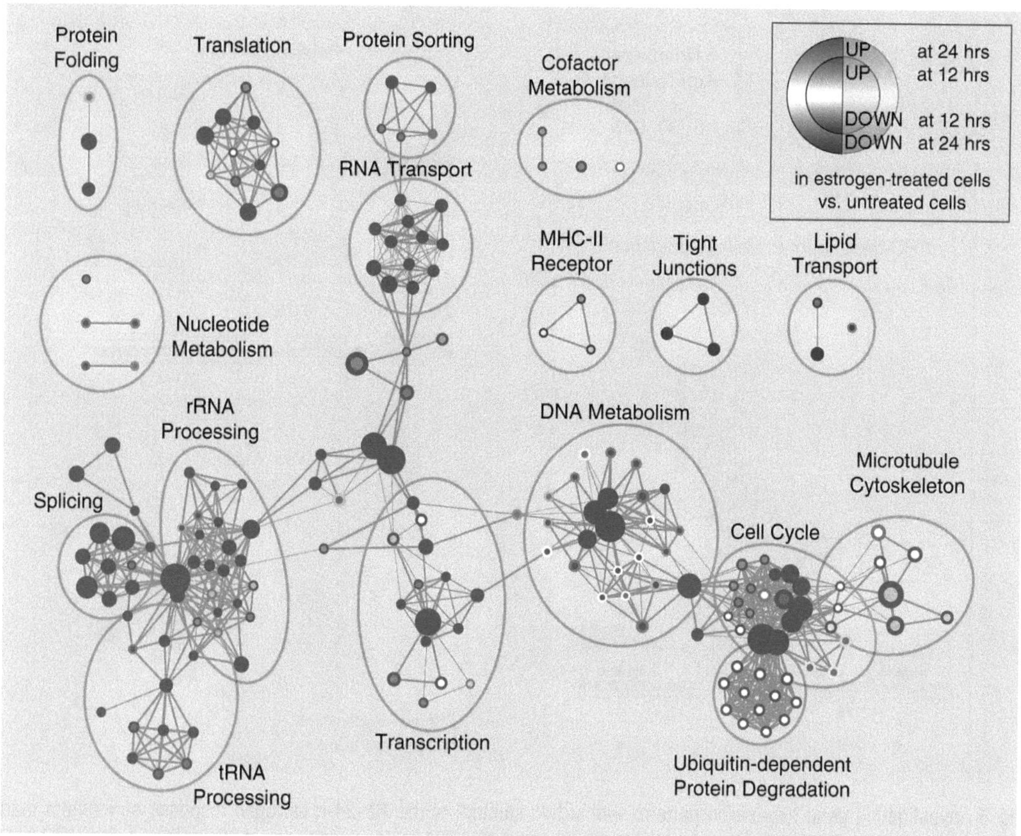

Fig. 4. Enrichment Map of the 12–24-h estrogen response enrichments. The network was manually rearranged to improve layout, and major clusters were manually labeled as in Fig. 3.

clusters of gene-sets belonging to the same functional theme were manually identified and annotated. Figure 5 shows the heat-map visualization of gene expression patterns for selected gene-sets.

3.6. Protocol 3: Query Set Analysis

Once an enrichment map has been generated, the *query set analysis* can be used to investigate the overlap between a *query gene-set*, not present in the map, and the enriched gene-sets in the map. In this protocol, we investigate the overlap between gene-sets enriched in the estrogen response (Protocol 2 example) and experimentally determined direct targets of the estrogen receptor (8). We are specifically interested in evaluating whether the transcriptional response to estrogen is dominated by indirect targets. The query set analysis can also be used to identify the relations between known disease genes and the gene expression or genetic alterations observed in that disease.

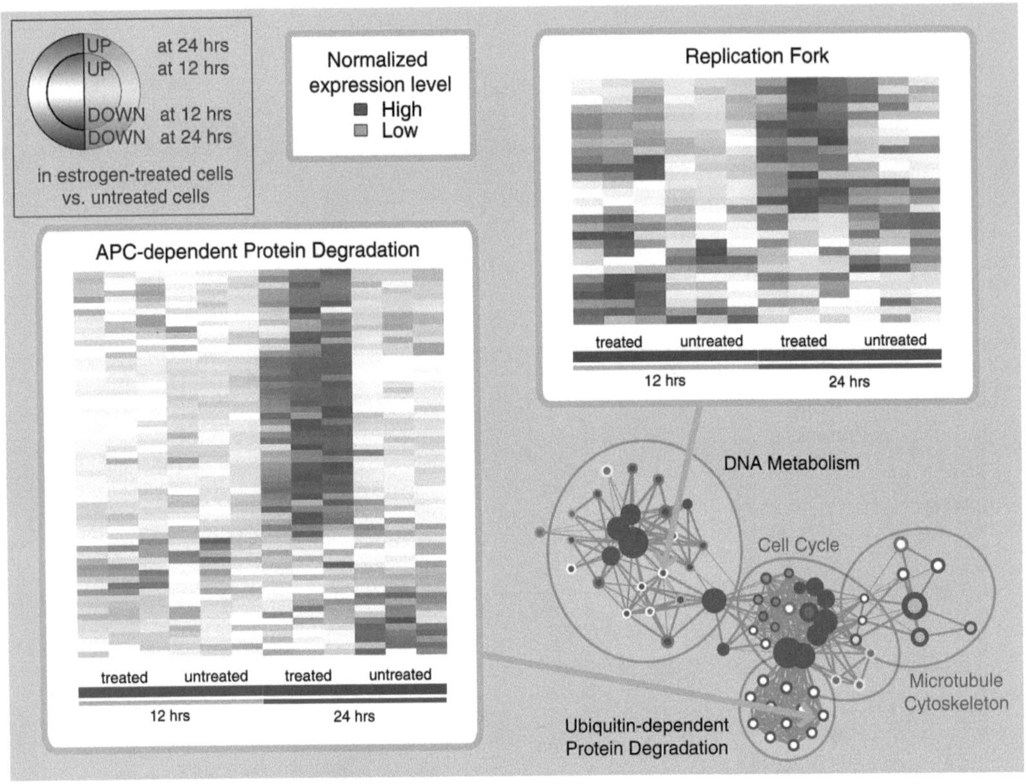

Fig. 5. Investigating gene expression patterns within two clusters of the 12–24-h estrogen response enrichment map. Enrichment of the *Replication Fork* gene-set is significant and associated to estrogen treatment (dark node center) at 12 h but not at 24 h (white node border); accordingly, the heat-map displays globally higher expression levels in estrogen-treated cells, although with a smaller difference at 24 h due to increased levels in the untreated cells, which is responsible for the absence of enrichment at that time point. Enrichment of the *APC-dependent Protein Degradation* gene-set is significant and associated to estrogen treatment at 24 h (dark node border) but not at 12 h (white node center); accordingly, the heat-map displays globally higher expression in estrogen-treated cells, although this pattern is much stronger at 24 h, explaining why a enrichment significance is observed only at that time point.

3.6.1. Load Data and Set Parameters

1. Select from the menu: *Plugins/Enrichment Map/Post Analysis.*
2. Make sure the gene-set file used to generate the current enrichment map is correctly loaded in the *GMT* box.
3. Load the query gene-set(s) in *SigGMT* box. Example data: load *estrogenTargetsLin2007.GMT*
4. Press the *Load Gene-sets* button.
5. Select the query gene-set(s) to use in the analysis and press the down arrow to add the selected gene-set(s) to the active selection. Example data: select *E2_TARGETS.*
6. Set the *Hypergeometric Test cutoff.* The overlap between the query gene-set(s) and preexisting gene-sets will be tested using the hypergeometric test (also known as Fisher's Exact Test) and only gene-set pairs with overlap *p*-value smaller than the

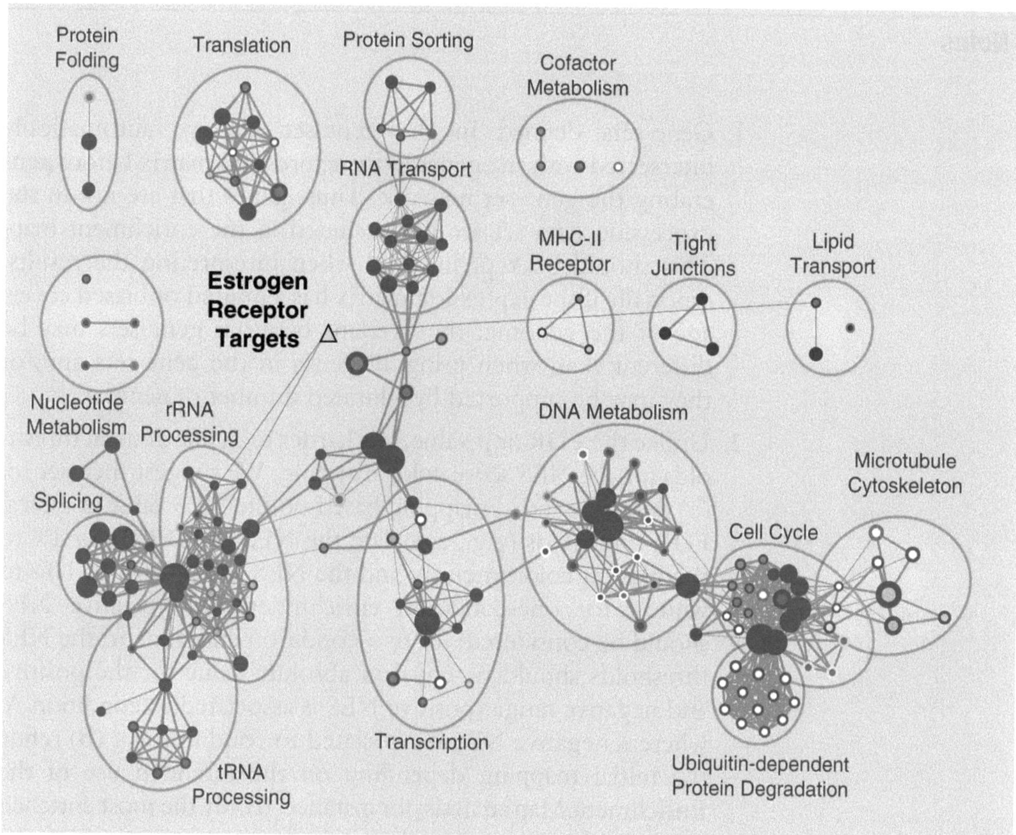

Fig. 6. Query set analysis investigating the relations between gene-sets enriched in the transcriptional response to estrogen treatment (12–24 h) and direct targets of estrogen receptor (query gene-set). The small triangle represents the query gene-set, and magenta links represent overlaps with uncorrected *p*-value <0.05 according to Fisher's Exact Test. Only a few links are present, even at a liberal *p*-value threshold. This suggests that either there are many false negatives in this target list, or the transcriptional response to estrogen is dominated by indirect targets.

cutoff will be connected by an edge. Example data: set to 0.05 (less stringent than default).

7. Press the *Run* button. The query gene-set will be displayed as a yellow triangle, connected to preexisting gene-sets by magenta edges. If no edges are present, the cutoff parameter may be too stringent.

3.6.2. Interpretation (Example Data)

1. The direct targets of the estrogen receptor and the gene-sets enriched by estrogen response have very little overlap: only three edges are displayed, even at a liberal Fisher's Exact Test *p*-value cutoff. Either the list of estrogen receptor targets is incomplete due to false negatives, or the transcriptional response observed after estrogen treatment is dominated by indirect targets.

Figure 6 displays the results of the query set analysis.

4. Notes

1. Gene-sets defined in the gene-set file are automatically intersected with the genes in the expression matrix before generating the gene-set network. Thus, genes that are not in the expression data set are not included in the enrichment map. This should be kept in mind when interpreting the results, especially if the expression matrix has a limited or biased coverage of the genome: the overlaps between gene-sets may be different than when using all genes in the gene-sets and/or they may be supported by a limited number of genes.
2. Unlike the FDR or *p*-value, it is harder to define general thresholds for the NES score color mapping. We suggest the user to: (a) define an initial mapping based on the NES value at typical FDR thresholds (e.g., associate the NES at FDR 0% or 1% to the highest color intensity and the NES at FDR 5% or 10% to white); for one-condition enrichment, only positive NES should be considered; for two-condition enrichment, the NES thresholds should be equal in absolute value for the positive and negative range (positive NES is associated to condition A, whereas negative NES is associated to condition B); (b) refine the initial mapping depending on the intended use of the Enrichment Map analysis; for instance, if only the most enriched gene-sets have intense color, it is very easy to prioritize the most enriched gene-sets within clusters; however, the other gene-sets may be less visible.
3. We recommend using Entrez-Gene as a gene identification system. Pathway Commons gene-sets are available with Entrez-Gene or official gene symbols. Several ID systems can be selected in WhichGenes, including Entrez-Gene. MSigDB gene-sets are available only with official symbols as gene IDs. The gene-sets at the Enrichment Map download page are available only with Entrez-Gene as gene IDs.
4. We recommend using capitalized alphanumeric IDs to identify gene-sets, rather than text labels (e.g., *GO:0007049* for GO *Cell cycle*). Capitalization is recommended as certain enrichment software packages, such as GSEA, use capitalized gene-set IDs in their results files.
5. The phenotype values represent the condition of gene-set enrichment. For one-condition enrichments, only one value will have to be used ("1"). For two-condition enrichments, two values should be used ("1" and "-1"). The choice of "-1", instead of other values, follows the sign convention of the NES score in the GSEA format (i.e., positive NES is associated to condition A and negative NES is associated to condition B).

References

1. Allison, D. B., Cui, X., Page, G. P., and Sabripour, M. (2006) Microarray data analysis: from disarray to consolidation and consensus., *Nature reviews. Genetics 7*, 55–65.
2. Huang da, W., Sherman, B. T., Tan, Q., Collins, J. R., Alvord, W. G., Roayaei, J., Stephens, R., Baseler, M. W., Lane, H. C., and Lempicki, R. A. (2007) The DAVID Gene Functional Classification Tool: a novel biological module-centric algorithm to functionally analyze large gene lists, *Genome Biol 8*, R183.
3. Huang da, W., Sherman, B. T., and Lempicki, R. A. (2009) Systematic and integrative analysis of large gene lists using DAVID bioinformatics resources, *Nat Protoc 4*, 44–57.
4. Subramanian, A., Tamayo, P., Mootha, V. K., Mukherjee, S., Ebert, B. L., Gillette, M. A., Paulovich, A., Pomeroy, S. L., Golub, T. R., Lander, E. S., and Mesirov, J. P. (2005) Gene set enrichment analysis: a knowledge-based approach for interpreting genome-wide expression profiles., *Proceedings of the National Academy of Sciences of the United States of America 102*, 15545–15550.
5. Isserlin, R., Merico, D., Alikhani-Koupaei, R., Gramolini, A., Bader, G. D., and Emili, A. (2010) Pathway analysis of dilated cardiomyopathy using global proteomic profiling and enrichment maps. *Proteomics* 1316–1327.
6. Shannon, P., Markiel, A., Ozier, O., Baliga, N. S., Wang, J. T., Ramage, D., Amin, N., Schwikowski, B., and Ideker, T. (2003) Cytoscape: a software environment for integrated models of biomolecular interaction networks, *Genome Res 13*, 2498–2504.
7. Cline, M. S., Smoot, M., Cerami, E., Kuchinsky, A., Landys, N., Workman, C., Christmas, R., Avila-Campilo, I., Creech, M., Gross, B., Hanspers, K., Isserlin, R., Kelley, R., Killcoyne, S., Lotia, S., Maere, S., Morris, J., Ono, K., Pavlovic, V., Pico, A. R., Vailaya, A., Wang, P.-L., Adler, A., Conklin, B. R., Hood, L., Kuiper, M., Sander, C., Schmulevich, I., Schwikowski, B., Warner, G. J., Ideker, T., and Bader, G. D. (2007) Integration of biological networks and gene expression data using Cytoscape., *Nature protocols 2*, 2366–2382.
8. Lin, C.-Y., Vega, V. B., Thomsen, J. S., Zhang, T., Kong, S. L., Xie, M., Chiu, K. P., Lipovich, L., Barnett, D. H., Stossi, F., Yeo, A., George, J., Kuznetsov, V. A., Lee, Y. K., Charn, T. H., Palanisamy, N., Miller, L. D., Cheung, E., Katzenellenbogen, B. S., Ruan, Y., Bourque, G., Wei, C.-L., and Liu, E. T. (2007) Whole-genome cartography of estrogen receptor alpha binding sites., *PLoS genetics 3*, e87.

Chapter 13

Quality Control Methodology for High-Throughput Protein–Protein Interaction Screening

Alexei Vazquez, Jean-François Rual, and Kavitha Venkatesan

Abstract

Protein–protein interactions are key to many aspects of the cell, including its cytoskeletal structure, the signaling processes in which it is involved, or its metabolism. Failure to form protein complexes or signaling cascades may sometimes translate into pathologic conditions such as cancer or neurodegenerative diseases. The set of all protein interactions between the proteins encoded by an organism constitutes its protein interaction network, representing a scaffold for biological function. Knowing the protein interaction network of an organism, combined with other sources of biological information, can unravel fundamental biological circuits and may help better understand the molecular basics of human diseases. The protein interaction network of an organism can be mapped by combining data obtained from both low-throughput screens, i.e., "one gene at a time" experiments and high-throughput screens, i.e., screens designed to interrogate large sets of proteins at once. In either case, quality controls are required to deal with the inherent imperfect nature of experimental assays. In this chapter, we discuss experimental and statistical methodologies to quantify error rates in high-throughput protein–protein interactions screens.

Key words: Protein interaction, High-throughput screen, False positives, False negatives, Sensitivity, Specificity, Biological networks

1. Introduction

Numerous cellular structures and processes are mediated by protein–protein interactions (PPIs) (1). High- and intermediate-affinity PPIs allow for the formation of protein complexes, resulting in multimeric receptors, enzymes, and other important molecular machines. On the contrary, intermediate- and low-affinity interactions are fundamental in signaling cascades, allowing for chemical transformation, e.g., phosphorylation, of a target protein by another protein or protein complex with enzymatic activity,

Gerard Cagney and Andrew Emili (eds.), *Network Biology: Methods and Applications*, Methods in Molecular Biology, vol. 781, DOI 10.1007/978-1-61779-276-2_13,

e.g., a protein kinase. Although mapping PPIs is subject to technical biases, e.g., scoring a PPI depends on the signal detection resolution of a given assay, and is imperfect, e.g., subject to experimental errors, it has helped biologists to unravel the macromolecular organization of cellular processes.

The set of all proteins encoded by the genome of an organism together with the set of all pairwise PPIs that can biophysically occur between them constitutes the protein interaction network (PIN) of the corresponding organism. PPIs can be characterized by their binding affinities or, simply, by the fact that they "do" or "do not" interact under physiologically relevant conditions. Furthermore, not all interactions in the PIN are biologically relevant in a given physiological context. For example, the interaction between two proteins may not be relevant because the proteins are never expressed at the same time in the same cell or tissue. We argue in favor of an iterative approach that consists in first, determining the entire PIN of an organism and second, evaluating their biological relevance using additional means.

A large proportion of currently available PPIs have been detected by coimmuno or coaffinity purifications followed by mass spectroscopy (coIP or coAP/MS) (2–7) or by yeast two-hybrid (Y2H) assays (8–17). These assays can be carried on at low-throughput, testing interactions between one or a few protein pairs at a time (12, 18), or they can be optimized for high-throughput screens, testing a large set of protein pairs (4, 14–17, 19). Despite considerable improvements in these experimental methodologies to probe protein interactions, false protein interactions (false positives) may still be reported, while true protein interactions may be missed (false negatives).

Several approaches have been implemented to estimate quality parameters and examine the biological relevance of PPI maps (20–23). For example, previous analyses to determine the quality of PPI maps relied on indirect supporting evidence by determining the extent to which interacting proteins share other biological attributes, such as expression patterns or functional annotation. However, in addition to relying on indirect evidence when using such an approach, one must assume that our knowledge of functional annotation is complete and unbiased, which is not the case. An alternative strategy relies on analyzing the extent of overlaps between PIN maps. However, such a strategy requires the use of datasets that are homogeneous in their composition, i.e. datasets that are generated using similar assays. Our scope here is to summarize a recent strategy that attempts to overcome some of these limitations and employs systematically designed experiments together with statistical methods to quantify the rate of false positives and false negatives in high-throughput PPI screens (24). Our focus is on pairwise biophysical interactions, leaving the discussion of how to determine biological relevance to other sources (25–30).

The layout will be more oriented to the practitioners of the field, providing technical details that were omitted or less emphasized in previous publications (24).

2. Parameters Definition

Various quality parameters associated with PPI mapping assays can be defined (21, 24).

Completeness refers to the fraction of all protein pairs that have been interrogated for physical interactions. For example, initial human PIN mapping efforts tested for pairwise interactions among ~4,000 × 4,000 (17) protein pairs or 8,000 × 8,000 protein pairs (16) out of a possible set of 24,000 × 24,000 annotated proteins, thus interrogating ~3–10% of all possible pairs. While this parameter is not directly a property of a PPI assay, the size of the search space is part of the design of a PIN mapping experiment and consequently contributes to missed interactions or false negatives.

Assay sensitivity is defined as the fraction of all biophysical interactions that can possibly be identified by an assay conducted under a specific set of experimental conditions. Each PPI mapping assay is usually characterized by a specific in vitro or in vivo experimental setting in which PPI are tested. To be detected and scored positive, a given PPI must, therefore, occur in the constrained environment defined by these experimental conditions. Such constraints may prevent the detection of a significant proportion of the interactions and lead to an inherent rate of false negatives and a limited sensitivity. Indeed, a recent comparative analysis of 4 independent PPI assays led to the conclusion that each PPI assay cannot detect more than 30% of all true interactions (31). A strategy to determine the assay sensitivity is described in Subheading 3.

Sampling sensitivity of an assay is defined as the probability that an interaction that is detectable by the assay is reported positive in one screen of this assay. The sampling sensitivity takes into account the fact that, due to stochastic errors, some detectable interactions may be reported as negative (termed *stochastic false negatives*) in one screen, but that they would eventually report positive after several screens are performed. Estimating the sampling sensitivity requires the analysis of repeated screens (see Subheading 4).

Precision is defined as the fraction of observed pairs in a PIN data set that are true positives. Indeed, each PPI assay is characterized by an inherent rate of false positives corresponding to "interactions" erroneously detected. For example, the output signal of a given assay may be turned on for reasons that are independent of the interactions between the pairs of protein tested, e.g., autoactivation in the Y2H system (11). Two strategies to determine the precision are described in Subheading 3. The false positives can be further divided into unsystematic and systematic false positives. *Unsystematic false positives* represent pairs of proteins that do not interact under the experimental conditions tested, but they are reported as positives stochastically without any particular distinction between the different noninteracting pairs. The latter emphasizes that the chance to observe an unsystematic false positive must be the same for all noninteracting pairs (see Subheading 4). *Systematic false positives* are protein pairs that do not interact but, when performing repeated screens, they are reported as positive in a similar manner as true interacting pairs (see Subheading 5).

PIN size is defined as the total number of interactions in the PIN. A methodology to estimate the PIN size that relies on combining all quality parameters described above is described in Subheading 6.

3. Assay Sensitivity and Precision

The assay sensitivity and precision can be estimated using well-defined reference sets of positive and negative interacting pairs.

Assay sensitivity/Positive Reference Set (PRS): The sensitivity of an assay can be estimated by measuring the ability of a given assay to detect interactions that are known to be true, i.e. a positive reference set of PPIs. Ideally, a PRS should be made exclusively of true, or highly likely to be true, interactions and be representative of all the different types of interactions in all their different "flavors." In a previous effort, we relied on interactions curated from the literature to generate a PRS. To avoid the inclusion of potential false positives described in the literature, we consider only interactions that were described in two independent curation databases and supported by at least two publications. A limitation of our PRS is that it is biased toward what is already described in the literature, e.g., toward human disease-related proteins and toward interactions detectable with most commonly used PPI assays. In other words, a PRS selected from the current literature probably does not provide a fair representation of the true PIN. Given the significant contribution of Y2H and co-IP for interaction mapping in the past decades, a novel assay that maps interactions in a subspace of the true PIN that does not overlap with either Y2H or co-IP subspaces would probably score poorly. Thus, PPI assays with high potential for mapping new territories of the PIN terra incognita could eventually score poorly in our approach given the biases of the PRS set. When testing a given PPI assay with a PRS, it is important to check how well the assay detects for example, Y2H-derived PRS versus non-Y2H-derived PRS. It would be interesting to develop a PPI assay that, despite an overall apparent low sensitivity, would have the ability to detect interactions that have been poorly explored so far, e.g., scores well with the ~40% of PRS that was found negative by 4 independent assays (31). For example, the split ubiquitin system (32) would score well with membrane-tethered proteins. Although one has to keep these limitations of a PRS in mind, this strategy still represents a better approach in comparison to previous strategies, which rely on testing only one or a few well-described protein–protein interactions.

Assay precision/Negative Reference Set (NRS): To assess the precision of an assay, one can test the ability of an assay to erroneously score interactions that are, *a priori*, true negative interactions. Several strategies can be implemented to define a set of pairs of proteins that have a high chance of not interacting. One strategy consists in selecting pairs of proteins that are anticorrelated for a particular functional attribute, e.g., tissue localization (33), and are therefore highly unlikely to interact in vivo. However, such set of protein pairs might not be representative of all possible pairs of proteins in the proteome. In a proteome scale experiment, the majority of the pairs tested are truly noninteracting pairs can include pairs of protein that are coexpressed. Thus, such a strategy might be biased and might limit the ability to measure the exact precision of a given assay. To avoid this bias, another strategy consists in selecting pairs of proteins at random (34, 35). This strategy relies on the assumption that among all proteins pairs that can be tested, e.g., ~500 million in human (22,000 × 22,000), only a small fraction of them are true positives, i.e., 0.1% if one assumes that there are ~500,000 true interactions (probably an over-estimate). In PIN mapping, the search space is vast and the number of true positives are small in comparison. Thus a good PPI assay must combine a reasonable sensitivity and a high precision. Indeed, even a small false-positive rate (<1%) would lead to a very low precision. Given the very small number of expected positive hits, one limitation when testing such a reference set is that one needs to test a large number of pairs in order to accurately estimate the precision of an assay.

Assay precision using an independent assay: A complementary, and more practically feasible approach for estimating assay precision is to test interactions found in a given assay A using another independent assay B. This first necessitates calibration of the false-positive and false-negative rate of assay B as a benchmark for the rate at which interactions observed using assay A are reproduced in assay B. The positive and negative reference sets (PRS and NRS sets) described above can be used for this purpose, where the fraction of PRS pairs scoring positive in assay B reflects that assay's sensitivity and the fraction of NRS pairs scoring positive in assay B reflects its false-positive rate. Once these parameters are determined experimentally, we can estimate the precision of assay A as follows.

Let I_{obsAB} be the percent of observed positives in assay A that are also observed positive in assay B. If $I_{+\text{A}}$ is the percent of true interactions among pairs reported positive in assay A, $f_{+\text{B}}$ is the false-positive rate of assay B, and $1-f_{-\text{B}}$ is the assay sensitivity of assay B (obtained as the fraction of PRS pairs scoring positive in assay B), then,

$$I_{\text{obsAB}} = (100 - I_{+\text{A}})f_{+\text{B}} + I_{+\text{A}}(I - f_{-\text{B}}) \tag{1}$$

Substituting the values for I_{obsAB}, $f_{+\text{B}}$ and $f_{-\text{B}}$ above, we can solve for $I_{+\text{A}}$, which represents the percent of true positives among positive pairs reported by assay A, and consequently provides an estimate of the precision or false discovery rate, i.e., $100 - I_{+\text{A}}$, generated by that assay.

$$I_{+\text{A}} = \frac{I_{\text{obsAB}} - 100 f_{+\text{B}}}{1 - f_{-\text{B}} - f_{+\text{B}}} \tag{2}$$

If assay B is independent from assay A, both systematic and unsystematic false positives arising from assay A should report negative to a similar extent in assay B. Therefore, the false discovery rate measured from experiments using assay B is a combined estimate of systematic and unsystematic false discovery rates. Methods to decompose their relative contribution are described in Subheadings 3 (unsystematic) and 4 (systematic).

Since we are always testing a finite number of protein pairs, the estimates of I_{obsAB} are imprecise and are better represented by a distribution. Indeed, the number of observed positives in assay A that are also observed positive in assay B (n_{obsA}) follows a binomial distribution Binomial (n_{obsA}; p_{obsAB}), where p_{obsAB} is the probability that an observed positive in assay A is also observed positive in assay B. p_{obsAB} is an unknown parameter that we want to estimate based on the observed data. Using a Bayesian approach (36), we obtain the p_{obsAB} estimated from our observations follows the beta distribution.

$$\text{Beta}(p_{\text{obsAB}}; n_{\text{obsA}} + \alpha, N_{\text{obsA}} - n_{\text{obsA}} + \beta) \tag{3}$$

where N_{obsA} is the number observed positives in assay A that were then tested using assay B. α and β are parameters characterizing the p_{obsAB} distribution in the absence of any observation, which are often assumed smaller or of the order of 1. Having derived Eq. 3, we can determine other statistical parameters as well. From example, the expected value of p_{obsAB} is given by $I_{\text{obsAB}} = 100(n_{\text{obsAB}} + \alpha)/(N_{\text{obsA}} + \beta)$ (in percent). Furthermore, the statistical errors on I_{obsAB} can be propagated to estimate those of $I_{+\text{A}}$. This can be done using Monte Carlo simulations (see Subheading 7).

4. Sampling Sensitivity and Unsystematic False-Positive Rate

The sampling sensitivity of an assay is associated with the existence of stochastic false negatives. In general the stochastic false-negative rate may be different for different interacting proteins and can be affected by the binding affinity or the protein expression levels achieved in the compartment where the interaction is tested (e.g., nucleus for Y2H), among other factors. For Y2H assays, one source of screen-to-screen variation is bait–prey protein expression fluctuations and other sources of cellular noise, or failure to identify interacting proteins by PCR/sequencing of interaction sequence tags (ISTs). In addition, there are unsystematic false positives. For example, in the Y2H assay, the promoters driving the expression of the reporter genes can be "leaky," resulting in variable levels of basal transcriptional activity expression irrespective of which protein pair is being tested. Owing to the stochasticity of gene expression, variations of this basal expression may become greater than the threshold needed to score positive. While the chance for this to happen in one screen of an experiment may be quite small, a significant number of unsystematic false positives can accumulate over the course of multiple screens. A systematic approach to boost the sensitivity of a high-throughput screen and to increase our ability to identify unsystematic false positives is to carry on repeated screens. For example, recent pilot high-throughput Y2H screens of the yeast (37) and human (24) protein interaction networks have shown that repeating the screen two times doubles the sensitivity, therefore justifying the additional cost involved in repeating the screen.

To model the stochastic variations, we introduce sampling sensitivity parameter p, the probability a tested protein pair is reported positive in one screen. After M independent repetitions of the screen, the probability to obtain n positives is given by the binomial distribution

$$\pi_n(p) = \binom{M}{n} p^n (1-p)^{M-n} \tag{4}$$

The sampling sensitivity may vary across protein pairs and have different mechanistic explanations. First, a distinction can be made between two classes of protein pairs: (1) one class that includes both true positives and systematic false positives (fraction e) and (2) another class that includes unsystematic false positives (fraction $1 - e$). We emphasize that, when performing repeated screens, the systematic false positives are reported as positive in a similar manner as true interacting pairs and are not distinguishable in this context.

The distinction between the two classes described above can be modeled assuming a constant sampling sensitivity across all interacting pairs and systematic false positives (p) and across all noninteracting pairs that do not lead to systematic false positives (q), replacing Eq. 4 by

$$\pi_n(e,p,q) = e\binom{M}{n}p^n(1-p)^{M-n} + (1-e)\binom{M}{n}q^n(1-q)^{M-n} \quad (5)$$

We can assume there is a range of various K sampling sensitivities in the first class (class that includes both interacting protein pairs and systematic false positives) due to their different binding affinities among interacting pairs or other factors, obtaining

$$\pi_n(e,p,q,K) = \sum_{i=1}^{K} e_i\binom{M}{n}p_i^n(1-p_i)^{M-n} + (1-e)\binom{M}{n}q^n(1-q)^{M-n}, \quad (6)$$

where $e=\Sigma_i e_i$. Note that, as discussed at the beginning of this section, the unsystematic false-positive rate is by definition uniform across all noninteracting pairs and, therefore, a single false-positive rate parameter (q) is sufficient to model them.

The next step requires us to extend the analysis to the case when several pairs are being tested. One screen consists of testing a search space S of protein pairs for protein–protein interactions. M screens consist of repeating the same screen over the search space M times. The outcome is N_0 protein pairs never found positive, N_1 pairs found positive once, N_2 pairs found positive twice, …, N_M pairs found positive M times, where $\sum_{n=0}^{M} N_n = S$. The likelihood of a particular outcome $N=(N_0, N_1,\ldots, N_M)$ is given by the multinomial distribution

$$\Pr(N|e,p,q,K) = \frac{S!}{N_0!\cdots N_M!}[\pi_n(e,p,q,K)]^{N_n} \quad (7)$$

We could proceed by computing the maximum likelihood estimate (MLE) for the model parameters given the data N. However, the MLE can bias the parameter estimation. A preferred strategy would be to also consider nearly optimal fits, producing intervals of confidence for the model parameters. Thus, we use a Bayesian approach (36) and compute the posterior distribution

$$\Pr(e,p,q,K \mid N) = \frac{\Pr(N|e,p,q,K)\Pr(e,p,q,K)}{\int \mathrm{d}e\mathrm{d}p\mathrm{d}q \Pr(N|e,p,q,K)\Pr(e,p,q,K)}, \quad (8)$$

of the model parameters, given the data N and a prior distribution $\Pr(e,p,q,K)$. For the prior distribution, we assume independence between the model parameters e, p, q, and K and a uniform distribution on the interval [0,1] for e, p, and q. The prior distribution for K takes into account the model complexity. For a given K, there are $2K+1$ independent parameters (e, p, q). Because a model with $K>1$ contains models with smaller K as particular examples, the larger the K value the better the model fits the data. On the contrary, larger K values greatly increase the model complexity. To account for this increase in model complexity, we use the Akaike information theoretical criterion (AIC) (38) and assume

$$\Pr(K) = Ae^{-\#\text{independentparameters}} = Ae^{-(2K+1)}, \quad (9)$$

where A is a normalization constant such that $\sum_{K=1}^{\infty} \Pr(K) = 1$. Following these assumptions for the prior distribution, from Eq. 9 we obtain

$$\Pr(e,p,q,K \mid N) = \frac{1}{Z(N)} \Pr(N|e,p,q,K)e^{-2K}. \quad (10)$$

where

$$Z(N) = \sum_{K=1}^{\infty} \int de \int dp \int dq P(N|e,p,q,K)e^{-2K} \quad (11)$$

is a normalization factor also known as the partition function. In essence, Eq. 10 represents a probability distribution in the space of parameters (e, p, q, K) given the data N and our assumptions for the prior distribution. This distribution can be used to determine the expectation of a certain variable $x(e, p, q, K)$, using the formula

$$E[x](N) = \sum_{K=1}^{\infty} \int de \int dp \int dq \Pr(e,p,q,K \mid N) x(e,p,q,K) \quad (12)$$

Some magnitudes of interest are the density of detected interactions after M screens $e_M = \Sigma_i e_i[1-(1-p_i)^M]$, the density of detectable interactions $e = e_\infty = \Sigma_i e_i$, the sampling sensitivity per screen $p = \Sigma_i e_i p_i / \Sigma_i e_i$, the fraction of pairs e_i and sampling sensitivity p_i on each class of true interacting proteins and systematic false positives, the rate of unsystematic false positives per screen q, and their standard deviations. Furthermore, the relative contribution of each model with K' classes is given by $E[\delta_{KK'}]$, where $\delta_{KK'} = 1$ when $K = K'$ and zero otherwise is the Kronecker delta symbol.

The calculation/computation of these averages are quite challenging in general. Some approximations can be used, however, to reduce computing times. For example, the following approximations were used to estimate these parameters for high-throughput Y2H screens (24): (1) Replace the integral over p_i by a sum with resolution $\Delta p = 0.01$. (2) For each K and set of p_i restrict the average over e_i and q to the MLE. (3) Truncate the sum over K at $K = 1$ or $K = 2$. Another alternative is to use Monte Carlo methods to sample the posterior distribution Eq. 10 (39).

Case study: a Y2H screen

The methodology described above has been used to characterize a Y2H screen (24). The Y2H screen was repeated four times in a defined search space of 1,822 DB-Xs ("baits" representing 1,744 unique genes) against 1,796 AD-Ys ("preys" representing 1,752 unique genes), representing ~3 million pairwise combinations. The expected sampling sensitivity per screen is 45%, the expected density of detectable interactions is 71 interactions per million tested and approximately six screens are needed to reach 90% saturation. Adjusting for these repeat screens being done in only one Y2H configuration (bait–prey versus prey–bait), we estimated that after testing both configurations, the sampling sensitivity per screen is 53%, and that after a saturating number of screens, Y2H can identify 118 interactions per million pairs tested. The analysis of repeated Y2H screens also revealed that, with 99% confidence, there are at least two sensitivity classes. More precisely, the density of detectable interactions reported above is distributed into a 72% of interactions with a low sampling sensitivity of 34% and the remaining 28% with a high sampling sensitivity of 72%. These estimates were also in agreement with the empirical evidence obtained in pilot studies, indicating that some interacting pairs score positive in most screens, while others manifest inconsistent scores from screen to screen. It should be emphasized, however, that the existence of two sampling sensitivity classes does not necessarily implies that the distribution of binding affinities of interacting proteins is bimodal. We cannot exclude the possibility that the two sensitivity classes is an experimental artifact of the threshold like nature of the scoring system, where interactions above certain affinity are scored positive with high probability while those below are scored positive with small probability. For example, Fig. 1 shows that, when the assay scoring scheme has a threshold-like behavior, a unimodal distribution of binding affinities can result in either a unimodal or a bimodal distribution of sampling sensitivities.

5. Systematic False Discovery Rate

The systematic false-positive rate can be obtained after subtracting the unsystematic false-positive rate (estimated in Subheading 4) from the overall false discovery rate (estimated in Subheading 3). Denoting by FD the combined estimate of systematic and unsystematic false discovery rates, by SFD the systematic false discovery rate, i.e., number of systematic false positives relative to the number of interactions reported positive, and by UFD the unsystematic

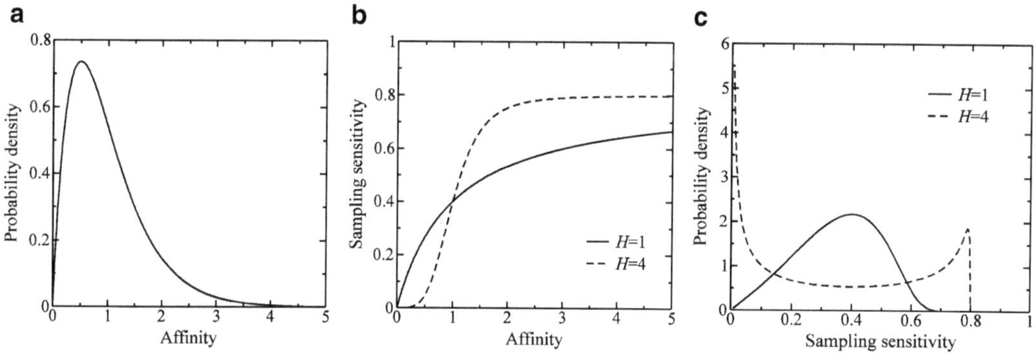

Fig. 1. Experimental artifacts can result in two sampling sensitivity classes. (**a**) Hypothetical unimodal distribution of binding affinities. (**b**) Hypothetical scoring curves transforming binding affinities into sampling sensitivities. The *solid line* represents a smooth scoring scheme, where the sampling sensitivity increases smoothly with increasing the binding affinity. The *dashed line* represents a threshold-like scoring scheme, where interactions with affinities below a certain threshold are detected with low probability, while those above a certain threshold are sampled with high probability. (**c**) The sampling sensitivity distribution resulting from the binding affinity distribution in (**a**) and the two scoring schemes in (**b**). The smooth scoring scheme (*solid line*) results in a unimodal sampling sensitivity distribution, while the threshold-like scoring scheme (*dashed line*) results in a bimodal distribution. These plots were obtained assuming the relationship $p(A) = p_{max}(A/K)^H/[1+(A/K)^H]$ between the binding affinity A and the sampling sensitivity p and a gamma probability density of binding affinities $f(A) = (1/A_0)(A/A_0)^{a-1}\exp(-A/A_0)$, with $p_{max} = 0.8$, $K = 1$, $A_0 = 0.5$, $a = 2$ and $H = 1$ or 4, where the affinities are expressed here in arbitrary units.

false discovery rate, i.e., the number of unsystematic false positives relative to the number of reported interactions,

$$\mathrm{FD} = \mathrm{UFD} + (1 - \mathrm{UFD})\mathrm{SFD} \tag{13}$$

FD is measured in the independent assay B validation experiments (see Eq. 2). The model of the repeat screens estimates the unsystematic false-positive rate q as the fraction of noninteracting pairs that are reported positive (see Subheading 4). From these, we obtain the unsystematic false discovery rate, UFD as

$$\mathrm{UFD} = \frac{q}{ep + q} \tag{14}$$

Thus, substituting the value of UFD from Eq. 14 and the value of FD from Eq. 2 into Eq. 13, we can obtain the value of SFD.

6. Estimation of the PIN Size

An important issue in the context of PIN mapping is estimating the PIN size. Knowledge of the PIN size indicates which fraction of the PIN has already been mapped and how many PPIs remain to be detected. Estimation of the various false-negative and false-positive rates of PPI mapping assays facilitates estimating the size of the entire PIN of a given organism. Having such an estimate can be very useful in planning what high-throughput assays would be best suited for exhaustively mapping a given PIN and thereby projecting the amount of effort, resources and time needed to completely map a given PIN. If N is the number of predicted genes in the genome, the size of the entire proteome search space, assuming one splice isoform per gene, is given by N^2.

We denote by e_0 the PIN size relative to the search space size. Among these interactions, only a fraction $e_0(1-\text{SFN})$ are expected to be uncovered by a given technology with assay sensitivity $1-\text{SFN}$. On the contrary, among the interactions reported positive after repeated screens using a given assay, only a fraction $e(1-\text{SFD})$ is expected to be true interactions, where SFD is the systematic false discovery rate of that assay. Because these two expected values should coincide we obtain

$$e_0(1-\text{SFN}) = e(1-\text{SFD}) \tag{15}$$

from which it follows

$$e_0 = e(1-\text{SFD})\,/\,(1-\text{SFN}) \tag{16}$$

Substituting the values of e (from Eq. 12), SFN (from the PRS experiments as the fraction of PRS pairs scoring positive in the assay under consideration) and SFD (from Eq. 13) into Eq. 15, we can compute the expected size of the PIN.

The PIN size computed is likely to be an under-estimate because of several reasons including but the fact that (1) our measurement of assay sensitivity may be an overestimate given the biased nature of the PRS and (2) we ignore splice variant complexity.

7. Estimation Errors: Monte Carlo Simulations of all Reported Parameters

The parameter estimates described above are subject to statistical errors, prompting us to estimate their confidence intervals. Specifically, the smaller the starting number of experimental data points used in estimating a parameter, the higher the variance of the estimate. Since our assay quality estimation framework involves modeling a phenomenon with significant uncertainty in various parameters, it lends itself well to the use of Monte Carlo simulations, rather than a more “deterministic” statistical approach, to compute the distribution of different magnitudes which are indirectly estimated from the experimental data, i.e., precision of a given assay, sampling sensitivity, assay sensitivity of a given assay, or the PIN size. These simulations can be facilitated by directly estimating the distribution of specific parameters whenever possible. For example, Eq. 3 provides the distribution of the probability that a reported positive in a screen A is reported positive in a retest screen B and Eq. 10 provides the distribution of parameters estimated from the repeated screen analysis.

In a given iteration of the simulation, we generate random numbers using those distributions that where directly estimated, e.g., Eqs. 3 and 10, and use those numbers to obtain estimates for the remaining dependent parameters. We can iteratively generate a large number L of random values for the various dependent parameters and thereby obtain an estimate of their distribution, allowing us to report mean, empirical standard deviation and empirical 95% confidence intervals. For example, the distribution of I_{+A} can be obtained using Eq. 2, sampling I_{obsAB} from the beta distribution in Eq. 3. Similarly, the distribution of parameters characterizing sampling sensitivity and stochastic false-positive rates are obtained by sampling them from the distribution Eq. 10 derived in Subheading 4. Thus, we can obtain error estimates for all of the assay quality parameters.

8. Outlook

Thanks to the use of stringent and systematic PPI assays and the introduction of repeated screens, high-throughput protein–protein interaction screens have been fine-tuned to the point that they currently exceed the quality their low-throughput counterparts (24, 37). Additional improvements could contribute to the generation of high-quality maps. Current screens generally test for only one splice variant per gene and, therefore, may be quite limiting when analyzing biological processes that require a distinction between them. This is particularly the case for several brain specific genes, which are often long genes encoding for several splice variants. Current assays are also limited in their ability to detect protein–protein interactions that require protein posttranslational modifications. The achievement of the technological and logistic developments that may address these limitations could require the introduction of some modifications to the methodology outlined here. The construction of PRS and NRS to estimate assay sensitivity and precision should be adapted to represent new search spaces or classes of interactions that are different from binary PPIs, e.g., cocomplex membership interactions that are defined by coAP/MS approach. The analysis of repeated screens may require additional sampling sensitivity classes to model interactions between different splice variants and protein modifications as well.

In the context of high-throughput Y2H screens, further work is required to explain the observation of two sampling sensitivity classes. The threshold nature of experimental methods to score positives in PPI assays can transform a unimodal distribution of binding affinities into a bimodal distribution of sampling sensitivities.

It remains to be determined whether by relaxing the scoring method one can fine-tune the assay to obtain only one sampling sensitivity class and, more important, whether this could increase the overall sampling sensitivity. On the contrary, a scoring method resulting in two sensitivity classes may be beneficial, provided we can correlate the classes with some biologically relevant property. A recent report indicates that the low sampling sensitivity class is enriched for transient interactions (40), suggesting that we could exploit this information to stratify PPIs into transient and nontransient interactions.

All PPI mapping assays are characterized by a specific in vitro or in vivo experimental setting in which PPIs are tested. To be detected and scored positive, a given PPI must, therefore, occur in the constrained environment defined by these experimental conditions. Such constraints may prevent the detection of a significant proportion of the interactions and lead to an inherent rate of false negatives and a limited sensitivity. Indeed, a recent comparative analysis of 4 independent PPI assays has led to the conclusion that each PPI assay cannot detect more than 30% of all true interactions (31). Thus, the complexity and heterogeneity of the PIN "space" cannot be captured by a single PPI mapping technology and requires the development of specific technologies, each PPI assay being amenable to detect a specific subspace of the true PIN.

Acknowledgments

AV is supported by the RWJ Foundation, Project title: "Strengthening the Cancer Institute of New Jersey in Cancer Prevention, Control and Population Science to Improve Cancer Care," Cancer Informatics Category. JFR is supported by a fellowship from the Deutsche José Carreras Leukämie-Stiftung e. V. We thank past and current members of the Barabasi, Roth, Tavernier, Vidal, and Wanker Labs for helpful discussions during the development of the empirical framework for binary PIN mapping.

References

1. Alberts B, Johnson A, Lewis J, Raff M, Roberts K, et al. (2002) Molecular Biology of the Cell New York: Garland Science.
2. Rigaut G, Shevchenko A, Rutz B, Wilm M, Mann M, et al. (1999) A generic protein purification method for protein complex characterization and proteome exploration. Nat Biotechnol 17: 1030–1032.
3. Bauer A, Kuster B (2003) Affinity purification-mass spectrometry. Powerful tools for the characterization of protein complexes. Eur J Biochem 270: 570–578.
4. Ewing RM, Chu P, Elisma F, Li H, Taylor P, et al. (2007) Large-scale mapping of human protein–protein interactions by mass spectrometry. Mol Syst Biol 3: 89.
5. Köcher T, Superti-Furga G (2007) Mass spectrometry-based functional proteomics: from molecular machines to protein networks. Nat Methods 4: 807–815.

6. Shevchenko A, Schaft D, Roguev A, Pijnappel WW, Stewart AF (2002) Deciphering protein complexes and protein interaction networks by tandem affinity purification and mass spectrometry: analytical perspective. Mol Cell Proteomics 1: 204–212.
7. Vasilescu J, Figeys D (2006) Mapping protein–protein interactions by mass spectrometry. Curr Opin Biotechnol 17: 394–399.
8. Fields S, Song O (1989) A novel genetic system to detect protein–protein interactions. Nature 340: 245–246.
9. Gan X, Kitakawa M, Yoshino K, Oshiro N, Yonezawa K, et al. (2002) Tag-mediated isolation of yeast mitochondrial ribosome and mass spectrometric identification of its new components. Eur J Biochem 269: 5203–5214.
10. Huang L, Baldwin MA, Maltby DA, Medzihradszky KF, Baker PR, et al. (2002) The identification of protein–protein interactions of the nuclear pore complex of *Saccharomyces cerevisiae* using high throughput matrix-assisted laser desorption ionization time-of-flight tandem mass spectrometry. Mol Cell Proteomics 1: 434–450.
11. Fields S (2005) High-throughput two-hybrid analysis. The promise and the peril. FEBS J 272: 5391–5399.
12. Fields S, Sternglanz R (1994) The two-hybrid system: an assay for protein–protein interactions. Trends Genet 10: 286–292.
13. Ito T, Tashiro K, Muta S, Ozawa R, Chiba T, et al. (2000) Toward a protein–protein interaction map of the budding yeast: A comprehensive system to examine two-hybrid interactions in all possible combinations between the yeast proteins. Proc Natl Acad Sci USA 97: 1143–1147.
14. Giot L, Bader JS, Brouwer C, Chaudhuri A, Kuang B, et al. (2003) A protein interaction map of *Drosophila melanogaster*. Science 302: 1727–1736.
15. Li S, Armstrong CM, Bertin N, Ge H, Milstein S, et al. (2004) A map of the interactome network of the metazoan *C. elegans*. Science 303: 540–543.
16. Rual JF, Venkatesan K, Hao T, Hirozane-Kishikawa T, Dricot A, et al. (2005) Towards a proteome-scale map of the human protein–protein interaction network. Nature 437: 1173–1178.
17. Stelzl U, Worm U, Lalowski M, Haenig C, Brembeck FH, et al. (2005) A human protein–protein interaction network: a resource for annotating the proteome. Cell 122: 957–968.
18. Cagney G, Uetz P, Fields S (2001) Two-hybrid analysis of the *Saccharomyces cerevisiae* 26S proteasome. Physiol Genomics 7: 27–34.
19. Ho Y, Gruhler A, Heilbut A, Bader GD, Moore L, et al. (2002) Systematic identification of protein complexes in *Saccharomyces cerevisiae* by mass spectrometry. Nature 415: 180–183.
20. von Mering C, Krause R, Snel B, Cornell M, Oliver SG, et al. (2002) Comparative assessment of large-scale data sets of protein–protein interactions. Nature 417: 399–403.
21. D'Haeseleer P, Church GM (2004) Estimating and improving protein interaction error rates. Proc IEEE Comput Syst Bioinform Conf: 216–223.
22. Hart GT, Ramani AK, Marcotte EM (2006) How complete are current yeast and human protein-interaction networks? Genome Biol 7: 120.
23. Stumpf MP, Thorne T, de Silva E, Stewart R, An HJ, et al. (2008) Estimating the size of the human interactome. Proc Natl Acad Sci USA 105: 6959–6964.
24. Venkatesan K, Rual JF, Vazquez A, Stelzl U, Lemmens I, et al. (2009) An empirical framework for binary interactome mapping. Nat Methods 6: 83–90.
25. Bader JS (2003) Greedily building protein networks with confidence. Bioinformatics 19: 1869–1874.
26. Bader JS, Chaudhuri A, Rothberg JM, Chant J (2004) Gaining confidence in high-throughput protein interaction networks. Nat Biotechnol 22: 78–85.
27. Han JD, Bertin N, Hao T, Goldberg DS, Berriz GF, et al. (2004) Evidence for dynamically organized modularity in the yeast protein–protein interaction network. Nature 430: 88–93.
28. Luscombe NM, Babu MM, Yu H, Snyder M, Teichmann SA, et al. (2004) Genomic analysis of regulatory network dynamics reveals large topological changes. Nature 431: 308–312.
29. Rachlin J, Cohen DD, Cantor C, Kasif S (2006) Biological context networks: a mosaic view of the interactome. Mol Syst Biol 2: 66.
30. Bossi A, Lehner B (2009) Tissue specificity and the human protein interaction network. Mol Syst Biol 5: 260.
31. Braun P, Tasan M, Dreze M, Barrios-Rodiles M, Lemmens I, et al. (2009) An experimentally derived confidence score for binary protein–protein interactions. Nat Methods 6: 91–97.
32. Stagljar I, Korostensky C, Johnsson N, te Heesen S (1998) A genetic system based on split-ubiquitin for the analysis of interactions between membrane proteins in vivo. Proc Natl Acad Sci USA 95: 5187–5192.
33. Jansen R, Yu H, Greenbaum D, Kluger Y, Krogan NJ, et al. (2003) A Bayesian networks approach

for predicting protein–protein interactions from genomic data. Science 302: 449–453.
34. Ben-Hur A, Noble WS (2006) Choosing negative examples for the prediction of protein–protein interactions. BMC Bioinformatics 7 Suppl 1: S2.
35. Gomez A, Domedel N, Cedano J, Pinol J, Querol E (2003) Do current sequence analysis algorithms disclose multifunctional (moonlighting) proteins? Bioinformatics 19: 895–896.
36. Hastie T, Tibshirani R, Friedman J (2001) The elements of statistical learning. New York: Springer.
37. Yu H, Braun P, Yildirim MA, Lemmens I, Venkatesan K, et al. (2008) High-quality binary protein interaction map of the yeast interactome network. Science 322: 104–110.
38. Akaike H (1974) A new look at the statistical model identification. IEEE Trans Aut Cont AC-19: 716–723.
39. Chen M-H, Shao Q-M, Ibrahim JG (2000) Monte Carlo methods in Bayesian computation. New York: Springer. xiii, pp. 386.
40. Vinayagam A, Stelzl U, Wanker EE (2010) Repeated two-hybrid screening detects transient protein–protein interactions Theoretical Chemistry Accounts 125: 613–619.

Chapter 14

Filtering and Interpreting Large-Scale Experimental Protein–Protein Interaction Data

Gabriel Musso, Andrew Emili, and Zhaolei Zhang

Abstract

Rarely acting in isolation, it is invariably the physical associations among proteins that define their biological activity, necessitating the study of the cellular meshwork of protein–protein interactions (PPI) before a full appreciation of gene function can be achieved. The past few years have seen a marked expansion in the both the sheer volume and number of organisms for which high-quality interaction data is available, with high-throughput interaction screening and detection techniques showing consistent improvement both in scale and sensitivity. Although techniques for large-scale PPI mapping are increasingly being applied to new organisms, including human, there is a corresponding need to rigorously evaluate, benchmark, and impartially filter the results. This chapter explores methods for PPI dataset evaluation, including a survey of previous techniques applied by landmark studies in the field and a discussion of promising new experimental approaches. We further outline practical suggestions and useful tools for interpreting newly generated PPI data. As the majority of large-scale experimental data has been generated for the budding yeast *S. cerevisiae*, most of the techniques and datasets described are from the perspective of this model unicellular eukaryote; however, extensions to other organisms including mammals are mentioned where possible.

Key words: Protein–protein interaction, Network, Affinity purification, Mass spectrometry, Yeast 2-hybrid, Large-scale assay, Systems biology

1. Introduction: The Need for Large-Scale Assay and Unbiased Filtering

Given the number of possible protein associations for even compact genomes such as the intensively studied model yeast *S. cerevisiae*, experimental techniques that can assay interactions among proteins on a large-scale are ultimately necessary to sufficiently survey the global interaction space (i.e., proteome-wide). Virtually, all high-throughput interaction detection techniques offer their increased survey capacity at a cost to assay accuracy and are

Gerard Cagney and Andrew Emili (eds.), *Network Biology: Methods and Applications*, Methods in Molecular Biology, vol. 781, DOI 10.1007/978-1-61779-276-2_14,

potentially subject to a substantial proportion of either missed (false negative) or incorrectly assessed (false positive) associations. Further, the major experimental frameworks in interaction detection (i.e., binary assays, complementation-based detection in vivo, or affinity purification-mass spectrometry of protein complexes) inherently vary in terms of specific biases and thus require different considerations when evaluating experimental findings. However, while approaches for assigning confidence in experimental predictions can vary based on the scope and nature of a particular study, the key parameters typically involve some combination of validation based on the assay itself (such as replicate or reciprocal testing for confirmation) and evaluation against an external reference (i.e., gold standard) to ensure general data reliability and to estimate systematic, experimental bias. Here, we describe how previous large-scale PPI studies have incorporated these two elements of validation to define high-confidence interaction networks.

This chapter first approaches the topic of PPI dataset scoring and filtering through a discussion of approaches taken by groundbreaking studies in the area (see Subheading 2). As the focus of this chapter is on analytical techniques used to assess and assign confidence to protein interactions, in-depth detail of the experimental methods is not given (however, for more detailed reviews on experimental procedures for large-scale protein–protein interaction detection methods, please see recent reviews by Musso et al. (1), Sanderson (2) and Cagney (3)). We then continue to outline a step-by-step approach to large-scale interaction filtering (see Subheading 3), providing anecdotal descriptions of common pitfalls and further considerations warranting close attention within Subheading 4.

2. Large-Scale Protein Interaction Detection Methods

2.1. Binary Interaction Detection Methods

2.1.1. Yeast 2-Hybrid

Binary interaction detection techniques typically employ transgenic modification of proteins of interest wherein each becomes tethered to one half of a functional reporter (e.g., a transcriptional readout). Sufficient proximity of the two proteins (or protein fragments) of interest causes activation of the reporter, generally allowing scoring of an interaction through monitored growth in a specific condition or in the presence of a fluorescent signal. Perhaps the most widely applied assay for binary protein interaction screening is the Yeast 2-Hybrid approach (Y2H), which as originally described involves reconstruction of the transcription factor GAL4 (4). Briefly, a bait protein of interest is fused to the DNA Binding Domain (BD) of GAL4 while a second prey protein is tethered to the GAL4 Activation Domain (AD); proximity of the two proteins allows binding of GAL4 to a specific DNA sequence and transcription of a selectable reporter.

In 2000 and 2001, two separate initiatives compiled Y2H-based global interaction maps in yeast (5, 6), noting a combined set of ~4,000 putative PPIs. Much as would later be the case with the other large-scale protein interaction studies (more below), the resulting interaction datasets showed surprisingly little overlap with each other or with annotated PPIs in public curation databases. Further, internal controls published by both studies showed low reproducibility of the Y2H assay. For example, Uetz et al. applied two independent approaches to identify PPI: the first, a systematic mating of 192 DNA-binding domain containing strains to an array of nearly 6,000 activation domain strains revealed an overall interaction reproducibility of only approximately 20%, while the second, involving a pooled assay of mated AD and BD strains followed by dual-marker selection and genome analysis, showed little agreement with the first assay. Compounding this, the results of a second, larger genome screen conducted by Ito et al. (6) using a similar pooling technique showed little in common with the previous work.

Early analyses attributed discrepancies in interaction networks generated using Y2H to a high-false positive and false-negative rate (7, 8); however, misestimations regarding the size of the yeast interactome (entire complement of protein interactions in the cell), as well as the use of comparative gold standards with inherent biases, might have led to unfair assumptions regarding the error rates of these screens. Fittingly, a more recently published second-generation Y2H map of protein interactions in yeast has revealed over 1,500 novel interactions demonstrated through rigorous testing to be of high quality (9). In the following section, we discuss common means for assessment and reduction of error in Y2H screens; however, for a more in-depth review, see Vidalain et al. (10).

While traditionally used to assay interactions among yeast proteins, ease of genetic recombination in *S. cerevisiae* has also allowed the screening of protein networks from other organisms including *Caenorhabditis elegans* (11), *Drosophila melanogaster* (12, 13), and human (14, 15). Further, a concerted effort now exists to recapitulate the principles of binary protein interaction detection directly in human cells, circumventing the criticism that cofactors necessary for a certain interaction may not be present when assaying in yeast. Owing to the lower throughput of mammalian Y2H techniques, however, they are best viewed as complementary to their yeast counterparts (for a more in-depth review and comment on mammalian two-hybrid systems, see Lievens et al. (16) and Suter et al. (17)).

2.1.2. Assessing Interaction Quality in Y2H

One source of experimental error during Y2H screening is autoactivation of the reporter, either by ectopic mutation or intrinsically by the fusion protein tethered to the DNA-binding domain alone.

If performing interaction analysis based on growth of colonies following reconstitution of the reporter, autoactivators can easily be identified by monitoring the growth of DB strains on the reporter selection media and manually removed. In their large-scale human protein interaction screen, Rual et al. (14) employed more rigorous testing for autoactivation of the DNA binding domain-containing proteins which included selection based on additional reporter genes, each controlled by a different GAL4-responsive promoter (described in detail in Walhout et al. (18)). Some proportion of DB-containing strains are also known to autoactivate during the course of screening, giving erroneous false positives despite this premating filtering (18); the lack of prey dependency can be eliminated by counterselection against markers contained in the AD plasmid. This filtering technique for identifying DB autoactivators was applied in a later large-scale yeast-interaction screen published by the same group (9) and resulted in the removal of 974 DB-containing yeast strains exhibiting spurious autoactivation. In both the yeast and human Y2H interactome studies, the authors then used experimental criteria such as reproducibility of interactions during multiple rounds of resampling to filter higher confidence PPI. However, while reproducibility of interactions may indicate lack of experimental error, the possibility still exists that a putative interaction does not have any biological relevance.

In addition to experimental error, like all high-throughput interaction detection methods, the Y2H assay has specific systematic biases that must be addressed. For example, due to the nature of the reporter systems commonly used, interactions are best observed when bait and prey are consistently present within the nucleus. Ectopic expression in a foreign cell may result either in missed interactions due to the absence of a particular cofactor or complex member (i.e., false-negative predictions), or conversely the observation of an interaction may be an artifact reflecting forced colocalization that does not normally occur in false positives. As both of these errors can be reproducible, determination of the overall prevalence of screen artifacts can only be achieved through evaluation against benchmark datasets (see Fig. 1). One typical approach is to compare experimental results against datasets either maintained in public databases or obtained through literature curation. However, Yu et al. argued that care must be taken when comparing against curated sets as selection of the "gold standards" can give misleading estimates of experimental error (9); while literature-curated datasets seemed to be appropriate for benchmarking Y2H data, interactions obtained from affinity purification studies (more below), while not necessarily less accurate, tend to overlap better with interactions that recapitulate stably bound protein complexes.

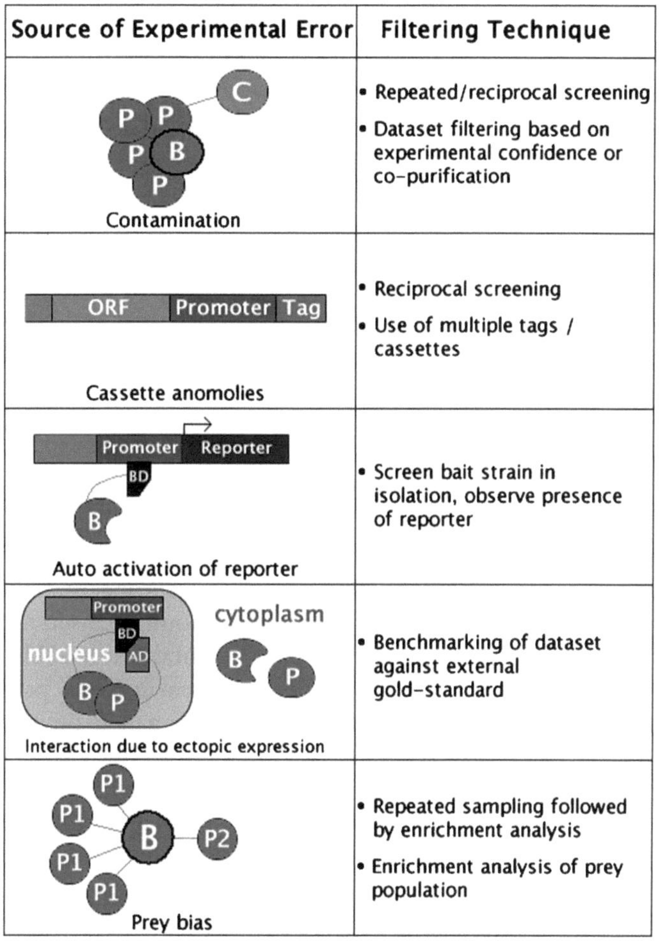

Fig. 1. Sources of experimental error. Presented are some of the main sources of experimental artifact in binary and complex-based PPI assay with suggested corrections. Within the figure, *B* represents a bait protein, *P* a prey (*P1* and *P2* represent specific preys), *C* a nonspecific contaminant, *BD* a DNA binding-domain, and *AD* an associated activation domain.

2.1.3. MYTH Screening

Owing to the need for colocalization of bait and prey proteins in yeast nuclei, Y2H is more effective at determining interactions among nuclear proteins and inherently has difficulty detecting interactions among proteins that are membrane-bound. Because of their role in cell permeability and transport, membrane proteins are frequently drug targets and make excellent biomarker candidates (19); thus, determination of their interactions is particularly useful. A specialized variant of Y2H utilizing reconstruction of ubiquitin and subsequent cleavage of an adjoined transcription factor (the so-called split-ubiquitin assay (20, 21)) has been developed to address this shortcoming. Application of this Membrane Yeast 2-Hybrid (MYTH) method resulted in the identification of nearly 2,000 putative interactions involving 536 transmembrane proteins in yeast (21).

Given the lack of existing interaction data for membrane proteins (only 7% of the 1,985 interactions had been previously reported), the authors turned to machine learning to rank and filter their interaction dataset. Specifically, the authors used functional associations contained in other sources as feature data in a Support Vector Machine (SVM) framework, predicting the validity of membrane PPI based on patterns in Gene Ontology (22) annotations, coexpression levels, and transcriptional coregulation. The SVM was trained 100 times, each time using the set of interactions confirmed by other studies as positive examples and randomly selected protein pairings as negative examples. Tested interactions were then assigned a confidence score based on their ranking in the 100 iterations (131 interactions were classified as being positive in all 100 SVM runs and thus were listed as the highest confidence interaction set (21)). As with any computational learning procedure, this approach assumed that there are inherent patterns for true interactors that could be used to independently determine the accuracy of experimental results. While evaluation techniques such as cross-validation (leaving one portion of training set unused, then evaluating performance on this withheld set) will indicate accuracy of predictions, the small number of known reference membrane PPI could lead to overfitting (modeling the training data too closely), giving biased prediction results. Thus, while learning techniques allow for filtering of experimental biases, they themselves should not be considered bias-free.

2.1.4. Protein Fragment Complementation Assay

Another technique highly analogous to more commonly applied Y2H methods that has recently been used to create a large binary protein interaction map is the protein fragment complementation assay (PCA) involving reconstruction of murine Dihydrofolate Reductase (mDHFR) (23). Unlike other techniques used to assay the yeast interactome, this assay was originally developed in mammalian cells (24) (although validated on a larger-scale in *E. coli* (25)), and involves the reconstruction of mDHFR in yeast cells expressing a DHFR variant with methotrexate sensitivity. Yeast cells are then exposed to methotrexate, allowing an instant growth-based reporting of protein interaction. Unlike traditional Y2H, which requires interaction within the nucleus, the reconstruction of DHFR can rescue cells if occurring in any cellular compartment, allowing endogenous expression of bait and prey. This additionally facilitates survey of physical associations between membrane proteins (23), although sensitivity is far greater for proteins with parallel membrane orientations due to the nature of the tethered reporter. In order to adapt this protocol to large-scale yeast screening, the authors optimized the experimental method by evaluating a subset of crosses with known interactions (as annotated by MIPS (26)).

They then mated thousands of test strains on solid agar plates and monitored growth in media containing methotrexate. The plates were photographed and image recognition was used to obtain a pixel intensity score corresponding to the size of each colony. An appropriate cutoff score for a metric that incorporated pixel intensity relative to both the mean and standard deviation for intensities on the plate was determined based on analysis of intensity values for those interactions also found in the MIPS database.

2.1.5. Summary

Virtually, all of the studies described above utilized methods to gauge both the experimental accuracy of the technique through repeated experimentation and the potential systematic biases through comparisons with outside data sources. Ultimately, there is a trade-off when addressing these two sources of error. While incorporating curated information into the filtering criteria may help to correct any method-specific error, it ultimately makes the assumption that curated data are of consistently high quality, and thus some valid, novel interactions may be discarded. This again underscores the importance of selecting appropriate external references for comparison and evaluation. Next, we review filtering techniques applied in high-throughput affinity purification techniques, highlighting both the similarities and the differences in the approaches to analyze data generated through disparate experimental means.

2.2. Complex-Based Detection Methods

In typical large-scale affinity purification studies, a universal affinity capture or epitope tag (27) is systemically expressed as a fusion to the query open reading frames of interest which allows for routine protein bait capture along with any stably associated interactors via one or more well-defined and often commercially available tag-specific capture reagents (e.g., antibody). Proteins bound to the bait protein can then be identified using mass spectrometry, which offers both high accuracy and sensitivity and reasonable throughput. However, the unfortunate drawback of the high sensitivity of mass spectrometry is that it is not uncommon for retrieved interactions to be riddled with nonspecific contaminants, even with the addition of multiple purification steps as in Tandem-Affinity-Purification (TAP) (28). For this reason, a postexperimental filtering technique is generally applied to identify and remove these nonspecific frequent flyers (see Fig. 1).

In an early large-scale affinity purification protein interaction screen in yeast, Ho et al. (29) empirically filtered copurifying proteins if they were frequently retrieved by other baits, or if they were a known constituent of high abundance cellular complexes, such as the ribosome. With later larger-scale screening, these empirical filtering techniques were not thought to be sufficient to remove lower abundance yet nonspecific interactions from the protein interaction dataset (30, 31); thus, more advanced filtering techniques were applied. Perhaps the most pertinent example of the importance of filtering technique on the outcome of a PPI dataset is that of two large-scale Affinity-Purification Mass Spectrometry (APMS) interaction studies published in 2006. Two independent publications simultaneously reported genome-wide surveys of soluble protein complexes in *S. cerevisiae* using the TAP procedure (30, 31). Yet, despite having substantial overlap in the total number of proteins detected (1,304 in common out of the 1,993 reported in Gavin et al. (31) study and the 2,388 found in Krogan et al. (30)), cross-comparisons initially revealed a surprisingly modest overlap in the respective reported high-quality interactions (<25%). A predominant cause for the disparity was later inferred to be differences in the computational scoring procedures (32) (also see subsequent analysis by Collins et al., which combines datasets using a unified filtering procedure (33)).

In one of the two studies, Gavin et al. filtered their interaction dataset based effectively on a metric of interaction detection fidelity and reproducibility. More specifically, the authors applied a technique called Socio-Affinity Indexing (SAI) that indicates the likelihood that two copurified proteins interact given their respective frequencies of occurrence and co-occurrence in the dataset (based on the work of Bader and Hogue (34)). SAI incorporates three factors: the likelihood of the bait retrieving a prey, the expectation of the prey reciprocally retrieving the same bait protein when itself is used as a bait, and the likelihood that both bait and prey would copurify with a third protein. A cutoff SAI score, defined by optimization of reproducibility, was used to filter the dataset based on reproducibility of interactions (evaluation against a gold standard was also used in an iterative procedure to derive protein clusters). This was a substantial departure from a previous (smaller-scale) interactome survey published by this same group in which experimental reproducibility was used to apply a general accuracy level to the interactome dataset (35).

Alternately, Krogan et al. incorporated the use of a gold standard into their filtering procedure instead of relying on interaction reproducibility alone. The Krogan study applied two independent mass spectrometry techniques to identify preys consistently copurifying with the retrieved bait proteins (MALDI-TOF and LC-MS/MS; each of which yields experimental confidence scores to tentative protein identifications). However, the authors noted that when relying on either the intersect of these two techniques alone, or even when including interactions validated multiple times using either method alone, the coverage of the resulting set was limited. Analogous to what had been observed previously in Y2H screening (see above), this lack of reproducibility may have been more due to sheer size of the interactome and subsequent sampling inefficiencies than due to false reporting of spurious interactions. Rather, the authors took the union of the two interaction datasets and then turned to other computational techniques to filter the combined network. In particular, they used the two experimental confidence scores and other properties such as reproducibility as features in a machine-learning procedure in an effort to capture a maximally comprehensive, but still accurate set of interactions. As with any supervised learning procedure, this process depended on an existing "gold-standard" set of PPI to inform its learners; thus, those complexes contained within the MIPS database (26) were taken as true. The result of this process was a filtered interaction dataset that was of high quality but with greater coverage than the intersect of the LCMS and MALDI-TOF detected interactions, and recapitulated the properties of the complexes within the MIPS database.

As has been shown in examples using both high-throughput Y2H and affinity profiling screening, interaction reproducibility ultimately is a strong indication of a true positive, but can be an overly stringent criterion for filtering an interaction dataset due to inherent assay variability and undersampling. Additionally, while learning procedures or comparisons with other published datasets have the advantage that they can both assess accuracy and address any systematic biases, very few existing datasets qualify as a bona fide gold standard for universal benchmarking. Therefore, even when comparative datasets are well selected, the possibility of trading one systematic bias for another is inevitable. Ultimately, the choice of filtering procedure can have some subjectivity and is influenced by the goal of the research project (a completely de novo assessment of interactions may rely solely on reproducibility, while a step forward in protein complex detection may filter based on known and validated complexes). As a result, the impact of filtering on the resulting data should not be underestimated and should be considered in subsequent network analysis. This again highlights the usefulness of approaches such as that taken by Yu et al. (9), in which several reference sets were evaluated using multiple techniques before selection of a gold standard.

3. Methods for PPI Filtering

Generating an extensive set of PPI through experimental methods ultimately leaves the researcher with some difficult decisions in terms of postanalysis. Here, we describe an approach we found helpful when managing newly generated PPI datasets so as to minimize the false discovery rate while maintaining maximal coverage. As such analyses can vary depending on the overall goal of a particular study and the cost of validation, the procedure described below is not intended to be universally applicable. Instead, we hope to provide a basic framework for processing data and indicate what we feel are priorities in the accurate reporting of a reliable PPI network.

3.1. Reciprocal Screening

The concept of reciprocal confirmation (confirming a bait-prey interaction by using the prey in subsequent bait experimentation) has become a hallmark of large-scale Y2H and TAP screens. As the majority of methods discussed in Subheading 2 depend on utilization of genetic constructs, reciprocal confirmations performed to determine that observed interactions are not merely the result of improper construct formation are a necessity. The concept of assay stochasticity is important here, as sampling rate will influence the number of interactions expected to be confirmed by reciprocal assay (to evaluate this, a subset of baits with known preys should be screened multiple times). Finding that you deviate from your expectation of reciprocal confirmations (using a Poisson test, Fisher's exact test, or a Chi-squared test, see Subheading 4) could indicate a technical problem with the bait, and additional experimental confirmations may be required before reporting retrieved preys. The proportion of reciprocally observed interactions can also serve as an indicator of assay specificity (more below).

3.2. Determination of Assay Sensitivity and Specificity

Perhaps the most intimidating aspect in filtering interaction data is the estimate of false positive and true positive rates, specifically because this typically involves comparison against gold-standard sets, which can be difficult to define. The importance of selecting these sets appropriately has been dealt with at length in the literature (see Yu et al. (9) and Jansen and Gerstein (36)) and will not be revisited here, except to state that an ideal gold-standard positive set would be one confirmed by multiple data sources and obtained using methodology similar to that being applied. Obviously, this is not something that is generally available; however, regardless of the quality of available data, any PPI study should compare its results to previous work to attempt to determine accuracy of the assay. Curated reference PPI can be obtained from one of many public repositories (see Table 1), or from heterogeneous data sources and integrated (see Franzosa et al. (37) for methods in combining PPI data sources).

True positive rate (assay sensitivity) is defined as the number of true positives divided by the sum of the number of true positives and false negatives. An estimation of the number of true positives and false negatives can be obtained through comparison against the gold-standard positive set. Alternately, the true negative rate (or assay specificity) is much more difficult to determine as it typically involves comparison against a gold-standard negative set and penalizing an assays results for detecting novel interactions that are presumed to be incorrect. As previously mentioned, a preferable way to gauge assay specificity may be through evaluation of the results of repeated screens (i.e., proportion of interactors detected multiple times, or via different assays, or the proportion of reciprocally retrieved interactors). These sensitivity and specificity values are useful for dataset benchmarking.

Table 1
List of publically available protein interaction data sources

Database name	Description	Statistics	Web address
BPID	*E. coli* interaction database containing both high and low quality experimental and predicted interactions	3,888 (high quality only) protein interactions from *E.coli*	http://www.bacteriome.org/
BioGRID	Compiled protein and genetic interactions. Literature curated, multiple species available	355,947 protein and genetic interactions among 17 species	http://thebiogrid.org/
DIP	Protein interactions both automatically and manually curated	71,274 protein interactions among 274 species	http://dip.doe-mbi.ucla.edu/dip/Main.cgi
HCPIN	Protein interactions of known cancer genes (7 major pathways)	9,784 interactions among 2,977 human proteins	http://nesg.org:9090/HCPIN/index.jsp
HPRD	Collection of manually curated human protein–protein interactions	39,194 protein interactions among 30,047 human proteins	http://www.hprd.org
IntAct	Collection of protein interaction data either from literature evidence or direct submission	228,262 protein interactions among dozens of species	http://www.ebi.ac.uk/intact/main.xhtml
MINT	Collection of experimentally verified protein interactions mined by curators	89,469 protein interactions among 30 species	http://mint.bio.uniroma2.it/mint/
Mpact	Curated yeast protein interaction data	4,300 protein interactions among 1,500 yeast proteins	http://mips.helmholtz-muenchen.de/genre/proj/mpact/

A brief description of some of the most commonly used protein interaction databases along with their respective URLs. Data generally obtained through a combination of literature and manual curation, although some sources integrate structural information and also indicate whether interactions were observed using multiple lines of evidence

3.3. Benchmarking

Finally, an effort should be made to limit the reported PPI to only those interactions deemed to be of high quality. Some metric of interaction reliability (i.e., experimental confidence score, socio-affinity index, reproducibility score for interactions, or a confidence score resulting from machine learning) should be evaluated empirically to determine an appropriate cutoff score. This process should involve gradual incremental threshold values and evaluation of accuracy at each iteration in terms of sensitivity and specificity (or if unavailable, coverage of the gold-standard positive set). If the reference set was already used in a computational learning process, a score corresponding to a preselected precision value may be used (with precision evaluated using cross-validation). While there is no general rule for selecting an appropriate score cutoff, the ideal confidence value is one that gives a reasonable trade-off between coverage and sensitivity. Ultimately, where this trade-off is most beneficial depends upon the intuition of the researchers and the goal of the project.

4. Notes

1. Assessing prey enrichment
   Appearance of preys in repeat sampling runs can be compared with random expectation to determine if the experimental method has an inherent preference for retrieval of certain proteins (either nonspecific contaminants or proteins with a particular property, e.g., high abundance). For example, assume bait protein *b* has 10 putative interactors and is rescreened five times by Y2H, each time producing 3 correct hits. If one prey protein appears in 4 of the 5 screens, is there a preferential enrichment for its detection? One approach to determining this is to apply the Poisson distribution:

   $$P(x;\mu) = (e^{-\mu})(\mu^{x}) / x!$$

   where x represents our observed number of retrievals of the prey in question and μ the expectation. In this case if baits are being sampled randomly, then our expectation is that we will see any given individual bait b can be approximated by:

   $$\sum_{i=1}^{n} s_i / S,$$

   where n is the number of sampling runs (5 in this case), *si* the number of retrieved baits in this run, and S the number of possible baits (10 here). Applying these numbers, the null hypothesis suggests the prey should appear less than twice, which means that our observation of this prey four times has a probability of occurrence <0.05 and thus is being favorably retrieved. Determining which proteins have prevalence in your retrieved preys could help you understand biases in your resulting interaction dataset (i.e., are you preferentially retrieving high-abundance proteins or members of a particular functional category). In addition to analyzing repeated runs using the Poisson distribution, one could examine their entire retrieved set of preys for evidence of enrichment using a comparison of categorizations such as Fisher's exact test.

2. Applying secondary characteristics to filter interaction data
   In the absence of available reference PPI datasets for comparison of experimental findings, other experimental properties (i.e., reproducibility) and secondary functional characteristics can at least provide a high accuracy subset with which to train a learning procedure. In this context, patterns in secondary genomic characteristics such as gene coexpression, coconservation, or presence of interaction domains can be defined for the "true" learning set and used to assign confidence in all other interactions. In prokaryotic species, the use of features such as operon structure and phylogenetic footprinting can be highly useful for predicting functional associations (so-called genome context methods; see Hu et al. (38) and Skrabanek et al. (39)). Again, we reiterate our caution though that reliance upon secondary characteristics will cause a filtered interaction dataset that recapitulates the properties of the selected gold standard, so both the "true" set and the secondary characteristics should be chosen carefully.

3. Useful software
   If unfamiliar with machine learning, Weka (40) offers a wide array of learning techniques packaged in a user-friendly GUI (http://www.cs.waikato.ac.nz/ml/weka/). Also, a very useful package for calculating precision, recall, and Receiver Operating Characteristic (ROC) curves for learning procedures is the ROCR package (41) for the R statistical language. Both ROCR and R are freely available (ROCR from http://rocr.bioinf.mpi-sb.mpg.de/, and R from http://www.r-project.org/). Finally, by far the most valuable single tool in network visualization and interpretation is Cytoscape (42), also freely available for academic use (http://www.cytoscape.org/). Cytoscape allows upload of protein networks as well as integrated data download from publically available databases. In addition to being visualized, third-party plug-ins allow networks to undergo functional enrichment analysis (43), gene expression comparisons (44), and clustering (45, 46). See Yeung et al. (47) for a detailed description of the basic use of Cytoscape.

Acknowledgments

AE and ZZ acknowledge a Team Grant from the Canadian Institute of Health Research (CIHR MOP#82940).

References

1. Musso GA, Z Zhang, et al. (2007). Experimental and computational procedures for the assessment of protein complexes on a genome-wide scale. Chem Rev 1078: 3585–3600.
2. Sanderson CM (2009). The Cartographers toolbox: building bigger and better human protein interaction networks. Brief Funct Genomic Proteomic 81: 1–11.
3. Cagney G (2009). Interaction networks: lessons from large-scale studies in yeast. Proteomics 920: 4799–4811.
4. Fields S and O Song (1989). A novel genetic system to detect protein–protein interactions. Nature 3406230: 245–246.
5. Uetz P, L Giot, et al. (2000). A comprehensive analysis of protein–protein interactions in Saccharomyces cerevisiae. Nature 4036770: 623–627.
6. Ito T, T Chiba, et al. (2001). A comprehensive two-hybrid analysis to explore the yeast protein interactome. Proc Natl Acad Sci USA 988: 4569–4574.
7. von Mering C, R Krause, et al. (2002). Comparative assessment of large-scale data sets of protein–protein interactions. Nature 4176887: 399–403.
8. Ito T, K Ota, et al. (2002). Roles for the two-hybrid system in exploration of the yeast protein interactome. Mol Cell Proteomics 18: 561–566.
9. Yu H, P Braun, et al. (2008). High-quality binary protein interaction map of the yeast interactome network. Science 3225898: 104–110.
10. Vidalain PO, M Boxem, et al. (2004). Increasing specificity in high-throughput yeast two-hybrid experiments. Methods 324: 363–370.
11. Li S, CM Armstrong, et al. (2004). A map of the interactome network of the metazoan C. elegans. Science 3035657: 540–543.
12. Stanyon CA, G Liu, et al. (2004). A Drosophila protein-interaction map centered on cell-cycle regulators. Genome Biol 512: R96.
13. Formstecher E, S Aresta, et al. (2005). Protein interaction mapping: a Drosophila case study. Genome Res 153: 376–384.
14. Rual JF, K Venkatesan, et al. (2005). Towards a proteome-scale map of the human protein–protein interaction network. Nature 4377062: 1173–1178.
15. Stelzl U, U Worm, et al. (2005). A human protein–protein interaction network: a resource for annotating the proteome. Cell 1226: 957–968.
16. Lievens S, I Lemmens, et al. (2009). Mammalian two-hybrids come of age. Trends Biochem Sci 3411: 579–588.
17. Suter B, S Kittanakom, et al. (2008). Two-hybrid technologies in proteomics research. Curr Opin Biotechnol 194: 316–323.
18. Walhout AJ and M Vidal (1999). A genetic strategy to eliminate self-activator baits prior to high-throughput yeast two-hybrid screens. Genome Res 911: 1128–1134.
19. Helbig AO, AJ Heck, et al. (2010). Exploring the membrane proteome – challenges and analytical strategies. J Proteomics 735: 868–878.
20. Johnsson N and A Varshavsky (1994). Split ubiquitin as a sensor of protein interactions in vivo. Proc Natl Acad Sci USA 9122: 10340–10344.
21. Miller JP, RS Lo, et al. (2005). Large-scale identification of yeast integral membrane protein interactions. Proc Natl Acad Sci USA 10234: 12123–12128.
22. Harris MA, J Clark, et al. (2004). The Gene Ontology (GO) database and informatics resource. Nucleic Acids Res 32Database issue: D258–261.
23. Tarassov K, V Messier, et al. (2008). An in vivo map of the yeast protein interactome. Science 3205882: 1465–1470.
24. Remy I and SW Michnick (1999). Clonal selection and in vivo quantitation of protein interactions with protein-fragment complementation assays. Proc Natl Acad Sci USA 9610: 5394–5399.
25. Pelletier JN, KM Arndt, et al. (1999). An in vivo library-versus-library selection of optimized protein–protein interactions. Nat Biotechnol 177: 683–690.
26. Mewes HW, C Amid, et al. (2004). MIPS: analysis and annotation of proteins from whole genomes. Nucleic Acids Res 32Database issue: D41–44.

27. Fritze CE and TR Anderson (2000). Epitope tagging: general method for tracking recombinant proteins. Methods Enzymol 327: 3–16.
28. Puig O, F Caspary, et al. (2001). The tandem affinity purification (TAP) method: a general procedure of protein complex purification. Methods 243: 218–229.
29. Ho Y, A Gruhler, et al. (2002). Systematic identification of protein complexes in Saccharomyces cerevisiae by mass spectrometry. Nature 4156868: 180–183.
30. Krogan NJ, G Cagney, et al. (2006). Global landscape of protein complexes in the yeast Saccharomyces cerevisiae. Nature 4407084: 637–643.
31. Gavin AC, P Aloy, et al. (2006). Proteome survey reveals modularity of the yeast cell machinery. Nature 4407084: 631–636.
32. Goll J and P Uetz (2006). The elusive yeast interactome. Genome Biol 76: 223.
33. Collins SR, P Kemmeren, et al. (2007). Toward a comprehensive atlas of the physical interactome of Saccharomyces cerevisiae. Mol Cell Proteomics 63: 439–450.
34. Bader GD and CW Hogue (2002). Analyzing yeast protein–protein interaction data obtained from different sources. Nat Biotechnol 2010: 991–997.
35. Gavin AC, M Bosche, et al. (2002). Functional organization of the yeast proteome by systematic analysis of protein complexes. Nature 4156868: 141–147.
36. Jansen R and M Gerstein (2004). Analyzing protein function on a genomic scale: the importance of gold-standard positives and negatives for network prediction. Curr Opin Microbiol 75: 535–545.
37. Franzosa E, B Linghu, et al. (2009). Computational reconstruction of protein–protein interaction networks: algorithms and issues. Methods Mol Biol 541: 89–100.
38. Hu P, SC Janga, et al. (2009). Global functional atlas of Escherichia coli encompassing previously uncharacterized proteins. PLoS Biol 74: e96.
39. Skrabanek L, HK Saini, et al. (2008). Computational prediction of protein–protein interactions. Mol Biotechnol 381: 1–17.
40. Frank E, M Hall, et al. (2004). Data mining in bioinformatics using Weka. Bioinformatics 2015: 2479–2481.
41. Sing T, O Sander, et al. (2005). ROCR: visualizing classifier performance in R. Bioinformatics 2120: 3940–3941.
42. Shannon P, A Markiel, et al. (2003). Cytoscape: a software environment for integrated models of biomolecular interaction networks. Genome Res 1311: 2498–2504.
43. Maere S, K Heymans, et al. (2005). BiNGO: a Cytoscape plugin to assess overrepresentation of gene ontology categories in biological networks. Bioinformatics 2116: 3448–3449.
44. Cline MS, M Smoot, et al. (2007). Integration of biological networks and gene expression data using Cytoscape. Nat Protoc 210: 2366–2382.
45. Vlasblom J, S Wu, et al. (2006). GenePro: a Cytoscape plug-in for advanced visualization and analysis of interaction networks. Bioinformatics 2217: 2178–2179.
46. Rivera CG, R Vakil, et al. (2010). NeMo: Network Module identification in Cytoscape. BMC Bioinformatics 11 Suppl 1: S61.
47. Yeung N, MS Cline, et al. (2008). Exploring biological networks with Cytoscape software. Curr Protoc Bioinformatics Chapter 8: Unit 8 13.

Chapter 15

Classification of Cancer Patients Using Pathway Analysis and Network Clustering

David C.Y. Fung, Amy Lo, Lucy Jankova, Stephan J. Clarke, Mark Molloy, Graham R. Robertson, and Marc R. Wilkins

Abstract

Molecular expression patterns have often been used for patient classification in oncology in an effort to improve prognostic prediction and treatment compatibility. This effort is, however, hampered by the highly heterogeneous data often seen in the molecular analysis of cancer. The lack of overall similarity between expression profiles makes it difficult to partition data using conventional data mining tools. In this chapter, the authors introduce a bioinformatics protocol that uses REACTOME pathways and patient–protein network structure (also called topology) as the basis for patient classification.

Key words: iTRAQ-mass spectrometry data, Network clustering, Patient classification, REACTOME pathways

1. Introduction

Molecular expression patterns have often been used for patient classification in oncology in an effort to improve prognostic prediction and treatment compatibility. This approach is currently driving the field of molecular pathology (1–3). The common practice has been to measure the similarity between expression profiles within a cohort using conventional data mining methods, e.g. principal component analysis and *k*-means clustering. However, these methods work poorly on highly heterogeneous data, which are often seen in the molecular analysis of cancer. The alternative is to apply a two-step approach by first comparing the expression state of the various pathways between samples, and then build a patient–protein network for identifying proteins that cluster with a particular subset of patients. This approach is based on the complex

Gerard Cagney and Andrew Emili (eds.), *Network Biology: Methods and Applications*, Methods in Molecular Biology, vol. 781, DOI 10.1007/978-1-61779-276-2_15, © Springer Science+Business Media, LLC 2011

systems theory that self-organization and preferential attachment are properties of biological systems (4). As such, certain proteins may tend to express in one group of patients but not another. Alternatively, a particular protein may express only in a particular individual. When the patient–protein relation is being modelled as a node-link network, the resulting structure of the network (also known as the network topology) should reflect the differential expression of proteins. In this chapter, we introduce a bioinformatics protocol that realizes the benefits of our two-step approach using an iTRAQ-mass spectrometry cancer dataset cohort. In the first step, we overlay the dataset onto the REACTOME map using SkyPainter (5), which assists us in identifying pathways of interest. In the second step, we construct a patient–protein network for network clustering. The procedure for carrying out these two steps forms the two parts of our protocol in this chapter.

2. Materials

2.1. iTRAQ-Mass Spectrometry Cancer Data

The protein ratios of 16 paired samples (cancer versus normal tissue) were measured using iTRAQ-mass spectrometry. The samples were surgically removed from patients in various disease stages. A set of 117 out of over 1,400 proteins has been selected for further analysis. All of which showed >1.5 fold change in expression ratios between paired cancer and the normal tissue. This data was provided in an Excel spreadsheet with the first three columns labelled Accession, Symbol and Description respectively.

2.2. Pathway Modelling Software

SkyPainter is an online software provided with the REACTOME map. It provides an easy-to-use interface for data input (Fig. 1). The user can import an expression dataset as a tab-delimited file with the list of gene or protein identifiers as the first column followed by succeeding columns of numerical values. Identifiers recognized by the software are accession numbers from UniProt and Entrez Gene, and probe identifiers of Affymetrix and Agilent probesets. SkyPainter automatically assigns a colour gradient to the numerical range of the input dataset. A time series or patient cohort can be displayed as an animation by providing multiple values (on the same line, separated by a single space or tab) for each identifier, thereby allowing the user to visually detect the varying expression profiles between time points or patients.

2.3. Network Modelling Software

We used the NetMap Analytics™ software suite (Netmap Analytics Proprietary Limited, Sydney, Australia) for network modelling and analysis. The suite contains three pieces of software: NetMap DecisionDirector™ for network design, NetMap StudyManager™ for designing the analytical plan, and NetMap Visualizer™ for network visualization and visual discovery.

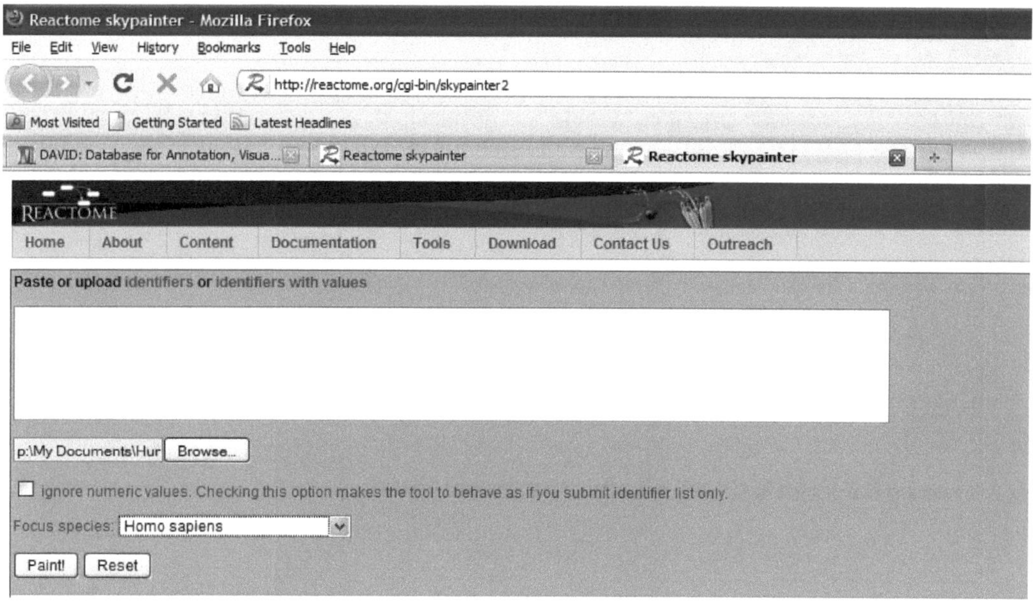

Fig. 1. The user interface of SkyPainter.

3. Methods

In the first part of our protocol, the REACTOME pathway map is used as the basis of pathway analysis. The map is a directed network in which the nodes represent proteins and the link between two nodes represents the direction of a cellular or metabolic reaction. Since REACTOME.org contains a manually curated compendium of human pathways, the overlaying of iTRAQ-mass spectrometry data can provide a global but low-resolution view on the expression state of a cellular network. This should inform us on the interpersonal variation of the different pathways in terms of their expression ratios.

3.1. REACTOME Pathway Analysis

1. Save the Excel spreadsheet as a tab-delimited file containing the UniProt Accession as the first column and the iTRAQ expression ratios as columns 2–17 (see Note 1). The first row contains the column titles preceded by a '#'. The output file should have the following format:

#Accession	PL002	PL003	PL001	PL006	PL007	PL008	PL004	PL005	PL009	PL010
P59666	16.08	10.67	11.98	1.48	2.79	2.59	9.69	2.89	…	…

While there is no limit to the number of data columns in the input, the data size limit for SkyPainter is 2,500 data points.

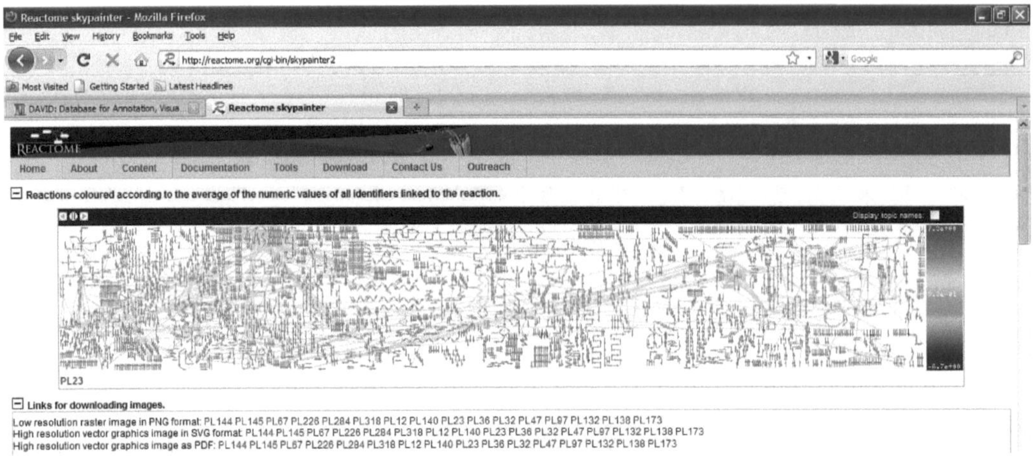

Fig. 2. Screenshot of a painted REACTOME map after expression data overlaid.

2. Enter the tab delimited file into the textbox and select **Homo sapiens** in the **Focus species** dropdown menu (Fig. 1). Click the **Paint** button to start the overlay. This will return an animated series of painted REACTOME maps. The colour bar on the right-hand side shows the colour gradient assigned to the data range (Fig. 2). If a UniProt accession number fails to map to a particular gene node or in the case of missing data, that node will be painted grey.
3. To identify which pathways the painted nodes belong to on the REACTOME map, check the box **Display topic name** on the top right-hand corner.
4. On identifying the pathway(s) a painted node set belongs to, enter the pathway name(s) into a five-column table as shown in Table 1 (see Note 2). The first two columns tabulate the disease stages of cancer and the patient identifiers respectively. The last three columns tabulate the pathways that contain genes with either no change, up-regulated, or down-regulated gene expression in cancer in reference to normal tissue.

3.2. Network Modeling and Analysis

Although the REACTOME pathway analysis can give us a preliminary concept on which pathways show variation in expression ratios in the patient cohort, it does not help us pinpoint genes that can classify patients according to their range of expression ratios. Hence, we need to construct a bipartite network that models the association between patients and proteins. The resulting structure of the network should inform us as to which proteins associate more with which patients at a certain range of expression ratios. In the second part of our protocol, we construct and analyze such a network using the NetMap Analytics™ software suite.

Table 1
Pathway expression profiles of four sample cancer patients in different disease stages

Stage	Patient ID	No change	Up-regulated	Down-regulated
A	PL002	• Integrin cell surface interactions • TCA cycle • ATP synthesis • G-protein complex • mRNA transport • Signaling in immune system • Metabolism of lipids and lipoproteins (partial) • Synaptic transmission • Haemostasis	• mRNA processing • Synthesis of bile acids and bile salts • Muscle contraction • Metabolism of carbohydrate	• Apoptosis • Glycolysis • Nucleocytoplasmic transport • Metabolism of lipids and lipoproteins (partial)
B	PL008	• mRNA processing • Synaptic transmission	• Integrin cell surface interactions • Synthesis of bile acids and bile salts • ATP synthesis • G-protein complex • Signaling in immune system • Metabolism of lipids and lipoproteins (partial) • Metabolism of carbohydrate • Haemostasis	• TCA cycle • Apoptosis • Metabolism of lipids and lipoproteins (partial) • Nucleocytoplasmic transport • Muscle contraction
C	PL009	• Integrin cell surface interactions • Synthesis of bile acids and bile salts • G-protein complex • Signaling in immune system • Haemostasis	• mRNA processing • Apoptosis • Metabolism of carbohydrate	• TCA cycle • ATP synthesis • Metabolism of lipids and lipoproteins • Nucleocytoplasmic transport • Muscle contraction • Synaptic transmission
C	PL012	• Integrin cell surface interactions • G-protein complex • Signaling in immune system • Synaptic transmission • Haemostasis	• mRNA processing • Synthesis of bile acids and bile salts • Metabolism of carbohydrate	• TCA cycle • ATP synthesis • Apoptosis • Metabolism of lipids and lipoproteins • Muscle contraction

(continued)

Table 1 (continued)

Stage	Patient ID	No change	Up-regulated	Down-regulated
D	PL016	• ATP synthesis	• Integrin cell surface interactions • Synthesis of bile acids and bile salts • G-protein complex • Apoptosis (partial) • Signaling in immune system • Metabolism of lipids and lipoproteins (partial) • Metabolism of carbohydrate • Haemostasis	• TCA cycle • Apoptosis (partial) • Metabolism of lipids and lipoproteins (partial) • Muscle contraction • Synaptic transmission

3.2.1. Data Discretization

1. Discretize iTRAQ expression data for each protein into seven bins, i.e. −3, −2, −1, 0, +1, +2, +3. The discretization is based on the deviation of an iTRAQ value from the sample mean by the number of standard deviation. For example, if the value is within three standard deviations above the mean, it will be assigned to bin +3. If the value deviates by the same amount but below the mean, it will be assigned to bin −3. The sample mean is computed as the average expression value of all the samples, in this case 16 patient samples, for a particular protein. In general, the bins with the positive sign indicate up-regulated expression while those with the negative sign indicate the opposite.
2. Flatten the Excel file to a three-column spreadsheet which contains the columns patient_id, gene_symbol, and bin_no. This can be done by copying and pasting columns from the original spreadsheet to a new one. Save the new spreadsheet as **cancer_data.xls**.

3.2.2. Develop a Database for Dataset Storage

After collecting the necessary datasets, a relational database has to be developed before using NetMap DecisionDirector™ for network construction. The purpose is to store the datasets in a relational database management system (DBMS) such that they can be extracted by running Structured Query Language (SQL) statements through NetMap DecisionDirector™. Because of its ease of use and availability, the database will be built with the use of Microsoft Access®. The SQL statements used in this protocol can be modified for use with other open-source, e.g. MySQL®, PostgreSQL®, and commercial DBMSes, e.g. Oracle®.

1. Open the Microsoft Access® DBMS and create a new database named **iTRAQ_expression**.
2. Create a new table called **cancer_expression** with the named columns **patient_id**, **protein_symbol** and **bin_no** by importing the Excel file **cancer_data.xls** (see Note 3).
3. Create a new table called **patient_list** with the named columns **patient_id** and **disease_stage**. This table will store a unique list of patient_ids that maps to different disease stages. Manually enter the patient_id and the appropriate disease stage as in Table 2.
4. Copy and paste the first three columns **Accession**, **Symbol**, and **Description** from the original Excel file to a new spreadsheet.
5. Change the column names in the new spreadsheet to **uniprot_id**, **protein_symbol**, and **protein_name**. Save the spreadsheet as **proteinlist.xls**.
6. Create a new table called **protein_list** with the named columns **uniprot_id**, **protein_symbol**, and **protein_name** by importing the Excel file **proteinlist.xls**.

Table 2
Data input for the PATIENT table

Patient_id	Disease_stage
PL001	A
PL002	A
PL003	A
PL004	B
PL005	B
PL006	B
PL007	B
PL008	B
PL009	C
PL010	C
PL011	C
PL012	C
PL013	C
PL014	D
PL015	D
PL016	D

3.3. Network Construction Using DecisionDirector™

With the database built, the next procedure is to construct the patient–gene network using DecisionDirector™. The procedure can be divided into four stages.

(a) Define Pass-Thru queries for mapping database tables to node and link attributes.

(b) Define the node and link types.

(c) Map the table columns to the appropriate node and link attributes.

(d) Build the network drawing application (called NetMap study).

3.3.1. Define Pass-Thru Queries

1. Add the **iTRAQ_expression** database to the ODBC data-source list. Open the **Control Panel** in the Windows Start Menu. Double click the **Administrative Tools** icon, and then double click the **Data Sources (ODBC)** icon.
2. In the **ODBC Data Source Administrator** window (Fig. 3), click the **Add** button.
3. In the **Create New Data Source** window, select the **Microsoft Access Driver (*.mdb, *.accdb)**.
4. In the **ODBC Microsoft Access Setup** window, enter **iTRAQ_Cancer** as the **Data Source Name**. Enter **iTRAQ database for Cancer** as the **Description** (Fig. 4b).
5. Click the Select button in the **Select Database** window and select the **iTRAQ_expression.mdb** database in the local directory, e.g. d:\Human_Molecular_System\Human_Disease_Network\TUTORIAL directory (Fig. 4c).

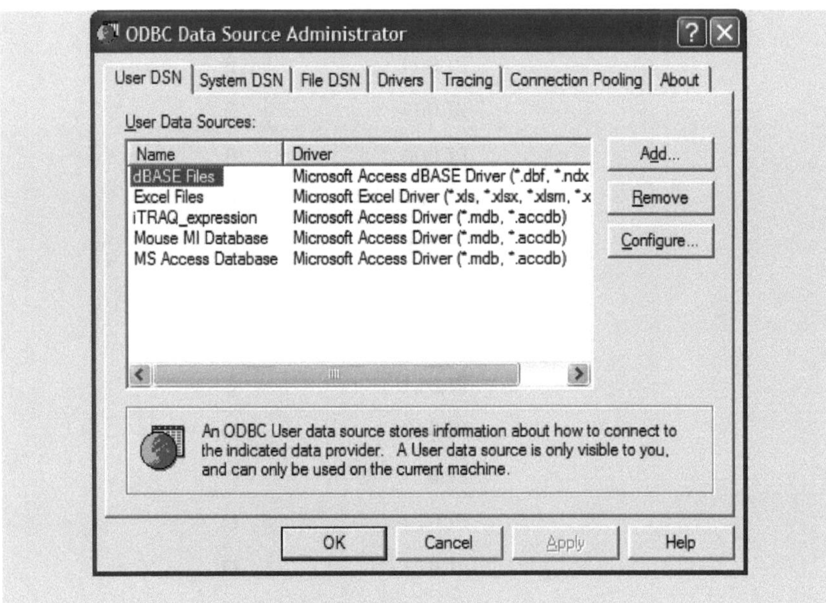

Fig. 3. Screenshot of the ODBC Data Source Administrator window.

6. Open the DecisionDirector™ in the Windows Start Menu. Select **File > New**. Create a datastore file named **human_cancer** under the TUTORIAL directory. The **Inspector** box should now appear on screen (Fig. 5).

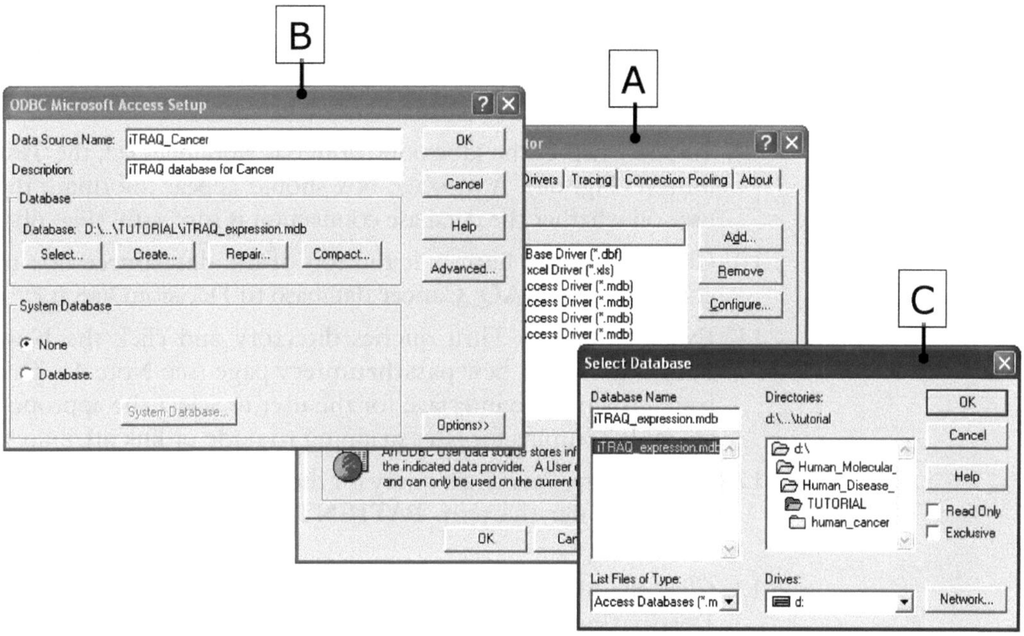

Fig. 4. Screenshot sequence for setting up the Microsoft Access® database iTRAQ_expression.mdb as the ODBC datasource. (**a**) The Create New Data Source window. (**b**) The ODBC Microsoft Access Setup window. (**c**) The Select Database window.

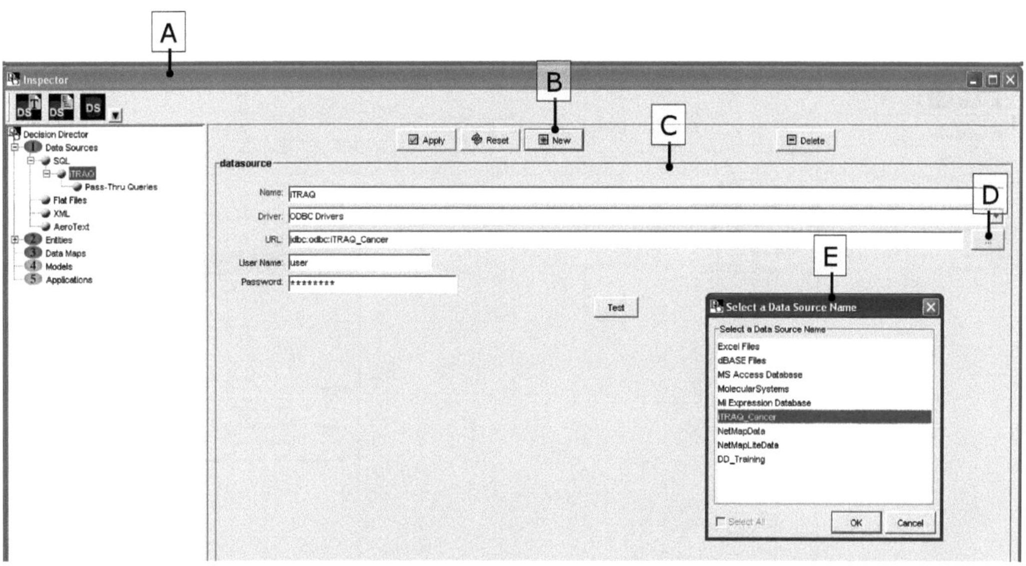

Fig. 5. Screenshot sequence for connecting the DecisionDirector™ to the iTRAQ_expression datasource via the jdbc:odbc bridge. (**a**) The Inspector box. (**b**) The New button for opening the (**c**) new datasource page. (**d**) The **|** … **|** button for opening the (**e**) Select a Data Source Name window.

7. Expand the Data Sources directory tree on the left navigation panel. Click on the SQL item. Then, click on the **New** button to open a new **datasource** page (Fig. 5b).
8. Enter **iTRAQ** as the Name. Click on the | … | button on the right of the URL textbox (Fig. 5d). Select the **iTRAQ_Cancer** item in the **Select a Data Source Name** window (Fig. 5e). Click the **OK** button to select. The URL textbox should automatically be filled with the jdbc:odbc address of the **iTRAQ_Cancer** database.
9. To test whether the jdbc:odbc bridge is working, click the **Test** button (Fig. 6a). A message box should appear informing the user on whether the database connection is successful (Fig. 6b).
10. Click the **Apply** button at the top of the **datasource** box to connect the **iTRAQ_Cancer** database to DecisionDirector™.
11. Expand the Pass-Thru queries directory and click the **New** button to open a new **passthruquery** page (see Note 4). This page provides the interface for the user to select the appropriate table columns for later mapping to node or link attributes.
12. To define the Pass-Thru query for the node named **PATIENT_NODE**, enter the term **PATIENT_NODE** into the **Name:** textbox.
13. Click the **More** button on the right-hand column next to the **Destination** window frame (Fig. 7b). In the **SQL statement** textbox, enter the following:

    SELECT PATIENT_LIST.PATIENT_ID, PATIENT_LIST.DISEASE_STAGE FROM PATIENT_LIST

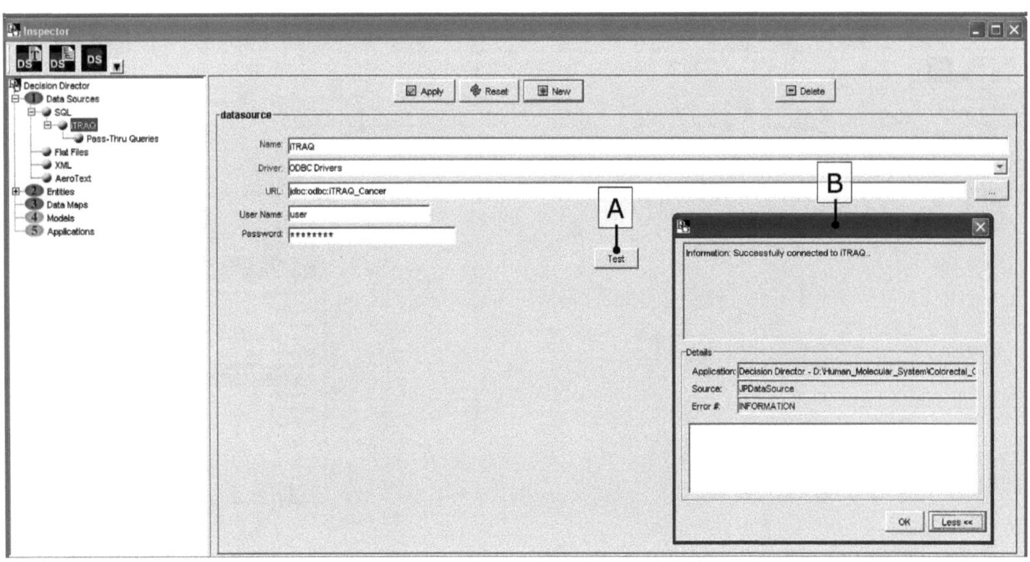

Fig. 6. Screenshot sequence for testing the jdbc:odbc bridge. (**a**) The Test button for opening the (**b**) message box.

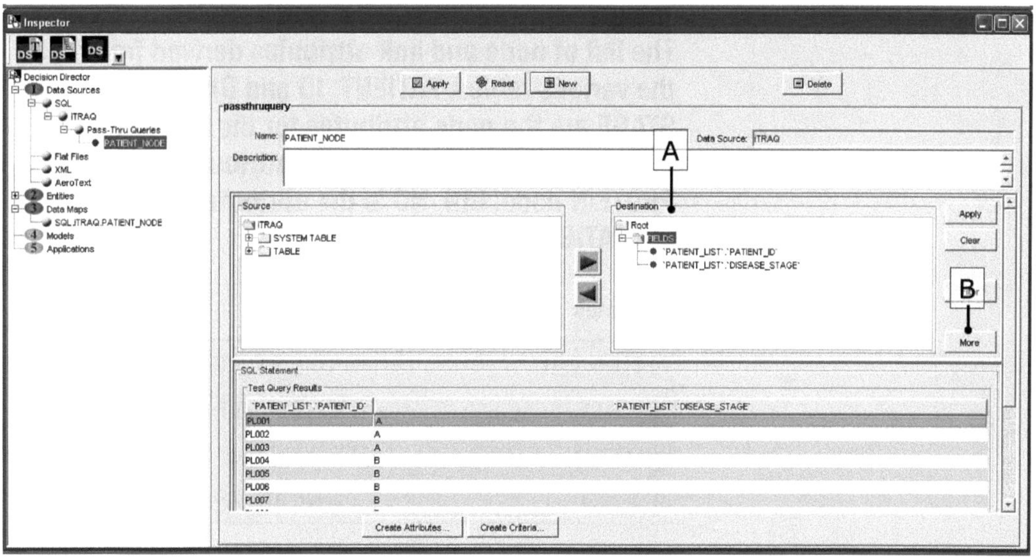

Fig. 7. Screenshot of the passthruquery page. (**a**) The Destination window frame. (**b**) The More button for opening the SQL statement box.

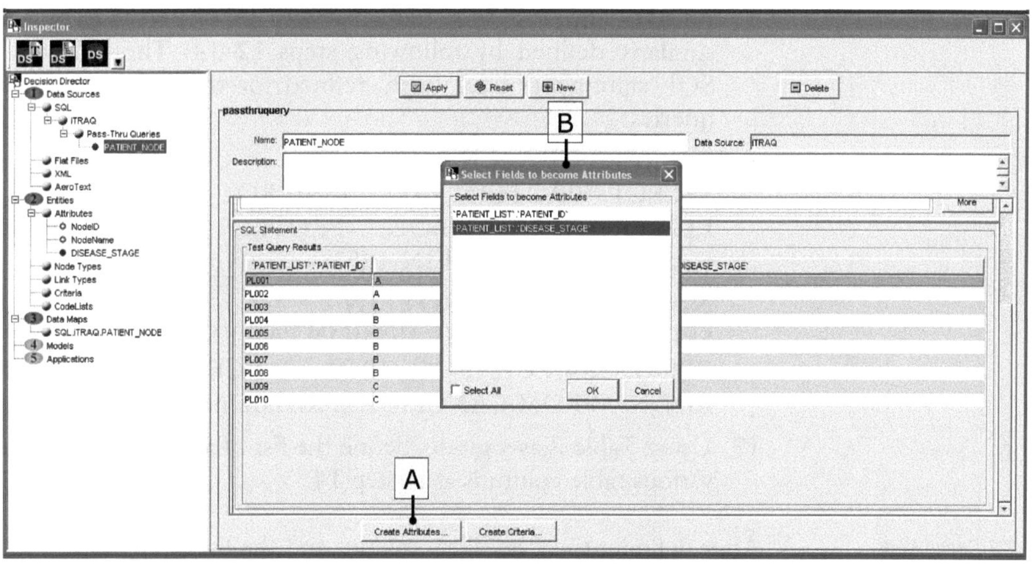

Fig. 8. Screenshot sequence for creating new attributes. (**a**) The Create Attributes... button for opening the (**b**) Select Fields to become Attributes window.

14. To define **DISEASE_STAGE** as one of the attributes for use in node and link mapping, click the **Create Attributes...** button at the bottom of the passthruquery page (Fig. 8a).
15. Select **'PATIENT_LIST'.'DISEASE_STAGE'** in the **Select Fields to become Attributes** window. Click the **OK** button to add the **DISEASE_STAGE** attribute to the **Attribute**

Table 3
The list of node and link attributes derived from the various tables. PATIENT_ID and DISEASE_STAGE are the node attributes for the PATIENT node. PROTEIN_SYMBOL is the attribute for the PROTEIN node. BIN_NO is the attribute for the PATIENT2PROTEIN link

Attributes	Tables
PATIENT_ID	PATIENT_LIST
DISEASE_STAGE	PATIENT_LIST
PROTEIN_SYMBOL	PROTEIN_LIST
BIN_NO	CANCER_EXPRESSION

subdirectory under the **Entities** directory in the left-hand navigation panel (Fig. 8).

16. The Pass-Thru queries for the node named **PROTEIN_NODE** and the link named **PATIENT2PROTEIN** are similarly defined by following steps 12–14. The following SQL statements have been defined for the two Pass-Thru queries:

    Name: PROTEIN NODE
    SELECT PROTEIN_LIST. UNIPROT_ID, PROTEIN_LIST.PROTEIN_SYMBOL, PROTEIN_LIST.PROTEIN_NAME FROM PROTEIN_LIST

    Name: PATIENT2PROTEIN
    SELECTCANCER_EXPRESSION.PATIENT_ID,CANCER_EXPRESSION.GENE_ SYMBOL, CANCER_EXPRESSION.BIN_NO FROM CANCER_EXPRESSION

17. Using Table 3 as a guide, define the list of attributes from the various table columns as in step 14.

3.3.2. Define Node and Link Types

After defining the Pass-Thru queries and the list of attributes, the node and link types to which the attributes will be assigned need to be defined. The Pass-Thru queries defined previously inform the user that there are two node types in the network, i.e. **PATIENT** and **PROTEIN**, and one link type, i.e. **PATIENT2PROTEIN** required for building the patient–protein network.

1. To define the node type **PATIENT** which attributes are **NodeID**, **NodeName**, **DISEASE_STAGE**, click on the **Node Types** subdirectory under the **Entities** directory.
2. Click the **New** button to open a new **nodetype** page. Enter the term **PATIENT** into the **Name:** textbox (Fig. 9a–c).

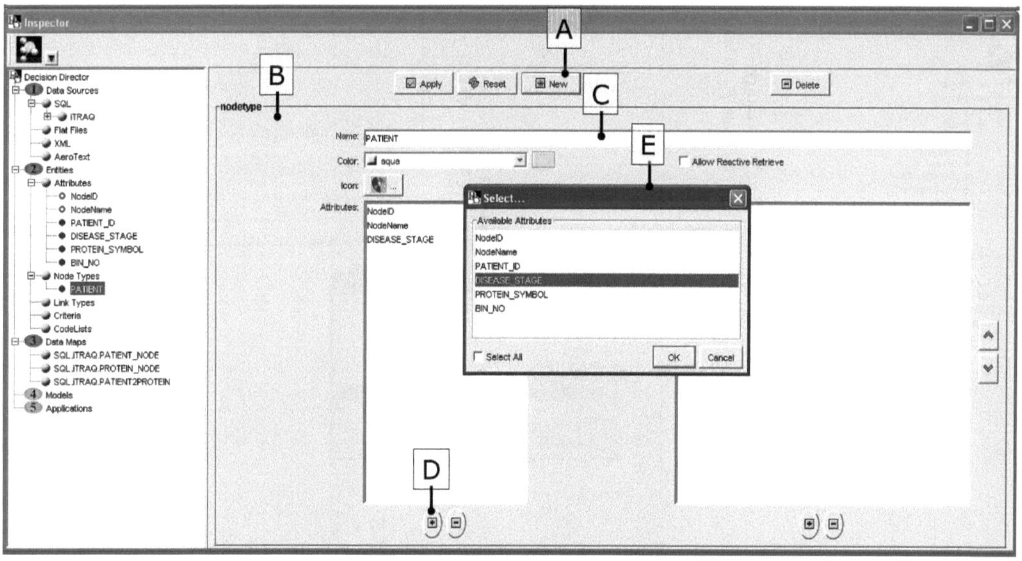

Fig. 9. Screenshot sequence for defining the PATIENT node type. (**a**) The New button for opening (**b**) the new nodetype page. (**c**) The Name: textbox. (**d**) The oval-shaped button for opening the (**e**) the Select… window.

3. Click the + signed oval-shaped button at the bottom of the **Attributes:** textbox and select **DISEASE_STAGE** in the **Select…** window. Click the **OK** button to add the attribute to the **Attribute** textbox (Fig. 9d, e).
4. Similarly, add the attributes **NodeID**, **NodeName**, **DISEASE_STAGE**, **Total Link Count**, and **Node Type** to the **Description** textbox.
5. Define the node type **PROTEIN**, for which attributes are **NodeID** and **NodeName**, by repeating steps 2 and 4. When repeating step 4, add the attributes **NodeID**, **NodeName**, **Total Link Count**, and **Node Type** to the **Description** textbox while leaving the list of attributes in the **Attribute** textbox unchanged.
6. To define the link type **PATIENT2PROTEIN**, for which the attribute is **BIN_NO**, click on the **Link Types** subdirectory under the **Entities** directory.
7. Click the **New** button to open a new **linktype** page. Enter the term **PATIENT2PROTEIN** into the **Name:** textbox (Fig. 10a–c).
8. Select **PATIENT** in the **Source:** dropdown menu and **PROTEIN** in the **Destination:** dropdown menu (Fig. 10d, e).
9. Click the + signed oval-shaped button at the bottom of the **Attributes:** textbox and select **BIN_NO** in the **Select…** window. Click the **OK** button to add the attribute to the **Attribute** textbox (Fig. 10f, g).

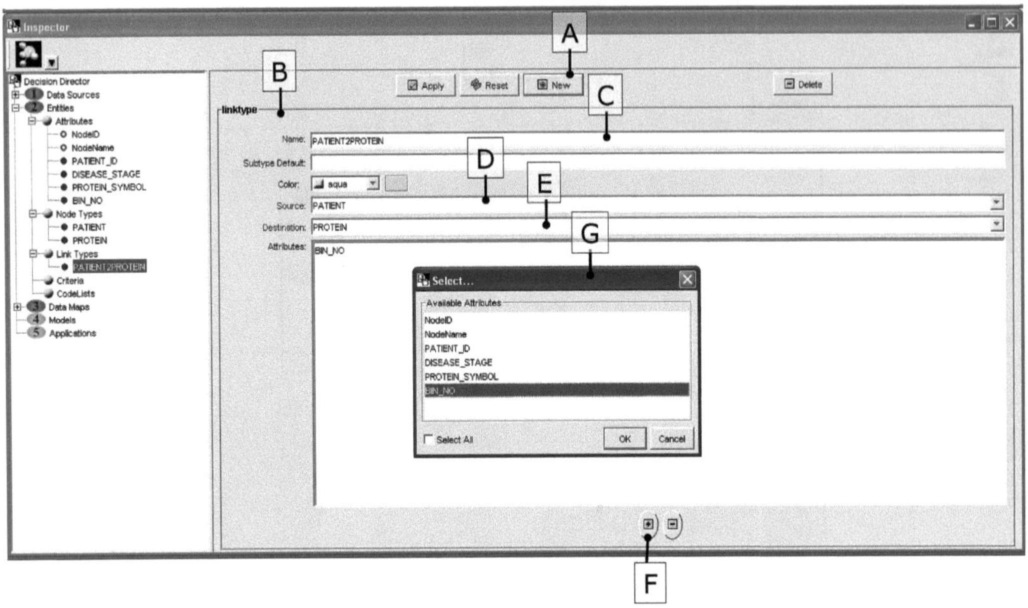

Fig. 10. Screenshot sequence for defining the PATIENT2PROTEIN link type. (**a**) The New button for opening (**b**) the new linktype page. (**c**) The Name: textbox. (**d**) The Source: dropdown menu. (**e**) The Destination: dropdown menu. (**f**) The oval-shaped button for opening the (**g**) the Select... window.

3.3.3. Data Mapping

With the node and link types defined, the list of attributes in every node and link type is ready to be mapped to the table columns in the **iTRAQ_expression** database. This mapping is defined in a data map and every node type or link type has its own data map.

1. To define the data map for the node type **PATIENT**, click on the **SQL.iTRAQ.PATIENT_NODE** subdirectory under the **Data Maps** directory.
2. Click the **New** button to open a new **datamap** page. Enter the term **PATIENT_nodemap** into the **Name:** textbox (Fig. 11a, b).
3. Select **PATIENT** as **Node Types**, **PATIENT_LIST.PATIENT_ID** as **NodeID**, and **PATIENT_LIST.DISEASE_STAGE** as **DISEASE_STAGE** (Fig. 11c–e).
4. Similarly, define the data map for the node type **PROTEIN** as follows:

   Name: PROTEIN_nodemap

   Node Types: PROTEIN

   NodeID: PROTEINLIST.PROTEIN_SYMBOL

   NodeName: PROTEINLIST.GENE_NAME

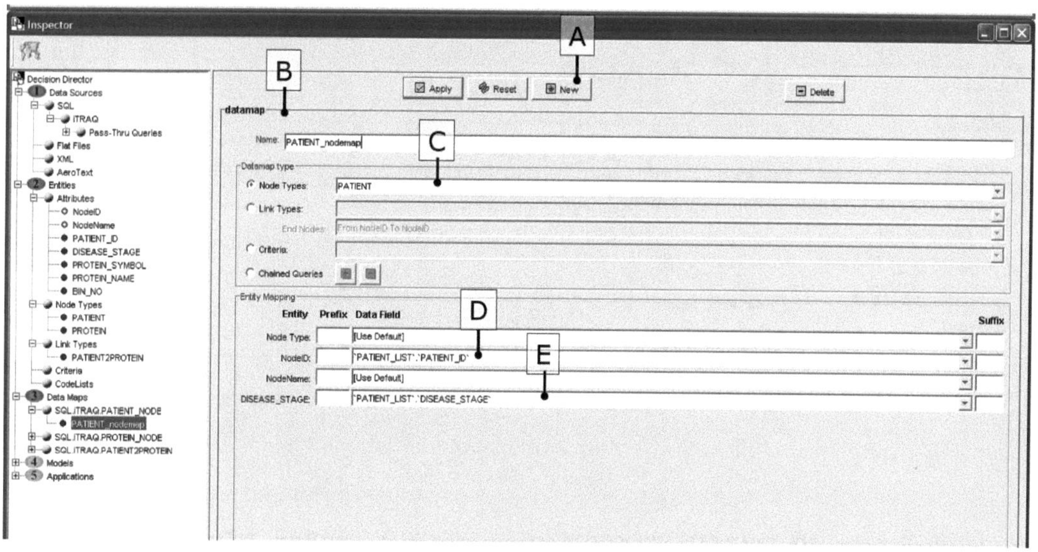

Fig. 11. Screenshot of the SQL.iTRAQ.PATIENT_NODE datamap page. (**a**) The New button for opening the new (**b**) datamap page. (**c**) The Node Types, (**d**) NodeID, and (**e**) DISEASE_STAGE dropdown menus.

5. Defining the data map for the link type **PATIENT2PROTEIN** is quite similar to the previous steps done for the node types. The inputs for the **datamap** page are as follows (Fig. 12):

   Name: PATIENT2PROTEIN_edgemap

   Link Types: PATIENT2PROTEIN

   End Nodes: From NodeID to NodeID

   From NodeID: CANCER_EXPRESSION.PATIENT_ID

   To NodeID: CANCER_EXPRESSION.PROTEIN_SYMBOL

   BIN_NO: CANCER_EXPRESSION.BIN_NO

3.3.4. Build the Network

With the data maps defined, the final stage is to construct the network model itself.

1. To create a new network model, click on the **Model** directory. Click the **New** button to open a new **model** page (Fig. 13).
2. Enter **CANCER_iTRAQ** into the **Name:** textbox. In the **Available Component** box, expand the **SQL Sources** directory and then the subdirectory **iTRAQ**. Under which, all the node and link types are being displayed (Fig. 13b). To specify the node and link types in the network model, check the checkboxes next to **PATIENT_NODE**, **PROTEIN_NODE**, and **PATIENT2PROTEIN** (see Note 5).
3. Click on the **Search Areas** directory under the **CANCER_iTRAQ** subdirectory. Click the **New** button to open a new **search** page on the fly. Enter default to the **Name:** textbox.

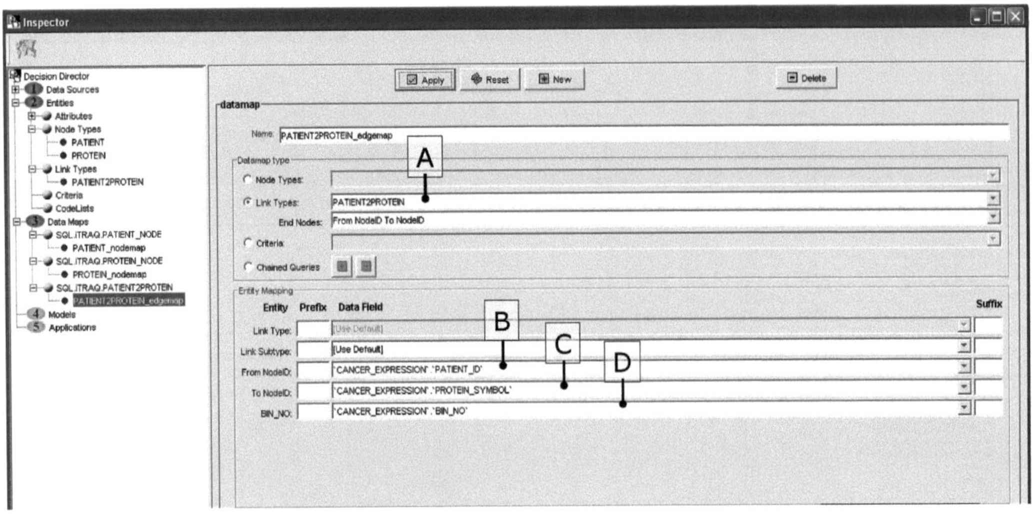

Fig. 12. Screenshot of the SQL.iTRAQ.PATIENT2PROTEIN datamap page. (**a**) The Link Types, (**b**) From NodeID, (**c**) To NodeID, and (**d**) BIN_NO dropdown menus.

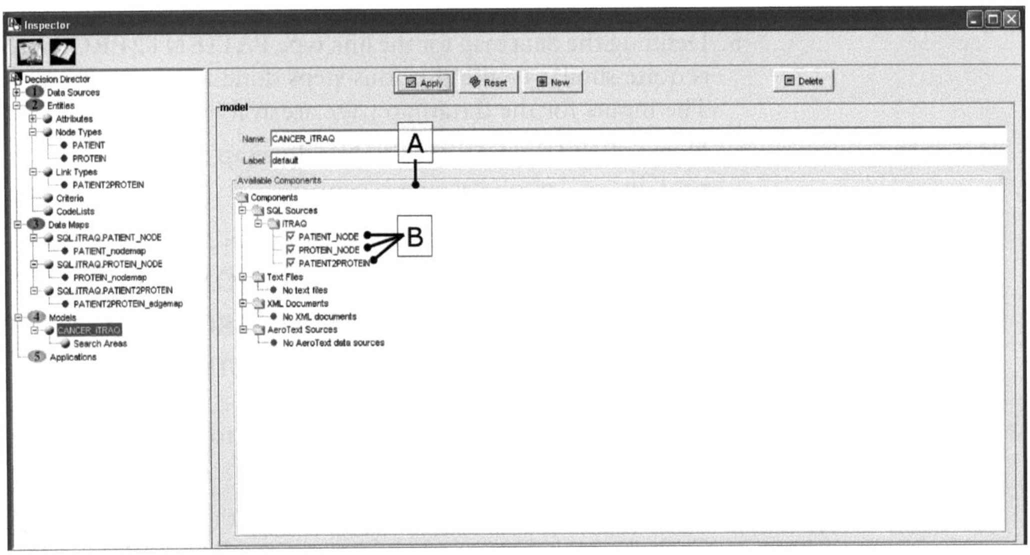

Fig. 13. Screenshot of the model page. (**a**) The Available Component box. (**b**) The node and link types.

4. Click on the **Applications** directory and then the **New** button to open a new **application** page on the fly (see Note 6).
5. Enter **CANCER_Patient** in the **Name:** textbox. This will become the name of the application. Specify the network model by selecting the checkbox next to **CANCER_iTRAQ** in the **Select Models** panel (Fig. 14a).

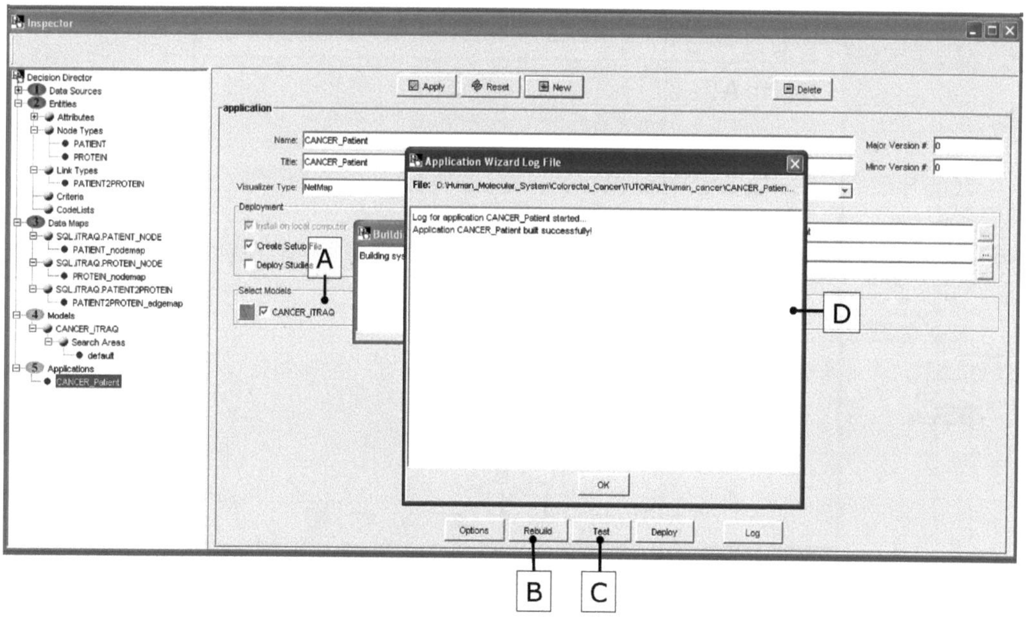

Fig. 14. Screenshot of the application page. (**a**) The Select Models panel. (**b**) The Rebuild button. (**c**) The Test button. (**d**) The **Application Wizard Log File** message box.

6. Click the **Build** button at the bottom of the **application** page to build the application. If the application has been successfully built, a notification will be display in the **Application Wizard Log File** message box. The **Build** button will automatically change to the **Rebuild** button (Fig. 14b).
7. To open the application, click the **Test** button at the bottom of the **application** page to open the StudyManager™ (Fig. 14c).

3.4. Network Visualization Using NetMap Visualizer™

1. In the **CANCER_Patient** window, select **File > New Study** (Fig. 15).
2. In the dialog box, enter **test** into the **Name:** textbox. This will be the name of the new network analysis plan which is also labelled as a *study* in the StudyManager™. After this step, the **test** icon should appear as a leaf under the root directory **CANCER_PATIENT** on the left-hand panel (Fig. 15).
3. Click the **Retrieve** button to retrieve the data from the database (Fig. 15b).
4. Click the **Visualise** button to draw the network (Fig. 15c). This will open the Netmap Visualizer™ as a new window.
5. To allow for filtering out patients of certain disease stages in later analysis, four patient groups according to their respective disease stages of A, B, C, or D need to be created. To do so, zoom in the patient node set to a scale that the node attributes are readable (Fig. 16).

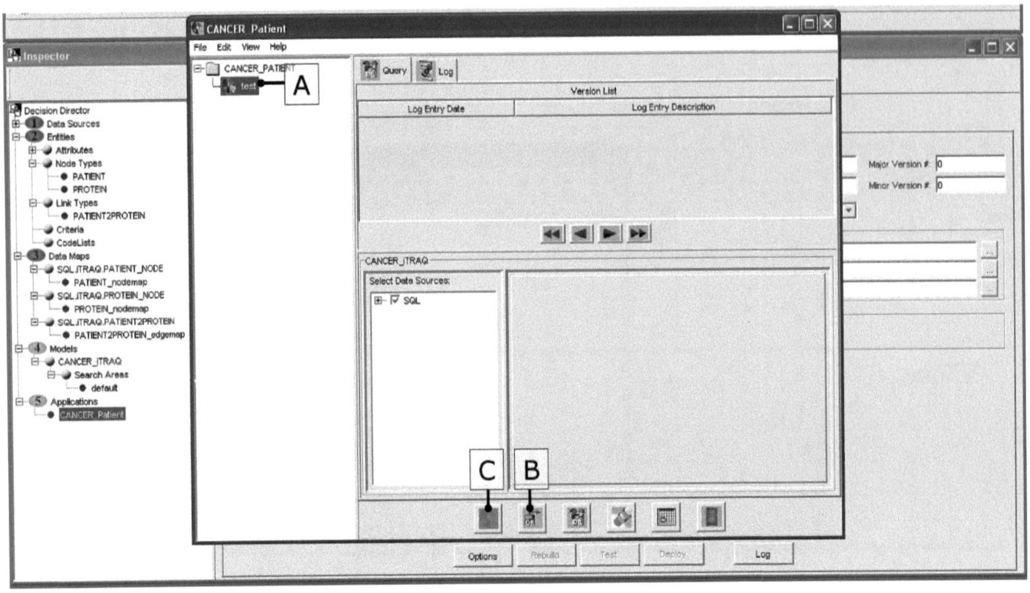

Fig. 15. Screenshot of the CANCER_Patient window opened by the StudyManager™. (**a**) The test icon in the root directory CANCER_PATIENT. (**b**) The Retrieve button. (**c**) The Visualise button.

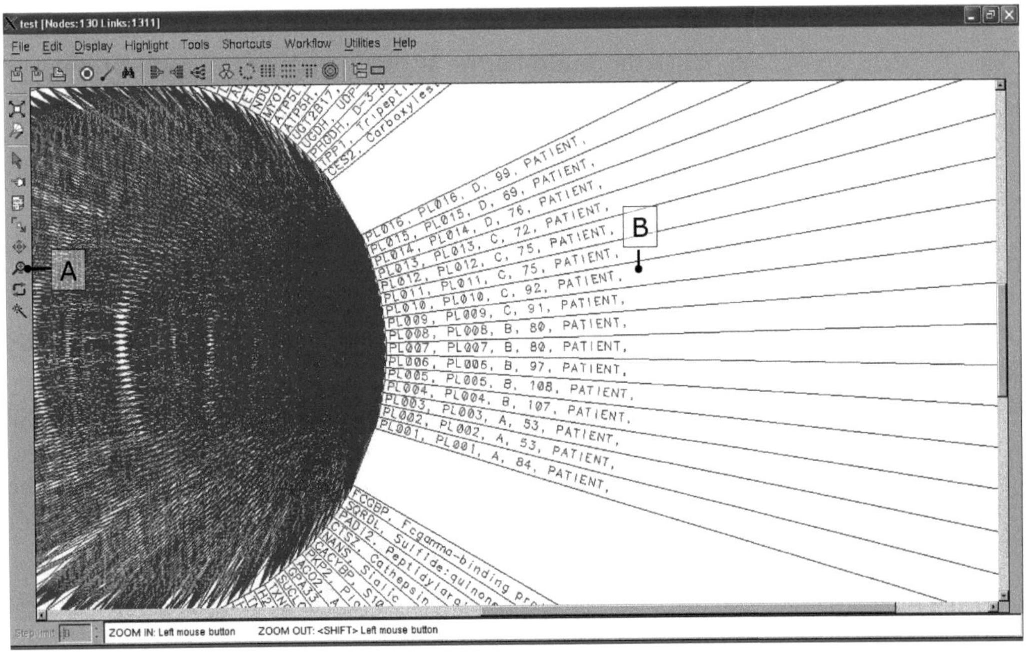

Fig. 16. Screenshot of the PATIENT nodes after scaling by zooming in. (**a**) The zoom icon on the Navigation bar. (**b**) One of the PATIENT nodes with the attributes NodeID = PL010, DISEASE_STAGE = C, Number of connected nodes = 92, and NodeType = PATIENT clearly displayed.

6. Select **Edit > Named Group Editor** in the menu bar (Fig. 17a). In the **Named Group Editor** window, select **Edit > New**. An empty directory named **New Named Group** will appear. Click on the directory name and type in **A** as the new name for the

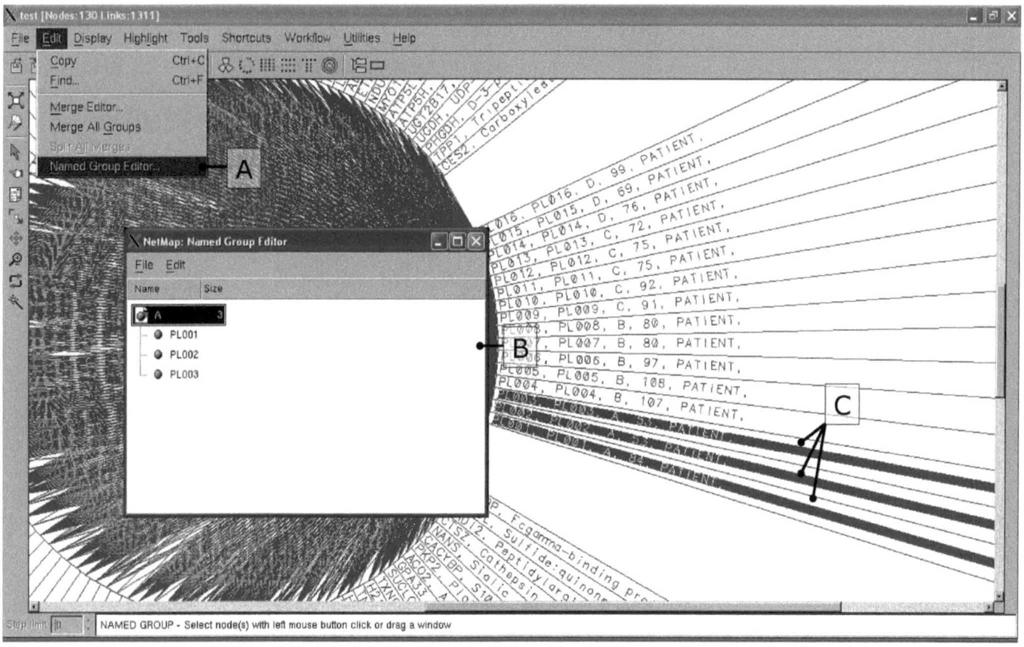

Fig. 17. Screenshot sequence for defining Named Groups. (**a**) The Edit > Named Group Editor selection in the menu bar. (**b**) The Named Group Editor window. (**c**) Highlighted PATIENT nodes.

named group. A square cursor will appear in the Visualizer™ window. By moving this cursor over and highlighting the patient nodes, the **PATIENT_IDs** will appear in the **Named Group Editor** window (Fig. 17b, c). Repeat this step for creating the named groups B, C, and D.

7. To check whether the **Named Groups** have been added to the **Node/Link Selection Menu**, select the **Node Menu** icon in the tool bar (Fig. 18a, b). At the lower right-hand section, the four **Named Groups A**, **B**, **C**, and **D** should appear (Fig. 18c).

8. To allow for filtering **PATIENT2PROTEIN** links by the **BIN_NO**, the links in the network need to be coloured by the attribute values of the **BIN_NO** instead of **Link Types**. Click the **Link Menu** icon in the tool bar to open the **Node/ Link Selection Menu** (Fig. 19a). In the **Link** Tab, there are three columns of link attributes being displayed, i.e. **Link Type**, **BIN_NO**, and **DATASOURCE**. Each column contains the values of the corresponding link attribute. The second column contains the values +3 to −3 of the link attribute **BIN_NO** (Fig. 19b). To assign the link colours to the **BIN_NO** values, select the item **BIN_NO** in the leftmost dropdown menu next to **Link color** (Fig. 19c). The network links are now re-coloured according to the **BIN_NO** values (Fig. 20).

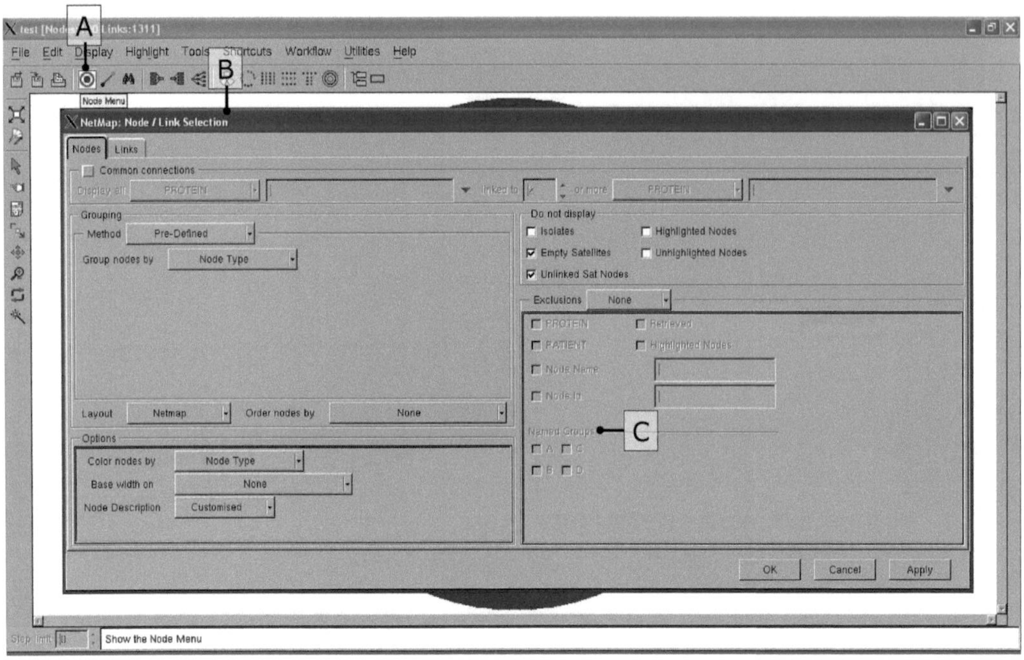

Fig. 18. Screenshot of the Node tab content in the Node/Link Selection Menu. (**a**) The Node Menu icon in the tool bar. (**b**) The Node Selection Menu window. (**c**) Named Groups section.

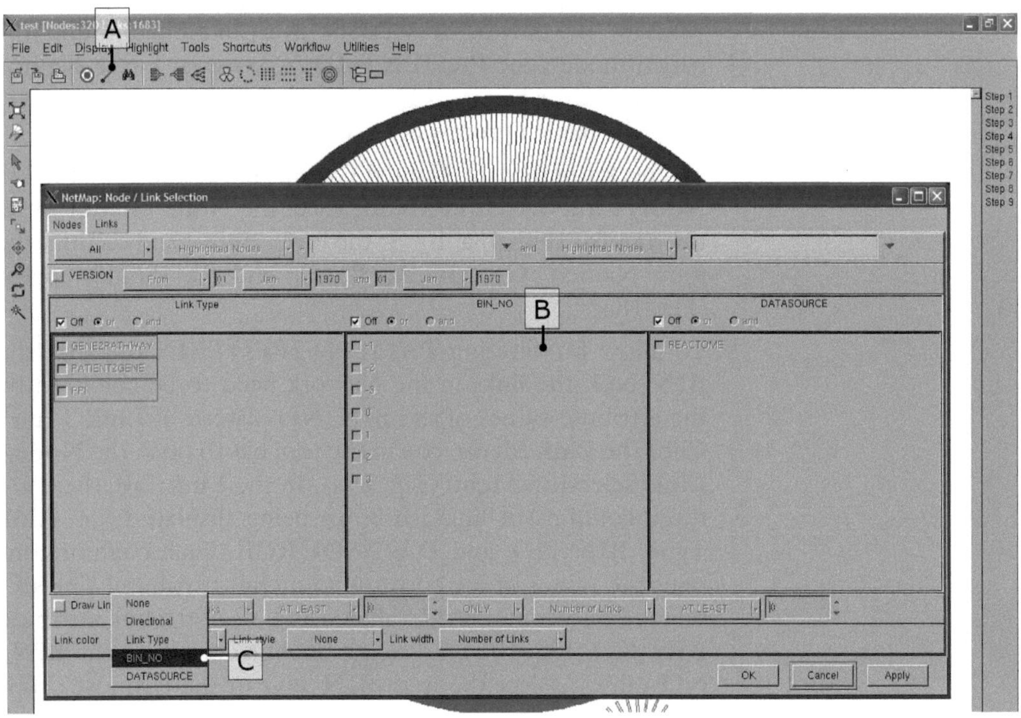

Fig. 19. Screenshot of the Link tab content in the Node/Link Selection Menu. (**a**) The Link Menu icon in the tool bar. (**b**) The BIN_NO panel. (**c**) The Link Color Selection Menu.

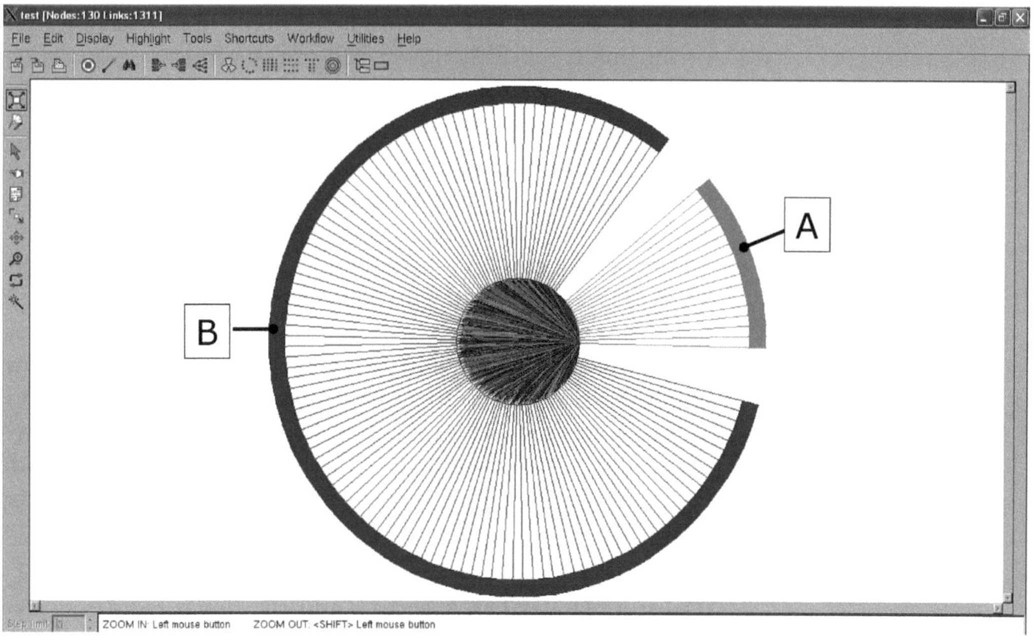

Fig. 20. The resulting patient–protein network with links re-coloured according to the BIN_NO values. (**a**) PATIENT nodes. (**b**) PROTEIN nodes.

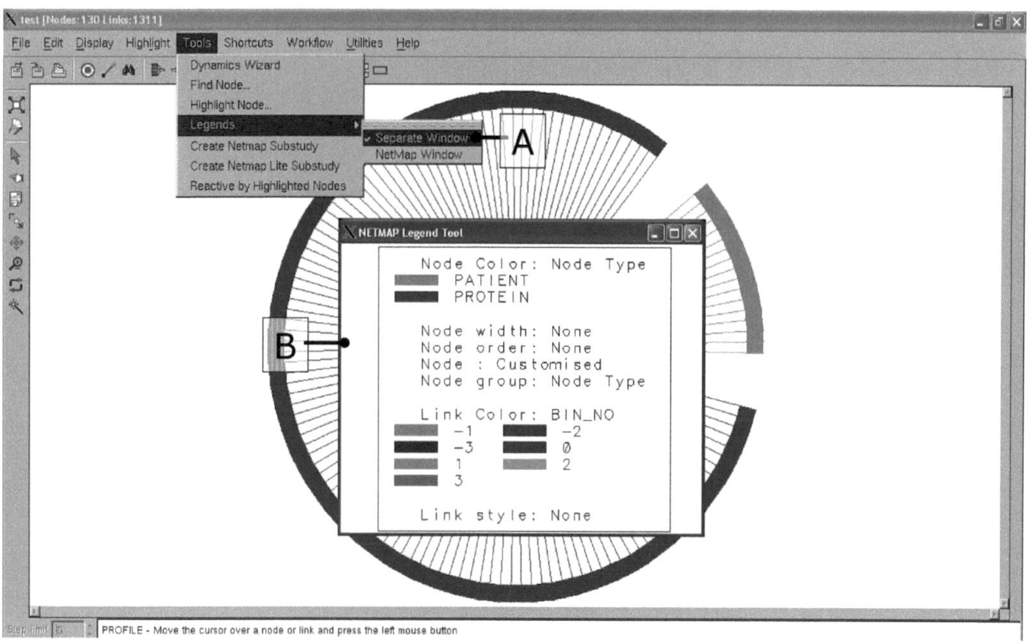

Fig. 21. Screenshot sequence for opening the Netmap Legend box. (**a**) The Legends Menu. (**b**) The NETMAP Legend Tool box.

9. To check the colour scheme assigned to the node and link types, select **Tools > Legends > Separate Window** in the menu bar (Fig. 21a). The Visualizer™ will display the **NETMAP Legend Tool box** as a separate window (Fig. 21b). The colour

coding of the node types is displayed in the upper half and that for the link types in the lower half of the window.

10. One useful function provided with NetMap Visualizer™ is emergent grouping. It is designed for finding sub-networks which consist of highly connected nodes. In an emergent group, the member nodes are more highly connected among themselves than with other nodes. Applied to the present case, the organization of emergent groups can help the user to identify proteins that are attached exclusively to a certain group of patients or those that are shared by patients of different disease stages. To apply emergent grouping, select the **Node Menu** icon in the tool bar to open the **Node/Link Selection Menu**. In the **Nodes** tab, select the item **Emergent** in the dropdown menu next to **Method** (Fig. 22a, b). Select the item **Node Type** in the dropdown menu next to **Order nodes by** (Fig. 22c).

11. As an example, only the emergent groups involving **PATIENT_IDs** of disease stages A and C, and the down-regulated proteins are shown. Select **Exclude** in the dropdown menu next to **Exclusion** (Fig. 22d). To exclude groups B and D, check the boxes **B** and **D** under the **Named Groups** (Fig. 22e).

12. To select the links containing only the **BIN_NO** values 2 and 3, click on the **Link** tab. Uncheck the box labelled **Off** in the

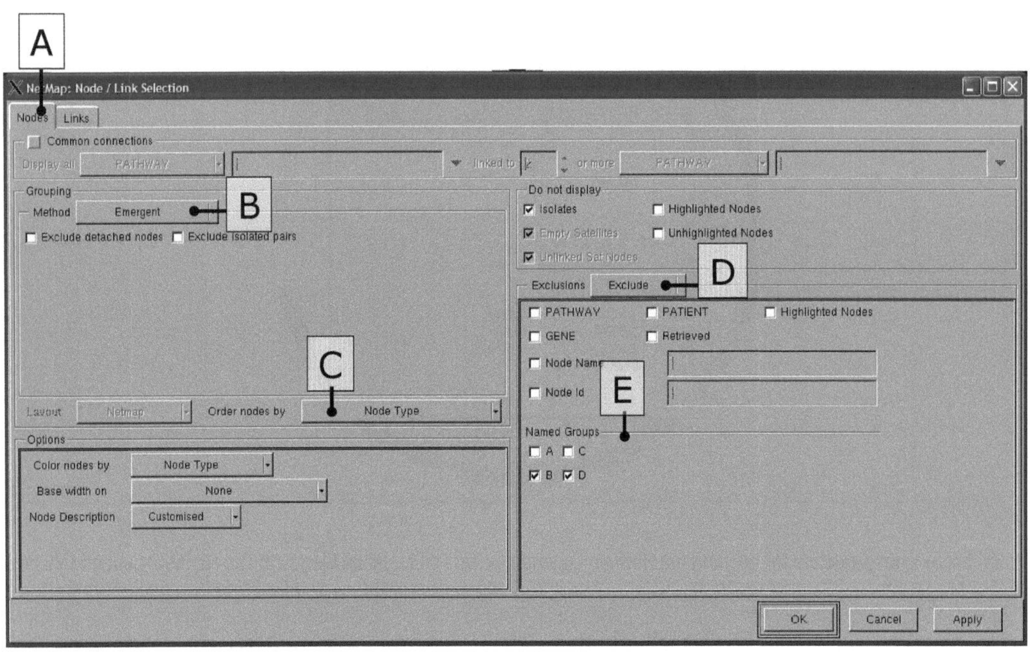

Fig. 22. Screenshot of the Node Selection Menu. (**a**) The Node tab. (**b**) The Method dropdown menu. (**c**) The 'Order nodes by' dropdown menu. (**d**) The Exclusions dropdown menu. (**e**) The Named Groups panel.

BIN_NO column and then check the boxes labelled **2** and **3**. Click the **OK** button to compute emergent groups (Fig. 19).

13. The resulting network shows that there are two clusters of **PATIENT2PROTEIN** links (Fig. 23). Both clusters contain four patients each. The patients in cluster A shared a set of 11 up-regulated proteins (Fig. 23a), whereas the patients in cluster B shared a set of 15 up-regulated proteins (Fig. 23b). The nodes making up the entire network are arranged in a circular layout. The width of each node is proportional to the number of links that it is connected to. The link colours represent the two levels of up-regulation. The links in the lighter shade represent the +2 bin, whereas those in the darkest shade represent the +3 bin. The links at the centre of the network shows that some of the proteins in cluster A are also up-regulated in patients belonging to cluster B (see Note 7). Therefore, the **PATIENT** nodes are classified by the degree that they share a subset of **PROTEIN** nodes that are up-regulated to the same range (see Note 8).
14. By repeating steps 11 and 12, the users can examine the network clusters produced by different combinations of **BIN_NO** values.

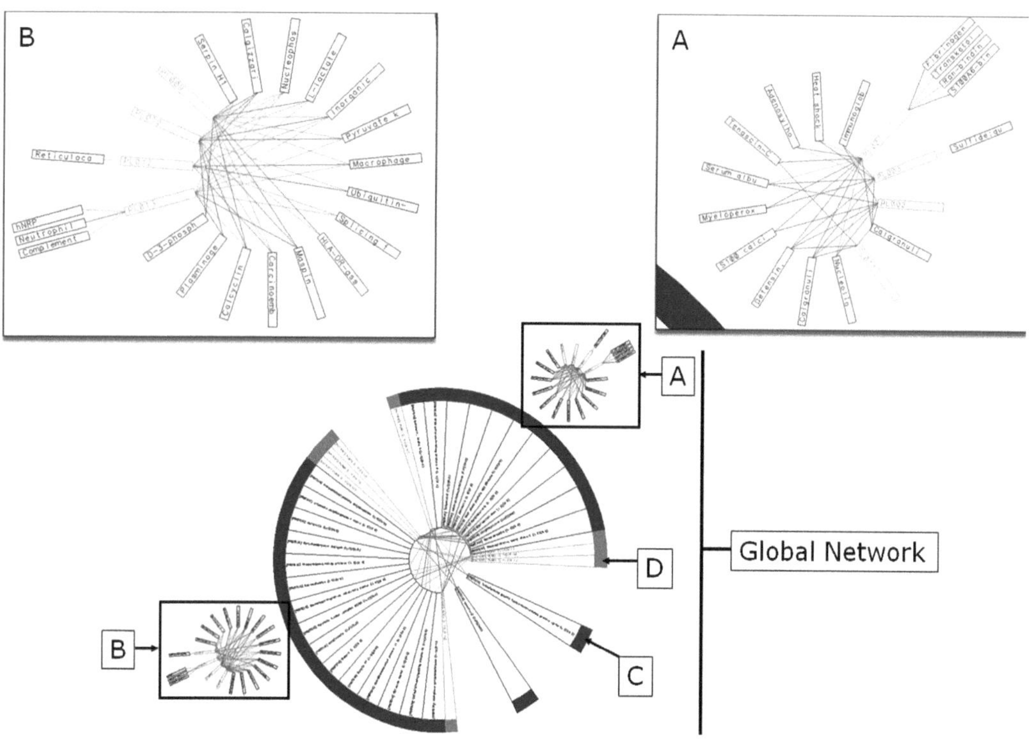

Fig. 23. Clustered patient–protein network. (**a**) Cluster A. (**b**) Cluster B. (**c**) PROTEIN node in dark shade. (**d**) PATIENT node in lighter shade.

3.5. Further Applications

The network construction method described so far involves only two node types and one link type. Additional node types, such as **PATHWAY** and **GO_PROCESS**, can be added to the patient–protein network to make the model more informative, thus enhancing biomarker discovery by identifying the pathway membership of protein candidates. Alternatively, additional edge types, such as protein–protein interactions, can be added to give more confidence to selective proteins as biomarkers.

4. Notes

1. For every new project, a good organization of data files and documents is part of good project management (6). A well-organized directory allows other succeeding users to understand the progress of the project. For the user, he/she can easily review past data and results when necessary. This could be for the purpose of conference presentation, manuscript authorship, or methodology revision. For example, a root directory called **Human_Cancer** has been created for the present project. Under which, four new folders named **data, doc, results**, and **related_work** have been created. All data, whether experimental or curated, are stored in the **data** directory. Conference presentation and yet to be published manuscripts are stored in the **doc** directory. Network models and outputs from statistical packages are stored in the **results** directory. Publications on cancer released by third parties are stored in the **related_work** directory.
2. As shown in Table 1, it is possible for different constituent proteins of the same pathway to exhibit varying expression dynamics. This could be a result of the heterogeneous cell population within the tumour sample or the inconsistent sampling of the tumour mass by different surgical practitioners.
3. There are two types of tables needed for network construction: tables for storing data that map to the nodes, and tables for storing data that map to the links. For example, **cancer_expression** is a table for storing the patient–gene relation and its attribute **bin_no**. The table **protein_list** is used for storing the node attributes **uniprot_id**, **protein_symbol**, and **protein_name**.
4. A Pass-Thru' query is actually a SELECT…FROM SQL statement that the DecisionDirector™ parses to the DBMS which in this case is the Microsoft Access®.
5. In the **SQL Sources** directory, the user can select different sets of node and link types for constructing a variety of network

models on demand. This design is especially useful for more complex networks that integrate more than five link types.

6. Under the DecisionDirector™, all network models are implemented as applications. Each application is a piece of visual network mining software for a particular model. As such, the user has to be equipped with an engineering mentality when constructing models out of large-scale microarray, transcriptome sequencing, and iTRAQ data.

7. The links between the two clusters indicate proteomic heterogeneity of these patient samples. In other words, patient samples from different disease stages do not cluster with distinct sets of down-regulated proteins.

8. The link pattern of each cluster also informs us on the variation between individual patients. For example, in cluster A, there are four **PROTEIN** nodes, ***fibrinogen alpha chain*** (Gene Symbol: *FGA*; Gene ID: 2243), ***transketolase*** (Gene Symbol: *TKT*; Gene ID: 7086), ***Ran-binding protein 1*** (Gene Symbol: *RANBP1*; Gene ID: 5902) and ***S100A6-binding protein*** (Gene Symbol: *CACYBP*; Gene ID: 27101), attached to the **PATIENT** node PL001 which means those proteins were up-regulated only in that particular patient. The neighbouring **PATIENT** node PL003 has only one **PROTEIN** node attached to it, i.e. ***sulfide:quinone oxidoreductase*** (Gene Symbol: *SQRDL*; Gene ID: 58472) indicating that the protein is up-regulated only in patient PL003 but not in patient PL001 (Fig. 24).

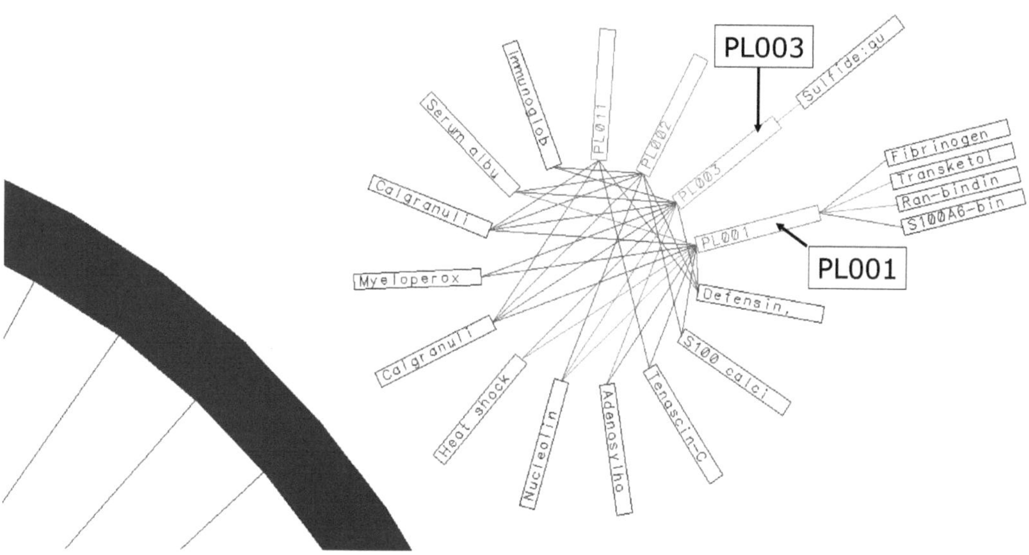

Fig. 24. Zoom-in view of cluster A in the patient–gene network shown in Fig. 23. The PATIENT nodes PL001 and PL003 are indicated with *arrows*.

Acknowledgement

This work is supported by the Cancer Institute New South Wales, Bioplatforms Australia, the New South Wales State Government Science Leveraging Fund, and EIF Super Science Scheme.

References

1. Mendes, A., Scott, R. J., Moscato, P. (2008) Microarrays—Identifying molecular portraits for prostate tumours with different Gleason patterns. In: Trent RJ (ed) *Methods in Molecular Medicine Vol.141: Clinical Bioinformatics*, Human Press, Totowa, New Jersey.
2. Dave, S. S., Fu, K., Wright, G. W., Lam, L. T., Kluin, P., Boerma, E. J., et al. (2006) Molecular diagnosis of Burkitt's lymphoma. *N. Eng. J. Med.* **354**, 2431–2442.
3. Hall, P., Ploner, A., Bjohle, J., Huang, F., Lin, C. Y., Liu, E. T., et al. (2006) Hormone-replacement therapy influences gene expression profiles and is associated with breast cancer prognosis: a cohort study. *BMJ Med.* **4**, 16.
4. Loscalzo, J., Kohane, I., Barabasi, A-L. (2007) Human disease classification in the postgenomic era: a complex systems approach to human pathobiology. *Mol. Syst. Biol.* **3**, 124.
5. REACTOME SkyPainter. (2010) http://www.reactome.org/skypainter
6. Noble, W. S. (2009) A quick guide to organizing computational biology projects. *PLoS Comput. Biol.* **5**, e1000424.

Chapter 16

Statistical Analysis of Dynamic Transcriptional Regulatory Network Structure

Jennifer J. Smith, Ramsey A. Saleem, and John D. Aitchison

Abstract

Here, we present a detailed method for generating a dynamic transcriptional regulatory network from large-scale chromatin immunoprecipitation data, and functional analysis of participating factors through the identification and characterization of significantly overrepresented multi-input motifs in the network. This is done by visualizing interactive data using a network analysis tool, such as Cytoscape, clustering DNA targets of the transcription factors based on their network topologies, and statistically analyzing each cluster based on its size and properties of its members. These analyses yield testable predictions about the conditional and cooperative functions of the factors. This is a versatile approach that allows the visualization of network architecture on a genome-wide level and is applicable to understanding combinatorial control mechanisms of DNA-binding regulators that conditionally cooperate in a wide variety of biological models.

Key words: Multi-input motif, Transcriptional regulatory network, Large-scale chromatin immunoprecipitation

1. Introduction

Overrepresentation of network motifs in an activated transcriptional regulatory network indicates mechanisms controlling transcriptional responses to stimuli. One such motif prevalent in many networks is a multi-input motif, whereby a combination of factors coincidently binds to a gene regulatory region (1). The motif implies complex mechanisms to control both expression profiles of target genes and the specificity of the response (2). This cooperation also enables multifunctionality of transcription factors, and therefore confers the ability for cells to respond to a wide variety of signals with relatively few regulators. In addition to these regulatory properties at the

Gerard Cagney and Andrew Emili (eds.), *Network Biology: Methods and Applications*, Methods in Molecular Biology, vol. 781, DOI 10.1007/978-1-61779-276-2_16, © Springer Science+Business Media, LLC 2011

gene level, multi-input motifs also enable network-level properties. For example, through the dynamic involvement of a factor in multiple multi-input motifs, it can divergently regulate and temporally synchronize different responses to the same stimulus (3). For these reasons, biological insights from the study of individual transcription factors in isolation are limited. A more promising approach is the study of dynamic networks involving multiple transcription factors under conditions relevant to their known functions. We have, therefore, developed an analysis strategy to identify and characterize combinatorial control mechanisms among factors that cooperate in response to a stimulus as outlined below.

This method was developed to characterize the conditional activities of four yeast transcription factors in response to fatty acid exposure (3). The analysis consists of the following steps: First, transcription factors that are known or predicted to cooperatively respond to the stimulus of interest are identified. Next, large-scale chromatin immunoprecipitation (ChIP) data are generated for the factors in both the presence and absence of stimulus and used to construct a dynamic transcription factor–DNA interaction network. Finally, condition-specific targets of the regulators are grouped based on their network topologies to identify multi-input motifs in the network. Each group is then characterized by statistical analysis of its size and properties of its members (including microarray gene expression data or other relevant data) to identify and characterize significantly overrepresented multi-input motifs in the network. The networks are visualized and analyzed using freely available network analysis software called Cytoscape.

The procedure presented here provides detailed protocols for experimental setup, and generation and analysis of a dynamic network. It does not cover generation of large-scale ChIP data or preliminary analysis of these data, including alignment and peak detection, as detailed protocols are readily available for these procedures from the platform manufacturers and elsewhere (4–6).

2. Experimental Setup

2.1. Choice of Factors/Transcriptional Response

Good model systems for this approach meet the following criteria:

1. The genomic sequence of the organism is known and there are supporting data suggesting that multiple transcription factors are involved and conditionally cooperate in response to the stimulus.
2. The transcription factors can each be labeled with the same molecular tag and expressed as functional proteins in vivo so that the tag can be used as a handle for ChIP. Note that using

the same tag for each factor reduces experimental noise due to otherwise variable affinities of antibodies used for ChIP (3).

3. Cell purity and abundance are sufficient for the large-scale ChIP analysis that will be used. About 10^9 cells and 10^7 cells are required for microarray and sequencing-based analysis, respectively (4, 6).

4. The positions of known or predicted transcriptional regulatory regions should be available for the organism. For functional analysis of the effect of network structure on gene expression in Subheading 3.5, the genomic coordinates of gene open reading frames and complementary large-scale gene expression data generated under the comparable experimental conditions should also be available or obtainable. Good gene expression data include a time course of wild-type cells, and transcription factor deletion or overexpression strains in the presence of the stimulus.

The example model system discussed throughout the protocol meets the criteria discussed above. The system is a group of four yeast transcription factors responsive to fatty acid exposure, including three factors, Oaf1p, Pip2p, and Adr1p, which are known to cooperate and upregulate genes involved in fatty acid metabolism (reviewed in (7)). In the presence of fatty acids, these genes are known to be upregulated by one of two multi-input motifs involving Oaf1p and Pip2p, or by Oaf1p, Pip2p, and Adr1p. Other responses to fatty acids include transient upregulation of oxidative stress response genes and a corresponding downregulation of general stress response genes, which were identified by large-scale gene expression studies (8), but at the time of the analysis, the mechanisms of regulation had not yet been elucidated. A fourth transcription factor, Oaf3p was also included in the analysis. The factor was uncharacterized at the time of the analysis, but transcriptome profiling and sequence analysis predicted that it is involved in the response and cooperates with fatty acid-responsive factor(s) (9). At the time of the analysis, time course gene expression data in response to fatty acid exposure were available (8, 9) and were complemented with gene expression data of transcription factor deletion strains under the same condition (3).

Genes in *Saccharomyces cerevisiae* can be tagged by homologous recombination with the c-myc epitope, which has been shown to be a suitable tag for ChIP analysis of various factors (10). The physical structure of the yeast genome is known and can be downloaded from Saccharomyces Genome Database (http://www.yeastgenome.org/). Yeast transcriptional regulatory regions are well-characterized; at low resolution, intergenic regions are considered to be capable of regulating genes with adjacent start sites (10, 11). If higher resolution data is desired, it has been predicted that the majority (74%) of transcription factor binding sites are from 100 to 500 bp upstream of protein-coding regions (10).

2.2. Choice of Experimental Conditions: System Dynamics

The exact experimental conditions depend on the hypotheses of the investigator; but they should be carefully designed to avoid extrinsic noise and to capture the transcriptional response of interest, which is usually on the order of minutes to hours. It is important to consider whether the datasets being used are compatible with the particular network state under analysis. A few specific points are listed below:

1. Apparent transcription factor binding and gene expression levels depend not only on the presence and duration of a stimulus, but also potentially on all other aspects of the environment (including cell density and media composition). In addition, they may be differently influenced by the environmental history of the cells. For example, if the expression level of a gene is high in the reference condition (i.e., before perturbation), mRNA levels may appear high during early time points of perturbation even if mRNA production stops and transcriptional activators are disassociated, but if the levels were low in the reference condition, levels of the same gene will appear low after perturbation.
2. Datasets of dynamic protein abundances can be used to infer transcriptional activity and are useful tools for design and analysis of transcriptional regulatory network models. For example, the abundance of a transcription factor can be an indicator of its activity level. In addition, protein level changes can be used as a readout for transcriptional regulation of a gene (offset slightly for the time of translation). However, datasets of mRNA levels reflect a different layer of cellular control than those of protein levels, and there may be lack of correlation between the two measurements for several reasons, including slow or variable protein turnover rate and translational regulation. One way to reduce inconsistency between the measurements is to use a well-characterized reporter gene like GFP in place of endogenous genes under control of the regulatory sequence in question.

3. Methods

The detailed method below describes dynamic network generation and analysis. Specifically, it includes comparing the targets of multiple transcription factors under different environmental conditions to determine the likelihood of conditional cooperation among the factors. The method assumes that large-scale ChIP experiments have been designed and conducted.

3.1. Preprocessing of Large-Scale DNA–Protein Interaction Data

For the purpose of this protocol, an experiment is the analysis of the DNA targets of one transcription factor under one experimental condition. This section describes formatting the data for each experiment into a tab-delimited table consisting of a column of unique identifiers, each representing a DNA regulatory region analyzed, followed by a column of statistical measurements [e.g., *p*-values or false discovery rates (FDRs)] reflecting the likelihood of interaction with the transcription factor tested. An important feature of the data is that the same set of DNA regions are analyzed for each experiment. Instructions for generating this format are below:

For PCR-based ChIP-CHIP data: The microarrays for these experiments consist of spots containing PCR products of broad DNA regulatory regions. For this type of data, interaction of a transcription factor with a DNA probe is positive if enrichment of IP versus whole cell extract DNA is above a confidence threshold (based on a *p*-value or other statistical measure generated from biological replicates of the data). For each experiment, make a tab-delimited text file of the names of all DNA probes on the array. For those probes passing the threshold of significance for interaction with the factor, enter a one letter unique identifier for the transcription factor in column 2 using the "if" function of Excel (see Fig. 1a). Continue to Subheading 3.2.

For tiling array and sequencing-based ChIP data: ChIP data can be analyzed at high resolution using either next-generation sequencing approaches, or tiled microarrays, for which a series of short probes spaced at regular intervals across the entire genome are spotted on the array (6). These platforms result in higher resolution data than PCR-based arrays; the binding locations of each factor are precisely localized to short DNA sequences rather than to broad regulatory regions. The data are often represented as a list of peak DNA regions with potential for transcription factor binding, along with corresponding statistical values representing the confidence of the interaction, which are calculated from peak attributes (rather than merging of biological replicates). This higher resolution can greatly enhance our understanding of the mechanisms of cooperation between factors, but it is not necessary for the generation and analysis of network motif graphs described here. In addition, higher resolution data are difficult to use for statistical analysis of cooperativity because although the sequences of DNA targets are known, the sequences that are tested but not bound by the factor are difficult to define and quantify, and it is difficult to compare data from different experiments. Therefore, to preprocess these data for network analysis, the data are standardized by first establishing a threshold for binding (based on the statistical value associated with the data), and then assigning each positive DNA segment to a broader regulatory region(s) in the genome using the following approach:

(a) For each experiment, generate a tab-delimited text file of the DNA coordinates of sequences that are bound by the factor (i.e., sequences of peaks meeting threshold of significance). The file should be in gff format.

(b) Generate an annotation file, which is a tab-delimited text file of DNA coordinates of a comprehensive set of broad regulatory regions in the genome (for example, intergenic regions or 5′ untranslated regions of a standard size). For *S. cerevisiae*, intergenic region names and coordinates have been determined previously (10) and are available as supplementary data file Table I of (3). This file should also be in gff format.

(c) Assign each peak to an overlapping regulatory region(s). This can be done by using the map peaks function of the Chip Analysis Package of NimbleScan software (Roche) to annotate each peak with overlapping regulatory region(s). Open NimbleScan software, choose Chip, and then Map Peaks under the analysis tab. Next, input the gff file created in (a) as the peak/data file, and input the gff file of regulatory regions as the annotation file. The program outputs a gff file of the peaks, each annotated with overlapping regulatory regions.

(d) In Microsoft Excel, edit each file output from step (c) into a tab-delimited text file with a unique identifier for each regulatory region in the first column, followed by a single letter in the second column (representing the transcription factor in the experiment). Do this by removing all columns except that with unique identifiers for regulatory

a

ID	topologies
iYPL144W	A
iYKL097W-A-1	A
iYNL132W	A
iYDR309C	A
itP(UGG)N1	A
iYNR036C	A
iYDR076W	A
iYKL110C-1	A
iYMR145C	A
iYDR134C	A
iYPR034W	A
iYML008C	A
iYOR056C	A
iYMR016C-1	A
iYDR040C	A
iYLR111W	A
iYPR148C-1	A

b

gene	topologies	ratio	significance	regulation	Excel reference
YDR406W	AOY	-0.6584	138.242745	up	AOYup
YDR524C	AOY	-0.0446	2.676476	nc	AOYnc
YDR525W-A	AOY	0.0135	0.265889	nc	AOYnc
YER189W	AOY	0.0145	0.409967	nc	AOYnc
YFL021W	AOY	0.0481	3.535472	nc	AOYnc
YFL022C	AOY	0.0432	3.365813	nc	AOYnc
YGL030W	AOY	0.0297	1.46524	nc	AOYnc
YGL031C	AOY	-0.0136	0.262611	nc	AOYnc
YGL171W	AOY	0.0072	0.05146	nc	AOYnc
YGR161C	AOY	0.2118	52.097604	down	AOYdown
YGR161W-A	AOY	0.0673	5.525777	nc	AOYnc
YGR161W-B	AOY	0.0249	0.889497	nc	AOYnc
YGR180C	AOY	0.1475	30.83924	nc	AOYnc
YGR249W	AOY	0.6813	7.361986	nc	AOYnc
YHR087W	AOY	0.4175	114.852157	down	AOYdown
YIL053W	AOY	0.2155	48.465971	down	AOYdown
YIL055C	AOY	-0.0248	0.569178	nc	AOYnc

Fig. 1. Two examples of DNA target property files. The tables in panels (**a**) and (**b**) are for statistical analysis of transcription factor cooperativity and gene properties, respectively.

regions, remove any duplicate entries using remove duplicates function (in data tools tab in data directory), and then add a column with a one-letter unique identifier for the transcription factor that was immunoprecipitated in the experiment.

(e) A note about biological replicates: If you have biological replicates of high-resolution ChIP data, but the platform vendor does not provide a method of merging the data, each biological replicate can be preprocessed separately through steps a–d and merged after preprocessing. For example, regulatory regions that were identified as bound by the factor in at least two of three replicates can be retained.

3.2. Merging of Preprocessed DNA–Protein Interaction Data into a Network Topology File

This section describes how to merge multiple interaction data files generated in Subheading 3.1 into one network topology file that can be used for network analysis in Subheading 3.3. A separate network topology file should be generated for each environmental condition and contain data for all factors tested under that condition.

1. For each environmental condition, start with a spreadsheet containing a list of all broad regulatory regions tested in the ChIP analysis. Regulatory region names (or other unique identifiers) should be listed, but not genomic coordinate data.
2. Add data for one factor tested under this condition in the next column. For Excel users, use the "vlookup" function to look up each regulatory region listed in the first column in another spreadsheet created in Subheading 3.1, and return the value in the second column. The function will return either the one-letter code for the transcription factor in the experiment, or the symbol #N/A if it is not found.
3. Repeat step 2 for each transcription factor tested under the experimental condition, creating a new column for each factor.
4. Replace all formulas in the spreadsheet with corresponding values, by selecting all cells and using the copy and paste special functions.
5. Remove all #N/A symbols by selecting the entire spreadsheet and using the find and replace tool to find #N/A and replace with nothing.
6. Merge all columns except the first, which can be done using the "concatenate" function of Excel. The spreadsheet should look similar to the network topology file in Fig. 2. In the example used here, each target in the network is annotated with a combination of the letters A, O, P, and Y, representing interactions with Adr1p, Oaf1p, Pip2p, and Oaf3p, respectively.

itA(AGC)D	OY
itD(GUC)G2	A
itF(GAA)N	AP
itG(GCC)F2	A
itG(GCC)P1	AO
itG(UCC)O	O
itK(CUU)F	A
itK(CUU)P	P
itK(UUU)O	A
itM(CAU)P	AO
itP(UGG)N1	AO
itS(UGA)I	AOPY
itV(UAC)B	P
itV(UAC)D	A
iYAL004W	A
iYAL039C-0	AP
iYAL040C	AOY
iYAL054C	OP
iYALCdelta2	Y
iYAR044W-1	P
iYAR047C-0	OP
iYAR047C-1	P
iYAR066W	O
iYARWdelta7	AOY
iYBL021C	O
iYBL028C	A
iYBL030C-0	OY
iYBL030C-1	AOY

Fig. 2. A network topology file.

3.3. Statistical Analysis of Transcription Factor Cooperativity in the Network

To characterize the structure of a network, a statistical function can be applied to determine if targets with a particular network topology are significantly overrepresented in the network. More specifically, it will determine if the number of genes targeted by each combination of factors is significantly more than expected by chance given the null hypothesis that the factors are not cooperative. Significantly overrepresented multi-input motifs can be identified using the cumulative binomial distribution function (CDF). Below is a description of how to do the statistical analysis using Microsoft Excel 2007 and R (12):

1. Open a network topology file generated in Subheading 3.2 containing a list of all regulatory regions tested and their network topologies for one experimental condition.

B	C	D	E	F	G	H	I	J	K	L	M	N
2	**Table I: Binding statistics for individual factors**											
3	values											
4	topology description	topology	number of targets				probability of success					
5	bound by Adr1p (A)	*A*	212				0.032924					
6	bound by Oaf1p (O)	*O*	394				0.06119					
7	bound by Pip2p (P)	*P*	137				0.021277					
8	bound by Oaf3p (Y)	*Y*	261				0.040534					
9	number of genes tested	*	6439									
10	formulas											
11	topology description	topology	number of targets				probability of success					
12	bound by Adr1p (A)	*A*	=COUNTIF(NW!$B$2:$B$6440,D5)				=E5/$E$9					
13	bound by Oaf1p (O)	*O*	=COUNTIF(NW!$B$2:$B$6440,D6)				=E6/$E$9					
14	bound by Pip2p (P)	*P*	=COUNTIF(NW!$B$2:$B$6440,D7)				=E7/$E$9					
15	bound by Oaf3p (Y)	*Y*	=COUNTIF(NW!$B$2:$B$6440,D8)				=E8/$E$9					
16	number of genes tested	*	=COUNTIF(NW!$B$2:$B$6440,D9)									
17												
18												
19	**Table II: Binding statistics for groups of factors**											
20	values											
21		AY	AO	AOY	OY	OP	OPY	AOP	AOPY	APY	AP	PY
22	actual number	11	46	72	44	18	25	7	28	5	6	6
23	probability of success	0.001226	0.001891841	7.99E-05	0.002348	0.001208	5.1E-05	4.1E-05	1.74E-06	2.67E-05	0.000631	0.000783
24	expected number	7.895794	12.18156566	0.514631	15.11606	7.778364	0.32861	0.26482	0.011188	0.171648	4.062986	5.041746
25	topology probability	1.74E-01	9.91E-14	3.8E-125	1.13E-09	1.17E-03	7.23E-38	1.43E-08	1.73E-84	1.07E-06	2.25E-01	3.91E-01
26	Bonferroni corrected	1.912	1.09E-12	4.18E-124	1.24E-08	0.0129	7.95E-37	1.58E-07	1.90E-83	1.18E-05	2.472	4.305
27	significance score	-0.2814	11.96234563	123.3788	7.90497	1.890652	36.09947	6.80218	82.72056	4.927219	-0.39308	-0.63398
28	formulas											
29		AY					AO					
30	actual number	=COUNTIF(NW!$B:$B,D$21)					=COUNTIF(NW!$B:$B,E$21)					
31	probability of success	=I$5*(I$8)*(1-I$6)*(1-I$7)					=I$5*I$6*(1-I$7)*(1-I$8)					
32	number expected	=D$23*$E$9					=E$23*$E$9					
33	topology probability	=1-(BINOMDIST(D$22-1,$E$9,D$23,TRUE))					=1-(BINOMDIST(G$13-1,$E$9,E$23,TRUE))					
34	Bonferroni corrected	=D25*11					=E25*11					
35	significance score	=-LOG10(D26)					=-LOG10(E26)					

Fig. 3. Example calculations for statistical analysis of network structure. Values in Excel tables are shown (*top*) as well as a few sample formulas (*bottom*).

2. On a new worksheet, make a table to calculate binding statistics for individual factors (as shown in the top table of Fig. 3 along with the corresponding formula table below). The table contains a description of each network topology, a corresponding Excel symbol for each, and the number of targets of each factor (calculated from the network topology file using the "countif" function). The data are used to calculate the probability of each factor binding to a particular target in the last column.
3. Make a second table to calculate the binding statistics for groups of factors (as shown in the lower table of Fig. 3 along with the corresponding formula table below). In this table, each possible transcription factor combination (of two or more factors) is listed in the first row. For each combination, the actual number of targets with each topology is calculated from the network topology file using the "countif" function, and listed in the second row. Note that for this function to work, the order of each factor listed must be the same as that in the network topology file. Next, the probability of success for each network topology (if the factors are not cooperative) is calculated using the binding probabilities of individual factors

calculated in the first table. For example, the probability of factors A and O binding to a target (without factors P and Y) is equal to the product of the probabilities of A and O binding, and probabilities of P and Y not binding. The expected number of targets with each topology (assuming that the factors are not cooperative) is then calculated in the fourth row from the probability of success values.

4. The values in the second table are used to calculate the likelihoods of observing the actual number of targets with each network topology (or more) given the null hypothesis that the factors do not cooperate, using CDF. Note that the cumulative version of the function is used to calculate the probability of observing the actual number of targets *or more* in each network topology, rather than the exact number of targets observed. CDF calculations are in the fifth row and Excel formulas are listed for AY and AO topologies in the lower table.
5. Note that with Excel 2007, it is not possible to precisely compute values less than 1E-25. For these calculations, the Stats package of R software can be used, which is freely available at http://www.r-project.org/. Description and usage of the CDF in R can be found at http://stat.ethz.ch/R-manual/R-devel/library/stats/html/00Index.html under the heading Binomial. For CDF, the following command is used: pbinom($q-1$, size, prob, lower.tail = FALSE), where q is the vector of quantiles, size is the number trials, prob is the probability of success on each trial, and lower.tail FALSE denotes calculation of $P[X > x]$, which is why $q-1$ is used as vector instead of q. For example, to calculate CDF of network cluster AO in Fig. 3, the following command is used: pbinom (45, 6,439, 0.001891841, lower.tail = FALSE), and a calculated value of 1.006137e-13 is returned, which closely matches that calculated by Excel.
6. Adjust probabilities to correct for multiple-hypothesis testing using Bonferroni correction. This is done by multiplying the topology probabilities in the fifth row by the number of hypotheses tested (which is 11 in Fig. 3, the number of combinations of multiple factors).
7. Calculate significance scores, which are negative log transformed p-values (shown in row 7 of the table). These are used for graphical representation of the data (as shown in Fig. 4).

3.4. Network Visualization Using Cytoscape Version 2.7.0

Cytoscape is freely available software for visualization and analysis of biological networks that can be downloaded at www.cytoscape.org (13). The program is very versatile and can be used for various quantitative analyses using plugin tools available at the Web site. Here, the program is used to generate graphical representations of the protein–DNA interaction networks for publication and

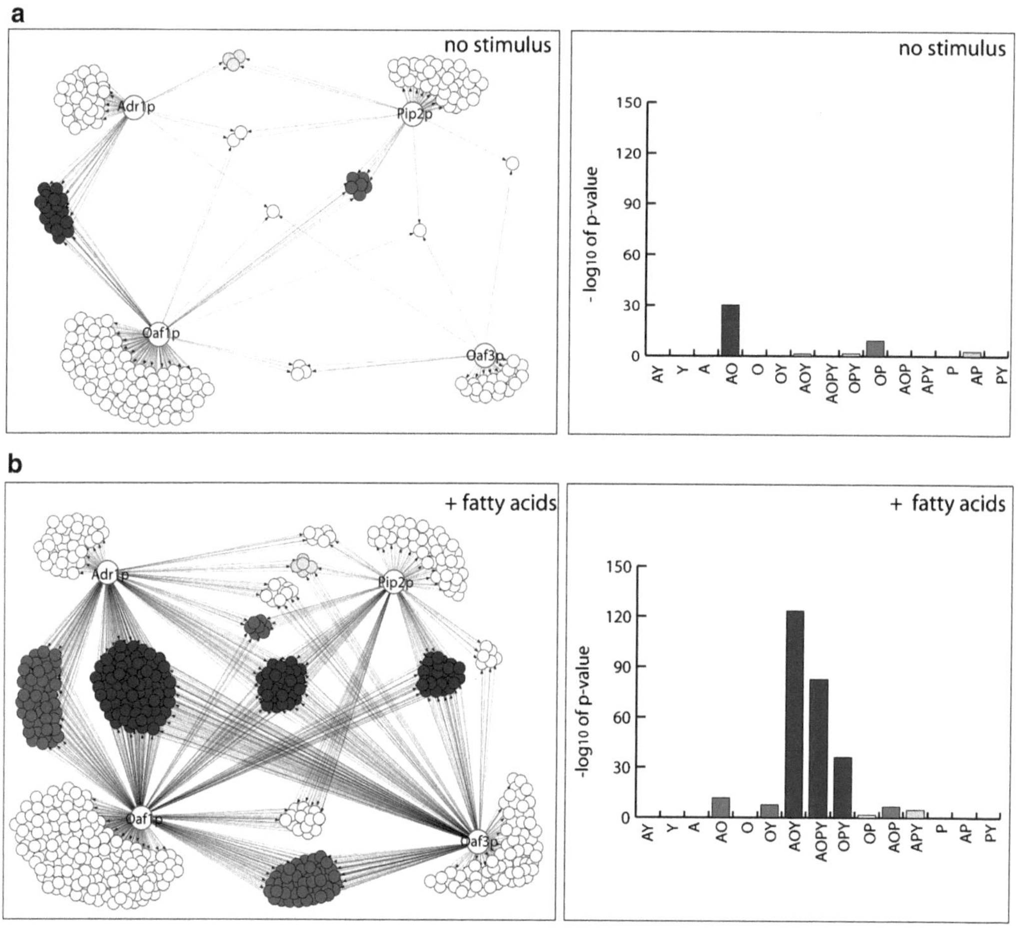

Fig. 4. Identification of overrepresented multiple input network motifs in the absence (**a**) and presence of fatty acids (**b**). *Left panels* are Cytoscape graphs showing condition-specific physical interaction networks between regulators (labeled nodes) and the intergenic regions with which they interact (unlabeled nodes). Protein–DNA interactions for the four transcriptional regulators are shown as directed edges. Intergenic regions with same network topology (i.e., targeted by the same group of factors) are clustered. The network expands, and there is more combinatorial control in the presence of the stimulus. *Right panels* are bar graphs showing the significance of overrepresentation for each cluster. On each graph, each cluster is labeled with up to four letters, representing the regulators targeting the cluster under that condition (A, Adr1p; O, Oaf1p; P, Pip2p; Y, Oaf3p). Overrepresented clusters have *bars*, the height of which reflects the significance of enrichment of that cluster in the network; note that as the *P*-values have been converted to significance scores [$-\log_{10}$(*P*-value)], 1 U on the scale corresponds to a tenfold difference in *P*-value. In all panels, significantly overrepresented topology clusters are *shaded*. Clusters with *P*-values of less than 1×10^{-25}, 1×10^{-5}, and 1×10^{-2} (significance scores greater than 25, 5, and 2) are *shaded dark*, *medium* and *light*, respectively.

qualitative analysis of the networks. The instructions below have been generated for Cytoscape version 2.7.0:

1. First, make Cytoscape network files by combining the interaction data generated in Subheading 3.2 so that there is one file for each experimental condition that contains the data for all transcription factors studied under that condition. The format is a tab-delimited text file with each protein–DNA interaction

represented in a separate row. Each row contains the transcription factor name, the text "pd," and the broad regulatory region name in the first, second, and third columns, respectively. "pd" indicates that interaction type is protein–DNA. This interaction type is represented as directed edges (from transcription factor to target). Other data/interaction types can be added to this spreadsheet if desired. For example, physical interactions among the transcription factors can be added as protein–protein interactions ("pp"), which are displayed as undirected edges.

2. To load a network file into Cytoscape, open Cytoscape and from the file menu choose import, then "network from table." From the select file(s) button, select file to load. Under interaction definition, set source interaction, interaction type, and target interaction to columns 1, 2, and 3, respectively. The first 100 entries of the spreadsheet are shown in the preview section. Select import from the lower right hand corner.
3. A node attribute file should be generated and loaded into Cytoscape that will be used for layout and visualization purposes. Open the network topology file generated in Subheading 3.2 in Excel, add headings "ID" and "topologies" to the first and second columns, respectively, and save as a tab-delimited text file. This file can be loaded into cytoscape by choosing File, import, "attribute/expression matrix."
4. Organize nodes by their network topologies. To do this, select layout from the pull down menu at the top, and then Cytoscape layouts, spring embedded, and all nodes. This layout loosely clusters targets based on their network topologies. Transcription factors and groups of targets can be moved to make topology groups more obvious. This can be done by selecting and moving them directly with the mouse. If the network is too dense for this approach, groups of nodes of the same topology can be selected for repositioning using the filters tool in the control panel at the left. Select the attribute/filter to be node. cytoscape.alias.list and select add. To select a topology group, type in the topology name and select apply filter. For example, to select genes only targeted by Oaf1p and Pip2 (but not Oaf3 or Adr1), the topology name is "OP." After application of the filter, genes of the selected topology are highlighted and can be moved as a group to a new location.
5. To color nodes based on *p*-value overrepresentation of the topology in the network (determined in Subheading 3.3), choose VizMapper from the control panel then node color. Chose the "topologies" attribute that represents node color and set mapping type to "discrete mapper." This allows coloring of nodes based on topology, and therefore allows topologies with different *p*-values to be colored differently.

6. To export the network to another program for publication, choose file/export/network as graphics. Select the desired format and choose okay. A dynamic network that was analyzed this way is shown in Fig. 4 in the absence and presence of fatty acids.

3.5. Statistical Analysis of Gene Properties in the Network

The above method can be modified to analyze other properties of the DNA targets in the network, including position in the genome (14), or gene expression profiles (3). To analyze significant overrepresentation of gene properties in topology clusters, first the transcriptional regulatory regions need to be converted to the names of the genes that they (potentially) regulate. If the regulatory regions are yeast intergenic regions, they can be converted to genes with adjacent start sites as annotated in the supplementary data of (15) which can be found at http://jura.wi.mit.edu/young_public/regulatory_code/GWLD.html in the binding ratios/Excel directory under the name fraenkel_ratios_abbr. Note that each intergenic region potentially controls multiple genes, and each gene can be regulated by multiple intergenic regions, thus care should be taken to preserve these relationships during the conversion. A suitable way to convert DNA targets to gene names in a network is to use "vlookup" function of Excel to lookup the network topologies (from the network topology file generated in Subheading 3.2) for each intergenic region in the fraenkel_ratios_abbr spreadsheet. Next, use the "copy" and "paste special" functions to replace all formulas in the spreadsheet with values, and then remove all columns of the spreadsheet except those with gene names or topologies. Finally, use the "remove duplicates" command to remove duplicate entries in the spreadsheet. After this conversion, one can use gene expression data to identify clusters significantly enriched for gene properties, and make predictions about the conditional influences of each transcription factor on each network cluster. The approach is similar to that used in Subheading 3.3, except a table of genes (annotated as significantly differentially up- or downregulated, or not changed) is used in place of the list of regulatory regions (annotated with network topology) (see Fig. 1b for example). This table of genes can be made using nested "if" or "and" functions in Excel from gene expression data with significance values. With this setup, one can determine if genes upregulated (or downregulated) by a perturbation are significantly enriched in any particular topology cluster. Figure 5 illustrates how this was done for the network example used here. The example formulas for the A and AOY topology clusters (*bottom panel*) refer to a spreadsheet called "properties_file," the layout of which is shown in Fig. 1b. The results of the analysis are shown in the table of cumulative distribution calculations (*top panel*). For example, the topology cluster OP is significantly enriched for genes upregulated by the stimulus

1	C	D	E	F	G	H	I	J	K	L	M	N	O	P	Q	R	S
2	Values							Table for Excel reference									
3		A	AO	AOP	AOPY	AOY	AP	APY	AY	O	OP	OPY	OY	P	PY	Y	
4	up	Aup	AOup	AOPup	AOPYup	AOYup	APup	APYup	AYup	Oup	OPup	OPYup	OYup	Pup	PYup	Yup	
5	down	Adown	AOdown	AOPdown	AOPYdown	AOYdown	APdown	APYdown	AYdown	Odown	OPdown	OPYdown	OYdown	Pdown	PYdown	Ydown	
6	TOTAL	38	54	9	39	78	4	3	13	176	23	32	50	32	11	61	
7								Table of actual values									
8		A	AO	AOP	AOPY	AOY	AP	APY	AY	O	OP	OPY	OY	P	PY	Y	
9	up	1	4	3	3	1	0	0	0	6	6	8	2	3	2	1	
10	down	1	4	0	2	6	1	0	1	7	2	0	5	0	0	2	
11								Table of expected values									
12		A	AO	AOP	AOPY	AOY	AP	APY	AY	O	OP	OPY	OY	P	PY	Y	TOTAL
13	up	0.19301587	0.2742857	0.04571429	0.198095238	0.3961905	0.02031746	0.015238	0.066032	0.89397	0.116825	0.16254	0.253968	0.16254	0.055873	0.309841	32
14	down	0.19301587	0.2742857	0.04571429	0.198095238	0.3961905	0.02031746	0.015238	0.066032	0.89397	0.116825	0.16254	0.253968	0.16254	0.055873	0.309841	32
15								Table of cumulative distribution calculations									
16		A	AO	AOP	AOPY	AOY	AP	APY	AY	O	OP	OPY	OY	P	PY	Y	
17	up	0.17594	0.00017	1.0759E-05	0.00104	0.32780	1	1	1	0.0003	1.61E-09	4.18E-12	0.027	0.001	0.001	0.267	
18	down	0.17594	0.00017	1	0.01688	3.225E-06	0.020163184	1	0.0641	3.8E-05	0.00608	1	5.92E-06	1	1	0.03875	

Formulas		Table for Excel reference	
	A	AOY	
up	=CONCATENATE(D$3,$C4)	=CONCATENATE(H$3,$C4)	
down	=CONCATENATE(D$3,$C5)	=CONCATENATE(H$3,$C5)	
TOTAL	=COUNTIF('properties_file'!$B:$B,D3)	=COUNTIF('properties_file'!$B:$B,H3)	
		Table of actual values	
	A	AOY	
up	=COUNTIF('properties_file'!$F:$F,D4)	=COUNTIF('properties_file'!$F:$F,H4)	
down	=COUNTIF('properties_file'!$F:$F,D5)	=COUNTIF('properties_file'!$F:$F,H5)	
		Table of expected values	
	A	AOY	TOTAL
up	=$S13/6300*D$6 (note: 6300 is number of genes on array)	=$S13/6300*H$6	Estimated number of falsely identified upregulated genes calculated from FDR
down	=$S14/6300*D$6	=$S14/6300*H$6	Estimated number of falsely identified downregulated genes calculated from FDR
		Table of cumulative distribution calculations	
	A	AOY	
up	=IF(D9>D13,1-BINOMDIST(D9-1,D$6,$S13/6300,TRUE),1	=IF(H9>H13,1-BINOMDIST(H9-1,H$6,$S13/6300,TRUE),1)	
down	=IF(D10>D14,1-BINOMDIST(D10-1,D$6,$S14/6300,TRUE),1	=IF(H10>H14,1-BINOMDIST(H10-1,H$6,$S14/6300,TRUE),1)	

Fig. 5. Example calculations for statistical analysis of gene expression in network clusters. Values in Excel tables are shown (*top*) as well as a few sample formulas (*bottom*).

(CDF $P = 1.61^{-9}$). To visualize gene expression values in a network using Cytoscape, network files can be made with gene names instead of broad regulatory region names, and a node attribute file can be made with gene names and gene expression values and loaded. Visualization is done as described in Subheading 3.4 except "continuous mapper" is chosen as the node color mapping type instead of "discrete mapper."

Acknowledgments

This work was funded by grants NIH/NIGMS R01 GM075152, NIH P50 GM076547, and NIH U54 RR022220. We also thank the Luxembourg Centre for Systems Biomedicine and the University of Luxembourg for support.

References

1. Shen-Orr SS, Milo R, Mangan S, Alon U. 2002. Network motifs in the transcriptional regulation network of Escherichia coli. *Nat Genet* 31: 64–8
2. Luscombe NM, Babu MM, Yu H, Snyder M, Teichmann SA, Gerstein M. 2004. Genomic analysis of regulatory network dynamics reveals large topological changes. *Nature* 431: 308–12
3. Smith JJ, Ramsey SA, Marelli M, Marzolf B, Hwang D, Saleem RA, Rachubinski RA, Aitchison JD. 2007. Transcriptional responses to fatty acid are coordinated by combinatorial control. *Mol Syst Biol* 3: 115
4. Ren B, Robert F, Wyrick JJ, Aparicio O, Jennings EG, Simon I, Zeitlinger J, Schreiber J, Hannett N, Kanin E, Volkert TL, Wilson CJ, Bell SP, Young RA. 2000. Genome-wide location and function of DNA binding proteins. *Science* 290: 2306–9
5. Nagalakshmi U, Wang Z, Waern K, Shou C, Raha D, Gerstein M, Snyder M. 2008. The transcriptional landscape of the yeast genome defined by RNA sequencing. *Science* 320: 1344–9
6. Park PJ. 2009. ChIP-seq: advantages and challenges of a maturing technology. *Nat Rev Genet* 10: 669–80
7. Gurvitz A, Rottensteiner H. 2006. The biochemistry of oleate induction: transcriptional upregulation and peroxisome proliferation. *Biochim Biophys Acta* 1763: 1392–402
8. Koerkamp MG, Rep M, Bussemaker HJ, Hardy GP, Mul A, Piekarska K, Szigyarto CA, De Mattos JM, Tabak HF. 2002. Dissection of transient oxidative stress response in Saccharomyces cerevisiae by using DNA microarrays. *Mol Biol Cell* 13: 2783–94
9. Smith JJ, Marelli M, Christmas RH, Vizeacoumar FJ, Dilworth DJ, Ideker T, Galitski T, Dimitrov K, Rachubinski RA, Aitchison JD. 2002. Transcriptome profiling to identify genes involved in peroxisome assembly and function. *J Cell Biol* 158: 259–71
10. Harbison CT, Gordon DB, Lee TI, Rinaldi NJ, Macisaac KD, Danford TW, Hannett NM, Tagne JB, Reynolds DB, Yoo J, Jennings EG, Zeitlinger J, Pokholok DK, Kellis M, Rolfe PA, Takusagawa KT, Lander ES, Gifford DK, Fraenkel E, Young RA. 2004. Transcriptional regulatory code of a eukaryotic genome. *Nature* 431: 99–104
11. Xu Z, Wei W, Gagneur J, Perocchi F, Clauder-Munster S, Camblong J, Guffanti E, Stutz F, Huber W, Steinmetz LM. 2009. Bidirectional promoters generate pervasive transcription in yeast. *Nature* 457: 1033–7
12. R Development Core Team. 2010. R: A language and environment for statistical computing. R Foundation for Statistical Computing, Vienna, Austria. ISBN 3-900051-07-0, URL http://www.R-project.org
13. Shannon P, Markiel A, Ozier O, Baliga NS, Wang JT, Ramage D, Amin N, Schwikowski B, Ideker T. 2003. Cytoscape: A software environment for integrated models of biomolecular interaction networks. *Genome Res* 13: 2498–504

14. Smith JJ, Miller LR, Kreisberg R, Vazquez L, Wan Y Aitchison JD. 2011. Environment-responsive transcription factors bind subtelomeric elements and regulate gene silencing. *Mol Syst Biol* 7: 455

15. MacIsaac KD, Wang T, Gordon DB, Gifford DK, Stormo GD, Fraenkel E. 2006. An improved map of conserved regulatory sites for Saccharomyces cerevisiae. *BMC Bioinformatics* 7: 113

Chapter 17

Imputing and Predicting Quantitative Genetic Interactions in Epistatic MAPs

Colm Ryan, Gerard Cagney, Nevan Krogan, Pádraig Cunningham, and Derek Greene

Abstract

Mapping epistatic (or genetic) interactions has emerged as an important network biology approach for establishing functional relationships among genes and proteins. Epistasis networks are complementary to physical protein interaction networks, providing valuable insight into both the function of individual genes and the overall wiring of the cell. A high-throughput method termed "epistatic mini array profiles" (E-MAPs) was recently developed in yeast to quantify alleviating or aggravating interactions between gene pairs. The typical output of an E-MAP experiment is a large symmetric matrix of interaction scores. One problem with this data is the large amount of missing values – interactions that cannot be measured during the high-throughput process or whose measurements were discarded due to quality filtering steps. These missing values can reduce the effectiveness of some data analysis techniques and prevent the use of others. Here, we discuss one solution to this problem, imputation using nearest neighbors, and give practical examples of the use of a freely available implementation of this method.

Key words: Protein interactions, Genetic interactions, Epistasis, Imputation, Biological networks

1. Introduction

1.1. Epistatic Mini Array Profiles

Genetic interactions are identified when mutations in multiple genes produce a phenotype which differs from what is expected based on individual mutations in each gene. Recently, high-throughput assays have been developed for the acquisition of large amounts of pairwise genetic interaction data in model organisms. Epistatic mini array profiles (E-MAPs) are one such approach capable of quantitatively measuring the strength of pairwise genetic interactions in yeast. The procedure supports the identification of both positive (alleviating) and negative (aggravating) interactions

Gerard Cagney and Andrew Emili (eds.), *Network Biology: Methods and Applications*, Methods in Molecular Biology, vol. 781, DOI 10.1007/978-1-61779-276-2_17, © Springer Science+Business Media, LLC 2011

between genes, assignments that are extremely valuable in interpreting the biological basis of the epistatic relationships (1).

An E-MAP is most commonly represented in the form of a symmetric matrix with real-valued entries indicating the type and strength of interaction between each pair of genes under consideration. These scores are calculated based on the divergence in growth of yeast strains with two disrupted genes from the expected growth rate. Typically, a normalization process is applied to the interaction scores so that positive matrix entries denote an alleviating interaction, negative matrix entries denote an aggravating interaction, and values close to zero indicate the probable absence of an interaction between two genes, i.e., they function in independent pathways in the cell. Full details of the experimental procedure and the normalization process are described in Collins et al. (2).

The standard analysis technique applied to the E-MAP score matrix is agglomerative hierarchical clustering using the the Cluster tool (3). This type of analysis often provides insight into the underlying biology. For example, subsets of genes with similar interaction profiles may signify complexes of proteins involved in common biological processes.

One common characteristic of E-MAPs is the high proportion of missing entries that they contain (up to ~35%). Missing entries correspond to pairs of genes for whom interaction strengths could not be measured during the high-throughput process or those that were subsequently filtered due to unreliability. These missing values can reduce the effectiveness of downstream analysis, e.g., introducing instability in clustering (4), and prevent the use of others, e.g., matrix factorization techniques such as SVD and PCA. As each genetic interaction implies a functional relationship between gene pairs, individual interactions themselves can provide valuable biological insight. Consequently, there is a need for imputation techniques to "fill in the blanks."

1.2. Imputation

A growing body of work focuses on the prediction of genetic interactions, but to date the emphasis has largely been on the prediction of binary links, i.e., synthetic lethality. These methods have had some success by integrating diverse biological data (5) or by focusing on network topology of the underlying protein interaction network (6, 7). Additionally, methods have been developed for predicting synthetic lethality using only the graph of synthetic lethal interactions (8, 9).

Recently, Ulitsky et al. (10) and Ryan et al. (11) have addressed the problem of imputing quantitative genetic interactions within E-MAPs, with both groups adopting related approaches.

Ryan et al. noted the similarities between E-MAP and gene expression datasets and adapted techniques originally developed for gene expression to work with the symmetric pairwise data found in

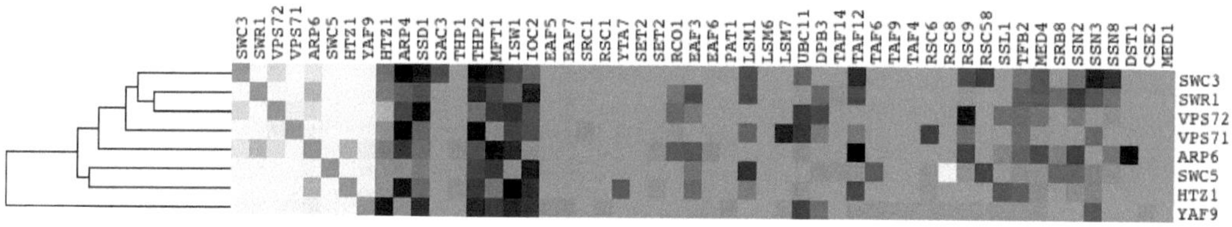

Fig. 1. An example of coherence taken from the Chromosome Biology E-MAP. Members of the Swr1 complex display similar interaction profiles, and as a result are clustered together.

E-MAPs. The goal in both cases is to construct a complete dataset by imputing quantitative measurements in order to improve the subsequent data analysis. Additionally, both E-MAPs and gene expression datasets display coherence among genes. For gene expression data, this is considered to be indicative of co-regulation while for E-MAPs, it is indicative of co-complex or pathway membership. An example of this coherence is shown in Fig. 1.

A number of gene expression imputation techniques were adapted to work with E-MAP data, including variations of the *K* nearest neighbors algorithm (12), local least squares (13), and Bayesian Principal Components Analysis (14). The performance of these techniques was assessed on a number of different datasets (including two different species) using a number of different metrics. Symmetric nearest neighbor-based approaches offered consistently accurate imputations in a tractable manner, and we focus here on one such approach – weighted symmetric nearest neighbors (wNN).

Nearest neighbor imputation is a simple strategy that uses genes with similar interaction profiles to impute missing values. Standard imputation algorithms based on nearest neighbors involve imputing values in feature-based asymmetric datasets (e.g., gene expression datasets). Ryan et al. proposed a simple modification to this approach designed to handle symmetric data. For each missing interaction (i, j), find the *K* nearest neighbor(s) for both gene *i* and gene *j*. Then, find the values for the interaction of *i* with *j*'s neighbors and *j* with *i*'s neighbors. These values are averaged to provide an imputed value for the missing entry (i, j). An illustration of this approach is shown in Fig. 2.

One issue with standard nearest neighbor-based approaches is that the accuracy of the method is very dependent on the choice of the parameter *K*. This can be alleviated by weighting the contribution of the neighbors based on their similarity to the query gene so that more similar genes make a greater contribution to the imputation. The degree of the contribution can be controlled by the choice of the weighting system. Ryan et al. employed the following weighting system (originally described in (15)), which ensures that closer neighbors are considerably more influential than more distant neighbors. Given a value *r* denoting the Pearson

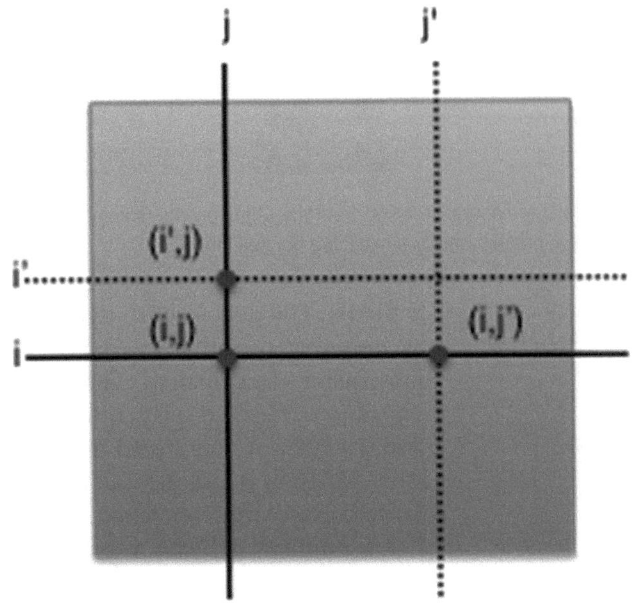

Fig. 2. Symmetric nearest neighbors with $K=1$. To estimate the missing value (i, j), the values given by (i', j) and (i, j') would be combined.

correlation between a gene i and its neighbor i', the weight $w(i, i')$ is calculated as follows:

$$w(i,i') = \left(\frac{r^2}{1 - r^2 + \varepsilon} \right)^2.$$

A benefit of this weighting system (determined empirically) is that once the choice of neighbors is sufficiently high ($K>20$), adding additional neighbors does not have a significant negative impact on the accuracy of the imputation. Near-optimal performance was obtained in all tested datasets with $K=50$, so this is the default for the implementation we discuss.

Ulitsky et al. used the nearest neighbors as the basis for a more advanced approach – performing a linear regression over 167 features, including nearest neighbors, interactions between neighbors, and diverse genomic information (e.g., protein–protein interactions, shared phenotypes, Gene Ontology terms). However, they noted that this additional information offered only a marginal improvement to the accuracy of imputation and that its absence would not seriously affect imputation.

Although the E-MAP approach was developed for the budding yeast *Saccharomyces cerivisiae* (16), it has recently been extended to the fission yeast *Schizosaccharomyces pombe* (17). An analogous technique has also been developed for *Escherichia coli* (18) while related approaches exist for other organisms (e.g., *Caenorhabditis elegans* (19)). The additional genomic data available for these organisms is more limited than that for *S. cerivisiae*. The method we describe here, the wNN method of Ryan et al., does not rely on the availability of external data, a factor that may be important as the range of epistasis studies increases. Furthermore, the implementation is freely available.

1.3. Assessing the Accuracy of Imputation

While methods for assessing the accuracy of imputation are primarily of interest to those developing imputation methods, they should also be of concern to those applying the results. No imputation method performs equally well on all datasets, so for each new dataset it is worth estimating the accuracy of imputation prior to analyzing the results.

We introduce in the methods section an implementation of *K* fold cross validation, a simple tool for estimating the accuracy of imputation on a given E-MAP. It works as follows – the measured interactions are randomly partitioned into ten equally sized sets (or folds). Imputation is then performed ten times – with the interactions from onefold hidden in each run. We can then compare the imputations for these artificially introduced missing values with their actual measured interactions.

The program uses two metrics for assessing the accuracy of imputation – Pearson's correlation and the Normalized Root Mean Squared Error (NRMSE). Pearson's correlation has a scale of −1 to 1, and a higher correlation indicates a more accurate imputation.

NRMSE is defined as follows:

$$\text{NRMSE} = \sqrt{\frac{\text{mean}[(ij_{\text{answer}} - ij_{\text{guess}})^2]}{\text{variance}\,[ij_{\text{answer}}]}},$$

where ij_{answer} denotes the set of known values and ij_{guess} denotes the corresponding set of imputed values. This is designed so that estimating the dataset mean results in an NRMSE of 1 and more accurate imputation results in a lower score.

1.4. Benefits of Imputation

There are two primary reasons for imputation – to improve downstream analysis, such as clustering, and to identify individual interactions which may be worth testing experimentally. Ryan et al. gave an example of how imputation could improve the use of average linkage hierarchical clustering, the standard analysis applied to E-MAPs. They identified clusters which had a statistically significant overlap with a known protein complex before and after imputation. They noted that after imputation in one E-MAP, a number of complexes were found with higher precision while in another E-MAP, an extra complex was identified.

Previous studies have used protein complexes as a way of interpreting genetic interactions – either within or between protein complexes. Many complexes were found to be "monochromatic"–enriched with either predominantly alleviating or predominantly aggravating interactions. Ulitsky et al. showed how the number of such complexes was increased significantly after imputation, suggesting a further benefit of imputation prior to downstream analysis.

Both groups showed that gene pairs predicted to have a strong genetic interaction were enriched with annotations similar to those of measured interactions (e.g., shared phenotypes, shared biological process, protein–protein interactions). This is an indication that imputation can generate reliable predictions for epistatic interactions, and thus may be worth validating using smaller scale experiments.

2. Materials

An implementation of the symmetric nearest neighbors algorithm can be obtained from http://www.bioinformatics.org/emapimputation.

This implementation requires an installation of Python 2.x preinstalled on many operating systems and also available at http://www.python.org/download/.

The expected input to the program is a symmetric score matrix in tab-delimited plain text with the first row and column containing gene names (see Note 1). For those using the E-MAP toolbox for MATLAB (20), such a file can be created using the "exportForCluster3_0" command. If the matrix is in Excel, the plain text matrix can be created by choosing "File"→"Save As"→"Tab Delimited Text." For reference, an example matrix is included with the implementation.

It may also be beneficial to prefilter the matrix (see Note 2).

3. Methods

Here, we give examples of three different use cases for our imputation software – obtaining a complete matrix for further analysis, obtaining a list of strong interactions for small-scale follow-up, and estimating the accuracy of imputation.

1. To obtain a complete matrix as output, run the program using the following command:

   ```
   python symmetricNN.py -i matrix_input.txt -o
   matrix_output.txt
   ```

   This matrix can then be used as an input to other programs, e.g., the *Cluster* program (3) (see Note 3) or the R environment (21). Additional input parameters can be specified for this command, e.g., to adjust the number of neighbors used for imputation, but in practice they are rarely needed (see Note 4).

2. To obtain a ranked list of strong genetic interactions, use the following command:

   ```
   python symmetricNN.py -i matrix_input.txt -r
   ranked_output.txt
   ```

Table 1
Strong epistatic interactions ranked based on the weight calculated from their neighbors

Gene A	Gene B	Score	Weight
YJL004C	YJL024C	-7.08	53.61
YGL005C	YGR261C	-6.02	37.31
YBR288C	YGL005C	-6.33	35.31
YGL005C	YHL031C	-11.40	34.83
YLR039C	YLR056W	4.33	33.24

The output of this command is a list of genetic interactions predicted to be strongly alleviating (score > 2) or strongly aggravating (score < -2.5). They are ranked according to the total weight of their neighbors – so pairs which were imputed using very similar neighbors are ranked higher than those with more distant neighbors. There is no direct interpretation to the weight column; however, the pairs at the top of the list can be considered a more reliable estimate, and thus a better candidate for a small-scale follow-up. Sample output is shown in Table 1.

3. The accuracy of imputation on a given dataset can be estimated using the following command:

```
python test_imputation.py -i matrix_input.txt
```

Note that this command takes approximately ten times longer to compute than standard imputation, as it is essentially performing the imputation ten times.

The output of this command is similar to the following:

```
Estimated correlation=0.63
Estimated NRMSE=0.77
```

Typically, for an E-MAP, the correlation is in the range 0.6–0.7 and the NRMSE is in the range 0.7–0.8.

4. Notes

1. Although tab-delimited text files are the default input file, other formats are acceptable using additional parameters. For example, to read a comma-separated matrix where missing values are identified using the term "NA," you would execute the following command:

```
python symmetricNN.py -i input.csv -o output.csv -s ',' -m 'NA'
```

This allows imputation to be carried out on matrices generated by other applications, e.g., R.

2. Genes for which there are very few measurements present can result in poor imputation because their correlation with other genes may be overestimated. These genes can also cause problems with traditional clustering methods, so it is advisable to remove them prior to analysis.
3. There is no guarantee that imputation will "improve" clustering, just as there is no single metric for assessing the quality of a clustering. One approach is to use external benchmarks – Ryan et al. use this approach to demonstrate that at least one use of clustering (identifying known protein complexes) can be improved by imputation. Stability analysis is an alternative approach – a way of assessing a clustering's robustness to slight perturbations – e.g., if a small percentage of missing values are added, does the resulting clustering appear significantly different? Our own internal experiments indicate that imputation can significantly improve clustering stability.
4. The number of neighbors used for imputation can be specified using the "K" parameter. As discussed, Ryan et al. previously showed that wNN is relatively insensitive to the choice of K and that near-optimal performance is achieved in every dataset using $K=50$. Consequently, 50 is the default in our implementation and does not need to be altered. Additionally, it is possible to use the application in an "unweighted" mode, where each of the K neighbors contributes equally to the imputation. In practice, this has never returned more accurate results than wNN and is only included for completeness.

References

1. Collins, S.R., Roguev, A. & Krogan, N.J. Quantitative Genetic Interaction Mapping Using the E-MAP Approach. *Guide to Yeast Genetics: Functional Genomics, Proteomics, and Other Systems Analysis* **Volume 470**, 205–231 (2010).
2. Collins, S.R., Schuldiner, M., Krogan, N.J. & Weissman, J.S. A strategy for extracting and analyzing large-scale quantitative epistatic interaction data. *Genome Biol* 7, R63 (2006).
3. Eisen, M.B., Spellman, P.T., Brown, P.O. & Botstein, D. Cluster analysis and display of genome-wide expression patterns. *Proc Natl Acad Sci USA* **95**, 14863–8 (1998).
4. Tuikkala, J., Elo, L., Nevalainen, O. & Aittokallio, T. Missing value imputation improves clustering and interpretation of gene expression microarray data. *BMC Bioinformatics* **9**, 202 (2008).
5. Wong, S.L. et al. Combining biological networks to predict genetic interactions. *Proceedings of the National Academy of Sciences of the United States of America* **101**, 15682–15687 (2004).
6. Kelley, R. & Ideker, T. Systematic interpretation of genetic interactions using protein networks. *Nat Biotech* **23**, 561–566 (2005).
7. Paladugu, S., Zhao, S., Ray, A. & Raval, A. Mining protein networks for synthetic genetic interactions. *BMC Bioinformatics* **9**, 426 (2008).
8. Qi, Y., Suhail, Y., Lin, Y., Boeke, J.D. & Bader, J.S. Finding friends and enemies in an enemies-only network: A graph diffusion kernel for predicting novel genetic interactions and co-complex membership from yeast genetic interactions. *Genome Research* **18**, 1991–2004 (2008).

9. Chipman, K. & Singh, A. Predicting genetic interactions with random walks on biological networks. *BMC Bioinformatics* **10**, 17 (2009).
10. Ulitsky, I., Krogan, N. & Shamir, R. Towards accurate imputation of quantitative genetic interactions. *Genome Biology* **10**, R140 (2009).
11. Ryan, C., Greene, D., Cagney, G. & Cunningham, P. Missing value imputation for epistatic MAPs. *BMC Bioinformatics* **11**, 197 (2010).
12. Troyanskaya, O. et al. Missing value estimation methods for DNA microarrays. *Bioinformatics* **17**, 520-525 (2001).
13. Kim, H., Golub, G.H. & Park, H. Missing value estimation for DNA microarray gene expression data: local least squares imputation. *Bioinformatics* **21**, 187–198 (2005).
14. Oba, S. et al. A Bayesian missing value estimation method for gene expression profile data. *Bioinformatics* **19**, 2088–2096 (2003).
15. Bø, T.H., Dysvik, B. & Jonassen, I. LSimpute: accurate estimation of missing values in microarray data with least squares methods. *Nucleic Acids Res.* **32**, e34 (2004).
16. Schuldiner, M. et al. Exploration of the Function and Organization of the Yeast Early Secretory Pathway through an Epistatic Miniarray Profile. *Cell* **123**, 507–519 (2005).
17. Roguev, A., Wiren, M., Weissman, J.S. & Krogan, N.J. High-throughput genetic interaction mapping in the fission yeast Schizosaccharomyces pombe. *Nat Meth* **4**, 861–866 (2007).
18. Typas, A. et al. High-throughput, quantitative analyses of genetic interactions in E. coli. *Nat Meth* **5**, 781–787 (2008).
19. Lehner, B., Crombie, C., Tischler, J., Fortunato, A. & Fraser, A.G. Systematic mapping of genetic interactions in Caenorhabditis elegans identifies common modifiers of diverse signaling pathways. *Nat Genet* **38**, 896–903 (2006).
20. *EMAP toolbox for MATLAB.* at <http://sourceforge.net/projects/emap-toolbox/>
21. Team, R.D.C. R: A Language and Environment for Statistical Computing. **3**, 2673.

Chapter 18

Displaying Chemical Information on a Biological Network Using Cytoscape

Iain M. Wallace, Gary D. Bader, Guri Giaever, and Corey Nislow

Abstract

Cytoscape is an open-source software package that is widely used to integrate and visualize diverse data sets in biology. This chapter explains how to use Cytoscape to integrate open-source chemical information with a biological network. By visualizing information about known compound–target interactions in the context of a biological network of interest, one can rapidly identify novel avenues to perturb the system with compounds and, for example, potentially identify therapeutically relevant targets. Herein, two different protocols are explained in detail, with no prior knowledge of Cytoscape assumed, which demonstrate how to incorporate data from the ChEMBL database with either a gene–gene or a protein–protein interaction network. ChEMBL is a very large, open-source repository of compound–target information available from the European Molecular Biology Laboratory.

Key words: Cytoscape, Network visualization, Druggable targets, Chemical biological networks, Chemical networks

1. Introduction

Cytoscape is an open-source software package that is used to visualize biological data as an interaction network (1). In such a biological network, nodes represent biological entities (e.g., proteins) and connections or edges between nodes represent a defined relationship between the biological entities (e.g., a direct physical interaction between proteins). This type of network can then be analyzed in a variety of ways to gain insight into the system of interest (2). For example, identifying clusters of highly connected proteins could indicate that they share function(s); alternatively, proteins that are connected to many other proteins could be a central node in the system.

Gerard Cagney and Andrew Emili (eds.), *Network Biology: Methods and Applications*, Methods in Molecular Biology, vol. 781, DOI 10.1007/978-1-61779-276-2_18, © Springer Science+Business Media, LLC 2011

One of the strengths of Cytoscape is its ability to easily allow a researcher to integrate and visualize different types of data, such as data from gene expression experiments or functional annotations, onto a biological network (3). This is accomplished by first loading the data onto a node or an edge attribute (such as gene expression data or interaction confidence) and then mapping these attributes onto visual attributes (such as node color or edge width). In this manner, diverse data sources and types can be visualized simultaneously in a user-defined manner, in a network context. For example, Cytoscape was used to visualize a network based on the correlation of yeast genetic interaction profiles (4) and the clustering of functionally related yeast genes was visualized by coloring the nodes based on gene function annotations.

Another source of information that can be easily added to a biological network in Cytoscape is compound–target information, for example the name or mode of action of a target's inhibitor. As we describe in the methods, this type of data can be annotated to nodes in a network and visualized by changing the color of nodes if a compound is known to inhibit the protein in question. By visualizing information about known compound–target interactions in the context of a biological network of interest, an experimentalist can rapidly identify novel avenues to perturb the system with compounds to obtain new mechanistic insight and potentially identify therapeutically relevant targets. For example, such a strategy can be employed using a synthetic lethal interaction network, a type of gene-gene network where nodes represent genes and edges indicate that if both of these genes are mutated the cell is sick or dead. Thus, coloring such nodes highlights tractable drug targets for inhibition. The prediction from such a visualization is that a cancer cell harboring a mutation in one of these genes will be hypersensitive to the chemical inhibition of its synthetic lethal partner. For instance, genes in a synthetic lethal network based on known oncogenes can be colored to indicate if there is a known inhibitor for the protein product of this gene. A strategy based on identifying druggable synthetic lethal interactions in cancer was demonstrated to be effective in a recent phase II clinical trial. In this study, Olaparib™, a poly(ADP-ribose) polymerase (PARP) inhibitor that induces synthetic lethality in homozygous BRCA-deficient cells in culture, was found to be effective in the clinic for treating advanced ovarian cancer in patients with *BRCA1* or *BRCA2* mutations (5).

There are a number of freely available databases that contain compound–target information which can be displayed on a biological network using Cytoscape. They range from databases focused on approved drugs and clinical candidates (6) to databases that attempt to extract all available information regarding named compounds and their targets with text mining algorithms available in the literature (7). One of the largest, freely available databases of this type is ChEMBL which is available from the European Molecular Biology Laboratory (EMBL). ChEMBL is a database of bioactive drug-like small molecules with associated bioactivities that were manually curated from the literature. It contains more than 500,000 compounds and more than 2.5 million bioactivities. We used ChEMBL to demonstrate how to display chemical data and activity information on a biological network because each compound–target interaction is supported by multiple levels of quality control. First, the source publication for each interaction is provided. Second, each interaction is given a confidence level to indicate how likely the named target is the true target based on the type of assay involved. Finally, the concentration of the compound is also provided.

2. Materials: Software and Databases

2.1. Cytoscape

Cytoscape is a freely available, open-source bioinformatics software platform for visualizing interaction networks and integrating theses networks with alternative sources of data. The core distribution of Cytoscape provides a basic set of features for network analysis and visualization. There is an active community of developers who expand the features of Cytoscape by creating freely available plug-ins for a wide variety of uses. For example, there is a plug-in available to predict gene function (8) or for searching a network for a subgraph matching a specific topology (9). There are extensive tutorials on how to use Cytoscape in different contexts, such as integrating networks with gene expression data (3).

To install Cytoscape, navigate to the Cytoscape homepage (10), click on the button "Download Cytoscape Now," and follow the installation instructions for the appropriate operating system. Cytoscape is available for all of the major platforms as a JAVA program.

2.2. Chemviz Plug-In

Chemviz is a Cytoscape plug-in that extends Cytoscape's capabilities into the realm of cheminformatics. It can display 2D chemical structures using "smiles" or InChI strings that are annotated to one or more nodes in a network. Chemviz can also perform calculations based on these attributes, such as creating a compound similarity network for a selection of nodes. More details are available on the Chemviz home page (11).

To install the plug-in:

1. In the Cytoscape menu bar, click on "Plugins" → "Manage Plugins."
2. Click on the "Analysis" tab in the "Available for Install" section.
3. Select the plug-in called "ChemViz" and press "Install." A status bar appears indicating the installation progress. When installed, the plug-in is immediately available from the menu bar under "Plugins" → "Cheminformatic tools." Default settings, such as which attributes contain the structure details, can be modified in the "Settings" option of the plug-in.

2.3. Sample Files

Two sample files, a network file and an attribute file, are required to run the first protocol described in Subheading 3.1. Both are available from our supplementary Web page (12).

2.3.1. Network File

"KRAS_SL_genes_from_publications.txt" is a tab-delimited file that describes a synthetic lethal interaction network between KRAS and 116 genes that was derived from two publications (13, 14). A synthetic lethal interaction is a type of genetic interaction, where cell death results from the combined mutations in two genes but not from mutations in each gene individually. The file contains five columns:

Screened Entrez Gene: The KRAS Entrez Gene number.

SL Entrez Gene: The Entrez Gene number for the gene that was reported to be synthetic lethal with the KRAS gene.

Screened Gene: Gene symbol for KRAS.

SL Gene: Gene symbol for the gene that was reported to be synthetically lethal with the KRAS gene.

PubMed Link: Link to the publication that reported this interaction.

2.3.2. Attribute File

"ChEMBL_targets_with_named_compounds.txt" is a tab-delimited file describing 2,104 proteins from a variety of species that are reported in the ChEMBL database (v6) to have at least one published, high-confidence (≥5 ChEMBL confidence level), potent interaction (reported potency in ChEMBL <1 μM) with a compound. This file is meant to be used as a guide, and the interested reader is encouraged to generate similar files to display different types of chemical data, such as compound–target interactions from different databases (see Note 1).

This sample file contains 14 columns:

UniProt identifier (ID): The UniProt ID for the protein, as listed in ChEMBL.

Entrez Gene ID: The Entrez Gene ID for this protein based on the ID mapping service provided by the UniProt Web site (see Note 2).

Protein name from ChEMBL: The common name for this protein as listed by ChEMBL.

Number of different compounds reported in ChEMBL: The number of distinct compounds that have a value of <1 μM associated with this protein at a ChEMBL confidence level of 5 or greater. Because commercial availability of particular compounds can vary over time, this number could be used as an estimate of how easy it is to source a compound to inhibit this protein. Additional availability information can be obtained from the ZINC database (15), a free database of commercially available compounds.

Number of publications reporting a compound–protein interaction: Based on the same criteria as above, this is the number of distinct publications that report at least one compound with an interaction with this protein. This number could be used as a proxy for how often different binding studies on this protein are carried out.

Top Named Compound: Based on the criteria above, this is the compound in ChEMBL that has the most publications demonstrating an interaction with this protein out of all the compounds in ChEMBL that have a name. Only ~4,000 compounds in ChEMBL (version 6) have a name associated with them. Names in ChEMBL are derived from a variety of sources, including commonly used names, research codes, trade names, and United States Adopted Names (USANs, unique nonproprietary names assigned to pharmaceuticals marketed in the USA).

SMILES for example compound: The SMILES string in this column represents the chemical structure for an example compound that has been reported in ChEMBL to interact with this protein. The example compound was chosen because it has the most publications demonstrating an interaction with this protein out of all the compounds in ChEMBL. Note: This compound is not necessarily a compound with a name associated with it in ChEMBL.

SMILES for named compound: Same as above, but for the named compound.

InChIKey for example compound: InChIKey is an alternative chemical representation of compounds that is optimized for searching for chemical information using text-based search engines, such as Internet searches and database queries.

InChIKey for named compound: Same as above, but for the named compound.

#Publications reporting interaction for example compound: This column contains the number of publications in ChEMBL in which the example compound has been shown to interact with this protein. It could be used as a measure of how reliable this particular compound interaction is based on the assumption that the more often the interaction is tested and reported the more reliable it is.

#Publications reporting interaction for named compound: Same as above, but for the named compound.

ChEMBL link: A link to the ChEMBL Web page for this protein. The Web page contains a wide variety of additional information regarding the reported interaction of this protein with different compounds.

Compound: A flag to identify which proteins/genes have been annotated in ChEMBL. This is useful for changing the visualization of the network in Cytoscape based on the presence or absence of compound information for a particular node.

3. Methods

In this section, we outline two different approaches to annotating biological networks with compound information available from the ChEMBL database. In the first protocol, the visualization of a node representing a gene in a gene–gene interaction network will be modified if an inhibitor is known for this gene product (Fig. 1). This protocol can be easily adapted to other types of networks, such as protein–protein interactions, provided an annotation file exists that can be used to map what is being visualized to an interaction with a compound. It can also be adapted to display compound information from any database that reports compound–target interaction, such as Drugbank (6) or the Therapeutic Target Database (16). All that is required is a Cytoscape attribute file, similar to the one used here for ChEMBL, which enumerates these interactions.

In the second protocol, a protein–protein interaction network is augmented with additional nodes that represent compounds known to inhibit a particular protein in the network. This type of network can be used to explore drug poly-pharmacology, i.e., to identify compounds that bind multiple proteins, as well as for identifying which such protein inhibitors have been published. It uses the PSICQUIC Web service (17), which provides a standard interface to access a variety of molecular interaction databases. In this

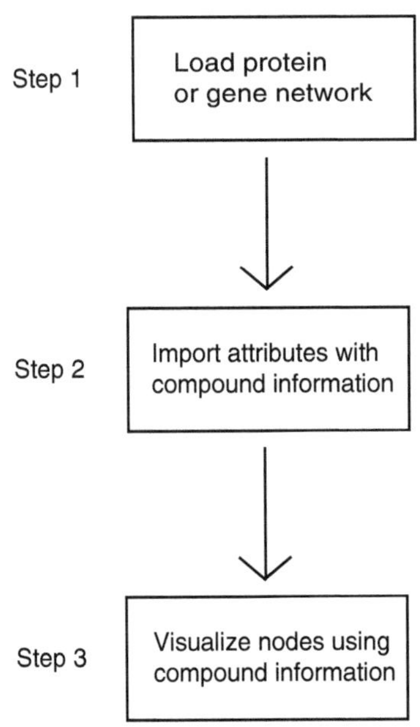

Fig. 1. A flowchart outlining the basic steps involved in annotating a biological network with chemical information in Cytoscape.

protocol, we focus on integrating compound data from ChEMBL with the protein–protein interaction database, MINT (18). The protocol also serves as a general guide on how to annotate any of the interaction networks that are available through the PSICQUIC service with compounds from the ChEMBL database. An updated list of available networks is available from the PSICQUIC Web page (19). The advantage of using this protocol over the first is that the network is always current because it is downloaded directly from the source repository and does not require an attribute file that would have to be updated with each new source repository update. A disadvantage is that the initial network visualization might get too crowded particularly if the protein of interest has an abundance of inhibitors present in ChEMBL.

3.1. Annotating a Gene–Gene Interaction Network with Compounds

In this protocol, a gene–gene interaction network is created and annotated with known compound interaction data. The gene–gene interaction network represents a synthetic lethal interaction network centered on the KRAS gene. Mutations in the KRAS gene, a member of the Ras family of small GTPases, are frequently found in pancreatic, thyroid, colon, lung, and liver cancers and are correlated with poor prognosis (14). In this network, nodes represent genes and edges represent a synthetic lethal interaction based on published interactions (13, 14). By overlaying compound information onto this network, we can rapidly identify proteins that could be of therapeutic relevance for cancers expressing the KRAS oncogene. Compounds that target a gene product that is synthetically lethal with KRAS should, in principle, affect only cancer cells and not normal cells.

3.1.1. Read In and Layout the Network

1. From the Cytoscape menu bar, click "File"→"Import"→ "Network from Table/MS Excel." This opens a new screen called "Import Network and Edge Attributes from Table."
2. Click on "Select File(s)" and open the file "KRAS_SL_genes_from_publications.txt." A preview of the contents of this file is displayed at the bottom of the "Import Network and Edge Attributes" screen. Note that the default importing options do not work for this file, as it appears that more than 14 attributes are being imported (only 14 attributes are defined in the file).
3. Click on the "Show Text File Import Options." This provides a number of options to allow Cytoscape to read in the data correctly. Deselect the box on the left-hand side of the "Space" delimiter option. Next, click on "Transfer first line as attribute names." Note how the preview of the data changes as each option is chosen. It is possible to check that all the text is being imported properly by using the scrollbar in the data preview window. Now, there are 14 attributes being imported which is the same number as in the original file.

4. In the interaction definition area of the import screen, choose "Column 1" from the "Select Source node Column" drop-down menu. Next, choose "Column 2" from the "Select Target node column." Then, in the data preview window, click on the column entitled "PubMed Link" and click on the "Import" button. Click "Close" once the network has been successfully loaded. The nodes in this network represent genes, which are labeled with Entrez Gene IDs, and edges represent synthetic lethal interactions.
5. From the menu bar, click "Layout" → "Y Files" → "Organic." This layout displays the nodes and edges in the network using a force-directed algorithm (3). Nodes are modeled with mutually exclusive forces while edges induce an attractive force between nodes that they connect. The algorithm places the nodes in such a manner as to minimize these forces. This layout facilitates the identification of "hub" nodes that are involved in many interactions and the identification of clusters of tightly connected nodes, which could indicate functional modules. In this example, the KRAS gene is connected to all of the other genes and is laid out in the middle of a circle.

3.1.2. Import Node Attributes with Compound Information

1. From the menu bar, click "File" → "Import" → "Attributes from Table (Text/MS Excel)." This opens a new screen called "Import Annotation File."
2. Click on the "Select Files" button and locate the file "ChEMBL_targets_with_named_compounds.txt." A preview of the contents of the file appears in the data preview box in a similar manner as when the network was imported at Subheading 3.1.1.
3. Click on the "Show Text File Import Options" check box and then select the "Transfer first line as attribute names" check box. The "UniProt ID" column is highlighted in blue because the default option to map attributes to a network is to use column 1 as the key ID. However, in the sample network, the key ID is Entrez Gene ID. This must be modified to read in the attributes correctly so that they match with the nodes, also identified using Entrez Gene IDs (see Note 2).
4. Click on the "Show Mapping Options" check box. Chose "Entrez Gene ID" from the "Primary Key" drop-down box. Click "Import" to annotate the nodes with these attributes.

3.1.3. Change the Visualization of Nodes that Contain Compound Information

1. Click the "VizMapper" tab. Next, click the "Node Color" label in the Visual Mapping Browser to highlight it. Then, double click on the text that reads "Double-Click to create." This moves the Node Color object from the "Unused Properties" to the "Node Visual Mapping" properties section.

2. Click the "Please select a value" beside the "Node Color" label and select the "compound" attribute created earlier. Next, click "Please select a mapping type!" below the Node Color label and select "Discrete Mapper." Choose a color to assign to the nodes that represent compounds.
3. Repeat the above two steps for the "Edge Color" attribute, choosing a color based on the "PubMed Link" attribute.
4. Repeat steps 1 and 2 for the "Node Size" attribute, choosing the size of the node based on the attribute "Number of different compounds reported in ChEMBL." In this example, we suggest binning the values such that a node with 0–10 compounds gets a size of 40, 11–100 compounds, a size of 60, 101–500, a size of 80, and if more than 500 compounds are associated with this node, it gets a size of 100.
5. Next, assign a visual mapping to the "Node Label" attribute based on the attribute "Top named compound." In this case, choose the "Passthrough Mapping" option. This labels each of the nodes with a compound name. It is possible to change the position of this label and font size used by clicking on the box labeled "Defaults." This box is located just above the label "Visual Mapping Browser." It is possible to override a default setting for any of the visual mapping attributes by right clicking on the node and selecting "Visual Mapping Bypass." This is particularly useful if labels overlap with other objects, such as other labels or edges.

3.1.4. Exploring Compounds Annotated to Nodes

In the final network (Fig. 2), there are 24 genes that are colored due to their having an inhibitor annotated in ChEMBL. Each of these genes has nine compound-related attributes annotated to them which are described in Subheading 2.3.2. To display them in the "Data Panel," click on the "Select Attributes" button, which is the left-most icon in the "Data Panel," and select each of them. Selecting a colored node on the network displays the attributes for that gene in the "Data Panel."

One intriguing observation about this network is that four nodes are annotated to the compound "Alvocidib." The largest node with this label is the node corresponding to Gene ID "983." It is one of the three largest nodes in the network, reflecting that compound–protein interactions are often reported for this protein. To select this node, just click on it. This highlights the node and displays its attributes in the data panel. This gene encodes the protein "Cyclin-dependent kinase 1." There are 77 publications in which 558 compounds are associated with this protein in ChEMBL. To go to the ChEMBL page for this protein, click on the "Chembl Link" attribute in the data panel. To view the chemical structure of the example compound, right click on this node and select the option "Cheminformatics Tools" → "Depict2DStructure" → "Show

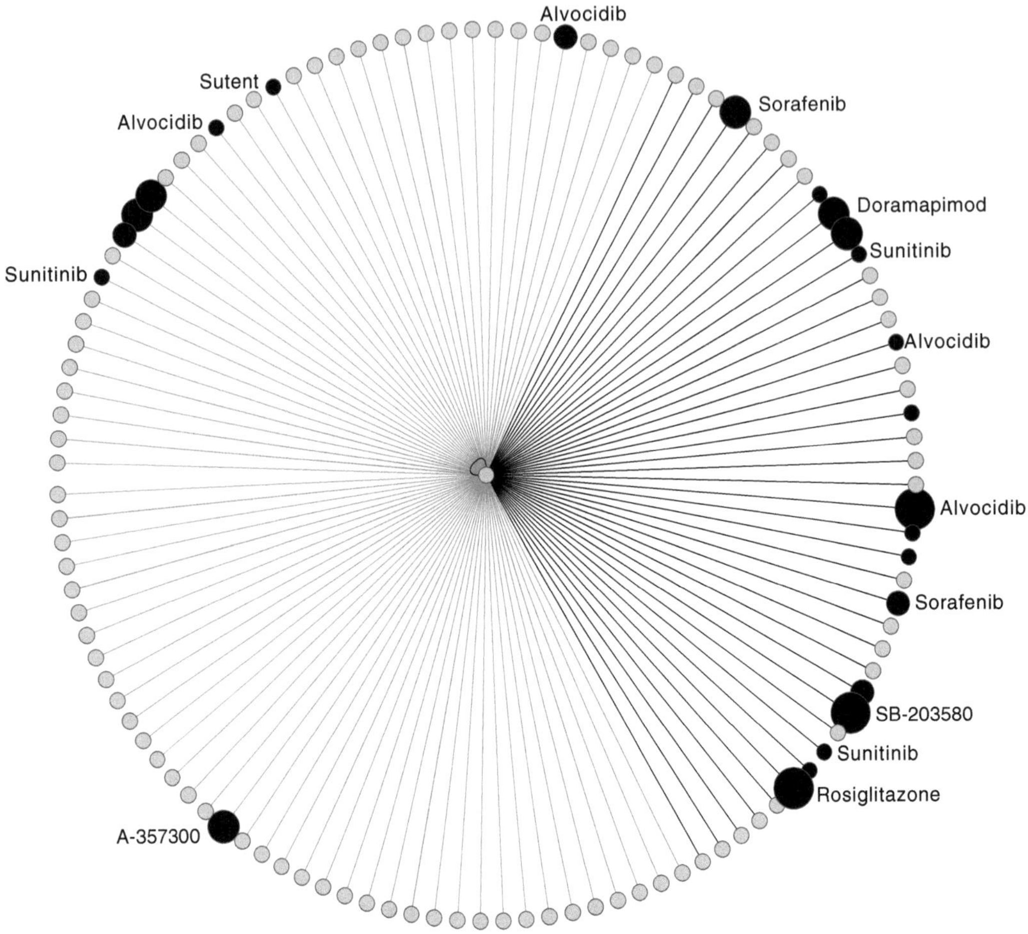

Fig. 2. A synthetic lethal network centered on the KRAS oncogene that is produced by following the steps outlined in Subheading 3.1. Nodes represent genes, and edges represent a synthetic lethal interaction between the two connected nodes. Nodes are colored *black* if an inhibitor is known for the gene product of this node. Node size represents the number of compounds that have been associated with this gene in ChEMBL. Nodes are labeled with a compound name if a suitable name exists in ChEMBL. Edges are colored depending on which one of two publications the synthetic lethal interaction is derived from.

2D structures from this Node." By right clicking on the InChIKey for this compound, one can search for it on the Web using the Google search engine. The top hit is from Drugbank and shows that an alternative name for this compound is flavopiridol and that it is a known cyclin-dependent kinase inhibitor. The other three proteins associated with this compound are cyclin A2, dual-specificity tyrosine-phosphorylation-regulated kinase 1B, and mitogen-activated protein kinase 7. Perhaps, the fact that it induces numerous synthetic lethal interactions with KRAS contributes to the effectiveness of alvocidib for treating certain cancer types. It is in a phase I trial (Clinical Trials ID: NCT00101231) for treating patients with relapsed or refractory acute myeloid leukemia (AML), acute lymphoblastic leukemia, (ALL), or chronic myelogenous leukemia.

AML and ALL are cancers associated with mutations in the KRAS oncogenes (20).

Clicking on "File → Save as" from the Cytoscape menu bar can save the network session. The network visualization can be exported to a variety of file formats, including PDF, by clicking on "File" → "Export" → "Network view as graphics."

3.2. Annotating a Protein–Protein Interaction Network with Compounds

This protocol allows one to annotate a protein–protein interaction network derived from the MINT database with compounds from the ChEMBL database. In this network, nodes represent both compounds and proteins. Edges between two protein nodes represent an experimentally verified protein–protein interaction that has been curated from the literature. Edges between a protein and compound nodes represent a high-confidence protein–compound interaction from ChEMBL as defined by the ChEMBL curators (see Note 1). A key step in this protocol is coloring the compound nodes differently from protein nodes. The advantage of this protocol over the one described in Subheading 3.1 is that the most up-to-date networks are downloaded dynamically from the relevant groups. It does not rely on a static attribute file, which has to be updated with every new release of ChEMBL. The disadvantage is that the final network might be cluttered, as some proteins have many known inhibitors annotated in ChEMBL.

3.2.1. Load Networks

1. From the Cytoscape menu bar, click on "File" → "Import" → "Import network from webservices." A new window appears entitled "Import network from database." This window allows Cytoscape to connect to a variety of resources on the Internet to download networks. Different data sources can be chosen from the drop-down list located at "Data Source," and information describing the selected data source can be obtained by clicking the "About" button.
2. Select the "PSICQUIC Universal Web Service Client" from the "Data Source" drop-down list.
3. In the search box, type "P19438 or Q13546" and press the Search button. P19438 and Q13546 are UniProt protein identifiers for the input proteins: tumor necrosis factor receptor 1 (P19438) and receptor-interacting serine/threonine-protein kinase 1 or (Q13546), both involved in the apoptosis pathway. Click "yes" when prompted to generate a new network from all the interactions that were found. Click "OK" when presented with the PSICQUIC data import result message box. This searches all of the molecular interaction databases available to PSICQUIC and identifies any of the interactions that involve any of these three proteins. The data import box displays all of the databases from which interactions were obtained. Clicking "OK" creates a separate network

for each of the different databases. Of particular note for this protocol are the ChEMBL and the MINT networks. The ChEMBL network contains compounds that have been shown to interact with the proteins that were input into the PSICQUIC search box and the input proteins themselves if they are present in ChEMBL. The MINT network contains protein–protein interactions involving the input proteins.

3.2.2. Color the Compound Nodes by Compound Property

1. Select the ChEMBL network from the network panel on the left.
2. From the menu bar, click "Layout" → "Y Files" → "Organic." In this example, the hub nodes are the input proteins and the rest of the nodes are the compounds retrieved from ChEMBL.
3. Open the "Create New Integer Attribute" by clicking on the middle icon in the top left-hand corner of the data panel and selecting "Integer Attribute."
4. Enter "compound" in the "Create New Integer Attribute" box, and select OK. This creates a new attribute named "compound" which is used to identify and color nodes that are compounds in a similar manner for protocol 1.
5. From the menu bar, click on "Select" → "Nodes" → "Select All Nodes."
6. Click on each of the hub genes while holding shift key. This step deselects each of the original input proteins and leaves all of the compounds selected. An alternative approach to deselect the original input proteins would be sorting the ID column in the data panel. The compound IDs start with either "chebi" or "ChEMBL" while the protein IDs start with "uniprotkb." Clicking on the protein IDs (holding shift key while clicking allows multiple IDs to be selected) in the data panel highlights the corresponding nodes in green on the network. Clicking on each of these nodes in turn deselects them.
7. Open the Node Attribute Batch Editor by clicking on the left-most icon on the top right-hand corner of the data panel. Select the "compound" attribute from the blank drop-down box, enter "1" into the text box, and press "GO." Close the node attribute batch editor by pressing the "x." This sets the color attribute to "1" for all the compounds.
8. Use the VizMapper protocol as outlined in protocol 1 to color the nodes based on the compound property. Coloring nodes in this manner means that compound nodes can be easily distinguished from protein nodes when the ChEBML network is merged with the protein–protein network.

3.2.3. Merge Networks

1. From the menu bar, click on "Plugins" → "Advanced Network Merge." This launches the "Advanced Network Merge"

plug-in that allows the user to combine two or more networks into one network. In this protocol, we merge the ChEMBL and MINT networks. Any of the networks downloaded can also be merged and annotated with compounds in a similar fashion.

2. In the "Networks to merge box," select the ChEMBL network and then click on the arrow pointing to the selected "networks box." Repeat this for the MINT network, and then click on the Merge button. This creates a new network called "Union" which contains all of the nodes and edges in both of the original networks. It also retains the node coloring that was generated based on the compound annotation. Compound nodes connecting multiple protein nodes indicate poly-pharmacology for compounds in the network. The user can rerun this protocol to expand the initial list of input proteins to annotate interesting regions of the protein–protein network with compound–protein interactions.

4. Notes

1. Care must be taken when creating an attribute file to use to map compound–target information onto a biological network from any compound–target database. While these databases contain vast amounts of useful data, they also contain unexpected data and potential experimental artifacts. For instance, a protein–compound network based on PubChem Web site, without careful filtering, contains many false positive interactions due to the presence of promiscuous compounds and other factors (21). In another database, Drugbank (6), there are two references that are annotated as providing evidence for over 7,000 compound–target interactions each. This is unlikely to be correct and will seriously bias downstream analysis if not filtered out.
2. It is important to take care when integrating multiple sources of data due to ambiguous naming conventions. In the case of mapping chemical information onto gene–gene networks, we have found that it is best to map the data using Entrez Gene IDs as they are unique and unambiguous. In contrast, gene names are not a useful for mapping information onto a network because multiple gene names can match to many genes. In the case of protein–protein networks, we have found it best to map compound interactions using UniProt IDs as they are the most widely used identifiers in the compound databases. To convert between various protein and gene identifiers, we have found the mapping service provided at the UniProt Web site (22), as well as the service provided by the EBI (23) to be particularly useful.

Acknowledgments

The second protocol discussed here is inspired by and adapted from a presentation by Anna Gaulton at the "Small molecule bioactivity resources at the EBI" course in January 2010. This work was supported by grants from the NHGRI to CN and GG, the CIHR to CN (MOPS-84305) and to GG (MOPS-81340), by NSERC support to GB and IMW, and from a Marie Curie fellowship to IMW. GG is a Canada Research Chair in Chemical Biology.

References

1. Shannon P, et al., *Cytoscape: a software environment for integrated models of biomolecular interaction networks*. Genome Res, 2003. **13**(11): p. 2498–504.
2. Merico D, D Gfeller, and GD Bader, *How to visually interpret biological data using networks*. Nat Biotechnol, 2009. **27**(10): p. 921–4.
3. Cline MS, et al., *Integration of biological networks and gene expression data using Cytoscape*. Nature protocols, 2007. **2**(10): p. 2366–82.
4. Costanzo M, et al., *The genetic landscape of a cell*. Science, 2010. **327**(5964): p. 425–31.
5. Audeh MW, et al., *Oral poly(ADP-ribose) polymerase inhibitor olaparib in patients with BRCA1 or BRCA2 mutations and recurrent ovarian cancer: a proof-of-concept trial*. Lancet, 2010. **376**(9737): p. 245–51.
6. Wishart DS, et al., *DrugBank: a comprehensive resource for in silico drug discovery and exploration*. Nucleic Acids Research, 2006. **34**(Database issue): p. D668–72.
7. Kuhn M, et al., *STITCH: interaction networks of chemicals and proteins*. Nucleic Acids Res, 2008. **36**(Database issue): p. D684–8.
8. Montojo J, et al., *GeneMANIA Cytoscape Plugin: Fast gene function predictions on the desktop*. Bioinformatics, 2010.
9. Ferro A, et al., *NetMatch: a Cytoscape plugin for searching biological networks*. Bioinformatics, 2007. **23**(7): p. 910–2.
10. *Cytoscape Home Page*. http://www.Cytoscape.org.
11. *ChemViz Plugin Home page*. http://www.cgl.ucsf.edu/Cytoscape/chemViz/.
12. *Supplementary Web Page*. http://baderlab.org/VisualizingChemicalInformation.
13. Barbie DA, et al., *Systematic RNA interference reveals that oncogenic KRAS-driven cancers require TBK1*. Nature, 2009. **462**(7269): p. 108–12.
14. Luo J, et al., *A genome-wide RNAi screen identifies multiple synthetic lethal interactions with the Ras oncogene*. Cell, 2009. **137**(5): p. 835–48.
15. Irwin JJ and BK Shoichet, *ZINC--a free database of commercially available compounds for virtual screening*. Journal of chemical information and modeling, 2005. **45**(1): p. 177–82.
16. Zhu F, et al., *Update of TTD: Therapeutic Target Database*. Nucleic Acids Research, 2010. **38**(Database issue): p. D787–91.
17. Orchard S, et al., *implementing data standards: a report on the HUPOPSI workshop September 2009, Toronto, Canada*. Proteomics, 2010. **10**(10): p. 1895–8.
18. Ceol A, et al., *MINT, the molecular interaction database: 2009 update*. Nucleic Acids Res, 2010. **38**(Database issue): p. D532–9.
19. *PSICQUIC databases*. http://www.ebi.ac.uk/Tools/webservices/psicquic/registry/registry?action=STATUS.
20. Liang D-C, et al., *K-Ras mutations and N-Ras mutations in childhood acute leukemias with or without mixed-lineage leukemia gene rearrangements*. Cancer, 2006. **106**(4): p. 950-6.
21. Han L, Y Wang, and SH Bryant, *A survey of across-target bioactivity results of small molecules in PubChem*. Bioinformatics, 2009. **25**(17): p. 2251–5.
22. *UniProt*. www.uniprot.org.
23. Côté RG, et al., *The Protein Identifier Cross-Referencing (PICR) service: reconciling protein identifiers across multiple source databases*. BMC Bioinformatics, 2007. **8**: p. 401.

Chapter 19

Modeling of Proteins and Their Assemblies with the Integrative Modeling Platform

Benjamin Webb, Keren Lasker, Dina Schneidman-Duhovny, Elina Tjioe, Jeremy Phillips, Seung Joong Kim, Javier Velázquez-Muriel, Daniel Russel, and Andrej Sali

Abstract

To understand the workings of the living cell, we need to characterize protein assemblies that constitute the cell (for example, the ribosome, 26S proteasome, and the nuclear pore complex). A reliable high-resolution structural characterization of these assemblies is frequently beyond the reach of current experimental methods, such as X-ray crystallography, NMR spectroscopy, electron microscopy, footprinting, chemical cross-linking, FRET spectroscopy, small-angle X-ray scattering, and proteomics. However, the information garnered from different methods can be combined and used to build computational models of the assembly structures that are consistent with all of the available datasets. Here, we describe a protocol for this integration, whereby the information is converted to a set of spatial restraints and a variety of optimization procedures can be used to generate models that satisfy the restraints as much as possible. These generated models can then potentially inform about the precision and accuracy of structure determination, the accuracy of the input datasets, and further data generation. We also demonstrate the Integrative Modeling Platform (IMP) software, which provides the necessary computational framework to implement this protocol, and several applications for specific-use cases.

Key words: Integrative modeling, Protein structure modeling, Proteomics of Macromolecular assemblies, X-ray crystallography, Electron microscopy, SAXS

1. Introduction

To understand the function of a macromolecular assembly, we must know the structure of its components and the interactions between them (1–4). However, direct experimental determination of such a structure is generally rather difficult. While multiple methods do exist for structure determination, each has a drawback.

Gerard Cagney and Andrew Emili (eds.), *Network Biology: Methods and Applications*, Methods in Molecular Biology, vol. 781, DOI 10.1007/978-1-61779-276-2_19, © Springer Science+Business Media, LLC 2011

For example, crystals suitable for X-ray crystallography cannot always be produced, especially for large assemblies of multiple components (5). Cryo-electron microscopy (cryo-EM), on the contrary, can be used to study large assemblies, but it is generally limited to worse than atomic resolution (6–8). Finally, proteomics techniques, such as yeast two-hybrid (9) and mass spectrometry (10), yield information about the interactions between proteins, but not the positions of these proteins within the assembly or the structures of the proteins themselves.

1.1. Integrative Modeling

One approach to solve the structures of proteins and their assemblies is by integrative modeling, in which information from different methods is considered simultaneously during the modeling procedure. The approach is briefly outlined here for clarity; it has been covered in greater detail previously (11–17). These methods can include experimental techniques, such as X-ray crystallography (5), nuclear magnetic resonance (NMR) spectroscopy (18–20), electron microscopy (EM) (6–8), footprinting (21, 22), chemical cross-linking (23–26), FRET spectroscopy (27), small-angle X-ray scattering (SAXS) (28, 29), and proteomics (30). Theoretical sources of information about the assembly can also be incorporated, such as template structures used in comparative modeling (31, 32), scoring functions used in molecular docking (33), and other statistical preferences (34, 35) and physics-based energy functions (36–38). Different methods yield information about different aspects of structure and at different levels of resolution. For example, atomic resolution structures may be available for individual proteins in the assembly; in other cases, only their approximate size, approximate shape, or interactions with other proteins may be known. Thus, integrative modeling techniques generate models at the resolution that is consistent with the input information. An example of a simple integrative approach is building a pseudo-atomic model of a large assembly, such as the 26S proteasome (39, 40), by fitting atomic structures of its subunits predicted by comparative protein structure modeling into a density map determined by cryo-EM (41, 42).

The integrative modeling procedure used here (12, 17) is shown in Fig. 1. The first step in the procedure is to collect all experimental, statistical, and physical information that describes the system of interest. A suitable representation for the system is then chosen, and the available information is translated to a set of spatial restraints on the components of the system. For example, in the case of characterizing the molecular architecture of the nuclear pore complex (NPC) (12, 13), atomic structures of the protein subunits were not available, but the approximate size and shape of each protein was known, so each protein was represented as a "string" of connected spheres consistent with the protein size and shape. A simple distance between two proteins can be restrained by

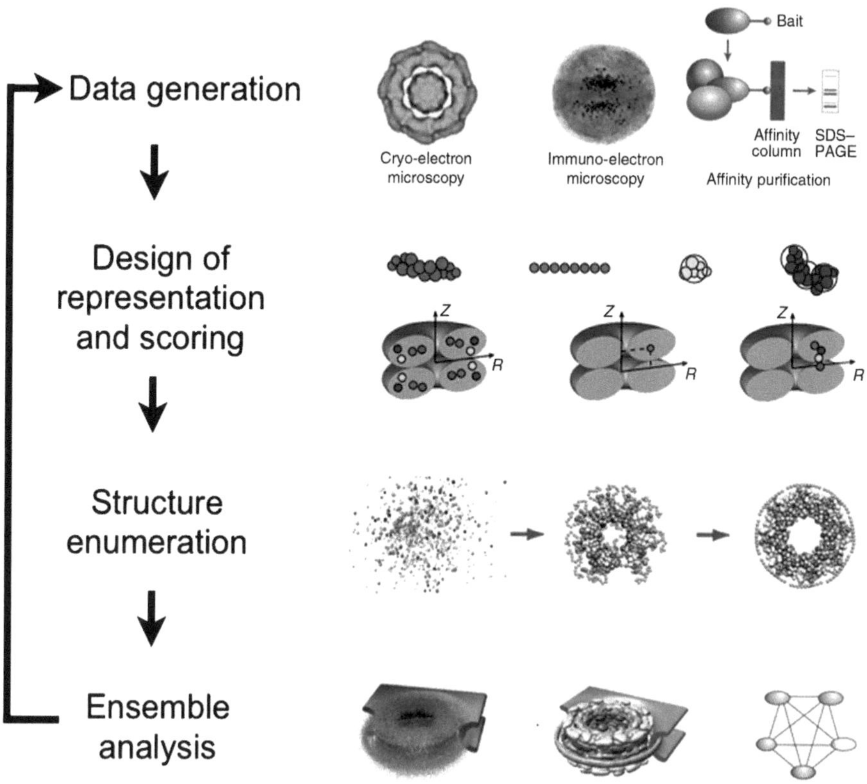

Fig. 1. Integrative modeling protocol. After the datasets to be used are enumerated, a suitable representation is chosen for the system, and the input information is converted into a set of spatial restraints. Models are generated that are optimally consistent with the input information by optimizing a function of these restraints. Analysis of the resulting models informs about the model and data accuracy and may help guide further experiments. The protocol is demonstrated with the construction of a bead model of the NPC (12).

a harmonic function of the distance, while the fit of a model into a three-dimensional (3D) cryo-EM density map can be restrained by the cross-correlation between the map and the computed density of the model. Next, the spatial restraints are summed into a single scoring function that can be sampled using a variety of optimizers, such as conjugate gradients, molecular dynamics, Monte Carlo, and inference-based methods (42). This sampling generates an ensemble of models that are as consistent with the input information as possible. In the final step, the ensemble is analyzed to determine, for example, whether all of the restraints have been satisfied or certain subsets of data conflict with others. The analysis may generate a consensus model, such as the probability density for the location of each subunit in the assembly.

1.2. Integrative Modeling Platform

We have developed the Integrative Modeling Platform (IMP) software (http://salilab.org/imp/) (12–15) to implement the integrative modeling procedure described above. Integrative modeling problems vary in size and scope, and thus, IMP offers a great deal

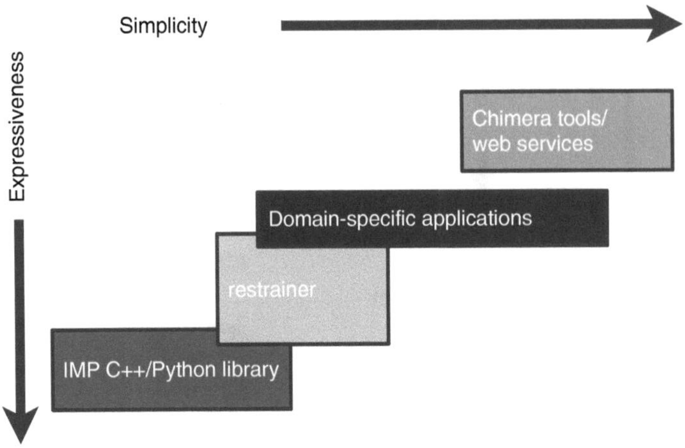

Fig. 2. Overview of the IMP software. Components are displayed by simplicity (or user-friendliness) and expressiveness (or power). The core C++/Python library allows protocols to be designed from scratch; higher-level modules and applications provide more user-friendly interfaces.

of flexibility and several abstraction levels as part of a multitiered platform (Fig. 2). At the lowest level, IMP provides building blocks and tools to allow methods developers to convert data from new experimental methods into spatial restraints, to implement optimization and analysis techniques, and to implement an integrative modeling procedure from scratch; the developer can use the C++ and Python programming languages to achieve these tasks. Higher abstraction levels, designed to be used by IMP users with no programming experience, provide less flexible but more user-friendly applications to handle specific tasks, such as fitting of proteins into a density map of their assembly, or comparing a structure with the corresponding SAXS profile. IMP is freely available as open source software under the terms of the GNU Lesser General Public License (LGPL). Integrative modeling, due to its use of multiple sources of information, is often a highly collaborative venture, and thus benefits from openness of the modeling protocols and the software itself.

2. Materials

To follow the examples in this discussion, both the IMP software itself and a set of suitable input files are needed. The IMP software can be downloaded from http://salilab.org/imp/download.html and is available in binary form for most common machine types and operating systems; alternatively, it can be rebuilt from the source code. The example files can be downloaded from http://salilab.org/imp/tutorials/basic.zip.

3. Methods

3.1. The IMP C++/ Python Library

The core of IMP is the C++/Python library, which provides all of the necessary components, as a set of classes and modules, to allow methods developers to build an integrative modeling protocol from scratch. Most users of IMP will use one of the higher-level interfaces described in later sections; however, we briefly demonstrate this library here to illustrate the core IMP concepts that these interfaces rely on.

The IMP library is split into a kernel and a set of extension modules (Table 1). The kernel is a small collection of classes that define the storage of information about the system and the main interfaces used to interact with that information. The information is stored in a set of `Particle` objects; these are flexible and abstract data containers, able to hold whatever information is necessary to represent the system. For example, a given `Particle` may be assigned *x*, *y*, and *z* attributes to store point coordinates, another may be assigned *x*, *y*, *z*, and a radius to represent a sphere, and another may contain two pointers to other `Particles` to represent a bond or another relationship. The kernel defines only the abstract interfaces to manipulate the data in the `Particles`, but does not provide implementations; these are provided in the extension modules. For example, it merely defines a `Restraint` as any object that, given a set of `Particles`, returns a score, and an `Optimizer` as an object that changes the attributes of all `Particles` to yield an optimized score over all restraints. It is the `core` module that provides, for example, a concrete `Restraint` acting like a harmonic "spring" between two point-like `Particles`, an `Optimizer` that utilizes the conjugate gradients minimization method, and much other functionality.

IMP includes a variety of modules (Table 1). Some modules provide the basic building blocks needed to construct a protocol, such as the `core` module that provides a functionality including harmonic restraints, point-like and spherical particles, and basic optimizers, and the `atom` module that provides atom-like particles, a molecular dynamics optimizer, etc. Other modules provide support for specific types of experimental data or specialized optimizers, such as the `em` module that supports electron microscopy data, and the `domino` module that provides an inference-based divide-and-conquer optimizer. IMP is designed so that it is easy to add a new module; for example, a developer working on incorporating data from a new experimental technique may add a new IMP module that translates the data from this technique into spatial restraints.

IMP is primarily implemented in C++ for speed; however, each of the classes is wrapped so that it can also be used from Python. A protocol can thus be developed from scratch by simply writing a Python script. As an example, we first look at the script `simple.py`

Table 1
IMP modules

Module name	Description
Basic modules	
`core`	Basic functionality commonly used in structural modeling, including representation of particles as rigid bodies, commonly used restraints such as distance, excluded volume, and connectivity (16), and frequently used optimizers such as Monte Carlo and conjugate gradients
`algebra`	General-purpose algebraic and geometric methods, including principal component analysis of attributes, geometric alignment between two sets of 3D coordinates and geometric manipulations of spheres, cones, cylinders, and cubes
`display`	Tools for displaying and exporting of IMP data, such as intermediate models in an optimization process, in PDB (50), Chimera (43), or PyMol (52) format
`statistics`	Basic statistics tools, including *k*-means clustering, Gaussian mixture model clustering and histogram calculation
`gsl`	Interfaces to allow algorithms from the GNU Scientific Library (53), including simplex, quasi-Newton and conjugate gradients optimizers, to be used in IMP
`container`	Tools and algorithms for manipulating subsets of the system's particles, such as maintaining a list of all pairs of particles that are spatially close
Structural modeling modules	
`atom`	Tools for manipulating atoms and proteins. The module allows molecules to be read or written in PDB format (50) and scored using force fields such as CHARMM (36). It also provides molecular dynamics and Brownian dynamics optimizers
`em`	Integration of 2D and 3D EM data into the integrative modeling procedure. The module provides functionality to read and write EM density maps in MRC, X-PLOR, Spider, and EM formats, to simulate density maps from a set of particles, and to represent the EM quality-of-fit as a restraint (54)

`modeller`	Interface to the MODELLER (31, 32) comparative modeling program. The module allows for MODELLER models and restraints to be imported into IMP, and for IMP restraints to be used with the MODELLER optimizers, or vice versa
`saxs`	Integration of SAXS data into the integrative modeling procedure. The module reads and writes SAXS profiles, and provides a restraint that scores a set of particles on their fit to an experimental profile (48, 49)
`restrainer`	High-level interface for setting up an integrative modeling procedure, reading the representation of the system and the sources of input information from a pair of XML files
`domino`	Implementation of an inferential message-passing optimization procedure (42). The module provides functionality to build a graphical model of the defined scoring function, and to decompose the graph into a tree on which a message-passing sampling procedure is performed.
`multifit`	Tools for fitting multiple proteins into their assembly density map. The main functionality includes the following: (1) fitting a single protein into its density based on point-alignment, principal component matching, or fast-Fourier transform search, (2) combinatorial consideration of fitting solutions of multiple components for generating an assembly model, and (3) modeling of cyclic symmetric assemblies (41)
Support modules	
`benchmark`	A set of benchmarks of the IMP software, to ensure that the algorithms perform optimally
`example`	Examples for developers on how to implement new IMP functionality
`helper`	High-level functionality to assist in setting up and manipulating a system, including simplified interfaces for creating restraints such as EM, connectivity, and excluded volume
`misc`	Miscellaneous and experimental functionality that has not been fully tested
`test`	Procedures to help in testing the IMP software itself

```
import IMP
import IMP.algebra
import IMP.core

m = IMP.Model()

# Create two "untyped" Particles
p1 = IMP.Particle(m)
p2 = IMP.Particle(m)

# "Decorate" the Particles with x,y,z attributes (point-like particles)
d1 = IMP.core.XYZ.setup_particle(p1)
d2 = IMP.core.XYZ.setup_particle(p2)

# Use some XYZ-specific functionality (set coordinates)
d1.set_coordinates(IMP.algebra.Vector3D(10.0, 10.0, 10.0))
d2.set_coordinates(IMP.algebra.Vector3D(-10.0, -10.0, -10.0))
print d1, d2

# Harmonically restrain p1 to be zero distance from the origin
f = IMP.core.Harmonic(0.0, 1.0)
s = IMP.core.DistanceToSingletonScore(f, IMP.algebra.Vector3D(0., 0., 0.))
r1 = IMP.core.SingletonRestraint(s, p1)
m.add_restraint(r1)

# Harmonically restrain p1 and p2 to be distance 5.0 apart
f = IMP.core.Harmonic(5.0, 1.0)
s = IMP.core.DistancePairScore(f)
r2 = IMP.core.PairRestraint(s, IMP.ParticlePair(p1, p2))
m.add_restraint(r2)

# Optimize the x,y,z coordinates of both particles with conjugate gradients
d1.set_coordinates_are_optimized(True)
d2.set_coordinates_are_optimized(True)
o = IMP.core.ConjugateGradients(m)
o.optimize(50)
print d1, d2
```

Fig. 3. simple.py, a simple Python script that uses IMP to build a model consisting of two particles satisfying a harmonic distance restraint.

in the "`library`" subdirectory of the zip file downloaded above (Fig. 3).

In the first part of the script, the IMP kernel and the `algebra` and `core` modules are loaded, as regular Python modules. We then proceed to set up the representation of the system, using the Model and `Particle` classes defined in the kernel. The `Model` class represents the entire system and keeps track of all the `Particles`, `Restraints`, and links between them. As mentioned earlier, the Particle class is a flexible container, but here we give the two `Particles` (p1 and p2) point-like attributes using the `XYZ` class defined in the core module. This `XYZ` class is known as a "decorator"; it does not create a new `Particle`, but merely presents a new interface to an existing `Particle`, in this case a point-like one (see Note 1). We can then treat each `Particle` like a point using methods in the `XYZ` class, here setting the *x*, *y*, and *z* coordinates to a provided vector.

In the second part, we set up the scoring of the system. We add two restraints to the `Model`, one of which harmonically restrains p1 to the origin and the other of which restrains p1 and p2 to be distance 5.0 apart (see Note 2). Note that the `core` module provides suitable building block restraints for this purpose. In the first case, we use the `SingletonRestraint` class that creates a restraint on a single particle (p1). It delegates the task of actually scoring the particle, however, to another class called `SingletonScore` that is simply given the `Particle` and asked for its score. In this example, we use a type of `SingletonScore` called a `DistanceToSingletonScore` that calculates the Cartesian distance between the point-like `Particle` and a fixed point (in this case the origin), and again delegates the task of scoring the distance to another class, a `UnaryFunction`. In this case, the `UnaryFunction` is a simple harmonic function with a mean of zero. Thus, the `Particle` p1 is harmonically restrained to be at the origin. The second restraint is set up similarly; however, in this case the restraints and scores act on a pair of particles. This building block functionality makes it easy to add a new type of restraint; for example, to implement a van der Waals potential, it is only necessary to provide a suitable `PairScore` that scores a single pair of particles; the functionality for efficiently enumerating all pairs of such particles is already provided in IMP.

Finally, in the third part of the script, we tell IMP that it can move the two point-like particles, and to build a system configuration that is consistent with all the restraints. In this example, a simple conjugate gradients optimization is used.

The script is a regular Python script. Thus, provided that both IMP and Python are installed (see Note 3), it can be run on any machine, by typing on a command line, in the same directory as the script:

```
python simple.py
```

The script will run the optimization, printing IMP log messages as it goes, and finally print the coordinates of the optimized particles.

IMP is designed such that the C++ and Python interfaces are similar to use. Thus, IMP applications or protocols can be constructed either in C++ or in Python, and new IMP functionality (for example, new types of `Restraint`) can be implemented in either language. For a comparison, please inspect the `simple.cpp` file. This file implements the same protocol as the first part of `simple.py`, but uses the IMP C++ classes rather than their Python equivalents. The two programs are very similar; the only differences are in the language syntax (e.g., the Python "`import IMP`" translates to "`#include <IMP.h>`" in C++) and in memory handling (Python handles memory automatically; in C++, memory handling must be done explicitly by using the `IMP::Pointer` class, which adds reference counting to automatically clean up after IMP objects when they are not used anymore).

3.2. `Restrainer`: A High-Level Interface for Integrative Modeling

The IMP C++/Python library offers a great deal of flexibility in setting up the system and restraints. However, in many cases, a simpler interface to solve modeling problems is preferable. The `restrainer` IMP module is one such interface that simplifies the setup of a complex system, generating the system representation and restraints from a pair of XML files. Optimization, however, may still need to be adjusted for specific cases.

As a simple demonstration of the module, we consider the construction of a model of a subcomplex of the NPC (12, 13). The yeast NPC is a large assembly of 50 MDa containing 456 proteins of 30 different types. The modeling of the entire assembly is beyond the scope of this tutorial; however, it has been observed that the NPC is made up of a set of smaller subcomplexes (Fig. 4). One of these complexes is the Nup84 complex, consisting of seven proteins, and the modeling of this complex is illustrated in this tutorial.

All of the XML and Python files necessary to perform the Nup84 modeling can be found in the "`restrainer`" subdirectory of the zip file downloaded above. The first of these XML files is `representation.xml`, which determines how the system is represented. IMP does not require every protein in the system to be modeled with the same representation; for example, some proteins could be modeled as sets of atoms and others at a lower resolution. As for the original NPC modeling, here we use a "bead model" for the Nup84 complex; each protein is represented as a sphere, or a pair of spheres (in the case of the more rod-like Nup133 and Nup120 proteins), with larger proteins using larger spheres. The second XML file encodes the input structural data as spatial restraints on the system. Here, we use two simple sources of information: First, excluded volume for each protein. Second, yeast two-hybrid results for some pairs of proteins. The third XML file is for visualization only, and assigns each sphere a different color. Finally, the Python script loads in all three of the XML files and performs a simple conjugate gradients optimization. This Python script can be executed just like any other Python script:

```
python nup84.py
```

`Restrainer` first generates a set of sphere-like particles to represent the system. It then converts the information in the restraints file into a set of IMP restraints. It generates an excluded volume restraint that prevents each protein sphere from penetrating any other sphere and a set of "connectivity" restraints (12) that force the protein particles to reproduce the interactions implied by the yeast two-hybrid experiments. The optimization generates a file `optimized.py` that is an input file for the molecular visualization program Chimera (43); when loaded into Chimera, it displays the final optimized configuration of the complex (Fig. 5).

In this example, the modeling problem is simple, and thus generating a single model is sufficient to find a solution that

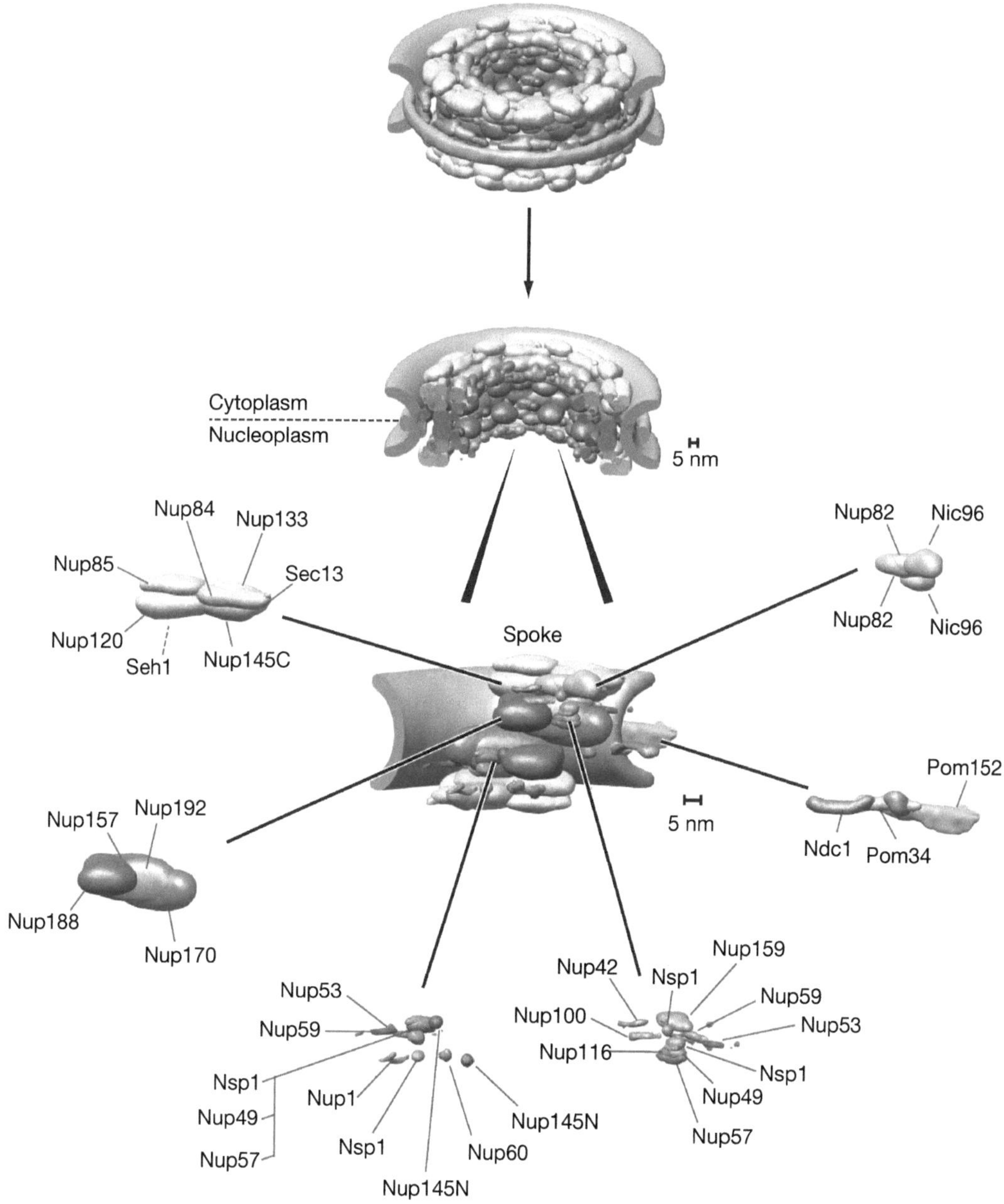

Fig. 4. Division of the yeast NPC into subcomplexes (13); one such subcomplex is the Nup84 complex of seven proteins.

satisfies all restraints. However, when all such models need to be found or, in more complex cases, when a global solution of the scoring function is hard to find (for example, because restraints are contradictory due to errors in experiments or experiment interpretations), the modeling procedure is repeated to generate an ensemble of models. When modeling the NPC, the top-scoring models were clustered and used to generate a probability density

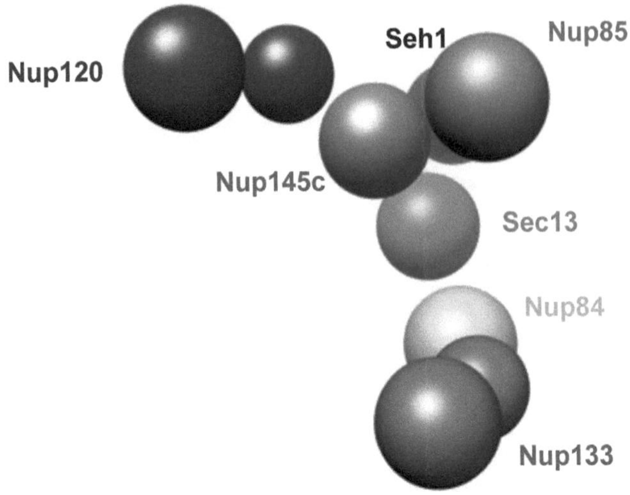

Fig. 5. Bead model of the Nup84 complex generated by `restrainer`, based on yeast two-hybrid system data and excluded volume considerations.

for each component within the complex (12). The envelope of this density defined the precision of the corresponding component localization. Only a single cluster of structures was found that satisfied all of the restraints. If contradictory information is presented, however, the optimization will be frustrated, unable to find solutions that simultaneously satisfy all restraints. The ensemble of solutions will exhibit more variability than that in a nonfrustrated case. Such frustration can be tested for in the iterative integrative modeling procedure by removing potentially conflicting restraints and repeating the modeling. Finally, the accuracy of the generated model(s) can be gauged by comparison with experimental data that were not used in the original modeling. For example, the generated bead model of the Nup84 complex has a characteristic Y-shape, which is consistent with electron micrographs of the complex (44), even though these data were not used in our example.

The `restrainer` XML and Python files, together with the experimental data, such as cryo-EM maps, constitute a complete modeling protocol. Thus, an assembly model built using this protocol can be published along with the input files to allow the model to be reproduced and easily updated. Such a model can thus act as a reference for future studies; for example, regions of the model that were poorly resolved can be investigated with new experiments, the resulting data incorporated into the protocol, and new models generated. Alternatively, existing unused experimental data can be added to the protocol to determine whether unused data is consistent with that used to build the model. The iterative nature of the protocol thus extends beyond the generation of the first "correct" model.

3.3. Integration of Comparative Modeling, X-ray Crystallography, and SAXS

The Nup84 complex structure determined above is consistent with all input information, but for a detailed understanding of its function, an accurate atomic structure is required. Two possible routes to such a structure, depending on the available information, are (1) fitting atomic structures of the individual protein subunits into a cryo-EM map of the assembly and (2) accurately placing pairs of subunits relative to each other using X-ray crystallography or molecular docking. For both routes, atomic structures of the subunits are required; these structures can be obtained *via* X-ray crystallography or comparative modeling.

One component of the Nup84 complex is the Nup133 protein; the structure of this protein has been characterized by both X-ray crystallography and SAXS (45). SAXS differs from X-ray crystallography in that it is applied to proteins in solution rather than crystals; thus, it can be applied to a much wider range of proteins in states more closely resembling their functional forms than X-ray crystallography, but the information is rotationally averaged and so the resulting SAXS profile gives less structural information (29, 46, 47). IMP contains a method that, given an atomic protein structure, can calculate its SAXS profile using the Debye formula, and then fit this profile against the experimentally determined one (48, 49). This method is implemented in the IMP `saxs` module and so can be used by writing a suitable Python script. However, because fitting against a SAXS profile is a common task, we provide an IMP application, FoXS, which automates this process. FoXS is available both as a command-line IMP application and as a Web service at http://salilab.org/foxs.

All input files for this demonstration are available in the "`saxs`" subdirectory of the downloaded zip file. The structure of the C-terminal domain of yeast Nup133 is available in the RCSB Protein Data Bank (PDB)(50) as code 3kfo (file `3KFO.pdb`), while the experimental SAXS profile is given in the `23922_merge.dat` file. The atomic structure can be fit against the SAXS profile by running FoXS in the directory containing both files:

```
foxs 3KFO.pdb 23922_merge.dat
```

Alternatively, the two files can be submitted to the FoXS Web server. FoXS compares the theoretical profile of the provided structure (solid line in Fig. 6) with the experimental profile (points), and calculates the quality of the fit, χ, with smaller values corresponding to closer fits.

The fit in this example is not a good one ($\chi = 2.96$). To understand why this is so, we examine the header of the 3kfo PDB file, which reveals two problems. Several residues at the N and C termini were not resolved in the X-ray experiment (eight in total, two at the N terminus and six at the C terminus), and the side chains of 16 other residues could also not be located (see Note 4).The missing 8 residues and 16 side chains need to be placed to create a

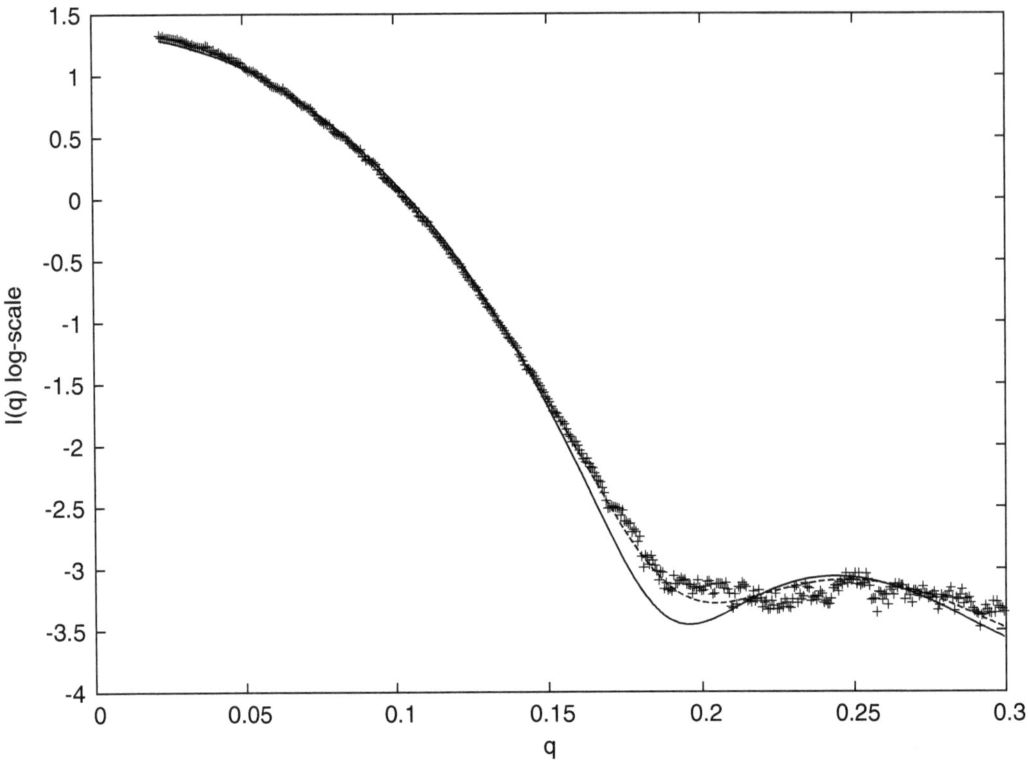

Fig. 6. Fit of the 3kfo PDB structure *(solid line)* against the experimentally determined SAXS profile of the same protein (points), using the FoXS Web service at http://salilab.org/foxs (48). A plot of a comparative model's profile is also shown *(dashed line)*.

complete atomic structure. One way to achieve this goal is to build a comparative model using a package such as MODELLER (http://salilab.org/modeller/) (31, 32) relying on the original 3kfo structure as a template and the full sequence as the target (see Note 5). The corresponding MODELLER alignment file (`3KFO-fill.ali`) and script file (`fill.py`) are provided in the downloaded zip file. Each candidate comparative model can be fitted against the SAXS profile using the FoXS command-line application or the Web service in exactly the same way as the original 3kfo structure; the best MODELLER model gives a significantly improved fit between the theoretical and experimental profiles (dashed line in Fig. 6; $\chi = 1.21$).

Given similar atomic structures of the subunits in the Nup84 complex, as either crystal structures or comparative models, `restrainer` can be used to build an atomic model of the complex. Note, however, that an accurate model of such a complex would require additional information beyond the proteomics data used above, since yeast two-hybrid data only show that proteins interact, but do not show the specific residues in the protein–protein interaction, thus not informing us about the relative orientations of the interacting proteins. Such information can be obtained, for example, from chemical cross-linking, molecular docking, or cryo-EM maps, as illustrated in the next section.

3.4. Determining Macromolecular Assembly Structures by Fitting Multiple Structures into an Electron Density Map

Often, we have available high-resolution (atomic) information for the subunits in an assembly, and low-resolution information for the assembly as a whole (a cryo-EM electron density map). A high-resolution model of the whole assembly can thus be constructed by simultaneously fitting the subunits into the density map. Fitting of a single protein into a density map is usually done by calculating the electron density of the protein followed by a search of the protein position in the cryo-EM map that maximizes the cross-correlation of the two maps. Simultaneously fitting multiple proteins into a given map is significantly more difficult, since an incorrect fit of one protein will also prevent other proteins from being placed correctly.

IMP contains a `multifit` (41, 42) module (http://salilab.org/multifit/) that can efficiently solve such multiple fitting problems for density-map resolutions as low as 25Å, relying on a general inferential optimizer DOMINO. The fitting protocol is a multistep procedure that proceeds *via* discretization of both the map and the proteins, local fitting of the proteins into the map, and an efficient combination of local fits into global solutions (Fig. 7). Here, we will demonstrate the use of `multifit` in building a model of the ARP2/3 complex (51) using crystal structures of its seven constituent proteins (ARP2, ARP3, and ARC1-5) and a 20Å density

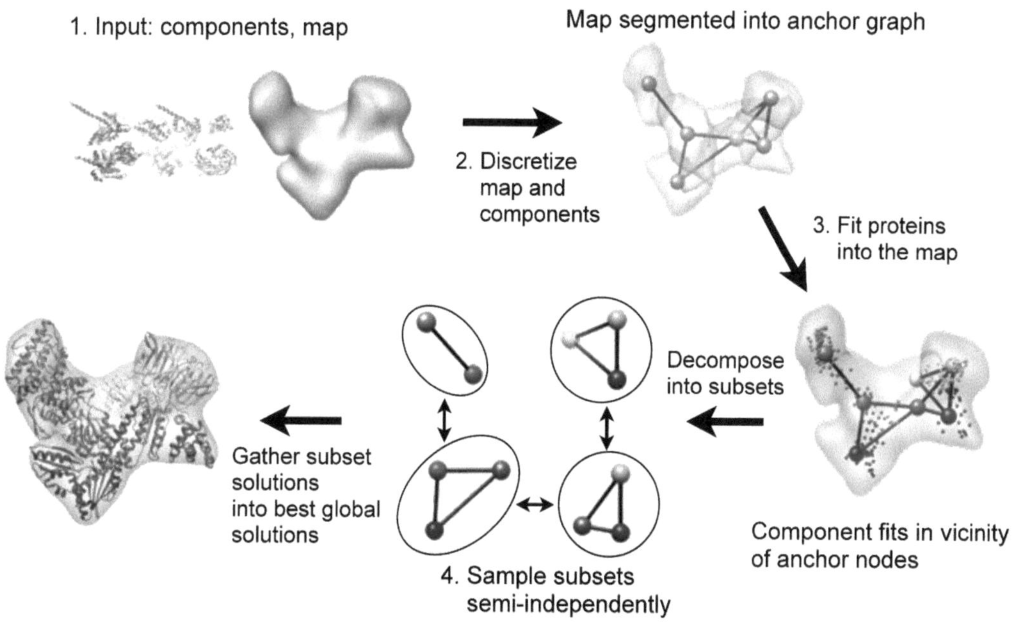

Fig. 7. The MultiFit protocol (42). Protein subunits are fitted into a density map of the assembly by discretizing both the map and the components, locally fitting each protein, and efficiently combining the local fits into global solutions.

map of the assembly. All input files for this procedure can be found in the "`multifit`" subdirectory of the downloaded zip file.

The first step in using `multifit` is to create input files that guide the protocol. The first of these files, `assembly.input`, lists each of the subunits and the density map, complete with the names of the files from which the input structures and map will be read, and those to which outputs from later steps will be written. In this case, we also know the native structure of the assembly (PDB code 1tyq), and so we add the subunit structures in native conformation to this input file (right-most column); `multifit` will use them to assess its accuracy. Normally, of course, the real native structure is not known, in which case this column in the input file is left blank. The second file, `multifit.par`, specifies various optimization parameters and is described in more detail on the `multifit` Web site (http://salilab.org/multifit/).

The second step is to determine a reduced representation for both the density map and the subunits, using the Gaussian Mixture Model. This task can be achieved by typing in the directory containing `assembly.input` as follows:

```
/opt/multifit/utils/run_anchor_points_detec-
   tion.py assembly.input 700
```

This run determines a reduced representation of the EM map that best reproduces the configuration of all voxels with density above 700, and a similar reduced representation of each subunit as a set of 3D Gaussian functions. The number of Gaussians is specified in `assembly.input` for each subunit. It should be at least three (the minimum required for fitting) and each Gaussian should cover approximately the same number of residues (for example, if you choose 50 residues per Gaussian, a 170-residue protein should use 3 Gaussians and a 260-residue protein should use 5 Gaussians). Each such reduced map representation can also be thought of as an anchor point graph, where each anchor point corresponds to the center of a 3D Gaussian, and the edges in the graph correspond to the connectivity between regions of the map or protein. These reduced representations are written out as PDB files containing fake Cα atoms, where each Cα corresponds to a single anchor point.

The third step is to fit each protein in the vicinity of the EM map's anchor points. This task is achieved by running the following:

```
/opt/multifit/utils/run_protein_fitting.py assem-
    bly.input multifit.par
```

The output is a set of candidate fits, where the subunit is rigidly rotated and translated to fit into the density map (see Note 6). Each fit is written as a PDB file in the "`fits`" subdirectory.

Finally, the fits are scored and then combined into a set of the best-scoring global configurations:

```
/opt/multifit/utils/run_all_scores.py  assembly.
   input > scores.log
```

```
/opt/multifit/utils/run_multifit.py assembly.input
   assembly.jt    assembly_configurations.output
   data/models/1tyq.fitted.pdb > multifit.log
```

The scoring function used to assess each fit includes the quality-of-fit of each subunit into the map, the protrusion of each subunit out of the map envelope, and the shape complementarity between pairs of neighboring subunits. The optimization avoids exhaustive enumeration of all possible mappings of subunits to anchor points by means of a branch-and-bound algorithm combined with the DOMINO divide-and-conquer message-passing optimizer using a discrete sampling space (42).

The final output from multifit is a file `assembly_configurations.output` that lists the best global solutions, ranked by their score, an excerpt of which is shown below:

```
ARP3,0|ARP2,14|ARC1,3|ARC2,24|ARC3,19|ARC4,11|
   ARC5,13|(17.5593729019)(rmsd:29.2637996674)
   (conf.0.pdb)
ARP3,5|ARP2,13|ARC1,9|ARC2,24|ARC3,19|ARC4,4|A
   RC5,13|(18.3258602619)(rmsd:11.997220993)
   (conf.1.pdb)
```

For each global solution, `multifit` lists the indices of the local fits for each subunit and the score. Each solution is also written out as a multichain PDB file of the assembly. In addition, because we also provided the native structure (`1tyq.fitted.pdb`), the RMSD between the native conformation and each solution is listed. In this case, the RMSD measure indicates that `multifit` has correctly determined the architecture of the assembly, placing each subunit in the approximately correct part of the map. However, the protein–protein interfaces are clearly not accurate at the atomic level. These models could thus be refined with a combination of pairwise computational docking and local sampling, ideally supported by additional experimental data, such as chemical cross-linking, various kinds of footprinting, and X-ray crystallography of binary subunit complexes.

3.5. Summary

The structures of protein assemblies can typically not be fully characterized with any individual computational or experimental method. Integrative modeling aims to solve this problem by combining information from multiple methods to generate structural models. Integrative modeling problems can be tackled using the method of satisfaction of spatial restraints. In this approach, a suitable representation for the system is chosen, the information is converted into a set of spatial restraints, the restraints are simultaneously satisfied as much as possible by optimizing a function that is the sum of all restraints, and the resulting models are analyzed. Further experiments as well as the precision and likely accuracy of both the model and the data can be informed. IMP is an open-source and flexible software package that provides all of the components needed to implement an integrative modeling protocol from scratch. It also contains higher-level applications and Web services that can tackle specific-use cases more conveniently.

4. Notes

1. Multiple decorators can be applied to a single `Particle`; for example, an atom-like `Particle` could be treated like a point, a sphere, an electrically charged particle, or an atom.
2. IMP does not enforce any units of distance; however, some physical optimizers, such as molecular dynamics, expect distances to be in angstroms.
3. The syntax for running Python scripts may vary slightly depending on where the IMP files are installed, and which version of Python is set up on your system. Please refer to the IMP Web site and documentation for system-specific syntax.
4. A PDB file is a plain text file that can be loaded in most text editors. The problems with the 3kfo structure are reported in the PDB file on the lines starting with "REMARK 465" and "REMARK 470".
5. The full sequence includes the eight missing N and C terminal residues. MODELLER constructs missing side chains automatically.
6. The fitting procedure is performed by either aligning a reduced representation of a protein to a reduced representation of the density map (41) or fitting the protein principal components to the principal components of a segmented region of the map.

Acknowledgments

We are grateful to all members of our research group, especially to Frank Alber, Friedrich Förster, and Bret Peterson who contributed to early versions of IMP. We also acknowledge support from National Institutes of Health (R01 GM54762, U54 RR022220, PN2 EY016525, and R01 GM083960) as well as computing hardware support from Ron Conway, Mike Homer, Hewlett-Packard, NetApp, IBM, and Intel.

References

1. Schmeing TM, and Ramakrishnan V (2009) What recent ribosome structures have revealed about the mechanism of translation, *Nature 461*, 1234–1242.
2. Sali A, Glaeser R, Earnest T, and Baumeister W (2003) From words to literature in structural proteomics, *Nature 422*, 216–225.
3. Mitra K, and Frank J (2006) Ribosome dynamics: insights from atomic structure modeling into cryo-electron microscopy maps, *Annu Rev Biophys Biomol Struct 35*, 299–317.
4. Robinson C, Sali A, and Baumeister W (2007) The molecular sociology of the cell, *Nature 450*, 973–982.
5. Blundell T, and Johnson L (1976) *Protein Crystallography*, Academic Press, New York.
6. Stahlberg H, and Walz T (2008) Molecular electron microscopy: state of the art and current challenges, *ACS Chem Biol 3*, 268–281.
7. Chiu W, Baker ML, Jiang W, Dougherty M, and Schmid MF (2005) Electron cryomicroscopy of biological machines at subnanometer resolution, *Structure 13*, 363–372.
8. Lucic V, Leis A, and Baumeister W (2008) Cryo-electron tomography of cells: connecting structure and function, *Histochem Cell Biol 130*, 185–196.
9. Parrish JR, Gulyas KD, and Finley RL Jr. (2006) Yeast two-hybrid contributions to interactome mapping, *Curr Opin Biotechnol 17*, 387–393.
10. Gingras AC, Gstaiger M, Raught B, and Aebersold R (2007) Analysis of protein complexes using mass spectrometry, *Nat Rev Mol Cell Biol 8*, 645–654.
11. Alber F, Kim M, and Sali A (2005) Structural characterization of assemblies from overall shape and subcomplex compositions, *Structure 13*, 435–445.
12. Alber F, Dokudovskaya S, Veenhoff L, Zhang W, Kipper J, Devos D, Suprapto A, Karni-Schmidt O, Williams R, Chait B, Rout M, and Sali A (2007) Determining the architectures of macromolecular assemblies, *Nature 450*, 683–694.
13. Alber F, Dokudovskaya S, Veenhoff L, Zhang W, Kipper J, Devos D, Suprapto A, Karni-Schmidt O, Williams R, Chait B, Sali A, and Rout M (2007) The molecular architecture of the nuclear pore complex, *Nature 450*, 695–701.
14. Lasker K, Phillips JL, Russel D, Velazquez-Muriel J, Schneidman-Duhovny D, Webb B, Schlessinger A, and Sali A (2010) Integrative Structure Modeling of Macromolecular Assemblies from Proteomics Data, *Mol Cell Proteomics 9*, 1689–1702.
15. Russel D, Lasker K, Phillips J, Schneidman-Duhovny D, Velazquez-Muriel J, and Sali A (2009) The structural dynamics of macromolecular processes, *Curr Opin Cell Biol 21*, 97–108.
16. Alber F, Forster F, Korkin D, Topf M, and Sali A (2008) Integrating diverse data for structure determination of macromolecular assemblies, *Annu Rev Biochem 77*, 443–477.
17. Alber F, Chait BT, Rout MP, and Sali A (2008) Integrative Structure Determination of Protein Assemblies by Satisfaction of Spatial Restraints, In *Protein-protein interactions and networks: identification, characterization and prediction.* (Panchenko, A., and Przytycka, T., Eds.), pp 99–114, Springer-Verlag, London, UK.
18. Bonvin AM, Boelens R, and Kaptein R (2005) NMR analysis of protein interactions, *Curr Opin Chem Biol 9*, 501–508.
19. Fiaux J, Bertelsen EB, Horwich AL, and Wuthrich K (2002) NMR analysis of a 900K GroEL GroES complex, *Nature 418*, 207–211.
20. Neudecker P, Lundstrom P, and Kay LE (2009) Relaxation dispersion NMR spectroscopy as a tool for detailed studies of protein folding, *Biophys J 96*, 2045–2054.
21. Takamoto K, and Chance MR (2006) Radiolytic protein footprinting with mass spectrometry to probe the structure of macromolecular complexes, *Annu Rev Biophys Biomol Struct 35*, 251–276.
22. Guan JQ, and Chance MR (2005) Structural proteomics of macromolecular assemblies using oxidative footprinting and mass spectrometry, *Trends Biochem Sci 30*, 583–592.
23. Taverner T, Hernandez H, Sharon M, Ruotolo BT, Matak-Vinkovic D, Devos D, Russell RB, and Robinson CV (2008) Subunit architecture of intact protein complexes from mass spectrometry and homology modeling, *Acc Chem Res 41*, 617–627.
24. Chen ZA, Jawhari A, Fischer L, Buchen C, Tahir S, Kamenski T, Rasmussen M, Lariviere L, Bukowski-Wills JC, Nilges M, Cramer P, and Rappsilber J (2010) Architecture of the RNA polymerase II-TFIIF complex revealed by cross-linking and mass spectrometry, *EMBO J 29*, 717–726.
25. Sinz A (2006) Chemical cross-linking and mass spectrometry to map three-dimensional protein structures and protein-protein interactions, *Mass Spectrom Rev 25*, 663–682.

26. Trester-Zedlitz M, Kamada K, Burley SK, Fenyo D, Chait BT, and Muir TW (2003) A modular cross-linking approach for exploring protein interactions, *J Am Chem Soc 125*, 2416–2425.
27. Joo C, Balci H, Ishitsuka Y, Buranachai C, and Ha T (2008) Advances in single-molecule fluorescence methods for molecular biology, *Annu Rev Biochem 77*, 51–76.
28. Mertens HD, and Svergun DI (2010) Structural characterization of proteins and complexes using small-angle X-ray solution scattering, *J Struct Biol 172*, 128–141.
29. Hura GL, Menon AL, Hammel M, Rambo RP, Poole FL 2nd, Tsutakawa SE, Jenney FE Jr, Classen S, Frankel KA, Hopkins RC, Yang SJ, Scott JW, Dillard BD, Adams MW, and Tainer JA (2009) Robust, high-throughput solution structural analyses by small angle X-ray scattering (SAXS), *Nat Methods 6*, 606–612.
30. Berggard T, Linse S, and James P (2007) Methods for the detection and analysis of protein-protein interactions, *Proteomics 7*, 2833–2842.
31. Sali A, and Blundell TL (1993) Comparative protein modelling by satisfaction of spatial restraints, *J Mol Biol 234*, 779–815.
32. Sali A, and Blundell TL (1994) Comparative protein modeling by statisfaction of spatial restraints, In *Protein Structure by Distance Analysis* (Bohr, H., and Brunak, S., Eds.), pp 64–86, TECH UNIV DENMARK, CTR BIOL SEQUENCE ANAL, LYNGBY, DENMARK.
33. Vajda S, and Kozakov D (2009) Convergence and combination of methods in protein-protein docking, *Curr Opin Struct Biol 19*, 164–170.
34. Shen MY, and Sali A (2006) Statistical potential for assessment and prediction of protein structures, *Protein Sci 15*, 2507–2524.
35. Melo F, Sanchez R, and Sali A (2002) Statistical potentials for fold assessment, *Protein Sci 11*, 430–448.
36. Brooks BR, Brooks CL 3rd, Mackerell AD Jr, Nilsson L, Petrella RJ, Roux B, Won Y, Archontis G, Bartels C, Boresch S, Caflisch A, Caves L, Cui Q, Dinner AR, Feig M, Fischer S, Gao J, Hodoscek M, Im W, Kuczera K, Lazaridis T, Ma J, Ovchinnikov V, Paci E, Pastor RW, Post CB, Pu JZ, Schaefer M, Tidor B, Venable RM, Woodcock HL, Wu X, Yang W, York DM, and Karplus M (2009) CHARMM: the biomolecular simulation program, *J Comput Chem 30*, 1545–1614.
37. Case DA, Cheatham TE 3rd, Darden T, Gohlke H, Luo R, Merz KM Jr, Onufriev A, Simmerling C, Wang B, and Woods RJ (2005) The Amber biomolecular simulation programs, *J Comput Chem 26*, 1668–1688.
38. Christen M, Hunenberger PH, Bakowies D, Baron R, Burgi R, Geerke DP, Heinz TN, Kastenholz MA, Krautler V, Oostenbrink C, Peter C, Trzesniak D, and van Gunsteren WF (2005) The GROMOS software for biomolecular simulation: GROMOS05, *J Comput Chem 26*, 1719–1751.
39. Forster F, Lasker K, Beck F, Nickell S, Sali A, and Baumeister W (2009) An Atomic Model AAA-ATPase/20S core particle sub-complex of the 26S proteasome, *Biochem Biophys Res Commun 388*, 228–233.
40. Nickell S, Beck F, Scheres SHW, Korinek A, Forster F, Lasker K, Mihalache O, Sun N, Nagy I, Sali A, Plitzko J, Carazo J, Mann M, and Baumeister W (2009) Insights into the Molecular Architecture of the 26S Proteasome, *Proc Natl Acad Sci USA 29*, 11943–11947.
41. Lasker K, Sali A, and Wolfson HJ (2010) Determining macromolecular assembly structures by molecular docking and fitting into an electron density map, *Proteins: Struct Funct Bioinform 78*, 3205–3211.
42. Lasker K, Topf M, Sali A, and Wolfson H (2009) Inferential optimization for simultaneous fitting of multiple components into a cryoEM map of their assembly, *J Mol Biol 388*, 180–194.
43. Pettersen EF, Goddard TD, Huang CC, Couch GS, Greenblatt DM, Meng EC, and Ferrin TE (2004) UCSF Chimera--a visualization system for exploratory research and analysis, *J Comput Chem 25*, 1605–1612.
44. Kampmann M, and Blobel G (2009) Three-dimensional structure and flexibility of a membrane-coating module of the nuclear pore complex, *Nat Struct Mol Biol 16*, 782–788.
45. Sampathkumar P, Gheyi T, Miller SA, Bain K, Dickey M, Bonanno J, Kim S, Phillips J, Pieper U, Fernandez-Martinez J, Franke JD, Martel A, Tsuruta H, Atwell S, Thompson D, Emtage JS, Wasserman S, Rout MP, Sali A, Sauder JM, and Burley SK (2011) Structure of the C-terminal domain of Saccharomyces cerevisiae Nup133, a component of the Nuclear Pore Complex, *Proteins: Struct Funct Bioinform 79*, 1672–1677.
46. Putnam CD, Hammel M, Hura GL, and Tainer JA (2007) X-ray solution scattering (SAXS) combined with crystallography and computation: defining accurate macromolecular structures, conformations and assemblies in solution, *Q Rev Biophys 40*, 191–285.
47. Petoukhov MV, and Svergun DI (2007) Analysis of X-ray and neutron scattering from

biomacromolecular solutions, *Curr Opin Struct Biol 17*, 562–571.

48. Schneidman-Duhovny D, Hammel A, and Sali A. (2010) FoXS: A Web Server for Rapid Computation and Fitting of SAXS Profiles, *Nucleic Acids Res 38*, W540–4.
49. Forster F, Webb B, Krukenberg KA, Tsuruta H, Agard DA, and Sali A (2008) Integration of small-angle X-ray scattering data into structural modeling of proteins and their assemblies, *J Mol Biol 382*, 1089–1106.
50. Berman HM, Battistuz T, Bhat TN, Bluhm WF, Bourne PE, Burkhardt K, Feng Z, Gilliland GL, Iype L, Jain S, Fagan P, Marvin J, Padilla D, Ravichandran V, Schneider B, Thanki N, Weissig H, Westbrook JD, and Zardecki C (2002) The Protein Data Bank, *Acta Crystallogr D Biol Crystallogr 58*, 899–907.
51. Robinson RC, Turbedsky K, Kaiser DA, Marchand JB, Higgs HN, Choe S, and Pollard TD (2001) Crystal structure of Arp2/3 complex, *Science 294*, 1679–1684.
52. DeLano WL (2002) The PyMOL molecular graphics system, *Version 1.2r3pre, Schrödinger, LLC.*
53. Galassi M, Davies J, Theiler J, Gough B, Jungman G, Booth M, and Rossi F (2002) *GNU Scientific Library.*
54. Topf M, Lasker K, Webb B, Wolfson H, Chiu W, and Sali A (2008) Protein structure fitting and refinement guided by cryo-EM density, *Structure 16*, 295–307.

Chapter 20

Predicting Node Characteristics from Molecular Networks

Sara Mostafavi, Anna Goldenberg, and Quaid Morris

Abstract

A large number of genome-scale networks, including protein–protein and genetic interaction networks, are now available for several organisms. In parallel, many studies have focused on analyzing, characterizing, and modeling these networks. Beyond investigating the topological characteristics such as degree distribution, clustering coefficient, and average shortest-path distance, another area of particular interest is the prediction of nodes (genes) with a given characteristic (labels) – for example prediction of genes that cause a particular phenotype or have a given function. In this chapter, we describe methods and algorithms for predicting node labels from network-based datasets with an emphasis on label propagation algorithms (LPAs) and their relation to local neighborhood methods.

Key words: Functional linkage networks, Gene function prediction, Label propagation

1. Introduction

Networks are commonly used to represent relationships between genes or proteins. For example, in a protein–protein interaction network, nodes represent proteins and the edges represent physical interactions between the connected nodes. Other widely available networks include genetic interaction networks and coexpression networks (see Table 1 for a summary of common network types). In parallel with the increase in number and size of interaction networks that are now available for various organisms, much work has been done to analyze, model, and make predictions from such network-based data.

An area of particular interest is prediction of node characteristics (or labels) in a network. For example, in gene function prediction, given a network and labeled genes that are involved in a function of interest (these input genes are referred to as "query genes" or "positives"), the goal is to predict other genes that are deemed to be involved in the same function (e.g., (1)). This is a

Gerard Cagney and Andrew Emili (eds.), *Network Biology: Methods and Applications*, Methods in Molecular Biology, vol. 781, DOI 10.1007/978-1-61779-276-2_20, © Springer Science+Business Media, LLC 2011

Table 1
This table summarizes some examples of widely used interaction networks whose edges are predictive of cofunctionality between the connected genes or proteins

Network	Experimental methodology	Description
Cocomplex network	Copurification (Affinity capture)	An interaction is inferred between members of subunits in a purified protein complex using affinity purification and one or more additional fractionation steps
Protein interaction networks	Protein-fragment complementation assay (PCA)	Two proteins of interest are fused to complementary fragments of a reporter protein; when proteins of interest react the reporter protein fluoresces
	Two-hybrid	Bait protein expressed as a DNA binding domain (DBD) fusion and prey expressed as a transcriptional activation domain (TAD) fusion; interaction measured by reporter gene activation
Genetic interaction networks	Synthetic genetic array (SGA) diploid based synthetic lethality analysis on microarrays (dSLAM)	Genetic interactions between two genes is inferred when the mutation (or deletion) of both genes results in a phenotype that is unexpected based on the single mutations of each of the genes. For example, in yeast, synthetic lethal genetic is inferred whenever a double mutation of two nonessential genes results in lethality
Colocalization networks	Green fluorescent protein (GFP) fusion	An interaction is inferred from colocalization of two proteins in the cell, including codependent association of proteins with promoter DNA in chromatin immunoprecipitation experiments
Coexpression networks	Microarray	Quantification of changes in expression level of genes by measuring the abundance of their corresponding mRNA in different conditions based on hybirdization of labeled mRNA to known probes
	Serial analysis of gene expression (SAGE)	Quantification of abundance of transcripts by cloning and sequencing of the extracted mRNA
Transcriptional regulatory networks	ChIP-on-Chip	Combines Chromatin immunoprecipitation (ChIP) with microarray (chip) to determine the binding sites of DNA-binding proteins on genome-wide basis
Coinheritance (shared phylogenetic profiles)		Coinheritance networks are derived from phylogenetic profiles that summarize the presence/absence of homologous proteins in various species

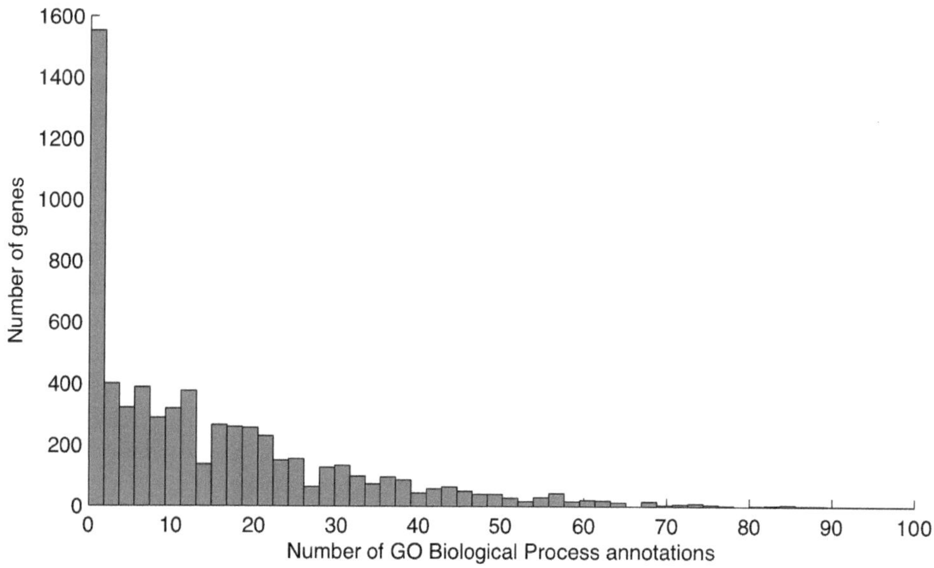

Fig. 1. This figure depicts our current knowledge of protein function in yeast. The number of annotations are based on *informative* GO Biological Process associations (downloaded April 2010). We define informative GO categories as those having fewer than 500 annotations.

well-motivated task: even in model organisms such as yeast, a large number of genes have not yet been annotated with a precise function (see Fig. 1). Similarly, in disease gene *prioritization*, given a network and a group of genes that are involved in a given disease, the goal is to *prioritize* other genes based on how likely they are to be involved in the same (or similar) disease(s) (e.g., (2, 3)). Often, the list of genes to be prioritized is a subset of the genes that are present in the network; for example, disease-associated chromosomal loci from Genome Wide Association studies (GWAS). Algorithms for solving these problems use the observation that genes (or proteins) that are coexpressed, or have similar physical or genetic interactions, tend to have similar functions (see ref. 4 for a review) or result in similar phenotypes (see ref. 5 for a review) – this principle is referred to as "guilt-by-association."

In the case of function prediction, positive gene labels can be obtained from databases such as Gene Ontology (GO) (6), KEGG (7), and MIPS (8). These databases provide both a controlled vocabulary for describing categories of gene function and curated lists of genes annotated to these functions. For disease prioritization, positive gene lists can be obtained from the Human Phenotype Ontology (9) and the OMIM database (10), which provide genes associated with various phenotypic abnormalities and genetic diseases.

Here, we focus on algorithms that solve a binary classification problem. If necessary, it is easy to generalize these approaches to predict multiple functions/diseases per gene. Formally, given a network over all entities (genes or proteins), and a set of binary labels where positives are labeled as +1, the goal is to predict which unlabeled nodes are likely to be positives. The "guilt-by-association" approach only considers direct neighbors of nodes when making predictions (1). However, often indirect interactions and global network topology can improve the prediction performance (11–13). As such, a large number of models and algorithms have been proposed that consider both direct and indirect interactions (12–18). For example, label propagation algorithms (LPAs) assign continuous scores (predicted labels) to all nodes in the network while considering both local and global network topology. In addition to offering a principled way of incorporating indirect interactions, the complexity of LPAs scales with the number of edges in the network and thus LPAs are computationally feasible for very large networks – empirically, less than 0.1% of total possible edges are often observed in real-world networks.

In this chapter, we focus on describing commonly used algorithms for predicting gene labels (gene function or involvement in a disease) from network-based data. In particular, as done in (19), we categorize such algorithms into two broad categories; those that use the node's local neighborhood and those that use global network topology when predicting node labels. For the latter category, we mainly focus on LPAs. As we show, LPAs are closely related to local neighborhood approaches; we describe an approximation framework to LPAs that will allow us to directly derive the local neighborhood methods.

The rest of this chapter is organized as follows: In Subheading 2.1, we describe how to construct networks from various high-throughput data sources, with the purpose of using these networks for prioritizing genes or prediction gene function; in Subheading 2.2, we review algorithms for predicting node labels from a network; in Subheading 2.3, we describe several methods for constructing networks from multiple high-throughput data sources; and in Subheading 2.4, we describe several online resources for gene prioritization and predicting gene and protein function.

2. Methods

The task of gene function prediction requires three components: A network that can be constructed from one or many different high-throughput data sources; a set of positive genes (see Note 1); and an algorithm for making prediction from the network. In the next section, we describe how to construct individual networks that represent the evidence for cofunctionality implied by individual high-throughput datasets.

2.1. Constructing Networks for Predicting Gene Function

To predict gene function, we assume that we are provided with a network whose nodes correspond to genes and whose edges represent the strength of the evidence for cofunctionality between the connected genes. These networks are called functional linkage networks (FLNs). There are several different types of FLNs that support good function prediction performance including those whose edge weights represent coexpression, genetic interaction, protein interaction, coinheritance, colocalization, or shared domain composition of the connected genes. A number of studies have demonstrated a drastic improvement in the accuracy of function prediction when multiple data sources are combined (13, 17, 18, 20–24). Below, we first review how to construct individual networks from high-throughput data sources; in Subheading 2.3, we describe how to combine multiple networks into one *composite* network for input to a label prediction algorithm.

We broadly classify networks into those that are derived from interaction-based data and those that are derived from profile-based data. The former includes networks derived from protein and genetic interaction datasets, and the latter includes those derived from gene expression profiles or patterns of protein localization.

For profile-based datasets, such as gene expression, the edges in the corresponding network are constructed from pairwise similarity scores. Determining the appropriate similarity metric for a given data type is an active area of research. For example, much work has been carried out on constructing coexpression networks (25, 26). Here, we present a simple method that performs well on a variety of data sources (as used in refs. 17, 27). In particular, for many types of profiled-based data, we have found that the Pearson Correlation Coefficient (PCC) results in networks with comparable or better performance than other similarity metrics. For binary profile-based data such as protein localization, prior to taking the PCC, we use the following background correction that significantly improves the resulting FLNs: given a binary matrix B with n rows (genes) and d columns (features), we set all 1's in column i to $-\log(p_i^{(1)})$ and 0's to $\log(1-p_i^{(1)})$, where $p_i^{(1)}$ is the probability of observing 1 in column i. In this way, genes that share "rare" features will have higher correlation than those that share "common" features (see Note 2).

For network-based data, we can always use the binary interactions alone as the final network. However, several studies have observed a drastic improvement in performance when constructing FLNs using the PCC between interaction partners of genes or proteins (28–30). Calculating PCC on frequency-corrected data as described above further improves the performance (28).

2.2. Predicting Node Labels from Networks

In this section, we review and discuss algorithms for predicting binary labels using networks. In particular, we focus on LPAs and their relationship to simpler direct and indirect neighbor methods.

2.2.1. Notation

Here, we assume we are given a network, represented as a symmetric matrix W. Assuming there are n nodes (genes) in the network, then W is an $n \times n$ matrix with entries $w_{ji} = w_{ij} \geq 0$ representing the weighted edge between nodes i and j. For example, in the case of protein–protein interaction, w_{ij} can be binary (indicating the absence or presence of a physical interaction) or weighted according to the score (e.g., –log of p-value) of the interaction between proteins i and j. We represent the labels using a vector $\vec{y} \in \{0,1\}^n$, where positive nodes (e.g., those involved in a given function of interest) are labeled as +1 and unlabeled nodes are labeled as 0. Ideally, in addition to the positives, some nodes should serve as negative examples and thus would be assigned a negative label ($y_i = -1$); however, such negative examples are rarely available – here, we assume we only have positive and unlabeled nodes.

2.2.2. Local Neighborhood Approaches

In guilt-by-association, genes are assigned a score based on their direct connections to the positive genes. For example, in (1), a gene is predicted to have the same function as those of the majority of its direct neighbors. More recent studies have extended guilt-by-association to include second-degree (indirect) interactions (13) or consider a small neighborhood around the positively labeled genes (31) when assigning scores to the unlabeled genes. Below, we present a single general framework that serves as the basis for deriving existing local neighborhood methods.

As a first attempt, we can calculate the gene scores for an unlabeled gene f_i as the weighted sum of the labels of its direct neighbors as $f_i = \sum_{j=1}^{n} w_{ij} y_j$ (e.g., (32)). Note that the summation is over all n genes; since the labels are binary, if W is also binary, this expression counts the number of neighbors of j that are labeled as +1. For a weighted network, this expression *weights* the label of i's positive neighbors according to their connection strength. However, it is standard practice to normalize the matrix W using weighted node degrees, as it results in better performance:

$$f_i = \frac{1}{d_i} \sum_{j=1}^{n} w_{ij} y_j,$$

where $d_i = \sum_{j=1}^{n} w_{ij}$ is the weighted degree of node i. After normalizing W, the score vector can be computed in matrix form as $\vec{f} = D^{-1} W \vec{y} = P \vec{y}$, where D is a diagonal matrix with diagonal elements d_i's i.e., $D = \text{diag}(d)$]. The matrix $P = D^{-1} W$ is known as the *Markov transition matrix* (or a *singly stochastic matrix*). Since all row sums of P equal 1, they are often interpreted as a probability distribution over random walks starting from a given node: for example, p_{ij} represents the probability of a random walk from node i to node j and we have $\sum_{j=1}^{n} p_{ij} = 1$.

Using the above formulation allows us to extend guilt-by-association to indirect neighbors. In particular, one can easily calculate the probability of a random walk of length 2 between all nodes in the network by computing P^2. The (i,j)-th entry of P^2 is given by $[P^2]_{ij} = \sum_{k=1}^{n} p_{ik} p_{kj}$ and represents the probability of a random walk of length 2 from node i to node j. In this way, we can include P^2 when calculating the node scores as:

$$f_i = \sum_{j=1}^{n} p_{ij} y_j + \sum_{j=1}^{n} [P^2]_{ij} y_j, \tag{1}$$

where the second term represents the probability of node j having the same label as its indirect neighbors (those neighbors that are two-steps away). Similarly, this approach can be extended to include other nodes at a distance of length r (usually $r<4$) by noting that $[P^r]_{ij}$ represents the probability of a random walk from i to j in r steps. We note that previous approaches have shown that increasing r beyond two often results in a degradation of the prediction performance (we elaborate on this point when presenting LPAs).

In the context of the above representation, several existing direct and indirect neighbor-based methods define node scores as $f_i = \sum_{j=1}^{n} p_{ij} y_j + \sum_{j=1}^{n} [\hat{P}^2]_{ij} y_j$, where $\hat{P}^2$ is a modified version of $\hat{P}^2$ with some entries set to zero (or modified). For example, in the BIOPIXIE graph search, the second summation only includes the top k genes that had the highest direct neighbor score; if node j is not among the top scoring direct neighbors of the positive genes, then $[\hat{P}^2]_{ij} = 0$.

In this section, we describe a simple framework for assigning node scores based on the label of other nodes in their local neighborhood. In the next section, we show that the local neighborhood approaches, as presented in Eq. 1, can be thought of as an approximation to LPAs.

2.2.3. Label Propagation Algorithms

Intuitively, LPAs can be thought of as an iterative scheme where positively labeled nodes propagate their labels to their neighbors and neighbors of neighbors and so on. At termination, the final score of a node is computed as the total amount of *propagation* it has received throughout this process. For example, Fig. 2 shows the solution to the LPA proposed by (33) when applied to a hypothetical network with 34 unlabeled and six positive nodes. As shown, the assigned scores are higher for nodes that have many short-length paths to the positives: nodes that are in the periphery of the cluster that contains the positives have lower scores than those in the center of the cluster.

Several variants of LPA have been used in gene function prediction including the works of (16, 34, 35). Here, we review an

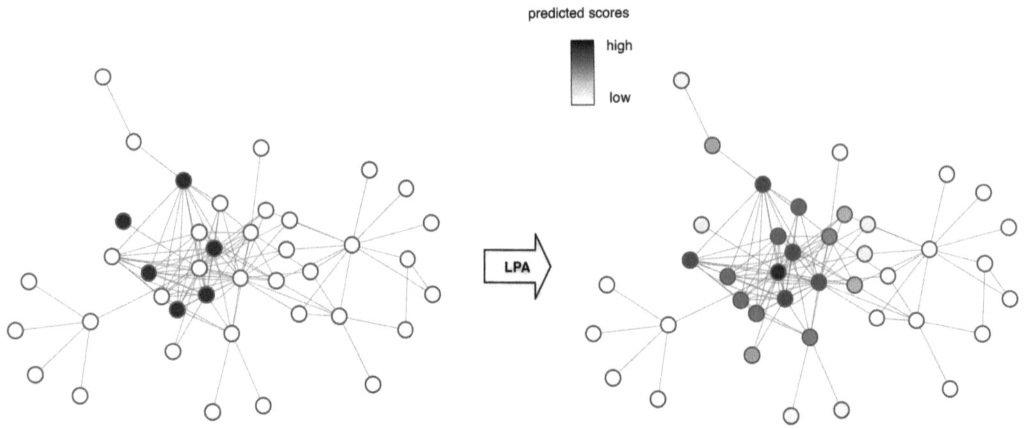

Fig. 2. Given six positive nodes (colored blue) on the left network, the network on the right shows the scores of all nodes assigned by LPA. The node colors on the right network depict the scores. A common approach to making predictions from the scores is to select the top *k* nodes according to their scores.

LPA derived from the Gaussian fields algorithm as it allows us to analytically calculate the final node scores (reviewed in ref. 36) and it has been shown to produce state-of-the-art performance in gene function prediction in yeast and mouse (17, 23).

The basic assumption of LPA is that the score of node i at iteration (or time) r can be computed from a weighted combination of score of its neighbors at the previous iteration and its initial label y_i. In its simplest form, we can state this principle as $f_i^{(r)} = \lambda \sum_{j=1}^{n} w_{ij} f_j^{(r-1)} + (1-\lambda) y_i$, where λ is a constant such that $0 < \lambda < 1$ thus making $f_i^{(r)}$ a convex combination of the scores of its neighbors and its initial label.

However, in the above formulation for node scores $f_i^{(r)}$, nodes with high weighted degree that have positive neighbors will end up influencing the scores of many nodes in their local neighborhood; it is standard practice to correct for this effect by normalizing W. In particular, we can normalize W in two different ways: (1) by dividing each row with its row sum and thus using the Markov Transition matrix in place of W: $P = D^{-1}W$ resulting in the expression $f_i^{(r)} = \lambda \sum_{j=1}^{n} \frac{w_{ij}}{d_i} f_j^{(r-1)} + (1-\lambda) y_i = \lambda \sum_{j=1}^{n} p_{ij} f_j^{(r-1)} + (1-\lambda) y_i$ or (2) performing a symmetric normalization $\dot{W} = D^{-1/2} W D^{-1/2}$ resulting in the expression $f_i^{(r)} = \lambda \sum_{j=1}^{n} \frac{w_{ij}}{\sqrt{d_i d_j}} f_j^{(r-1)} + (1-\lambda) y_i$. These two choices of normalization result in slightly different node scores. We first pursue the asymmetric normalization, thus using the Markov Transition Matrix P, as it will allow us to directly compare LPA to the local neighborhood method presented in Eq. 1.

Using P instead of W, we can write the node scores, at iteration r, in matrix form as $\vec{f}^{(r)} = \lambda P \vec{f}^{(r-1)} + (1-\lambda)\vec{y}$. A nice property of LPA is that we can analytically compute the final (steady-state) score by first writing the scores as $r \to \infty : \vec{f} = \lambda P \vec{f} + (1-\lambda)\vec{y}$, where, for simplicity, we represent $\vec{f}^{(\infty)}$ as $\vec{f}$. By re-arranging this equality we can directly calculate the final score vector as:

$$\vec{f} = (1-\lambda)(I - \lambda P)^{-1}\vec{y}. \tag{2}$$

Below we further simplify the solution to LPA using Taylor's matrix expansion. In particular, we make use of the following equality $(I - \lambda P)^{-1} = \sum_{r=0}^{\infty} (\lambda P)^r$. This equality holds in our case because we have $0 < \lambda < 1$, which makes the largest eigenvalue of λP less than 1 – a condition needed for using the Taylor's expansion (see Note 3). Thus, we can write the final score vector as:

$$\vec{f} = (1-\lambda)\sum_{r=0}^{\infty} \lambda^r P^r \vec{y}. \tag{3}$$

The above representation clarifies the connection between LPA and the local neighborhood method as presented in Eq. 1. However, there is a major difference: in LPA, because we multiply P^r with $\lambda^r < 0$, propagation from the positive nodes declines rapidly with paths of increasing length, whereas there are no such guarantees when using Eq. 1. This fact offers an explanation for the rapid decline in performance of neighborhood methods with path length increasing beyond two (e.g., as observed in (11)).

Similar to the derivation above, the solution to LPA with symmetrically normalized W can be written as:

$$\vec{f} = (1-\lambda)(I - \lambda \dot{W})^{-1}\vec{y}. \tag{4}$$

Again, using the Taylor's theorem, we can correspondingly derive the following form: $\vec{f} = (1-\lambda)\sum_{j=1}^{\infty} \lambda^r \dot{W}^r \vec{y}$ since we also have that the largest magnitude eigenvalue of $\lambda \dot{W}$ is again less than 1 (37).

The solution to LPA depends on a parameter $0 < \lambda < 1$. Note that smaller λ allow for more influence from paths of increasing length, in contrast, with large λ, only local paths (e.g., those with path lengths of 2 or 3) influence the final solution. In practice, λ can be set using cross-validation (see Note 4).

In this section, we present a formulation of LPA that is based on iterative propagation of scores to direct neighbors; it is also possible to derive the solution to LPA by solving a convex optimization problem (33) as we describe in Note 5. In particular, we have derived LPA for two different normalization of the matrix W: by using the asymmetric matrix P or symmetrically normalized matrix $\dot{W}$

(see Note 6). This formulation allows us to generalize several other LPAs: for example, the *RankProp* algorithm (34) uses asymmetrically normalized P (as in Eq. 3). On the contrary, the *functional Flow* algorithm (16) does not explicitly set a decay parameter λ or downweigh the influence of hubs by normalizations: these criterions are implicitly enforced by always propagating to shortest distance neighbors first and subtracting *out-flow* from *in-flow*. We also have shown that approximating the LPA solution using Taylor's matrix expansion results in an algorithm very similar to the local neighborhood method.

In addition to LPAs, discrete Markov Random Fields (DMRFs) present another class of methods that consider the global network topology when making predictions. In particular, DMRFs can be viewed as a discrete version of LPA where the predicted scores are constrained to be binary: i.e., $\vec{f} = [0,1]^n$. However, the integer constraint makes the DMRFs intractable. Nevertheless, previous studies have used DMRFs for predicting gene function either using simulated annealing (38) or approximating the solution using coordinate descent (39). However, typically solving for the node scores in a DMRF requires considerable computational effort. We note that several studies have found that methods based on discrete MRFs do not perform any better than LPAs or neighborhood-based methods (16, 19).

2.3. Constructing a Composite Network from Multiple Data Sources

As we discussed in Subheading 1, previous studies have shown that combining multiple high-throughput data sources into a single functional linkage network (FLN) results in better prediction performance. There are two broad categories of methods for constructing FLNs: (1) probabilistic FLNs, where edges between two genes represent the probability of their cofunctionality (functional *coupling*) and (2) FLNs constructed by weighted summation of the underlying networks, each constructed from a different data source. Probabilistic FLNs are commonly constructed using Bayesian network. In fact, most existing methods are similar to Naïve Bayes (13, 40–43). In the following, we describe the second approach for constructing FLNs. For a more extensive review on the subject, we refer the reader to (44).

A simple and widely used approach for combining multiple networks constructs a composite network, denoted by a weighted matrix W^*, as the average of the individual networks: e.g., $W^* = \frac{1}{D}\sum_{d=1}^{D} W_d$, where we have D networks represented as W_d (24). We can extend the above approach by taking a weighted sum: $W^* = \sum_{d=1}^{D} \alpha_d W_d$, where each network W_d is weighted according to the coefficient α_d (17, 18, 21, 22). For example, we can set these coefficients to downweigh redundant network and ignore irrelevant ones. These coefficients α_d are often set to optimize the performance of the composite network in predicting a single or a group of gene functions. As an example, (17) used linear regression to determine the coefficients $\vec{\alpha}$ (see Note 7).

To summarize, combining multiple networks into a single graph creates a better platform for the label propagation performance. In this section, we briefly described two simple and widely used methods for combining multiple networks. In the next section, we list online resources for prioritizing genes and predicting gene function from multiple heterogeneous data sources.

2.4. Online Resources for Predicting Gene Function

Table 2 summarizes several online resources that allow for on-demand gene prioritization and prediction of gene function: given a set of query (positive) genes, these resources use internally collected FLNs to predict other genes that are likely to share the same function or phenotype. Generally, there are three steps to making predictions: (1) collection and construction of multiple FLNs (offline), (2) integration of multiple FLNs into a single combined network (offline or online), and (3) predicting genes that are *similar* to the query set of genes (online).

3. Conclusions

In this chapter, we describe how to construct networks from high-throughput data sources and how to combine multiple networks into a composite functional linkage network. In addition, we provide a review of algorithms for predicting node labels from networks; such algorithms are often applied to predict genes that are involved in a given function or result in a specific phenotype. In particular, we focus on describing the label propagation algorithms and their relation to simpler neighborhood-based methods. As we show, these two types of algorithms are closely related. This observation may explain why several studies have found that neighborhood methods result in similar performance to LPAs (16, 45) – noting that average shortest distance in molecular interaction networks often follows the same distribution as small-world networks; in other words, most nodes are connected by small number of steps (e.g., 3 or 4). Finally, we summarize several online resources for gene prioritizing and predicting gene function.

4. Notes

1. *Positive Labels for Gene Function Prediction.* As mentioned in the Introduction, the set of positive labels can be derived from online databases such as Gene Ontology (GO) (6), MIPS (8),

Table 2
Summary of Web-based resources for on-demand prediction of gene function from FLNs

	Organisms	Algorithm	Flexibility in choosing FLNs
GeneMANIA (49)	Yeast, mouse, human, fly, worm, *Arabidopsis thaliana*	*Network integration*. Using the choice of (a) linear regression, (b) averaging, or (c) averaging while accounting for redundancy *Label prediction*. LPA using the combined network	Allows users to choose any combination of the available FLNs (by type or based on a publication) and network integration method. Additionally, users can upload their own networks
FunCoup (50)	Yeast, mouse, human, fly, worm, rat, *Arabidopsis thaliana*, *Ciona intestinalis*	*Network integration*. Naïve Bayes *Label prediction*. Uses Naïve Bayes classifiers, which make prediction based on direct neighbors in the combined network	Allows users to choose a network *type* (for example coexpression) but no flexibility on using specific networks based on a given publication
STRING (51)	630 Organisms	*Network integration*. Individual scoring of FLNs prior to construction of a network combination using a *Noisy OR* model *Label prediction*. Considers direct neighbors only	All predictions are made from a fixed network
BioPIXIE (13)	Yeast	*Network integration*. Naïve Bayes *Label prediction*. Uses direct and second-order neighbors heuristic (BioPIXIE Graph Search)	All predictions are made from a fixed network
MouseNet (52)	Mouse	*Network integration*. Naïve Bayes *Label prediction*. Uses direct and second-order neighbors heuristic (BioPIXIE Graph Search)	All predictions are made from a fixed network
HFalMp (43)	Human	*Network integration*. Regularized Naïve Bayes *Label prediction*. Uses direct and second-order neighbors heuristic	All predictions are made from a fixed network

and KEGG (7). GO is one the most widely used annotation databases covering a large number of organisms. GO defines three hierarchies for describing properties of gene products: Biological Process, Molecular Function, and Cellular Component. The categories defined in GO range from very broad properties that consist of hundreds of genes (e.g., biological regulation) to very specific properties that consist of only a few genes (e.g., positive regulation of mitosis). Algorithms for predicting gene function are often tested on categories that have between 10 and 300 annotations; these categories have enough positives for training without being too broad (23). For a discussion on choosing informative GO categories *see* ref. 46.

2. *Sparsification of FLNs.* To construct an FLN from profile-based data, we can use a similarity metric such as the PCC; for example, in a coexpression network, the edge between protein *i* and *j* is the PCC between their expression profiles. However, since the PCC is often nonzero for many pairs of genes, we can often improve performance (both in accuracy and computational time requirement) by sparsifying the PCC derived networks. A common way for doing this is to keep the top *m* interacting partners for each gene and set the rest to zero; *m* can range from 50 to 100 as done in (17, 21, 23). A second approach is to set a threshold value *t*, where all PCC's smaller than *t* are set to zero (25).
3. *Taylors Matrix Expansion Theorem.* To be able to use the Matrix Expansion Theorem we need to ensure that the largest magnitude eigenvalue of P is less than 1: $\max_i |\sigma_i| < 1$. We use the *Perron-Frobenius Theorem* (PFT) to show that this condition holds: from PFT we have that $\min_i \sum_{j=1}^{n} \lambda p_{ij}\ \lambda p_{ij} \le \max_i |\sigma_i| \le \max_i \sum_{j=1}^{n} \lambda p_{ij}$ note that $\sum_{j=1}^{n} p_{ij} = 1$ for all *i*, since we have a multiplication with the constant $0 < \lambda < 1$, then the maximum value of all row sums is less than 1.
4. *Setting the Parameter λ in LPA.* We can set the parameter λ using cross-validation. To do so, we investigate the performance of various settings of λ on the *validation set* (a fraction of training data not used to in training). In practice, we have found that setting $\lambda \approx 0.5$ (when solving LPA in Eq. 4) results in good performance in a wide variety of prediction tasks.
5. *LPA as the solution to a convex optimization problem.* LPA as proposed by (33, 47) is derived using the following objective function:

$$\arg\min_{f} c\sum_{i=1}^{n}(f_i - y_i)^2 + \frac{1}{2}\sum_{i,j=1}^{n} w_{ij}\left(\frac{f_i}{\sqrt{d_i}} - \frac{f_j}{\sqrt{d_j}}\right)^2, \qquad (5)$$

where $c>0$ is a constant. This equation can be written in matrix form as $\arg\min_{f} c(\vec{f}-\vec{y})^T(\vec{f}-\vec{y})+\vec{f}^T \dot{L}\vec{f}$ with $L = I - \dot{W}$ known as the normalized graph Laplacian (recall that $W^{\Upsilon} = D^{-1/2}WD^{-1/2}$ is the symmetrically normalized weight matrix W). Differentiating and setting the derivative to zero, we get $\vec{f} = c\left((1+c)I - \dot{W}\right)^{-1}\vec{y}$. Using $\lambda = \frac{1}{1+c}$ we can write as follows: $\vec{f} = (1-\lambda)(I-\lambda\dot{W})^{-1}\vec{y}$. Thus, this version of LPA uses the symmetrically normalized W that we used to derive Eq. 4.

6. *LPA with symmetrically and asymmetrically normalized weight matrix.* Empirically, we have observed better performance when using the symmetrically normalized $\dot{W}$ and so we suggest using this form of LPA in practice. Furthermore, because $\dot{W}$ is symmetric, we can solve Eq. 4 efficiently using Conjugate Gradient, which scales with the number of nonzero elements in $\dot{W}$.

7. *Constructing a Composite Network by Averaging FLNs.* Several previous studies have found that a simple averaging of underlying networks often results in composite networks that have comparable performance to those constructed with more sophisticated methods, for example, those that assign network weights α_d (24, 48). However, simple averaging will suffer when many of the underlying networks are redundant (or represent similar information); for example, there are often many more coexpression networks than other data types. One simple way to correct for redundancy is to simply group networks together based on their type and assign each type an equal weight: e.g., $W^* = \frac{1}{3}W_{\exp} + \frac{1}{3}W_{gi} + \frac{1}{3}W_{pi}$, where $W_{\exp}$, W_{gi}, and *Wpi* represent the *average network* of all coexpression, genetic interactions, and protein interactions networks, respectively.

References

1. Marcotte, E.M., et al., *Detecting protein function and protein-protein interactions from genome sequences.* Science, 1999. **285**(5428): p. 751–3.
2. Wu, X., et al., *Network-based global inference of human disease genes.* Mol Syst Biol, 2008. **4**: p. 189.
3. Aerts, S., et al., *Gene prioritization through genomic data fusion.* Nat Biotechnol, 2006. **24**(5): p. 537–44.
4. Sharan, R., I. Ulitsky, and R. Shamir, *Network-based prediction of protein function.* Mol Syst Biol, 2007. **3**: p. 88.
5. Oti, M. and H.G. Brunner, *The modular nature of genetic diseases.* Clin Genet, 2007. **71**(1): p. 1–11.
6. Ashburner, M., et al., *Gene ontology: tool for the unification of biology. The Gene Ontology Consortium.* Nat Genet, 2000. **25**(1): p. 25–9.
7. Ogata, H., et al., *KEGG: Kyoto Encyclopedia of Genes and Genomes.* Nucleic Acids Res, 1999. **27**(1): p. 29–34.
8. Ruepp, A., et al., *The FunCat, a functional annotation scheme for systematic classification of proteins from whole genomes.* Nucleic Acids Res, 2004. **32**(18): p. 5539–45.
9. Robinson, P.N., et al., *The Human Phenotype Ontology: a tool for annotating and analyzing human hereditary disease.* Am J Hum Genet, 2008. **83**(5): p. 610–5.
10. Hamosh, A., et al., *Online Mendelian Inheritance in Man (OMIM), a knowledgebase of human genes and genetic disorders.* Nucleic Acids Res, 2005. **33**(Database issue): p. D514–7.
11. Chua, H.N., W.K. Sung, and L. Wong, *Exploiting indirect neighbours and topological*

weight to predict protein function from protein-protein interactions. Bioinformatics, 2006. **22**(13): p. 1623–30.

12. Zhou, X., M.C. Kao, and W.H. Wong, *Transitive functional annotation by shortest-path analysis of gene expression data.* Proc Natl Acad Sci USA, 2002. **99**(20): p. 12783–8.
13. Myers, C.L., et al., *Discovery of biological networks from diverse functional genomic data.* Genome Biol, 2005. **6**(13): p. R114.
14. Karaoz, E., et al., *Protective role of melatonin and a combination of vitamin C and vitamin E on lung toxicity induced by chlorpyrifos-ethyl in rats.* Exp Toxicol Pathol, 2002. **54**(2): p. 97–108.
15. Deng, M., et al., *Prediction of protein function using protein-protein interaction data.* J Comput Biol, 2003. **10**(6): p. 947–60.
16. Nabieva, E., et al., *Whole-proteome prediction of protein function via graph-theoretic analysis of interaction maps.* Bioinformatics, 2005. **21 Suppl 1**: p. i302–10.
17. Mostafavi, S., et al., *GeneMANIA: a real-time multiple association network integration algorithm for predicting gene function.* Genome Biol, 2008. **9 Suppl 1**: p. S4.
18. Tsuda, K., H. Shin, and B. Scholkopf, *Fast protein classification with multiple networks.* Bioinformatics, 2005. **21 Suppl 2**: p. ii59-65.
19. Murali, T.M., C.J. Wu, and S. Kasif, *The art of gene function prediction.* Nat Biotechnol, 2006. **24**(12): p. 1474–5; author reply 1475–6.
20. Deng, M., T. Chen, and F. Sun, *An integrated probabilistic model for functional prediction of proteins.* J Comput Biol, 2004. **11**(2–3): p. 463–75.
21. Mostafavi, S. and Q. Morris, *Fast Integration of Heterogeneous Data Sources for Predicting Gene Function with Limited Annotation.* Bioinformatics, 2010.
22. Lanckriet, G.R., et al., *A statistical framework for genomic data fusion.* Bioinformatics, 2004. **20**(16): p. 2626–35.
23. Pena-Castillo, L., et al., *A critical assessment of Mus musculus gene function prediction using integrated genomic evidence.* Genome Biol, 2008. **9 Suppl 1**: p. S2.
24. Pavlidis, P., et al., *Learning gene functional classifications from multiple data types.* J Comput Biol, 2002. **9**(2): p. 401–11.
25. Zhang, B. and S. Horvath, *A general framework for weighted gene co-expression network analysis.* Stat Appl Genet Mol Biol, 2005. **4**: p. Article17.
26. Yona, G., et al., *Effective similarity measures for expression profiles.* Bioinformatics, 2006. **22**(13): p. 1616–22.
27. Warde-Farley, D., et al., *The GeneMANIA prediction server: biological network integration for gene prioritization and predicting gene function.* Nucleic Acids Res, 2010. **Accepted**(Webserver Issue).
28. Costanzo, M., et al., *The genetic landscape of a cell.* Science, 2010. **327**(5964): p. 425–31.
29. Tong, A.H., et al., *Global mapping of the yeast genetic interaction network.* Science, 2004. **303**(5659): p. 808–13.
30. Weirauch, M.T., et al., *Information-based methods for predicting gene function from systematic gene knock-downs.* BMC Bioinformatics, 2008. **9**: p. 463.
31. Hishigaki, H., et al., *Assessment of prediction accuracy of protein function from protein--protein interaction data.* Yeast, 2001. **18**(6): p. 523–31.
32. Schwikowski, B., P. Uetz, and S. Fields, *A network of protein-protein interactions in yeast.* Nat Biotechnol, 2000. **18**(12): p. 1257–61.
33. Zhou, D., et al., *Learning with Local and Global Consistency*, in *Neural Information Processing Systems.* 2003, MIT Press: Vancouver, BC, Canada.
34. Weston, J., et al., *Protein ranking: from local to global structure in the protein similarity network.* Proc Natl Acad Sci USA, 2004. **101**(17): p. 6559–63.
35. Hu, P., H. Jiang, and A. Emili, *Predicting protein functions by relaxation labelling protein interaction network.* BMC Bioinformatics, 2010. **11 Suppl 1**: p. S64.
36. Bengio, Y., O. Delalleau, and N. Le Roux, *Label Propagation and Quadratic Criterion*, in *Semi-Supervised Learning*, O. Chapelle, B. Scholkopf, and A. Zien, Editors. 2006, MIT Press.
37. Chung, F., *Spectral Graph Theory.* Number 92 in CBMS Regional Conference Series in Mathematics. 1999: American Mathematical Society.
38. Vazquez, A., et al., *Global protein function prediction from protein-protein interaction networks.* Nat Biotechnol, 2003. **21**(6): p. 697–700.
39. Karaoz, U., et al., *Whole-genome annotation by using evidence integration in functional-linkage networks.* Proc Natl Acad Sci USA, 2004. **101**(9): p. 2888–93.
40. Fraser, A.G. and E.M. Marcotte, *A probabilistic view of gene function.* Nat Genet, 2004. **36**(6): p. 559–64.
41. Lee, I., et al., *A probabilistic functional network of yeast genes.* Science, 2004. **306**(5701): p. 1555–8.
42. Myers, C.L. and O.G. Troyanskaya, *Context-sensitive data integration and prediction of*

biological networks. Bioinformatics, 2007. **23**(17): p. 2322–30.

43. Huttenhower, C., et al., *Exploring the human genome with functional maps*. Genome Res, 2009. **19**(6): p. 1093–106.
44. Noble, W.S. and A. Ben-Hur, *Integrating Information for Protein Function Prediction*, in *Bioinformatics-From Genomes to Therapies*, T. Lengauer, Editor. 2007, Wiley-VCH Verlag GmbH & Co KGaA: Weinheim, Germany.
45. Song, J. and M. Singh, *How and when should interactome-derived clusters be used to predict functional modules and protein function?* Bioinformatics, 2009. **25**(23): p. 3143–50.
46. Myers, C.L., et al., *Finding function: evaluation methods for functional genomic data*. BMC Genomics, 2006. 7: p. 187.
47. Zhu, X., J. Lafferty, and Z. Ghahramani. *Semi-supervised learning using Gaussian fields and harmonic functions*. in *International Conference on Machine Learning*. 2003. Washington DC, USA.
48. Lewis, D.P., T. Jebara, and W.S. Noble, *Support vector machine learning from heterogeneous data: an empirical analysis using protein sequence and structure*. Bioinformatics, 2006. **22**(22): p. 2753–60.
49. Warde-Farley, D., et al., *The GeneMANIA prediction server: biological network integration for gene prioritization and predicting gene function*. Nucleic Acids Res, 2010. **38 Suppl**: p. W214–20.
50. Alexeyenko, A. and E.L. Sonnhammer, *Global networks of functional coupling in eukaryotes from comprehensive data integration*. Genome Res, 2009. **19**(6): p. 1107–16.
51. von Mering, C., et al., *STRING: known and predicted protein-protein associations, integrated and transferred across organisms*. Nucleic Acids Res, 2005. **33**(Database issue): p. D433–7.
52. Guan, Y., et al., *A genomewide functional network for the laboratory mouse*. PLoS Comput Biol, 2008. **4**(9): p. e1000165.

Chapter 21

Mathematical Modeling of Biomolecular Network Dynamics

Alexander V. Ratushny, Stephen A. Ramsey, and John D. Aitchison

Abstract

Mathematical and computational models have become indispensable tools for integrating and interpreting heterogeneous biological data, understanding fundamental principles of biological system functions, generating reliable testable hypotheses, and identifying potential diagnostic markers and therapeutic targets. Thus, such tools are now routinely used in the theoretical and experimental systematic investigation of biological system dynamics. Here, we discuss model building as an essential part of the theoretical and experimental analysis of biomolecular network dynamics. Specifically, we describe a procedure for defining kinetic equations and parameters of biomolecular processes, and we illustrate the use of fractional activity functions for modeling gene expression regulation by single and multiple regulators. We further discuss the evaluation of model complexity and the selection of an optimal model based on information criteria. Finally, we discuss the critical roles of sensitivity, robustness analysis, and optimal experiment design in the model building cycle.

Key words: Biomolecular network, Differential equation, Dynamical system, Inverse problem, Mathematical model, Systems biology

1. Introduction

Molecular biology research has been profoundly impacted by the development of high-throughput measurement technologies such as next-generation sequencing, DNA microarrays, and mass spectrometry, and by the application of these technologies in large-scale functional genomic, proteomic, metabolomic, protein–protein interaction, and protein–DNA interaction assays. These advances are enabling researchers to comprehensively map the molecular components, processes, networks, and functions that underlie biological processes and human diseases. A systems biology approach integrating these heterogeneous molecular data in quantitative network models promises to facilitate the comprehensive understanding of biological systems, the identification of novel diagnostic markers and drug targets, and rational intervention into disease processes.

Gerard Cagney and Andrew Emili (eds.), *Network Biology: Methods and Applications*, Methods in Molecular Biology, vol. 781, DOI 10.1007/978-1-61779-276-2_21, © Springer Science+Business Media, LLC 2011

The dynamic biomolecular network model is a powerful framework for integrating heterogeneous experimental data (1, 2). It allows us to interpret the behavior of complex biological systems both quantitatively and mechanistically, and to generate targeted experimentally testable hypotheses. Thus, modeling is an essentially component of the systems biology cycle – "experiment–model–hypothesis–experiment" (Fig. 1). To construct models with predictive utility, it is critical to describe cellular mechanisms with

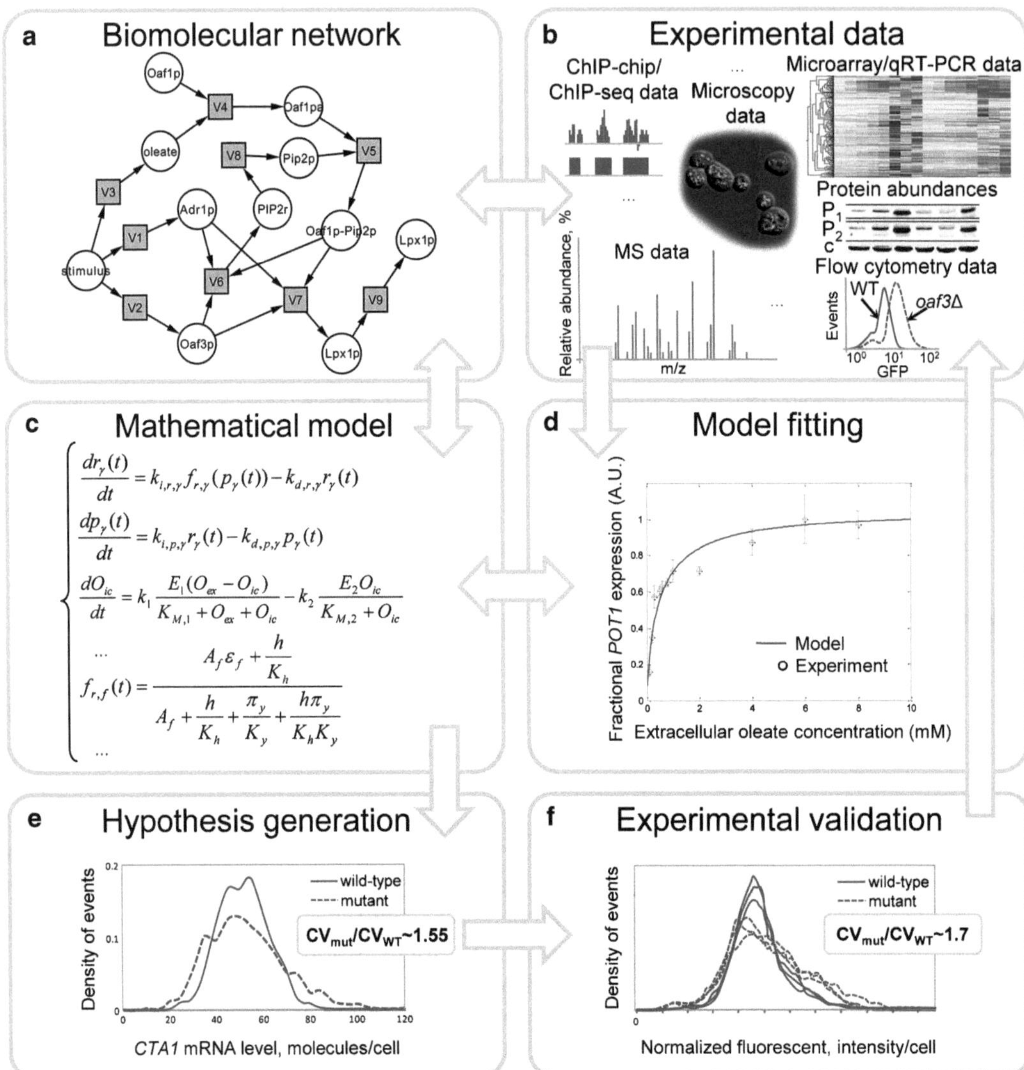

Fig. 1. Workflow of the theoretical and experimental analysis of biomolecular network dynamics (the model building cycle). *Arrows* denote information flow, illustrating how the model can lead to the formulation of hypotheses that can be experimentally tested. The results from experimental validation of the model are used to improve the model, and simulations from the refined model can lead to new predictions. (**a**) Diagram showing regulatory network controlling oleate response in yeast. (**b**) Types of data used in systems biology. (**c**) Equations showing the structure of the ordinary differential equation (ODE) model of the oleate response network. (**d**) Predicted and measured steady-state dose-response of the oleate-inducible

the appropriate level of detail. In the predictive kinetic biomolecular models we describe below, we seek to compile information on the system process details including reaction rates and concentrations of molecular constituents. Estimating reaction rates and other parameters is often challenging due to the heterogeneity of experimental data (e.g., different source origins, setups, cell types, experimental conditions, etc.) and the difficulty of obtaining data relevant to the specific in vivo conditions.

Despite the tremendous data-generating capabilities of new high-throughput technologies, the compendium of available measurements of many cellular component levels and processes in space and time remains sparse. Often, the available data are not sufficient to infer the detailed molecular mechanisms underlying a given cellular process. Furthermore, current knowledge of molecular mechanisms is highly nonuniform, varying from the well supported and very detailed to the hypothetical and poorly described. Thus, it is impossible to describe all processes in the model equally comprehensively. The ideal modeling method should allow rational selection of the most appropriate level of detail in the model based on prior knowledge of relevant pathways and the resolution of relevant measurements (3, 4).

In engineering terminology, one could describe the problem of inferring regulatory mechanisms within biological systems from prior pathway/interaction knowledge and measurements of system behavior as a "structural inverse problem with prior information." Numerous theoretical and experimental approaches have been developed to attempt to solve this biological inverse problem (5–12), and the investigation of integrative methods to identify biomolecular interactions and mechanisms from heterogeneous data remains an active area of research (13). This problem naturally divides into two components: mapping the network of molecular components and processes relevant to the system, and constructing a kinetic and quantitative model of the network behavior. It is the latter component that is the specific area of focus of this chapter.

Broadly speaking, mathematical approaches to modeling biological networks can be grouped by the type of mathematical abstraction used to represent the state vector of the system, such as discrete-valued (14–16), continuous-valued (17–19), stochastic (20–22), and various combinations of the three types (hybrid models) (18, 19, 23–26). In this chapter, we focus primarily on a continuous-valued approach for modeling biomolecular network dynamics using nonlinear ordinary or delay differential equations and the possible combination with the discrete approach for mod-

Fig. 1. (continued) target gene, POT1. (**e**) Predicted cell population distributions of the mRNA level of an oleate-inducible gene in the wild-type oleate network and in a perturbed ("mutant") network. (**f**) Measured cell population distributions of the level of a GFP-tagged, oleate-inducible protein in wild-type yeast and in a yeast strain bearing a mutation corresponding to the network perturbation in the "mutant" model.

eling such events as transitions from different environments, cell divisions, etc. We use the fatty-acid (FA)-induced transcriptional regulatory network model (27) as a running example to illustrate different steps of the workflow of the theoretical and experimental analysis of the molecular network dynamics.

2. Methods

2.1. Construction of a Mathematical Model

1. Construction of a mathematical model of a biomolecular network of interest (the model building cycle) starts from stating the purpose and objectives of the model (see Note 1). For example, the main purpose of the FA-responsive transcriptional regulatory network model is to understand the carbon source-sensing mechanism of the core regulatory network as well as the interplay between transcription factors (TFs) that combinatorially regulate of the expression of target genes. Ideally, the FA-responsive network model should help us understand how the variations in the input stimulus (FA concentration) and the molecular system parameters (e.g., the FA transport efficiency, the ligand–sensor affinity, TF activities, transcription and translation initiation rates, component degradation rates, etc.) affect the network response.
2. Collect/gather all available information on the structural and functional organization of the biomolecular network of interest. Define the list of processes and variables (molecular components: genes, RNAs, proteins, small molecules, intermediate complexes, etc.) that should be included in the model. For example, the core FA-responsive transcriptional network consists of carbon source-sensing transcription factors that regulate key target genes through an overlapping feed-forward network motif. The model describes the transport of extracellular oleate into the cell, which activates a molecular network governing peroxisome protein production and organelle proliferation. The model describes the oleate-dependent expression and function of four TF genes (*ADR1*, *PIP2*, *OAF1*, and *OAF3*), as well as the expression of three oleate-inducible reporter genes that are downstream targets of these TFs (*POT1*, *CTA1*, and *LPX1*). For each gene in the model, both the gene-specific mRNA and total protein are accounted for. The model takes into account the synthesis and degradation of each species of mRNA and protein (27).
3. The structural organization of the biomolecular network of interest can be illustrated as a bipartite graph (Fig. 1a), which presents two independent disjoint sets of vertices. Within the graph, circular nodes represent variables of the model, i.e., levels of molecular species (RNA and protein components). Square

nodes represent molecular processes that together define the dynamic relationships among physical entities (e.g., translation of RNA to protein). Edges of the graph define the relationships between the participants in each process of the model. The bipartite graph of the model at initial steps helps to reveal the model structure, its inputs and end points and to detect possible gaps and incompleteness, which is particularly important for large-scale models. For example, the graph of the oleate response network (Fig. 1a) shows that transcription factors regulate oleate-responsive genes in a feed-forward network topology, which has important implications for the stability and robustness of control of oleate-responsive genes (27).

4. Collect/gather all available quantitative and qualitative experimental data for molecular components and processes of the model (concentrations of molecular components at different states (e.g., basal, induced, repressed, relative concentrations in different conditions); kinetics of molecular components during transitions between different conditions/environments; synthesis and degradation rates of molecular components of the model; interaction affinities; dependencies of process rates on the level of regulator(s), any available kinetic constants and their estimates of the processes of interest, etc.) (Fig. 1b). Examples of experimental techniques and quantitative measurements for kinetic models are listed in Table 1.
5. Define possible mechanisms and formally describe each process in the model. For example, gene expression in the FA-induced transcription model (hereafter, FAIT model) is represented using a *fractional activity function* for each gene, which is a function of the protein concentrations in the model (for a description of how to derive a fractional activity function from the mechanistic details of the molecular process, see Subheading 2.2). Protein synthesis is modeled as proportional to mRNA concentration. Often, mechanisms that occur rapidly relative to the time scales of interest in the model can be simplified using the assumption of quasi-steady-state (28). For example, activation of the oleate-sensing transcription factor Oaf1p is assumed to occur rapidly so that activation and complex dissociation are at quasi-steady-state with respect to the slowly varying total Oaf1p concentration (moreover, the extracellular oleate concentration is taken to be constant, and depletion of oleate from the media is not modeled). The transport of FA across the plasma membrane and subsequent esterification with coenzyme A (CoA) are modeled by assuming a hyperbolic saturating function of extracellular oleate concentration for the rate of transport and Michaelis–Menten kinetics for fatty acyl-CoA synthesis. Turnover is modeled using first-order kinetics (27).
6. Each component of the system might participate simultaneously in several processes with nonzero stoichiometry.

Table 1
Examples of experimental techniques and quantitative measurements for kinetic models

Quantity in the kinetic model	Experimental technique
Concentrations of molecular components	qRT-PCR; microarray; quantitative northern blot; fluorescence microscopy; quantitative mass spectrometry; flow cytometry; etc.
Enzymatic activity (initial rate, transient and relaxation kinetics, etc.)	Enzyme assays (spectrophotometric, fluorometric, calorimetric, chemiluminescent, light scattering, chromatographic and other assays)
Kinetics of molecular interactions	In vitro measurements; time-course measurements of protein activation; etc.
Time-course of protein–DNA binding	DNA-binding protein immunoprecipitation combined with microarray (ChIP-chip), sequencing (ChIP-seq) or PCR; in vitro binding assays for purified interacting partners; etc.
Time-course of protein–RNA binding	RNA-binding protein immunoprecipitation combined with microarray, sequencing or PCR; in vitro binding assays for purified interacting partners; etc.
Kinetics of protein–protein binding	Affinity electrophoresis; co-immunoprecipitation combined with quantitative mass spectrometry; quantitative immunoprecipitation combined with knock-downs; fluorescence correlation spectroscopy (FCS); surface plasmon resonance (SPR); static and dynamic light scattering; etc.
Kinetics of protein–ligand binding	Ligand-binding assays; radiolabeling, affinity electrophoresis; SPR; light scattering; etc.
Kinetics of molecular transport	Imaging; fluorescence microscopy; radiolabeling, pulse chase; etc.
Synthesis and degradation rates	Time-course abundance data (during induction or deactivation of system, or after inhibition of transcription); pulse chase; etc.

For example, the kinetics of changes in the abundance of the *i*th mRNA species can be described based on its time and state-dependent overall synthesis ($V_{\text{syntesis}}^{\text{mRNA}_i}$) and degradation ($V_{\text{degradation}}^{\text{mRNA}_i}$) rates as the following differential equation

$$\frac{\mathrm{d}[\text{mRNA}_i]}{\mathrm{d}t} = V_{\text{syntesis}}^{\text{mRNA}_i} - V_{\text{degradation}}^{\text{mRNA}_i}. \tag{1}$$

In the more specific case, when the synthesis of the ith mRNA is described in terms of a fractional activity f_i (see Subheading 2.2) and the mRNA degradation is proportional to the mRNA concentration in the cell, the kinetics of the mRNA changes can be described as follows:

$$\frac{d[\mathrm{mRNA}_i]}{dt} = k_i^{\max} f_i - k_i^{\mathrm{d}} [\mathrm{mRNA}_i], \qquad (2)$$

where $k_i^{\max}$ is the maximum rate constant for the mRNA synthesis and k_i^{d} is the mRNA degradation rate constant.

7. For each dynamic species in the model, a kinetic equation governing its time rate of change is derived by combining contributions from all relevant dynamic processes (from step 5), and taking into account the stoichiometry of the species in each process. When completed, this yields a system of differential equations

$$\frac{dx_i}{dt} = \sum_{n=1}^{N} V_n^{\mathrm{in}} - \sum_{m=1}^{M} V_m^{\mathrm{out}}, \qquad (3)$$

where x_i is the ith dynamic species of the kinetic model; V_n^{in} is a rate of the nth process in which x_i is produced ($n = \overline{1 \cdot\cdot N}$) and V_m^{out} is a rate of the mth process in which x_i is consumed, where ($m = \overline{1 \cdot\cdot M}$).
Examples of differential equations of the oleate model are shown in Fig. 1c.

8. Once the equations governing the model's dynamics (Eq. 3) are defined, the model is then specified by determining initial species concentrations and rate parameter values. Usually, some model parameter values are known from the literature or have been measured in direct assays and, therefore, can be fixed in the model. The values of other parameters may only be known to fall within a certain range. Parameters or species concentrations whose values are not known or adequately constrained by previous measurements need to be estimated (within physiologically relevant ranges), using the model, from the available measurements of whole-system response (see Note 2). Formally, the parameter estimation procedure can be described as solving the parametric inverse problem. In brief, a cost function representing the total deviation of the model predictions from the experimental data is defined, and the model parameters are then varied to identify the parameter set that minimizes the cost function. For example, the cost function (D) can be calculated as a sum of the squares of the model prediction error defined as a deviation of the quantitative characteristics calculated by the model (e.g., concentration of molecular components in different conditions) ($y^{\mathrm{predicted}}$) from the relevant experimental measurements (y^{measured})

$$D = \sum_{i=1}^{N} \frac{(y_i^{\mathrm{predicted}} - y_i^{\mathrm{measured}})^2}{\sigma_i^2}, \qquad (4)$$

where σ_i^2 is the variance in the ith measurement.

Numerous optimization algorithms have been developed to solve the parametric inverse problem in many fields of science (29). The gradient-based optimization algorithms are often used for the model parameter optimization. These algorithms are well suited to the task of finding a local minimum of an optimization function. Nonlinear mathematical models usually have multiple local minima and applying only gradient-based optimization methods is not effective for finding the global minimum of the cost function or the globally optimal solution. To overcome this problem, one can combine the traditional gradient methods with Monte Carlo methods (i.e., methods with random sampling). Evolutionary and genetic algorithms are often used as Monte Carlo heuristic methods. These algorithms incorporate into the optimization procedure basic principles of evolution such as mutation, recombination, selection, inheritance, etc. There are also other Monte Carlo optimization methods such as simulated annealing, stochastic optimization, etc. (see Notes 3 and 4).

In the case of the FAIT model, optimization was carried out using the constrained optimizers *ga* (genetic algorithm) and *fmincon* (constrained nonlinear optimization) in MATLAB. The undetermined model parameters were optimized within the physiologically relevant constraints to minimize the sum-squared of the model prediction error for the time-course and steady-state gene expression measurements. A comparison of the optimized FAIT model characteristics with experimental data is shown in Fig. 1d.

9. Steps 5 to 8 can be repeated for the models with different mechanistic representations of the processes in the model. For example, in the model of transcriptional induction of the yeast galactose metabolic pathway in yeast, two different model scenarios were studied. These two scenarios represented two different hypotheses regarding the mechanism of galactose-dependent de-repression of the *GAL* regulon by Gal3p, namely, direct nuclear binding of Gal3p and indirect derepression. Comparison of the fitness of the two model scenarios favored the indirect mechanism (30, as well as *unpublished data*).

10. Analyze/estimate the model complexity and/or choose the optimal model. Ideally, the model complexity should be commensurate with the amount and granularity of available experimental data. For example, given a body of highly fine-grained time-course data and data from a diverse set of system perturbations, a very simplistic model might be expected to poorly recapitulate the body of data. On the contrary, an inappropriately complex model with many undetermined parameters will

likely have poor performance recapitulating experimental measurements that were held out from the body of measurements used to train the model, a problem known as *overfitting*. Usually, in both cases the model would be said to have a low predictive power (see Note 5). To compare two or more models with differing levels of parametric complexity, one can apply the Bayesian or Akaike information criteria (BIC (31) and AIC (32), respectively). These criteria quantify the trade-off relationship between model prediction performance and the number of the model parameters. For example, the potential for overfitting in the FAIT model was estimated by quantifying the bias-versus-variance trade-off. The small-sample-corrected Akaike Information Criterion (c-AIC) (33) was used for this purpose. The Akaike penalty term was computed using the following formula:

$$\text{Penalty} = 2k + \frac{2k(k+1)}{N-k-1}, \qquad (5)$$

where k is a number of model parameters that were determined by parameter estimation and N is the number of experimental data points. The variance was computed as the sum-squared model prediction error using Eq. 4.

The Penalty/D ratio for the model was found to have a value of 1.4. Given that c-AIC is a conservative estimate of model fitness (i.e., the c-AIC involves a quite strong penalty for complexity), this penalty/error ratio in conjunction with results of the sensitivity analysis (see step 12) suggested that the available number and diversity of measurements used for model training are adequate to discern the best-fit model in the space of parameter values, within the model class necessitated by prior biological knowledge of the underlying molecular interactions.

At this point, we have an optimized model that can be used as a tool for exploring quantitative and qualitative characteristics of the molecular system of interest.

11. For both uncertainty analysis of model predictions and identifying potential targets for perturbation analysis in the biological system, a sensitivity analysis of the model output predictions can be beneficial. For example, it is often useful to perform a sensitivity analysis of the model output characteristics to variable inputs (stimuli). In general, sensitivity analysis allows one to explore quantitatively the relationship between input and output characteristics and detect the ranges of the input signal variations to which the model responses are most sensitive as well as the ranges to which the system is not responsive or saturated (see Note 6).

The sensitivity matrix (S^{io}) of the model response (r_i, $i = \overline{1 \cdot\cdot n}$) changes to input signal (s_j, $j = \overline{1 \cdot\cdot k}$) variations is defined as follows:

$$S^{io} = \begin{bmatrix} \frac{\partial r_1 / r_1}{\partial s_1 / s_1} & \cdots & \frac{\partial r_1 / r_1}{\partial s_k / s_k} \\ \cdots & \cdots & \cdots \\ \frac{\partial r_n / r_n}{\partial s_1 / s_1} & \cdots & \frac{\partial r_n / r_n}{\partial s_k / s_k} \end{bmatrix}. \tag{6}$$

For example, the model of the FAIT regulatory network was used to analyze whether the Oaf3p regulator acts to buffer the induced genetic switch against variations in the level of FA in the environment (27). The expression kinetics of the target gene (*POT1*) were investigated in WT and *oaf3*Δ model strains exposed to an oscillating oleate concentration. The kinetic model predicted that the transcriptional repressor Oaf3p modulates the amplitude of variation of expression levels of oleate-responsive element (ORE)-driven genes in a fluctuating environment. The target gene undergoes larger-amplitude variations in the *oaf3*Δ model than in the WT model, indicating that the loss of Oaf3p destabilizes the genetic switch with respect to transient oleate variations (27). The dependence of the target gene expression amplitudes on the oleate variation frequency and amplitude was also systematically explored (27). The model simulations show that the *POT1* amplitude difference between the *oaf3*Δ model and WT model increases with decreasing frequency of an oleate pulse, suggesting that the *oaf3*Δ strain is less able than the WT, to filter out oleate variations on a time scale of >40 min. Varying both the amplitude and period of the oleate concentration oscillations reveals a nonlinear relationship between amplitude and period. The maximal differences between *oaf3*Δ and WT strains shift toward greater period as the amplitude increases (27), which illustrates the potential of mathematical models of complex biomolecular networks to reveal nontrivial system-level properties.

12. Perform parameter sensitivity analysis (PSA). PSA can be used to systematically investigate how perturbations/changes in the model parameters impact on the system outcomes. PSA is very useful for analyzing the robustness of the behavior of the system with respect to perturbations of its components (e.g., pharmacological inhibition, knockouts, etc.), as well as for identifying the limiting parameters for system functioning. This analysis is also helpful for estimating the required precision levels of the parameter values that are necessary to achieve

a given precision level in a prediction of system output/behavior (see Note 6 and 7). There are two major types of PSA: local and global.

(a) *Local PSA:* The sensitivity matrix (S^{p}) of the model response (r_i, $i = \overline{1 \cdot\cdot n}$) to changes to the model parameters (k_j, $j = \overline{1 \cdot\cdot z}$) can be defined as follows

$$S^{p} = \begin{bmatrix} \dfrac{\partial r_1 / r_1}{\partial k_1 / k_1} & \cdots & \dfrac{\partial r_1 / r_1}{\partial k_z / k_z} \\ \cdots & \cdots & \cdots \\ \dfrac{\partial r_n / r_n}{\partial k_1 / k_1} & \cdots & \dfrac{\partial r_n / r_n}{\partial k_z / k_z} \end{bmatrix}. \tag{7}$$

Local PSA allows exploring the model response changes to one parameter variation at a time.

(b) *Global PSA* allows varying multiple model parameters simultaneously to analyze their effect on the model outputs. Global PSA methods usually apply random sampling techniques because it is impossible to explore systematically the influence of all possible combinations of the model parameter changes of the model outputs. PSA is extensively described and applied in a number of publications (34–39).

As an example application of PSA, the relative model prediction error for alternative values of the 14 fitted parameters was estimated in the FAIT regulatory network. The relative model prediction error (i.e., the optimization cost function) was calculated using Eq. 4. The model error was analyzed as a function of varying a single model parameter over a range of approximately eightfold up or down relative to the best-fit parameter value. In the variation of all 14 parameters except $K_{M,s}$, the Michaelis constant for the activation of transcriptional response of *ADR1* and *OAF3* by FA (27), a significant increase in the cost function value (i.e., decrease in the model fitness) was seen over the range of alternative parameter values. In the case of $K_{M,s}$, the relative fitness-insensitivity to the parameter value may indicate that the expression data, in conjunction with the model, are most consistent with FA-dependent activation of Adr1p and Oaf3p being essentially on-or-off (27).

13. *Optimal experiment design.* The systematic analysis of the mathematical model allows one to detect the most sensitive processes/parameters to a particular environmental/experimental condition that is relevant to a particular function of the system. For example, in the case of the FAIT model, the role of the TF Adr1p in reducing the cell population heterogeneity

of expression of the target genes (i.e., *CTA1*, *LPX1*, etc.) has been investigated. The model was modified to simulate a hypothetical Adr1p-independent mutant strain in which Adr1p regulates neither *PIP2* nor downstream target targets, but for which these genes fully induce in the presence of oleate (which might be envisioned as an *adr1*Δ strain with elevated ORE-binding affinity). To compare the cell population heterogeneity of expression (arising from the intrinsic stochasticity of gene expression) of the target gene *CTA1* in the presence of oleate in the two models, the steady-state stochastic dynamics in both models were simulated using the Gibson–Bruck stochastic chemical kinetic algorithm (21). The simulations revealed that the histogram of *CTA1* transcript levels has a broader distribution in the mutant model than in the WT model (Fig. 1e), indicating a greater heterogeneity of gene expression (approximately 1.5-fold, as measured by coefficient of variation (CV)) (27), which provides an explanation for the network structure and suggests experiments to test the predictions.

To investigate the FAIT model prediction that Adr1p serves as a noise reducer in the oleate-responsive network, the variability of expression of *LPX1*, was examined experimentally. The level of *LPX1* gene expression was measured in WT and *adr1*Δ yeast strains in the presence of oleate using flow cytometry and a GFP-tagged Lpx1p reporter. Consistent with simulation results, the CV of Lpx1-GFP in *adr1*Δ cells was 1.7-fold higher than in WT cells (Fig. 1f).

2.2. Formal Description of Fractional Activity Functions of Molecular Processes

The purpose of this section is to illustrate how to derive the so-called *fractional activity functions* for a formal description of the dependence of quasi steady-state rates of molecular processes on component concentrations of the model (e.g., TFs or other types of regulators). To give a flavor how to describe formally the fractional activity functions of molecular processes, which would, therefore, form the right parts of differential equations of the model (see Subheading 2.1, steps 6 and 7), we consider below some cases of a gene expression regulation by single and multiple TFs. The approach described here borrows from methods originally developed for modeling enzyme kinetics, and for this reason, it is helpful to have some familiarity with the King–Altman method (40, 41) and the generalized Hill function approach (this method is very flexible and does not require prior knowledge of the detailed mechanism of the molecular process of interest) (4). The reader is also referred to (17–19), which provide additional relevant background. Here, we outline the procedure for deriving the fractional activity function for a complex biological process (e.g., transcription) controlled by multiple regulators:

1. Constructing a fractional activity function for the rate of initiation of transcription of a gene y (hereafter written T_y) regulated

by a single TF x requires initially answering a few questions. First, is x an activator or a repressor in the context of the model (sometimes the regulator may have a dual role)? What is the constitutive level of the target gene expression? At which concentration of x does T_y reach one half of its maximum level? Is the regulation of T_y by x linear, or does x act in a cooperative fashion as it regulates T_y? Nonlinear regulation of T_y by x can be captured by a "Hill coefficient" – see below.

The fractional activity function of transcriptional initiation for a gene y, denoted by f_y, that is upregulated by x can be generally described as follows

$$f_y = \frac{k_0 + (x / k_1)^{h_1}}{1 + (x / k_2)^{h_2}}, \tag{8}$$

where x is a concentration of the regulator, k_0 is the basal expression level of y; k_1 and k_2 are efficiency constants for the regulator x, and h_1 and h_2 are Hill coefficients of the regulator x. Examples of the fractional activity function (Eq. 8) dependence

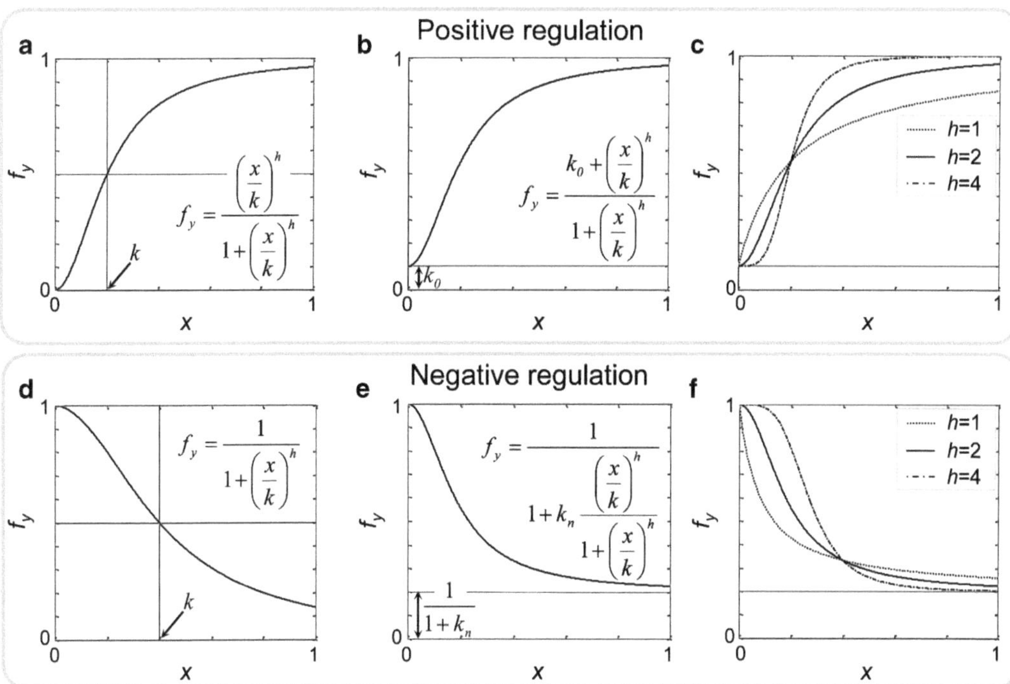

Fig. 2. Examples of regulation of transcription of a gene y by a single regulator x and their formal descriptions using a fractional activity function $f_y(x)$. Plots show the simulated dependence of the fractional activity on the level of regulator x, for different regulatory scenarios (positive and negative regulation) and for different kinetic parameter values. (**a**) Gene y has no basal level of expression. (**b**) Gene y has a constitutive level of expression (k_0). (**c**) Dependence of f_y on the level of activator (x) with the different cooperative activities (Hill coefficients, h). (**d**) Simplest formal description of negative regulation. (**e**) Partial negative regulation (k_n). (**f**) Dependence of f_y on the level of the inhibitor (x) with different cooperative activities. The following parameters were used for the simulations: $k_0 = 0.1$; $k = 0.2$ (**a–c**); $k = 0.4$ (**d–f**); $k_n = 4$; $h = 2$.

on the concentration of x with different parameter values are illustrated in Fig. 2a–c.

The fractional activity function for transcription of a gene y that is downregulated by x can be modeled as follows

$$f_y = \frac{1}{1+\left(\frac{x}{k}\right)^h}. \tag{9}$$

The partial inhibition of y by x can be described as follows

$$f_y = \frac{1}{1+k_n\left((x/k)^h/1+(x/k)^h\right)}. \tag{10}$$

where $1/(1+k_n)$ represents the level to which the expression of y can be downregulated by x when $x \to \infty$. Examples of how the fractional activity functions (Eqs. 9 and 10) depend on the concentration of x with different parameter values are illustrated in Fig. 2d–f.

2. The expression level of a gene y upregulated by two TFs (x_1 and x_2) acting combinatorially can be described as follows

$$f_y = \frac{k_0+\left(\frac{x_1}{k_1}\right)^{h_1}+\left(\frac{x_2}{k_2}\right)^{h_2}+q\left(\frac{x_1}{k_1}\right)^{h_1}\left(\frac{x_2}{k_2}\right)^{h_2}}{1+\left(\frac{x_1}{k_1}\right)^{h_1}+\left(\frac{x_2}{k_2}\right)^{h_2}+q\left(\frac{x_1}{k_1}\right)^{h_1}\left(\frac{x_2}{k_2}\right)^{h_2}}, \tag{11}$$

where x_1 and x_2 are concentrations of corresponding regulators, k_0 is the basal expression level of y, k_1 and k_2 are efficiency constants for x_1 and x_2, respectively, h_1 and h_2 are Hill coefficients for x_1 and x_2, respectively, and q is a cooperativity/synergistic constant for the joint regulation of y by x_1 and x_2. An example of the fractional activity function shown in Eq. 11 as a function of concentrations of x_1 and x_2 is illustrated in Fig. 3c.

Equation 11 can be modified for different mechanisms of combinatorial up- down- and mixed regulation (e.g., some terms in numerator and/or in denominator can be set to zero). Examples of using such modified functions for a competitive activation of y by x_1 and x_2 and the upregulation of y only when both factors are presented (e.g., x_1 and x_2 bind the promoter of y as a heterocomplex (x_1/x_2) to regulate the transcription initiation of gene y) are illustrated in Fig. 3a–b, respectively. Other examples are presented in Fig. 3d–i.

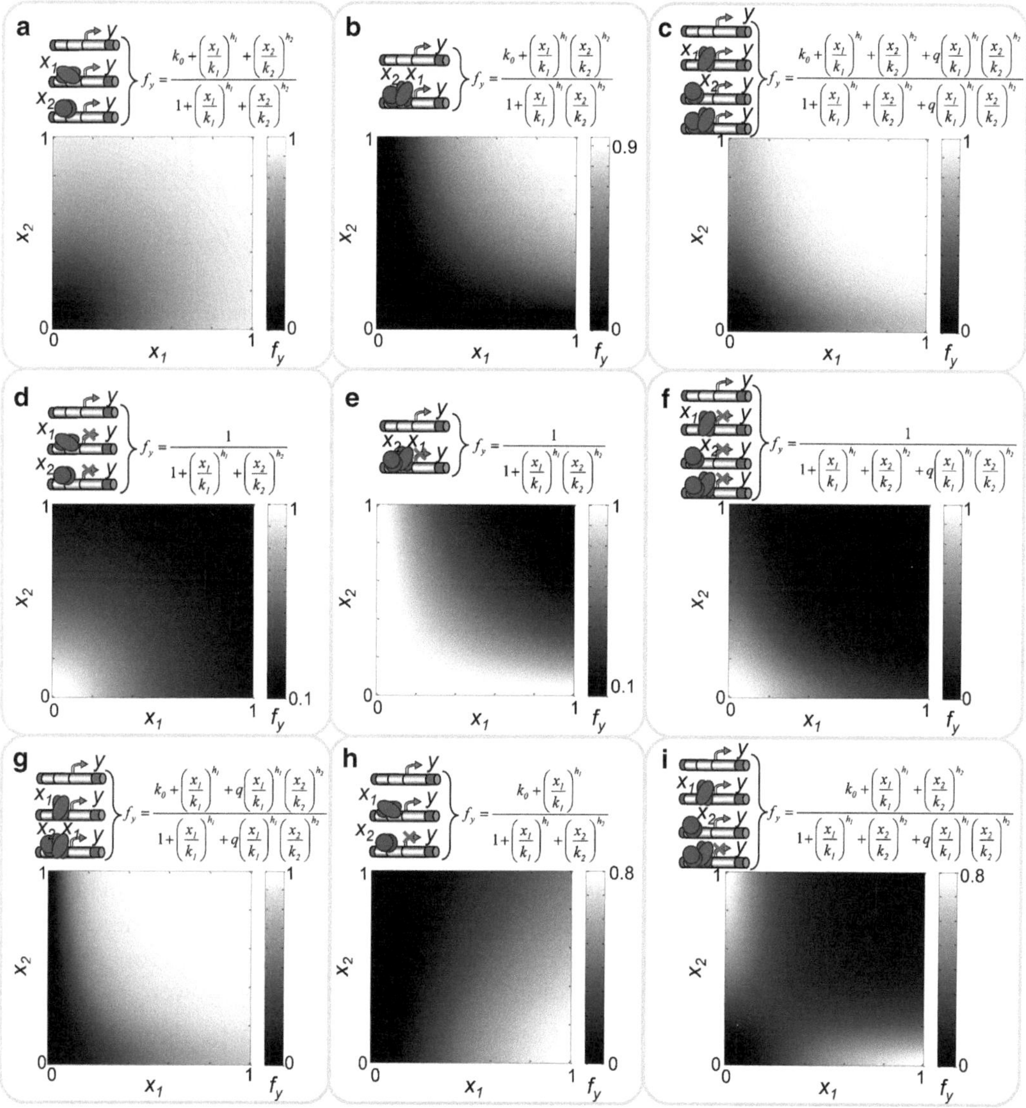

Fig. 3. Examples of combinatorial regulation of the expression level (initiation of transcription) of gene y (f_y) by two regulators (x_1 and x_2) and their formal descriptions. (**a**) Competitive activation of y by x_1 and x_2. (**b**) Upregulation of y only when both factors bind the promoter (e.g., by x_1/x_2 heterocomplex). (**c**) Independent activation of y by x_1 and x_2 with the synergistic effect (q) when both factors bind the promoter. (**d**) Competitive inhibition of y by x_1 and x_2. (**e**) Downregulation of y when only both factors bind the promoter (e.g., by x_1/x_2 heterocomplex). (**f**) Independent inhibition of y by x_1 and x_2 with synergistic (q) downregulation when both factors bind the promoter. (**g**) Sequential activation of y by x_1 and x_2 when x_1 binds the promoter first and both factors synergistically upregulate the target gene expression (**h**) Competitive activation and inhibition of y by x_1 and x_2, respectively. (**i**) Activation of y by x_1 and x_2 when they bind the promoter separately and synergistic inhibition when they bind the promoter together. The following parameters were used for the simulations: $k_0 = 0$; $k_1 = k_2 = 0.5$; $q = 10$; $h_1 = h_2 = 2$.

3. Equation 11 can be easily generalized for an arbitrary number of TFs with mixed types and mechanisms of the regulation, as follows

$$f_y = \frac{k_0 + \sum_{i_1}^{C_{A,1}} \left(\frac{x_{i_1}}{k_{i_1}}\right)^{h_{i_1}} + \sum_{i_1, i_2}^{C_{A,2}} q_{i_{12}} \left(\frac{x_{i_1}}{k_{i_1}}\right)^{h_{i_1}} \left(\frac{x_{i_2}}{k_{i_2}}\right)^{h_{i_2}} + \cdots + \sum_{i_1, \cdots, i_M}^{C_{A,M}} q_{i_{1 \cdots M}} \prod_{k=1}^{M} \left(\frac{x_{i_k}}{k_{i_k}}\right)^{h_{i_k}}}{1 + \sum_{j_1}^{C_{I,A,1}} \left(\frac{x_{j_1}}{k_{j_1}}\right)^{h_{j_1}} + \sum_{j_1, j_2}^{C_{I,A,2}} q_{j_{12}} \left(\frac{x_{j_1}}{k_{j_1}}\right)^{h_{i_1}} \left(\frac{x_{j_2}}{k_{j_2}}\right)^{h_{i_2}} + \cdots + \sum_{j_1, \cdots, j_N}^{C_{I,A,N}} q_{j_{1 \cdots M}} \prod_{w=1}^{N} \left(\frac{x_{j_w}}{k_{j_w}}\right)^{h_{j_w}}}, \quad (12)$$

where x_i is a concentration of the ith regulator in the system, k_i is an efficiency constant for x_i, $C_{A,m}$ is a number of different combinations of independently acting upregulators ($m = \overline{1 \cdot\cdot M}$); and $C_{I,A,n}$ is a number of different combinations of independently acting up- and downregulators ($n = \overline{1 \cdot\cdot N}$). In the more general case, k_i and h_i parameter values can vary with time and depend on the regulator levels as well as other environmental and intracellular factors (4).

3. Notes

1. To better formulate the purpose and objectives of the model, one should first state a set of questions/problems about the behavior of the system of interest which the model may help to answer.
2. Ideally, the available experimental data should be divided into at least two subsets, with one subset used for parameter estimation and the other subset used for the model cross-validation or the estimation of the model prediction power.
3. An important common feature of stochastic algorithms is that for complex cost functions, they are, in general, not guaranteed to find the globally optimal solution; instead, they will identify a solution that is very close to globally optimal, where the degree of difference is controlled by the criteria used to cease the optimization process (stopping criteria; e.g., number of generations, time limits, fitness limits, etc.) and other parameters of the optimization (42).
4. Occasionally, to solve the parametric inverse problem it is necessary to divide the model into small parts or submodels and fit each submodel to an appropriate set of experimental data. Such subdivisions/submodels and the corresponding experimental data represent a set of scenarios that might describe different experimental setups, conditions, etc. For instance, the framework of modeling in terms of scenarios is helpful for an optimal experiment design, hypothesis generation, etc.

5. Generally, the potential for overfitting in the model depends not only on the number of parameters but also on the model structure that may inappropriately reduce the model prediction error in comparison with the noise level in the experimental data.
6. The sensitivity analysis of the model input/output characteristics (see Subheading 2.1, step 11) can be repeatedly performed in the context of the parameter sensitivity analysis (see Subheading 2.1, step 12). This approach allows for the systematic investigation and quantitative comparative analysis of the biomolecular network dynamics relative to system parameters.
7. The sensitivity analysis can also precede the parameter estimation procedure (solving the inverse problem) and be used to rank parameters based on a sensitivity measure. The low-rank parameters can be roughly estimated or even eliminated from the models by changing relevant parts of the model (e.g., equations for fractional activities, etc.), whereas high-rank parameters must be estimated with an appropriate accuracy/precision.

Acknowledgments

This work was supported by the National Institutes of Health through grants from the National Institute of General Medical Sciences (R01-GM075152 and P50-GM076547), the National Technology Centers for Networks and Pathways (U54-RR022220), and the National Heart, Lung, and Blood Institute (K25-HL098807 to S.A.R).

References

1. Bornholdt S (2005) Systems biology. Less is more in modeling large genetic networks. *Science* **310**, 449–451.
2. Kim HD, Shay T, O'Shea EK, Regev A (2009) Transcriptional regulatory circuits: predicting numbers from alphabets. *Science* **325**, 429–432.
3. van Riel NA (2006) Dynamic modelling and analysis of biochemical networks: mechanism-based models and model-based experiments. *Briefings in Bioinformatics* 7, 364–374.
4. Likhoshvai V, Ratushny A (2007) Generalized Hill function method for modeling molecular processes. *J Bioinform Comput Biol* **5**, 521–531.
5. Wagner A (2001) How to reconstruct a large genetic network from n gene perturbations in fewer than n(2) easy steps. *Bioinformatics* **17**, 1183–1197.
6. Liao JC, Boscolo R, Yang YL, Tran LM, Sabatti C, Roychowdhury VP (2003) Network component analysis: reconstruction of regulatory signals in biological systems. *Proc Natl Acad Sci USA* **100**, 15522–15527.
7. Harbison CT, Gordon DB, Lee TI, Rinaldi NJ, Macisaac KD, Danford TW, Hannett NM, Tagne JB, Reynolds DB, Yoo J, Jennings EG, Zeitlinger J, Pokholok DK, Kellis M, Rolfe PA, Takusagawa KT, Lander ES, Gifford DK, Fraenkel E, Young RA (2004) Transcriptional regulatory code of a eukaryotic genome. *Nature* **431**, 99–104.
8. Herrgard MJ, Covert MW, Palsson BO (2004) Reconstruction of microbial transcriptional regulatory networks. *Current opinion in biotechnology* **15**, 70–77.
9. Basso K, Margolin AA, Stolovitzky G, Klein U, Dalla-Favera R, Califano A (2005) Reverse

engineering of regulatory networks in human B cells. *Nat Genet* **37**, 382–390.
10. Boyer LA, Lee TI, Cole MF, Johnstone SE, Levine SS, Zucker JP, Guenther MG, Kumar RM, Murray HL, Jenner RG, Gifford DK, Melton DA, Jaenisch R, Young RA (2005) Core transcriptional regulatory circuitry in human embryonic stem cells. *Cell* **122**, 947–956.
11. Hu Z, Killion PJ, Iyer VR (2007) Genetic reconstruction of a functional transcriptional regulatory network. *Nat Genet* **39**, 683–687.
12. Amit I, Garber M, Chevrier N, Leite AP, Donner Y, Eisenhaure T, Guttman M, Grenier JK, Li W, Zuk O, Schubert LA, Birditt B, Shay T, Goren A, Zhang X, Smith Z, Deering R, McDonald RC, Cabili M, Bernstein BE, Rinn JL, Meissner A, Root DE, Hacohen N, Regev A (2009) Unbiased reconstruction of a mammalian transcriptional network mediating pathogen responses. *Science* **326**, 257–263.
13. Marbach D, Prill RJ, Schaffter T, Mattiussi C, Floreano D, Stolovitzky G (2010) Revealing strengths and weaknesses of methods for gene network inference. *Proc Natl Acad Sci USA* **107**, 6286–6291.
14. Kauffman SA (1969) Metabolic stability and epigenesis in randomly constructed genetic nets. *J Theor Biol* **22**, 437–467.
15. Thomas R (1973) Boolean formalization of genetic control circuits. *J Theor Biol* **42**, 563–585.
16. Kauffman SA (1993) The Origins of Order: Self-Organization and Selection in Evolution. Oxford Univ. Press, New York.
17. Edelstein-Keshet L (2005) Mathematical Models in Biology. SIAM: Society for Industrial and Applied Mathematics, New York.
18. Bolouri H (2008) Computational Modelling Of Gene Regulatory Networks - A Primer. Imperial College Press, London.
19. Klipp E, Liebermeister W, Wierling C, Kowald A, Lehrach H, Herwig R (2009) Systems Biology: A Textbook. WILEY-VCH, Weinheim.
20. Gillespie DT (1976) A General Method for Numerically Simulating the Stochastic Time Evolution of Coupled Chemical Reactions. *Journal of Computational Physics* **22**, 403–434.
21. Gibson MA, Bruck J (2000) Efficient exact stochastic simulation of chemical systems with many species and many channels. *J Phys Chem A* **104**, 1876–1889.
22. Wilkinson DJ (2006) Stochastic Modelling for Systems Biology. Chapman & Hall/CRC, Boca Raton, FL.
23. McAdams HH, Shapiro L (1995) Circuit simulation of genetic networks, *Science* **269**, 650–656.
24. Likhoshvai VA, Matushkin IuG, Ratushnyi AV, Anan'ko EA, Ignat'eva EV, Podkolodnaia OA (2001) [A generalized chemical-kinetic method for modeling gene networks], *Molekuliarnaia Biologiia* **35**, 1072–1079.
25. Shmulevich I, Dougherty ER, Kim S, Zhang W (2002) Probabilistic Boolean Networks: a rule-based uncertainty model for gene regulatory networks. *Bioinformatics* **18**, 261–274.
26. de Jong H (2002) Modeling and simulation of genetic regulatory systems: a literature review. *J Comput Biol* **9**, 67–103.
27. Ratushny AV, Ramsey SA, Roda O, Wan Y, Smith JJ, Aitchison JD (2008) Control of transcriptional variability by overlapping feed-forward regulatory motifs. *Biophysical Journal* **95**, 3715–3723.
28. Rao CV, Arkin AP (2003) Stochastic chemical kinetics and the quasi-steady-state assumption: Application to the Gillespie algorithm. *J Chem Phys* **118**, 4999–5010.
29. Banga JR (2008) Optimization in computational systems biology. *BMC systems biology* **2**, 47.
30. de Atauri P, Orrell D, Ramsey S, Bolouri H (2004) Evolution of "design" principles in biochemical networks. *IET Sys Biol* **1**, 28–40.
31. Schwarz G (1978) Estimating the Dimension of a Model, *The Annals of Statistics* **6**, 461–464.
32. Akaike H (1974) A New Look at the Statistical Model Identfication, *IEEE Trans. Automatic Control* **19**, 716–723.
33. Burnham KP, Anderson DR (2002) Model Selection and Multi-Model Inference, Springer-Verlag New York, LLC.
34. Rabitz H (1989) Systems analysis at the molecular scale. *Science 246*, 221–226.
35. Ratushny AV, Likhoshvai VA, Ignatieva EV, Goryanin II, Kolchanov NA (2003) Resilience of Cholesterol Concentration to a Wide Range of Mutations in the Cell. *Complexus* **1**, 142–148.
36. Ratushnyi AV, Likhoshvai VA, Ignat'eva EV, Matushkin YG, Goryanin II, Kolchanov NA (2003) A computer model of the gene network of the cholesterol biosynthesis regulation in the cell: analysis of the effect of mutations. *Doklady Biochemistry and Biophysics* **389**, 90–93.
37. Stites EC, Trampont PC, Ma Z, Ravichandran KS (2007) Network analysis of oncogenic Ras activation in cancer. *Science* **318**, 463–467.
38. Saltelli A, Ratto M, Andres T, Campolongo F, Cariboni J, Gatelli D, Saisana M, Tarantola S (2008) Global sensitivity analysis: the primer. John Wiley & Sons Ltd, Chichester, UK.
39. Chen WW, Schoeberl B, Jasper PJ, Niepel M, Nielsen UB, Lauffenburger DA, Sorger PK

(2009) Input-output behavior of ErbB signaling pathways as revealed by a mass action model trained against dynamic data. *Mol Syst Biol* **5**, 239.

40. King EL, Altman C (1956) A Schematic method of deriving the rate laws for enzyme-catalyzed reactions. *The Journal of physical chemistry* **60**, 1375–1378.
41. Cornish-Bowden A (1977) An automatic method for deriving steady-state rate equations. *Biochem J* **165**, 55–59.
42. Press WH, Teukolsky SA, Vetterling WT, Flannery BP (2007) Numerical Recipes with Source Code CD-ROM 3rd Edition: The Art of Scientific Computing. Cambridge University Press; 3rd edn., New York.

Index

Gerard Cagney and Andrew Emili (eds.), *Network Biology: Methods and Applications*, Methods in Molecular Biology, vol. 781, DOI 10.1007/978-1-61779-276-2, © Springer Science+Business Media, LLC 2011

MIX
Papier aus verantwortungsvollen Quellen
Paper from responsible sources
FSC® C105338

If you have any concerns about our products,
you can contact us on
ProductSafety@springernature.com

In case Publisher is established outside the EU,
the EU authorized representative is:
Springer Nature Customer Service Center GmbH
Europaplatz 3, 69115 Heidelberg, Germany

Printed by Libri Plureos GmbH
in Hamburg, Germany